D. Pearce G. Wagner (Eds

Logics in AI

European Workshop JELIA '92
Berlin, Germany, September 7-10, 1992
Proceedings

Springer-Verlag
Berlin Heidelberg New York
London Paris Tokyo
Hong Kong Barcelona
Budapest

Series Editor

Jörg Siekmann
University of Saarland
German Research Center for Artificial Intelligence (DFKI)
Stuhlsatzenhausweg 3, W-6600 Saarbrücken 11, FRG

Volume Editors

David Pearce
Gerd Wagner
Free University Berlin, Department of Philosophy
Group of Logic, Epistemics and Information
Habelschwerdter Allee 30, W-1000 Berlin 33, FRG

CR Subject Classification (1991): I.2, F.3-4

ISBN 3-540-55887-X Springer-Verlag Berlin Heidelberg New York
ISBN 0-387-55887-X Springer-Verlag New York Berlin Heidelberg

Printed in Germany

Typesetting: Camera ready by author/editor
Printing and binding: Druckhaus Beltz, Hemsbach/Bergstr.
45/3140-543210 - Printed on acid-free paper

Lecture Notes in Artificial Intelligence 633

Subseries of Lecture Notes in Computer Science
Edited by J. Siekmann

Lecture Notes in Computer Science
Edited by G. Goos and J. Hartmanis

Preface

The present volume contains the proceedings of JELIA '92, les Journées Européennes sur la Logique en Intelligence Artificielle, or the Third European Workshop on Logics in AI.

Earlier meetings were held in France (1988) and the Netherlands (1990). JELIA '92 is taking place in Berlin (Gosen) from 7-10 September, 1992, organised by the Group *Logik, Wissenstheorie und Information* (LWI) of the Free University Berlin in co-operation with the German Informatics Society (GI - Fachausschuß 1.2 - Inferenzsysteme). As in previous workshops, the aim was to bring together researchers involved in all aspects of logic in artificial intelligence.

The volume contains 2 invited addresses and 21 selected papers covering such topics as

- Logical Foundations of Logic Programming and Knowledge-Based Systems
- Automated Theorem Proving
- Partial and Dynamic Logics
- Systems of Nonmonotonic Reasoning
- Temporal and Epistemic Logics
- Belief Revision

The papers appear here in their planned order of presentation at the conference. No fixed sections are employed, but thematically related contributions have, where possible, been grouped together.

The organising committee of J. van Eijck, L. Fariñas del Cerro, E. Orlowska, and D. Pearce was assisted by a programme committee comprising additionally

J. van Benthem, C. Cellucci, P. Enjalbert, M. Eytan, A. Fuhrmann, U. Furbach, D. Gabbay, P. Gärdenfors, H. Herre, M. Kanovich, M. Kracht, B. Nebel, P. Schroeder-Heister and F. Veltman.

Further help in the refereeing of submitted papers was provided by M. Kalsweek, F. Klusniak, G. Miskowska, D. Roorda, H. Rott, U. Scheffler, A. Skowron, F. Voorbraak, G. Wagner and H. Wansing.

We should like to thank all the referees for their prompt and valuable responses.

We are also grateful to the GI for their partial sponsorship of the meeting, to the Free University Berlin for further support, as well as to Jörg Siekmann for recommending these proceedings for the Lecture Notes series and to Springer-Verlag for agreeing to publish them in the shortest possible time.

Berlin, June 1992

David Pearce and Gerd Wagner

Contents

A Modal Theory of Arrows.
Arrow Logics I. (Invited Paper)
D. Vakarelov 1

Knowledge Without Modality: A Simplified Framework
for Chronological Ignorance
C. MacNish 25

Design Complete Sequential Calculus for Continuous Fixpoint
Temporal Logic
R. Pliuskevicius 36

Logical Omniscience and Classical Logic
R. Muskens 52

Weak Implication: Theory and Applications
K.L. Kwast and S. van Denneheuvel 65

Deriving Inference Rules for Terminological Logics
V. Royer and J.J. Quantz 84

Linear Proofs and Linear Logic
B. Fronhöfer 106

Relevance and Revision –
About Generalizing Syntax-based Belief Revision
E. Weydert 126

Modellings for Belief Change: Base Contraction, Multiple Contraction,
and Epistemic Entrenchment
H. Rott 139

A Framework for Default Logics
C. Froidevaux and J. Mengin 154

A Conceptualization of Preferences in Non-Monotonic Proof Theory
A. Hunter 174

Reasoning with Defeasible Arguments: Examples and Applications
G. Vreeswijk 189

About Deductive Generalization
Ph. Besnard and E. Grégoire 212

Transition Systems and Dynamic Semantics
T. Fernando 232

Declarative Semantics for Inconsistent Database Programs
M.S. Mircheva 252

Tableau-Based Theorem Proving and Synthesis of Lambda-Terms in the Intuitionistic Logic
O. Bittel 262

A Constructive Type System Based on Data Terms
H.-J. Goltz 279

An Ordered Resolution and Paramodulation Calculus for Finite Many-Valued Logics
N. Zabel 304

An Efficient Constraint Language for Polymorphic Order-Sorted Resolution
C. Prehofer 319

Default Theory for Well Founded Semantics with Explicit Negation (Invited Paper)
L.M. Pereira, J.J. Alferes and J.N. Aparício 339

Computing Answers for Disjunctive Logic Programs
U. Furbach 357

Expanding Logic Programs
C. Witteveen 373

Disjunctive Logic Programming, Constructivity and Strong Negation
H. Herre and D. Pearce 391

A MODAL THEORY OF ARROWS. ARROW LOGICS I

Dimiter Vakarelov

Department of Mathematical Logic
with Laboratory for Applied Logic,
Faculty of Mathematics and Informatics,
Sofia University, Boul. Anton Ivanov 5,
Sofia, Bulgaria

Keywords. Arrow structures, Modal logics.

Abstract. The notion of arrow structure /a.s./ is introduced as an algebraic version of the notion of directed multi graph. By means of a special kind of a representation theorem for arrow structures it is shown that the whole information of an a.s. is contained in the set of his arrows equipped with four binary relations describing the four possibilities for two arrows to have a common point. This makes possible to use arrow structures as a semantic base for a special polymodal logic, called in the paper BAL /Basic Arrow Logic/. BAL and various kinds of his extensions are used for expressing in a modal setting different properties of arrow structures. Several kinds of completeness theorems for BAL and some other arrow logics are proved, including completeness with respect to classes of finite models. And the end some open problems and possibilities for further development of the "arrow" approach are formulated.

Introduction

There exist many formal schemes and tools for representing knowledge about different types of data. Sometimes we can better understand this knowledge if it has some graphical representation. In many cases arrows are very suitable visual objects for representing various data structures: different kinds of graphs, binary relations, mappings, categories and so on. An abstract form of this representation scheme is the notion of arrow structure /a.s./, which in this paper is an algebraic version of the notion of directed multi graph. Simply speaking, a.s. is a two sorted algebraic system, consisting of a set of arrows Ar, a set of points Po and two functions 1 and 2 from arrows to points, assigning to each arrow x the point 1(x) - the beginning of x, and the point 2(x) - the end of x. By means of 1 and 2 we define four relations R_{ij} $i,j=1,2$ such that $xR_{ij}y$ iff $i(x)=j(y)$. These relations define the four possibilities for a two arrows to have a common point. So each a.s. S determine a relational system $W(S)=(Ar, \{R_{ij}/ij=1,2\})$ called arrow frame /a.f./. It is shown that the whole information of an a.s. S is contained in the arrow frame W(S). Arrow frames as relational systems with binary relations are suitable for interpreting polymodal logics, having modal operations, corresponding to each binary relation in the frame. So we introduce a modal language $\mathcal{L}$ with four boxes [ij] with standard Kripke semantics in arrow frames . We show how different properties of arrow frames are modally definable by means of modal formulas of $\mathcal{L}$. The logic of all arrow frames is axiomatized and called BAL - the Basic Arrow Logic. This paper is mainly devoted to study BAL and some of their extensions.

The paper is organized as follows.

Section 1 is devoted to arrow structures and arrow frames.

In section 2 we introduce semantically the notion of arrow logic as the class of all formulas true in a given class of arrow frames. Some definability and undefinability results are proved there. For instance, applying some special techniques, called "copying", we show that the logic of all arrow frames coincides with the logic of all normal arrow frames, which correspond to directed graphs, admitting no more than one arrow between an ordered pair of points.

In section 3 we give axiomatization of the logic of all arrow frames - BAL and prove several completeness theorems.

In section 4, applying the filtration technic from ordinary modal logic we prove that BAL and some other arrow logics possess finite model property and are decidable.

In section 5 we study an extension of BAL with a new connective interpreted by an equivalence relation between arrows, stating that two arrows are equivalent if they have common begins and common ends.

In section 6 we study another extension with a modal constant Loop, which is true in an arrow if it has common begin and end, i.e. if it forms a loop.

Section 7 is devoted to a short survey for possible directions for further development, including extensions with different polyadic modalities, corresponding to some typical relations between arrows as $Path_n$, $Path_\omega$, $Loop_n$, $Trapezium_n$, Triangle and others. There are some natural generalizations of modal logic of binary relations and Lambek Calculus. Finally it is shown a way of many dimensional generalization of arrow structures, which makes possible to consider an n-ary relation in a set as an n-dimensional arrow structure. Among the logics based on n-dimensional arrow frames are some natural generalizations of the so called cylindric modal Logics.

The idea to look for a logic based on two sorted structures having points and arrows, was suggested to me by Johan Van Benthem [BEN 90]. The first results were included in the manuscript [VAK 90] and the many-dimensional generalization in the abstracts [VAK 91a] and [VAK 92]. The terms arrow frame and arrow logic were introduced by Van Benthem [BEN 89] in connection with some generalizations of the modal logic of algebra of relations. Van Bethem's arrow frames consist of a set of objects with composition as a ternary relation, converse as a binary relation and a set of identity arrows. These relational structures are so abstract that there is no any representation theorem stating that the arrows look indeed as arrows, with beginning and end. We adopt Van Benthem's terminology, because it fits very well to the subject of this paper.

1. Arrow structures and arrow frames

By arrow structure (a.s.) we shall mean any system S=(Ar, Po, 1, 2), where

- Ar is a nonempty set, whose elements are called arrows,
- Po is a nonempty set, whose elements are called points. We assume also that $Ar \cap Po = \emptyset$.
- 1 and 2 are total functions from Ar to Po associated to each arrow x the following two points: 1(x) - the first point of x (beginning, source, domain), and 2(x) - the last point of x (end, target, codomain). Graphically:

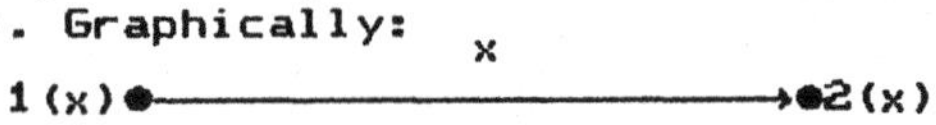

If A=1(x) and B=2(x) we say that x connects A with B, or, that (A,B) is a connected pair of points. It is possible for a pair of points (A,B) to be connected by different arrows.

• For some technical reasons we assume the following axiom for arrow structures:

(Ax) For each point A there exists an arrow x such that A=1(x) or A=2(x). In other words, each point is ether the first or the last point of some arrow.

An a.s. S is called normal if it satisfies the following condition of normality

(Nor) If 1(x)=1(y) and 2(x)=2(y) then x=y.

Sometimes, to denote that Ar, Po, 1, 2 are from a given a.s. S, we will write Ar_S, Po_S, 1_S and 2_S.

The main examples of a.s. structures are directed multi-graphs, and for normal a.s. - directed graphs without isolated points. These are notions studied in Graph theory where graphs are visualized, or sometimes defined, by geometrical notions of a point and arrow. In graph intuition arrow is a part of a line with some direction, connected two points. Formally, the notion of an arrow structure coincides with the notion of directed multi-graph without isolated points. We will prefer, however, the term "arrow structure" as more neutral, having models, not only connected with graph intuition, as for example, categories and binary relations.

The example of a.s. constructed from a binary relation can be defined as follows. Let R be a nonempty binary relation in a nonempty set W. Define Ar=R, Po={x∈W/(∃y∈W)(xRy or yRx)} and for (x,y)∈Ar define 1((x,y))=x and 2((x,y))=y. Then, obviously (Ar, Po, 1, 2) is a normal a.s. In some sense this example is typical, because each normal a.s. can be represented as an a.s. determined by a non-empty binary relation.

Let S be an a.s. The following binary relation $\rho=\rho_S$ can be defined in the set Po_S. For each $A,B\in Po_S$:

$A\rho B$ iff $(\exists x\in Ar_S)(1(x)=A$ and $2(x)=B)$

According to some properties of ρ_S we can consider different kinds of arrow structures:

• S is serial a.s. if ρ_S is a serial relation (i.e. $\forall A\ \exists B\ A\rho B$),

• S is reflexive a.s. if ρ_S is a reflexive relation,

• S is symmetric a.s. if ρ_S is a symmetric relation,

• S is transitive a.s. if ρ_S is a transitive relation,

• S is total a.s. if ρ_S is a total relation, i.e. $\rho_S=Po_S\times Po_S$.

Let S be a given a.s. The following four relations $R_{ij}=R^S_{ij}$, $i,j\in\{1,2\}$ in Ar_S (called incidence relations in S), will play a fundamental role in this paper:

$xR_{ij}y$ iff $i(x)=j(y)$

The following pictures illustrated the introduced relations:

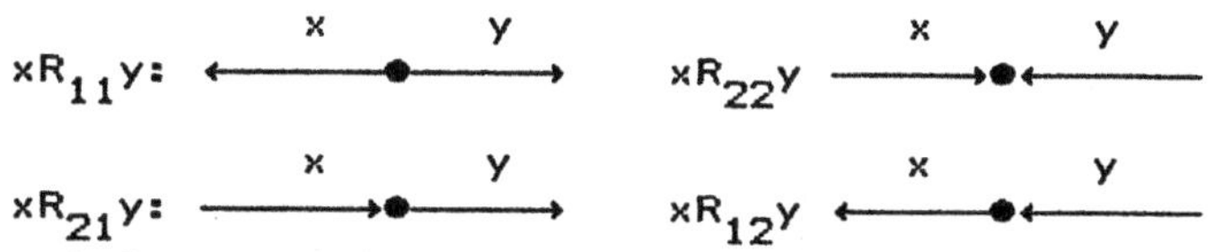

Lemma 1.1. The relations R_{ij} satisfy the following conditions for

any $x,y,z \in Ar_S$ and $i,j,k \in \{1,2\}$:

(ρii) $xR_{ij}x$,

(σij) If $xR_{ij}y$ then $yR^S_{ji}x$,

(τijk) If $xR_{ij}y$ and $yR_{jk}z$ then $xR_{ik}z$.

Proof. By an easy verification.■

Let $\underline{W}=(W, R_{11}, R_{22}, R_{12}, R_{21})$, $W\neq\emptyset$, be a relational system. $\underline{W}$ will be called arrow frame (a.f.) if it satisfies the axioms (ρii), (σij) and (τijk) for any $i,j,k\in\{1,2\}$and $x,y,z\in W$. The class of all arrow frames will be denoted by ARROW. If S is an a.s. then the a.f. $SAF(S)=(Ar_S, R^S_{11}, R^S_{22}, R^S_{12}, R^S_{21})$ will be called a standard a.f. over S. The class of all standard a.f. will be denoted by (standard)ARROW. One of the main results of this section will be the proof that each a.f. is a standard a.f. over some a.s., i.e. (standard)ARROW=ARROW.

Lemma 1.2. Let S be an a.s. Then the following equivalences are true, where x,y,z range over Ar_S:

(i) S is normal a.s.iff $\forall xy(xR_{11}y \;\&\; xR_{22}y\rightarrow x=y)$,

(ii) S is serial a.s. iff $\forall x\exists y\; xR_{21}y$,

(iii) S is reflexive a.s. iff $\forall x\exists y(xR_{11}y \;\&\; yR_{21}x)$ and $\forall x\exists y(xR_{21}y \;\&\; yR_{22}x)$,

(iv) S is symmetric a.s. iff $\forall x\exists y(xR_{12}y \;\&\; yR_{12}x)$,

(v) S is transitive a.s. iff $\forall xy\exists z(xR_{21}y\rightarrow xR_{11}z \;\&\; yR_{22}z)$.

(vi) S is total a.s. iff $\forall ij\in\{1,2\}\forall xy\exists z(xR_{i1}z \;\&\; zR_{2j}y)$

Proof. As an example we shall prove (ii).

($\rightarrow$) Suppose S is serial and let $x\in Ar_S$ and $2(x)=A$. By seriality there exists $B\in Po_S$ such that $A\rho B$. Then for some $y\in Ar_S$ we have $1(y)=A$ and $2(y)=B$, so $2(x)=1(y)$, which yields $xR_{21}y$. Thus $\forall x\exists y\; xR_{21}y$.

($\leftarrow$) Suppose $\forall x\exists y\; xR_{21}y$ and let $A\in Po_S$. Then by (Ax) there exists $x\in Ar_S$ such that $A=1(x)$ or $A=2(x)$.

Case 1: $A=1(x)$. Let $B=2(x)$, then $A\rho B$.

Case 2: $A=2(x)$. Take y such that $xR_{21}y$. From here we get $2(x)=1(y)$ and $A=1(y)$, Take $B=2(y)$, then we get $A\rho B$. So in both cases $\forall A\exists B\; A\rho B$.

The remaining conditions can be proved in a similar way. ■

This lemma suggests the following definition concerning arrow frames. Let $\underline{W}=(W, R_{11}, R_{22}, R_{12}, R_{21})$ be an a.f., then $\underline{W}$ is called:

- $\underline{W}$ is normal a.f.iff $\forall xy(xR_{11}y \;\&\; xR_{22}y\rightarrow x=y)$,
- $\underline{W}$ is serial a.f. iff 9$\forall x\exists y\; xR_{21}y$,
- $\underline{W}$ is reflexive a.f. iff $\forall x\exists y(xR_{21}y \;\&\; xR_{22}y)$ and $\forall x\exists y(xR_{12}y \;\&\; xR_{11}y)$,
- $\underline{W}$ is symmetric a.f. iff $\forall x\exists y(xR_{12}y \;\&\; xR_{21}y)$,
- $\underline{W}$ is transitive a.f. iff $\forall xy\exists z(xR_{21}y\rightarrow xR_{11}z \;\&\; yR_{22}z)$,
- $\underline{W}$ is total a.f. iff $\forall ij\in\{1,2\}\forall xy\exists z(xR_{i1}z \;\&\; zR_{2j}y)$, where the variables x,y,z range over the set W.

The class of all normal arrow frames will be denoted by (nor)ARROW. Analogously we introduce the notations (ser)ARROW, (ref)ARROW, (sym)ARROW, (tr)ARROW and (total)ARROW for the classes of all serial a.f., reflexive a.f., symmetric a.f., transitive a.f. and total a.f. respectively. We will use also a notation as (ref)(sym)ARROW denoting the class of all reflexive and symmetric arrow frames.

Obviously, if $\underline{W}$ is total a.f. then $\underline{W}$ is reflexive, symmetric and transitive a.f. An a.f. is called pretotal if it is reflexive, symmetric and transitive. The class of all pretotal a.f. is denoted by (pretotal)ARROW. Using combined notations we have that (pretotal)ARROW=(ref)(sym)(tr)ARROW.

Let $\underline{W}=(W, R_{11}, R_{22}, R_{12}, R_{21})$ be an a.f. and $W'\subseteq W$, $W'\neq\emptyset$ and R'_{ij} are the relations R_{ij} restricted over W'. Then obviously the system $\underline{W}'=(W', R'_{11}, R'_{22}, R'_{12}, R'_{21})$ is an a.f. called subframe of $\underline{W}$. The frame $\underline{W}'$ is called generated subframe of $\underline{W}$ if $\forall i,j\in\{1,2\}\forall x\in W'\forall y\in W(xR_{ij}y\Rightarrow y\in W')$. If $a\in W$ then by $\underline{W}_a$ will be denoted the smallest generated subframe of $\underline{W}$, containig a. $\underline{W}_a$ is called an arrow subframe of $\underline{W}$ generated by a. If $\underline{W}$ is an a.f. and there exists some $a\in W$ such that $\underline{W}=\underline{W}_a$ then $\underline{W}$ is called a generated a.f. (by a). If Σ is a class of arrow frames then by Σ_{gen} we denote the class of all generated frames of Σ.

Lemma 1.3. (i) $((\text{pretotal})\text{ARROW})_{gen}\subseteq(\text{total})\text{ARROW}$,

(ii) $((\text{pretotal})\text{ARROW})_{gen}=(\text{total})\text{ARROW}$

Proof. By an easy verification.■

Let S be an a.s. and for $i\in\{1,2\}$ and $A\in Po_S$ define:

$i(A)=\{x\in Ar_S/i(x)=A\}$, $g(A)=(1(A),2(A))$.

Lemma 1.4. The following is true for each $x,y\in Ar_S$ and $i,j\in\{1,2\}$:

(1) If $x\in i(A)$ and $y\in j(A)$ then $xR^S_{ij}y$,

(2) If $xR^S_{ij}y$ then $x\in i(A)$ iff $y\in j(A)$,

(3) $1(A)\cup 2(A)\neq\emptyset$.

Proof. By an easy verification.■

Lemma 1.4 suggests the following definition. Let $\underline{W}=(W, R_{11}, R_{22}, R_{12}, R_{21})$ be an a.f. and α_1 and α_2 be subsets of W. The pair (α_1,α_2) will be called a generalized point in $\underline{W}$ if it satisfies the following conditions for each $x,y\in W$ and $i,j\in\{1,2\}$:

(1) If $x\in\alpha_i$ and $y\in\alpha_j$ then $xR_{ij}y$,

(2) If $xR_{ij}y$ then $x\in\alpha_i$ iff $y\in\alpha_j$,

(3) $\alpha_1\cup\alpha_2\neq\emptyset$.

The set of generalized points of an a.f. $\underline{W}$ will be denoted by $Po(\underline{W})$. Lemma 1.2 now means that $g(A)=(1(A),2(A))$ is a generalized point in the standard a.f. SAF(S) over S.

For any binary relation R in W and $x\in W$ define $R(x)=\{y\in W/xRy\}$.

Lemma 1.5. Let $\underline{U}=(U, U_{11}, R_{22}, R_{12}, R_{21})$ be an a.f.. Then for any $x,y\in U$ and $i,j\in\{1,2\}$: $xR_{ij}y$ iff $R_{i1}(x)=R_{j1}(y)$ and $R_{i2}(x)=R_{j2}(y)$

Proof. By an easy calculation, using the axioms of a.f.■

Lemma 1.6. Let $\underline{W}$ be an a.f. Then for any $x,y,z\in W$ and $i,j,k\in\{1,2\}$:

(i) The pair $k(z)=(R_{k1}(z), R_{k2}(z))$ is a generalized point in $\underline{W}$.

(ii) For each generalized point (α_1,α_2) there exists $z\in W$ and $k\in\{1,2\}$ such that $k(z)=(\alpha_1,\alpha_2)$.

(iii) $xR_{ij}y$ iff $i(x)=j(y)$.

Proof. (i) Let $i,j\in\{1,2\}$ and $x\in R_{ki}(z)$ and $y\in R_{kj}(z)$. Then we have $zR_{ki}x$ and $zR_{kj}y$. Then by (σki) we obtain $xR_{ik}z$ and by (τikj) we get $xR_{ij}y$. This proves condition (1) from the definition of generalized point. In a similar way one can verify condition (2). By (ρkk) we have $xR_{kk}x$, so $R_{kk}(x)\neq\emptyset$. This shows that $R_{k1}(x)\cup R_{k2}(x)\neq\emptyset$, which proves condition (3).

(ii) Let (α_1,α_2) be a generalized point in $\underline{W}$. Then there exists $z\in W$ such that $z\in\alpha_1\cup\alpha_2$.

Case 1: $z\in\alpha_1$. In this case we will show that $k=1$, i.e. that $(\alpha_1,\alpha_2)=1(z)=(R_{11}(z),R_{12}(z))$ i.e. that $\alpha_1=R_{11}(z)$ and that $\alpha_2=R_{12}(z)$.

Let $x\in\alpha_1$. Since $z\in\alpha_1$, then by (1) of the definition of generalized point we get $xR_{11}z$. So by $(\sigma 11)$ we obtain $zR_{11}x$, which shows that $x\in R_{11}(z)$.

Now let $x\in R_{11}(z)$. Then $zR_{11}x$ and since $z\in\alpha_1$, by (2) of the definition of generalized point we get $x\in\alpha_1$. This proves the equality $\alpha_1=R_{11}$. In a similar way on can prove that $\alpha_2=R_{12}(z)$.

Case 2: $z\in\alpha_2$. In this case $k=2$ and we can proceed as in case 1.

(iii) By lemma 1.5. we have: $xR_{ij}y$ iff $R_{i1}(x)=R_{i2}(y)$ and $R_{j1}(x)=R_{j2}(y)$ iff $(R_{i1}(x),R_{i2}(x))=(R_{j1}(y),{}_{j2}(y))$ iff $i(x)=j(y)$.■

Now we shall give a construction of arrow structures from arrow frames. Let $\underline{W}=(W, R_{11}, R_{22}, R_{12}, R_{21})$. Define a system $S=S(\underline{W})$ as follows: $Ar_S=W$, $Po_S=Po(\underline{W})$ – the set of general points of $\underline{W}$, for $k=1,2$ and $z\in W$ let $k_S(z)=k(z)=(R_{k1}(z),R_{k2}(z))$ as in lemma 1.6. In the next theorem we shall show that $S(\underline{W})$ is an a.s. called the a.s. over $\underline{W}$.

Theorem 1.7. (i) The system $S(\underline{W})$ defined above is an a.s. More over:

(ii) The standard a.f. $SAF(S(\underline{W}))$ over $S(\underline{W})$ coincides with $\underline{W}$.

(iii) $S(\underline{W})$ is normal (serial, reflexive, symmetric, transitive, total) a.s. iff $\underline{W}$ is normal (serial, reflexive, .. and so on) a.f.

Proof. (i) By lemma 1.6.(i) and (ii) we obtain that the system $S(\underline{W})$ is an a.s.

(ii) By lemma 1.1 and lemma 1.6.(iii) $SAF(S(\underline{W}))$ is a standard a.f. such that for any $x,y\in W$ and $i,j\in\{1,2\}$: $xR_{ij}y$ iff $i(x)=j(y)$ iff $xR^S_{ij}y$, which shows that $SAF(S(\underline{W}))=\underline{W}$.

(iii) By lemma 1.2. $S(\underline{W})$ is normal (serial,...) a.s. iff the corresponding standard a.f. $SAF(S(\underline{W}))$ over $S(\underline{W})$ is normal (serial,...). By (ii) $SAF(S(\underline{W}))=\underline{W}$, which proves the assertion. ■

Corollary 1.8. (standard)ARROW=ARROW

Proof. From theorem 1.7.■

Let S and S' be two arrow structures. A pair (f,g) of

11-functions $f:Ar_S \to Ar_{S'}$ and $g:Po_S \to Po_{S'}$ is called an isomorphism from S onto S' if for any $x \in Ar_S$ and $i=1,2$ we have $g(i_S(x)) = i_{S'}(f(x))$.

Lemma 1.9. Let S be an a.s. $\underline{W}$=SAF(S) be the standard a.f. over S, Po($\underline{W}$) be the set of generalized points of $\underline{W}$, and S'=S($\underline{W}$) be the a.s. over $\underline{W}$. Let for $A \in Po_S$ $g(A)=(1(A),2(A))$ be the function defined before lemma 1.4. and for $x \in Ar_S$ and $i=1,2$ $i_{S'}(x) = (R^{S'}_{i1}(x), R^{S'}_{i2}(x))$ be the function defined as in lemma 1.6.(i). Then:

(i) g is a 11-function from Po_S onto Po($\underline{W}$).

(ii) For any $x \in Ar_S$ and $i=1,2$: $g(i(x))=i_{S'}(x)$.

Proof. Obviously g(A) is a generalized point in $\underline{W}$. Let g(A)= g(B). Then 1(A)=1(B) and 2(A)=2(B). For A we can find $x \in Ar_S$ such that 1(x)=A or 2(x)=A. Then $x \in 1(A)$ or $x \in 2(A)$. Suppose $x \in 1(A)$. Then $x \in 1(B)$, so 1(x)=B. From 1(x)=A and 1(x)=B we get A=B. In the case $x \in 2(A)$ we proceed in the same way and get A=B. This shows that the mapping is injective. To show that it is "onto" suppose that (α_1,α_2) is a generalized point in $\underline{W}$. We shall show that for some $A \in Po_S$ $g(A)=(1(A),2(A))=(\alpha_1,\alpha_2)$. Since $\alpha_1 \cup \alpha_2 \neq \emptyset$ there exists $z \in \alpha_1$ or $z \in \alpha_2$.

Case1: $z \in \alpha_1$. Let 1(z)=A, so $z \in 1(A)$. We shall show that $1(A)=\alpha_1$ and that $2(A)=\alpha_2$. Suppose $x \in 1(A)$. Then 1(x)=A, so 1(x)=1(z) which yields $xR^S_{11}z$. Since $z \in \alpha_1$, then, by the properties of generalized points we get $x \in \alpha_1$, so $1(A) \subseteq \alpha_1$. Suppose now that $x \in \alpha_1$. Then, since $z \in \alpha_1$, we get $xR_{11}z$, so 1(x)=1(z)=A. Then $x \in 1(A)$, so $\alpha_1 \subseteq 1(A)$. Consequently $1(A)=\alpha_1$. In a similar way one can show that $2(A)=\alpha_2$. Hence, in this case $g(A)=(\alpha_1,\alpha_2)$.

Case 2: $z \in \alpha_2$. The proof is similar to that of case 1.

(ii) Let $x \in Ar_S$ and $i=1,2$. Since $g(i(x))=(1(i(x)),2(i(x))$ and $i_{S'}(x)=(R^S_{i1}(x),R^S_{i2}(x))$, To show that $g(i(x))=i_{S'}(x)$, we have to prove that $1(i(x))=R^S_{i1}(x)$ and that $2(i(x))=R^S_{i1}(x)$. For that purpose suppose that $y \in 1((i(x))$, so $1(y)=i(x)$. Thus $yR^S_{i1}x$, which yields $y \in R^S_{i1}(x)$. Consequently $1(i(x)) \subseteq R^S_{i1}(x)$. The converse inclusion and the second equality can be proved in a similar way. ∎

Theorem 1.10. Let S be an a.s., $\underline{W}$=SAF(S) be the standard a.f. over S and S($\underline{W}$) be the a.s. over $\underline{W}$. Then S is isomorphic with S($\underline{W}$).

Proof. Let for $x \in Ar_S$ $f(x)=x$ and for $A \in Po_S$ $g(A)=(1(A),2(A))$. Then lemma 1.8 shows that the pair (f,g) is the required isomorphism. ∎

Theorems 1.10 and 1.7 show that the whole information of an a.s. S is contained in the standard a.f. SAF(S) over S and can be expressed in terms of arrows and the relations R^S_{ij}. An example of such a correspondence is lemma 1.2. As for first order conditions about the relation ρ this correspondence can be defined in an effective way. The intuitive idea of this translation can be explained in the following way. By the axiom (Ax) for each point A there exists $i \in \{1,2\}$ such that A=i(x). So each variable A for a point is translated by a pair (i,x), where x denotes an arrow and i denotes one of the numbers 1 and 2. Suppose now that we have $A\rho B$, $A=i(x)$ and $B=j(y)$. Then by the definition of ρ we have: $(\exists u)(1(u)=i(x)$ & $2(u)=j(y))$ which is equivalent to $(\exists u)(xR_{i1}u$ & $uR_{2j}y)$. So if A is translated by (i,x) and B by (j,y), then the corresponding translation of $A\rho B$ will be the formula $\varphi = xS_{ij}y = (\exists u)(xR_{i1}u$ & $uR_{2j}y)$. Here obviously $S_{ij}=R_{i1}\circ R_{2j}$. The parameters i and j in φ can be eliminated according to under what kind of quatifiers are A and B. If for example A is under the scope of $(\forall A)$, we change this quantifier by $(\forall i)(\forall x)$ and accordingly for $(\exists A)$. Then quantifiers of the type $(\forall i)$ and $(\exists i)$ can be eliminated in a standard way by conjunctions and disjunctions of formulas, putting on the place of i 1 and 2. Let us take for example the formula $(\forall A)(A\rho A)$. First this formula is translated by $(\forall i)(\forall x)xS_{ii}x$. Eliminating $(\forall i)$ we obtain

$(\forall x)(xS_{11}x)$ & $(\forall x)(xS_{22}x)$, which is equivalent to

$(\forall x)(\exists y)(xR_{11}y$ & $yR_{21}x)$ & $(\forall x)(\exists y)(xR_{21}y$& $yR_{22}x)$

which is exactly the condition of reflexivity of ρ from lemma 1.2. The translation of the formula $(\forall A)(\exists B)(A\rho B)$ is the following:

$(\forall i)(\forall x)(\exists j)(\exists y)(xS_{ij}y)$

Eliminating $(\forall i)$ we obtain the conjunction of the following two formulas:

$\varphi_{1j}=(\forall x)(\exists j)(\exists y)(xS_{1j}y)$,

$\varphi_{2j}=(\forall x)(\exists j)(\exists y)xS_{2j}y)$.

Eliminating $(\exists j)$ from φ_{1j} and φ_{2j} we obtain the following formulas φ_1 and φ_2:

$\varphi_1=(\forall x)((\exists y)(xS_{11}y)\vee(\exists y)(xS_{12}y))$,

$\varphi_2=(\forall x)((\exists y)(xS_{21}y)\vee(\exists y)(xS_{22}y))$. Substituting here S_{ij} we obtain

$\varphi_1=(\forall x)((\exists y)(\exists z)(xR_{11}z$ & $zR_{21}y)\vee(\exists y)(\exists z)(xR_{11}z$ & $zR_{22}y))$,

$\varphi_2=(\forall x)((\exists y)(\exists z)(xR_{21}z$ & $zR_{21}y)\vee(\exists y)(\exists z)(xR_{21}z$ & $zR_{22}y)$

The formula φ_1 is always true in a.s. because in the second disjunct we can put $y=z=x$. It follows logically from φ_2 the following formula $\varphi=(\forall x)(\exists z)(xR_{21}z)$, which is exactly the condition of seriality from lemma 1.2. It is easy to see that φ implies in a.s. the formula φ_2.

The described intuitive idea of translating first order sentences for points in terms of ρ and = in arrow structures into equivalent sentences for arrows in terms of the relations R_{ij} can be given in precise terms, but we will do not that in this paper.

2. Arrow logics - semantic definitions and some definability and nondefinability results

In this section we shall give a semantic definition of modal logics, called arrow logics. For that purpose we introduce the following modal language $\mathcal{L}$. It contains the following symbols:

- VAR - a denumerable set of proposition variables,
- $\neg$, $\wedge$, $\vee$ - classical propositional connectives,
- $[ij]$, $i,j=1,2$ - four modal operations,

• (,) – parentheses.

The definition of the set of all formulas FOR for $\mathcal{L}$ is defined in the usual way.

Abbreviations: $A \to B = \neg A \vee B$, $A \leftrightarrow B = (A \to B) \wedge (B \to A)$, $1 = A \vee \neg A$, $0 = \neg 1$, $\langle ij \rangle A = \neg [ij] \neg A$.

The general semantics of $\mathcal{L}$ is a Kripke semantics over relational structures of the type $\underline{W} = (W, R_{11}, R_{22}, R_{12}, R_{21})$ with $W \neq \emptyset$, called frames. The standard semantics of $\mathcal{L}$ is over the class ARROW of all arrow frames. Let us remind the basic semantic definitions and notations, which we will use /for more details about Kripke semantics and related notions we refer Segerberg [SEG 71], Hughes & Cresswell [H&C 84] and Van Benthem [Ben 86]/.

Let $\underline{W} = (W, R_{11}, R_{22}, R_{12}, R_{21})$ be a frame. A function $v: VAR \to 2^W$ assigning to each variable $p \in VAR$ a subset $v(p)$ of W is called a valuation and the pair $M = (\underline{W}, v)$ is called a model over $\underline{W}$. For $x \in W$ and $A \in FOR$ we define a satisfiability relation $x \Vdash_v A$ in M /to be read "A is true in x at the valuation v"/ by induction on the complexity of the formula A as in the usual Kripke definition:

$x \Vdash_v A$ iff $A \in v(A)$ for $A \in VAR$,

$x \Vdash_v \neg A$ iff $x \nVdash_v A$ /not $x \Vdash_v A$/,

$x \Vdash_v A \wedge B$ iff $x \Vdash_v A$ and $x \Vdash_v B$,

$x \Vdash_v A \vee B$ iff $x \Vdash_v A$ or $x \Vdash_v B$,

$x \Vdash_v [ij]A$ iff $(\forall y \in W)(x R_{ij} y \to y \Vdash_v A)$.

We say that A is true in the model $M = (\underline{W}, v)$, or that M is a model for A, if for any $x \in W$ we have $x \Vdash_v A$. A is true in the frame $\underline{W}$, or that W is a frame for A, if A is true in any model over $\underline{W}$. A is true in a class Σ of frames if A is true in any member of Σ. A class of formulas L is true in a model M, or M is a model for L, if any member of L is true in M. L is true in a class of frames Σ if any formula from L is true in Σ. L is called the logic of Σ and denoted by $L(\Sigma)$ if it contains all formulas true in Σ. Obviously, this operation of assigning sets of formulas to classes of frames is antymonotonic in the following sense:

If $\Sigma \subseteq \Sigma'$ then $L(\Sigma') \subseteq L(\Sigma)$.

In this paper we will study the logics L((standard)ARROW), L(ARROW), L((nor)ARROW), L((ser)ARROW), L((ref)ARROW), L((sym)ARROW), L((tr)ARROW), L((pretotal)ARROW), L((total)ARROW). The most important logic from this list is L((standard)ARROW). The first result which, can be stated for L(standard)ARROW) and which follows immediately from corollary 1.8. is that

L((standard)ARROW)=L(ARROW).

We say that a condition φ for R_{ij} is modally definable in a class Σ of frames if there exists a formula A such that for any frame $\underline{W} \in \Sigma$: A is true in $\underline{W}$ iff φ holds in $\underline{W}$. If a class of frames is characterized by a condition φ which is modally definable in the class of all frames, then we say that Σ is modally definable class of frames. The following lemma is a standard results in modal definability theory.

Lemma 2.1. /Modal definability of arrow frames/ Let Σ be the class of all frames and $A \in VAR$. Then in the next table the conditions from the left side are modally definable in Σ by the formulas from the right side: (i,j,k=1,2)

(ρii) $(\forall x) x R_{ii} x$,	(Pii) $[ii]A \to A$,
(σij) $(\forall xy)(x R_{ij} y \to y R_{ji} x)$	(Σij) $A \vee [ij] \neg [ji] A$,

(τijk) $(\forall xyz)(xR_{ij}y$ and $yR_{jk}z \longrightarrow xR_{ik}z)$ $\quad$ (Tijk) $[ik]A \to [ij][jk]A$.

Corollary 2.2. The class ARROW is modally definable.

Lemma 2.3. Let Σ=ARROW and $A\in$VAR. Then in the next table the conditions from the left side are modally definable in Σ by the formulas from the right side:

seriality of an a.f. (ser) $\langle 21\rangle 1$,
reflexivity of an a.f. (ref) $([11][21]A \to A) \wedge ([21][22]A \to A)$,
symmetricity of an a.f. (sym) $[12][12]A \to A$,
transitivity of an a.f. (tr) $[11][22]A \to [21]A$.

Proof. As an example we shall show the validity of (tr) in an a.f. $\underline{W}$ implies that $\underline{W}$ is a transitive a.f. For the sake of contradiction, suppose that (tr) is true in $\underline{W}$ and that $\underline{W}$ is not transitive a.f. Then for some $x,y,z\in W$ we have $xR_{21}y$ and not$(\exists z\in W)(xR_{11}z$ & $zR_{22}y)$. Define $v(A)=W\setminus\{y\}$. Then $y \not\Vdash_v A$ and since $xR_{21}y$ we get $x \not\Vdash_v [21]A$. We shall show that $x \Vdash_v [11][22]A$. Suppose that this is not true. Then for some $z,t\in W$ we have $xR_{11}z$, $zR_{22}t$ and $t \not\Vdash_v A$, hence $t=y$. So $(\exists z)(xR_{11}z$ & $zR_{22}y)$, which is a contradiction. ∎

Corollary 2.4. The classes (ser)ARROW, (ref)ARROW, (sym)ARROW, (tr)ARROW and (pretotal)ARROW are modally definable.

We shall show that the condition of normality of an a.f. is not modally definable and consequently that the class (nor)ARROW is not modally definable. We shall show first that the logic L((nor)ARROW) coincides with the logic L(ARROW). For that purpose we shall use a special construction called copying, adapted here for relational structures in the type of arrow frames.

Let $\underline{W}=(W, R_{11}, R_{22}, R_{12}, R_{21})$ and $\underline{W}'=(W', R'_{11}, R'_{22}, R'_{12}, R'_{21})$ be two frames and $M=(\underline{W}, v)$, $M'=(\underline{W}', v')$ be models over $\underline{W}$ and $\underline{W}'$ respectively. Let I be a nonempty set of mappings from W into W'. We say that I is a copying from $\underline{W}$ to $\underline{W}'$ if the following conditions are satisfied for any $i,j\in\{1,2\}$, $x,y\in W$and $f,g\in I$:

(I1) $(\forall y'\in W')(\exists y\in W)(\exists g\in I)g(y)=y'$
(I2) If $f(x)=g(y)$ then $x=y$,
(R_{ij}1) If $xR_{ij}y$ then $(\forall f\in I)(\exists g\in I)f(x)R'_{ij}g(y)$,
(R_{ij}2) If $f(x)R'_{ij}g(y)$ then $xR_{ij}y$.

We say that I is a copying from M to M' if in addition the following condition is satisfied for any $p\in$VAR, $x\in W$ and $f\in I$:

(V) $x\in v(p)$ iff $f(x)\in v'(p)$.

For $x\in W$ and $f\in I$ $f(x)$ is called f-th copy of x and $f(W)=\{f(x)/x\in W\}$ is called f-th copy of W. By (I1) we obtain that $W'=\bigcup\{f(W)/f\in I\}$, so W' is a sum of his copies. If I is one element set {f} then f is an isomorphism from $\underline{W}$ onto $\underline{W}'$.

The importance of the copying construction is in the following

Lemma 2.5. (i) (Copying lemma) Let I be a copying from the model M to the model M'. Then for any formula $A\in\mathcal{L}$, $x\in W$ and $f\in I$ the following equivalence holds:

$x \Vdash_v A$ iff $f(x) \Vdash_{v'} A$,

(ii) If I is a copying from the frame $\underline{W}$ to the frame $\underline{W}'$ and v is a valuation, then there exists a valuation v' such that I is a copying from the model $M=(\underline{W}, v)$ to the model $M'=(\underline{W}', v')$.

Proof. (i) The proof is by induction on the complexity of the formula A. For $A\in$VAR the assertion holds by the condition (V) of copying. If A is a Boolean combination of formulas the proof is straightforward. Let $A=[ij]B$ and by the induction hypothesis (i.h.) suppose that the assertion for B holds.

($\rightarrow$) Suppose $x \Vdash_v [ij]B$ and $f \in I$. To show that $f(x) \Vdash_{v'} [ij]B$ suppose $f(x)R'_{ij}y'$ and proceed to show that $y' \Vdash_v B$. By (I1) $(\exists y \in W)(\exists g \in I)g(y)=y'$, so $f(x)R'_{ij}g(y)$ and by ($R_{ij}2$) we get $xR_{ij}y$. From $xR_{ij}y$ and $x \Vdash_v [ij]B$ we get $y \Vdash_v B$. Then by the i.h. we get $g(y) \Vdash_{v'} B$, so $y' \Vdash_{v'} B$.

($\leftarrow$) Suppose $f(x) \Vdash_{v'} [ij]B$. To show that $x \Vdash_v [ij]B$ suppose $xR_{ij}y$ and proceed to show that $y \Vdash_v B$. From $xR_{ij}y$ we obtain by ($R_{ij}1$) that there exists $g \in W$ such that $f(x)R'_{ij}g(y)$. Then, since $f(x) \Vdash_{v'} [ij]B$, we get $g(y) \Vdash_{v'} B$ and by the i.h. $y \Vdash_v B$.

(ii) Define for $p \in VAR$:

$v'(p)=\{x' \in W' / (\exists x \in W)(\exists f \in I) f(x)=x' \text{ and } x \in v(p)\}$

We shall show that the condition (V) of copying is fulfilled. Let $x \in W$ and $f \in I$ and suppose $x \in v(p)$. Then by the definition of v' we have $f(x) \in v'(p)$. For the converse implication suppose $f(x) \in v'(p)$. Then there exists $y \in W$ and $g \in I$ such that $f(x)=g(y)$ and $y \in v(p)$. By (I2) we get $x=y$, so $x \in v(p)$.■

Lemma 2.6. Let $\underline{W}=(W, R_{11}, R_{22}, R_{12}, R_{21})$ be an arrow frame. Then there exists a normal arrow frame $\underline{W}'=(W', R'_{11}, R'_{22}, R'_{12}, R'_{21})$ and a copying I from $\underline{W}$ to $\underline{W}'$ and if $\underline{W}$ is a finite a.f. the same is $\underline{W}'$.

Proof. Let $\underline{B}(W)=(B(W), 0, 1, +, .)$ be the Boolean ring over the set W, namely B(W) is the set of all subsets of W, $0=\emptyset$, $1=W$, $A+B=(A\setminus B)\cup(B\setminus A)$ and $A.B=AB=A\cap B$. Note that in Boolean rings $a-b=a+b$.

We put $W'=W\times B(W)$, $I=B(W)$ and for $f \in I$ and $x \in W$ we define $f(x)=(x,f)$. Obviously the conditions (I1) and (I2) from the definition of copying are fulfilled and each element of W' is in the form of $f(x)$ for some $f \in I$ and $x \in W$.

For the relations R'_{ij} we have the following definition:

$f(x)R'_{ij}g(y)$ iff $xR_{ij}y$ & $(f+i.\{x\}=g+j.\{y\})$. Here the indices i, $j \in \{1,2\}$ are considered as elements of B(W): 1 is the unit of B(W) and $2=1+1=1-1=0$.

To verify the condition ($R_{ij}1$) suppose $xR_{ij}y$ and $f \in I$. Put $g=f+i.\{x\}-j.\{y\}$. Then $f+i.\{x\}=g+j.\{y\}$, which implies $f(x)R'_{ij}g(y)$. Condition ($R_{ij}2$) follows directly from the definition of R'_{ij}. So I is a copying.

The proof that W' with the relations R'_{ij} is an arrow frame is straightforward. For the condition of normality suppose $f(x)R'_{11}g(y)$ and $f(x)R'_{22}g(y)$. Then we obtain $xR_{11}y$ & $(f+1.\{x\}=g+1.\{y\})$ and $xR_{22}y$ & $(f+2.\{x\}=g+2.\{y\})$. From here, since $2=0$, we get $f=g$ and $f+\{x\}=g+\{y\}$, which implies $\{x\}=\{y\}$, hence $x=y$ and $f(x)=g(x)$. Thus W' is a normal a.f.

Suppose now that $\underline{W}$ is a finite a.f. Then the Boolean ring over W is finite too and hence $\underline{W}'$ is a finite a.f.■

If Σ is a class of a.f. then the class of all finite a.f. from Σ is denoted by Σ_{fin}.

Theorem 2.7. (i) L((nor)ARROW)=L(ARROW).

(ii) $L(((nor)ARROW)_{fin})=L(ARROW_{fin})$.

Proof. (i) Since (nor)ARROW$\subseteq$ARROW we get L(ARROW)$\subseteq$ L((nor)ARROW). To prove that L((nor)ARROW)$\subseteq$L(ARROW) suppose $A\notin L(ARROW)$. Then there exists an a.s. $\underline{W}$, $x\in W$ and a valuation v such that $x \not\Vdash_v A$. By lemma 2.6. there exists a normal a.s. $\underline{W}'$ and a copying I from $\underline{W}$ to W'. By lemma 2.5.(ii) there exists a valuation v' in W' such that I is a copying from the model ($\underline{W}$, v) to the model ($\underline{W}'$, v'). Then by the copying lemma we get for any $f\in I$ that $f(x) \not\Vdash_{v'} A$. So A is not true in $\underline{W}'$ and hence $A\notin L((nor)ARROW)$. So $L((nor)ARROW)\subseteq L(ARROW)$.

(ii) The proof is the same as the proof of (i), using the fact that lemma 2.6 guaranties that $\underline{W}'$ is a finite a.f. ∎

Corollary 2.8. The condition of normality of an a.f. is not modally definable.

Proof. Suppose that there exists a formula φ such that for any a.f. $\underline{W}$: φ is true in $\underline{W}$ iff $\underline{W}$ is normal. So $\varphi\in L((nor)ARROW)$. Let $\underline{W}_0$ be an a.f. which is not normal. Then φ is not true in $\underline{W}_0$, so $\varphi\notin L(ARROW)$, hence by theorem 2.7 $\varphi\notin L((nor)ARROW)$, which is a contradiction.∎

Another example of modally undefinable condition is totality. First we need the following standard result from modal logic.

Lemma 2.9. Let Σ be a nonempty class of a.f. closed under subframes and let Σ_{gen} be the class of generated frames of Σ. Then $L(\Sigma)=L(\Sigma_{gen})$.

Corollary 2.10. (i) $L((pretotal)ARROW)=L(((pretotal)ARROW)_{gen})=L((total)ARROW$.

(ii) $L((pretotal)ARROW)_{fin})=L((((pretotal)ARROW)_{gen})_{fin})=L(((total)ARROW)_{fin})$.

Proof. (i) The first equality follows from lemma 2.9 and the second - from lemma 1.3.

(ii) Use the fact that generated frame of a finite frame is a finite frame too. ∎

Corollary 2.11. The condition of totality of an a.f. is not modally definable.

Proof. Suppose that there exists a formula φ such that for any a.f. $\underline{W}$: φ is true in $\underline{W}$ iff $\underline{W}$ is total a.f. Then $\varphi\in L((total)ARROW)$ and by corollary 2.10 $\varphi\in L((pretotal)ARROW)$. Let $\underline{W}_0$ be a pretotal a.f. which is not total (such frames obviously exist). Then φ is not true in $\underline{W}_0$, so $\varphi\notin L((pretotal)ARROW)$ - a contradiction.∎

3. Axiomatization of some arrow logics

In this section we introduce a syntactical definition of arrow logic as sets of formulas containing some formulas as axioms and closed under some rules. The minimal set of axioms which we shall use, contains those from the minimal modal logic for each modality [ij] and the formulas, which modally define arrow frames. The formal system, obtained in this way is denoted by BAL and called Basic Arrow Logic.

Axioms and rules for BAL.

(Bool) All or enough Boolean tautologies,

(K[ij]) $[ij](A\Rightarrow B)\Rightarrow([ij]A\Rightarrow[ij]B)$,

(Pii) $[ii]A\Rightarrow A$,

(Σij) $A\vee[ij]\neg[ji]A$,

(Tijk) $[ik]A\Rightarrow[ij][jk]A$,

(MP) $\dfrac{A, A\Rightarrow B}{B}$, (N[ij]) $\dfrac{A}{[ij]A}$, i, j are anymembers of{1,2} and

A and B are arbitrary formulas.

We identify BAL with the set of its theorems.

By an arrow logic (a.l.) we mean any set L of formulas containing BAL and closed under the rules (MP), (N[ij]) and the rule of substitution of propositional variables. So BAL is the smallest arrow logic. We adopt the following notation. If X is a finite sequence of formulas , (taken as a new axiom) then by BAL+X we denote the smallest arrow logic containing all formulas from X. We shall us the following formulas as additional axioms:

(ser) <21>1,

(ref) ([11][21]A→A)∧([21][22]A→A),

(sym) [12][12]A→A,

(tr) [11][22]A→[21]A.

Let X⊆{ser, ref, sym, tr} and let for instance X={ser, tr}. Then BAL+X= BAL+ser+tr. We will use also the notation (X)ARROW and for that concrete X (X)ARROW= (ser)(tr)ARROW.

Let L be an a.l. and Σ be a class of arrow frames. We say that L is sound in Σ if L⊆L(Σ), L is complete in Σ if L(Σ)⊆L, and that L is characterized by Σ, or that L(Σ) is axiomatized by L, if L is sound and complete in Σ, i.e. if L=L(Σ).

In the completeness proofs we shall use the standard method of canonical models. We shall give a brief description of the method. For more details and some definitions we refer Segerberg [SEG 71] or Hughes & Cresswell [H&C 84].

Let L be an a.l. The frame $\underline{W}_L=(W_L, R^L_{11}, R^L_{22}, R^L_{12}, R^L_{21})$ will be called canonical frame for the logic L if W_L is the set of all maximal consistent sets in L and the relations R^L_{ij} are defined in W_L as follows: $xR^L_{ij}y$ iff {A∈FOR/[ij]A∈x}⊆y. For p∈VAR the function $v_L(p)=\{x\in W_L/p\in x\}$ is called canonical valuation and the pair $M_L=(\underline{W}_L,v_L)$ is called the canonical model for L. The following is a standard result from modal logic.

Lemma 3.1. (i) Truth lemma for the canonical model for L. The following is true for any formula A and $x\in W_L$: $x\Vdash_{v_L} A$ iff A∈x.

(ii) If A∉L then there exists $x\in W_L$ such that A∉x.

Lemma 3.2. Let L be an a.l. Then the canonical frame $\underline{W}_L$ of L is an a.f.

Proof. It is well known fact from the standard modal logic that the axiom (Pii) yields the condition (ρii) for the canonical frame. In the same way the axioms (Σij) and (Tijk) yield the conditions (σij) and (τijk) for $\underline{W}_L$. Thus $\underline{W}_L$ is an a.f.∎

Theorem 3.3. BAL is sound and complete in the class of all arrow frames.

Proof. Soundness follows by lemma 2.1 and the completeness can be proved by the method of canonical models. Let L=BAL. By lemma 3.2 the canonical frame for L is an a.f. To show that L(ARROW)⊆L suppose that A∉L. Then by lemma 3.1.(ii) there exists $x\in W_L$ such that A∉x. Then by the truth lemma we have $x\nVdash_{v_L} A$, so A is not true in the a.f. $\underline{W}_L$. Then A∉L(ARROW), which proves the theorem.∎

Corollary 3.4. BAL=L(ARROW)=L((nor)ARROW).

Proof - from theorem 3.3 and theorem 2.7. ∎

Lemma 3.5. Let L be an a.l. Then the following conditions are true:

(i) (ser)∈L iff $\underline{W}_L$ is a serial a.f.,

(ii) (ref)∈L iff $\underline{W}_L$ is a reflexive a.f.,

(iii) (sym)∈L iff $\underline{W}_L$ is a symmetric a.f.,

(iv) (tr)∈L iff $\underline{W}_L$ is a transitive a.f.

Proof. As an example we shall show (iv)(⟶). Suppose (tr)= [11][22]A→[21]A∈L. and proceed to show the condition of transitivity of $\underline{W}_L$: $(\forall xy\in W_L)(\exists z\in W_L)(xR^L_{21}y \longrightarrow xR^L_{11}z \ \&\ zR^L_{22}y)$.

Let M_1={A/[11]A∈x}, M_2={A/(∃B∈y)(¬A→[22]¬B∈L)} and $M=M_1 \cup M_2$. Then the following assertion is true:

Assertion. (i) If $A_1,\dots,A_n \in M_i$ then $A_1 \wedge \dots \wedge A_n \in M_i$, i=1,2,

(ii) If $A \in M_i$ and A→B∈L then $B \in M_i$, i=1,2,

(iii) $M_1 \cup M_2$ is L-inconsistent set iff ∃A∈FOR: $A \in M_1$ and $\neg A \in M_2$,

(iv) If $xR^L_{21}y$ then M is L-consistent set of formulas.

(v) Let z be a maximal consistent set. Then $M_2 \subseteq z$ implies $zR^L_{22}y$, and $M_1 \subseteq z$ implies $xR^L_{11}z$.

(vi) If xRy then $(\exists z\in W_L)(xR^L_{11}z \ \&\ zR^L_{22}y)$.

Proof. The proof of (i) and (ii) is straightforward and (iii) follows from (i) and (ii).

Let us proof (iv). Suppose $xR^L_{21}y$ and that M is not L-consistent. Then by (iii) there exists a formula A such that $A \in M_1$ and $\neg A \in M_2$. Then [11]A∈x and ∃B∈y such that ¬¬A→[22]¬B∈L, hence A→[22]¬B∈L. Then by the rule (N[11]) we get [11](A→[22]¬B)∈L and by axiom (K[11]) and (MP) we obtain that [11]A→[11][22]¬B∈L. But [11]A∈x, so [11][22]¬B∈x, Then by the axiom (tr): [11][22]¬B→[21]¬B and (MP) we get [21]¬B∈x and since $xR^L_{21}y$ we get ¬B∈y. Since B∈x we obtain a contradiction.

(v) Suppose $M_2 \subseteq z \in W_L$. Suppose that not $zR^L_{22}y$. Then for some formula A we have: [22]B∈z and B∉y, so ¬B∈y. Since ¬¬[22]B→[22]¬¬B∈L, then by the definition of M_2 we get that ¬[22]B∈M_2, hence ¬[22]B∈z – a contradiction. The second part of (v) follows by the definition of R^L_{11}.

(vi) Suppose $xR^L_{21}y$. Then by (iv) M is an L-consistent set. Then there exists a maximal consistent set z such that M⊆z and by (v) we have $xR^L_{11}z$ and $zR^L_{22}y$. Now the proof of (iv)(⟶) follows directly from assertion (vi). ■

Theorem 3.6. Let X⊆{ser, ref, sym, tr}. Then BAL+X=L((X)ARROW).

Proof. The consistency part of the theorem follows from lemma 2.3 and the completeness part can be obtained from lemma 3.5. as in the proof of theorem 3.3. ■

Corollary 3.7. (i) BAL+ref+sym+tr=L((pretotal)ARROW),

(ii) BAL+ref+sym+tr=L((total)ARROW).

Proof. (i) is a direct consequence of theorem 3.6 and (ii) follows from corollary 2.10. ■

4. Filtration and finite model property

In this section, applying the techniques of filtration coming from classical modal logic, we shall show that BAL and some of its extensions posses finite model property and are decidable. We adopt the Segerberg's definition of filtration, adapted for the language $\mathcal{L}$ of arrow logics (see [SEG 71]).

Let $\underline{W}=(W, R_{11}, R_{22}, R_{12}, R_{21})$ be an a.f. and $M=(\underline{W}, v)$ be a model over $\underline{W}$. Let Ψ be a finite set of formulas, closed under subformulas. For $x,y\in W$ define:

$x\sim y$ iff $9(\forall A\in\Psi)(x\Vdash_{v} A$ iff $y\Vdash_{v} A)$, $|x|=\{y\in W/x\sim y\}$,

$W'=\{|x|/x\in W\}$, for $p\in VAR$ $v'(p)=\{|x|/x\in v(p)\}$.

Let R'_{ij} $i,j=1,2$ be any binary relations in W' such that $\underline{W}'=(W', R'_{11}, R'_{22}, R'_{12}, R'_{21})$ be an a.f. We say that the model $M'=(W', v')$ is a filtration of the model M trough Ψ if following conditions are satisfied for any $i,j=1,2$ and $x,y\in W$:

($FR_{ij}1$) If $xR_{ij}y$ then $|x|R'_{ij}|y|$,

($FR_{ij}2$) If $|x|R|y|$ then $(\forall[ij]A\in\Psi)(x\Vdash_{v}[ij]A \rightarrow y\Vdash_{v} A)$.

The following lemma is a standard result in filtration theory.

Lemma 4.1. ([Seg 71])(i)/ Filtration lemma/ For any formula $A\in\Psi$ and $x\in W$ the following is true: $x\Vdash_{v} A$ iff $|x|\Vdash_{v'} A$.

(ii) $CardW'\leq 2^{n}$, where $n=Card\Psi$.

Let L be an a.l. We say that L admits a filtration if for any frame $\underline{W}$ for L and a model $M=(\underline{W},v)$ over $\underline{W}$ and for any formula A there exist a finite set of formulas Ψ containing A and closed under subformulas and a filtration $M'=(\underline{W}',v')$ of M trough Ψ, such that $\underline{W}'$ is a frame for L.

Corollary 4.2. (i) Let Σ be a class of arrow frames, let Σ_{fin} be the class of all finite arrow frames from Σ and let $L(\Sigma)$ admits a filtration. Then $L(\Sigma)=L(\Sigma_{fin})$.

(ii) If $L(\Sigma)$ is finitely axiomatizable then it is decidable.

Lemma 4.3. Let $\underline{W}$ be an a.f., $M=(\underline{W}, v)$ be a model over $\underline{W}$ and $M'=(\underline{W}',v,)$ be a filtration of M trough Ψ. Then:

(i) If $\underline{W}$ is a serial a.f. then $\underline{W}'$is a serial a.f.,

(ii) If $\underline{W}$ is a reflexive a.f. then $\underline{W}'$ is a reflexive a.f.,

(iii) If $\underline{W}$ is a symmetric a.f. the $\underline{W}'$ is a symmetric a.f.,

(iv) If $\underline{W}$ is a total a.f. then $\underline{W}'$is a total a.f.

Proof. As an example we shall prove (iii). We have to show that $(\forall|x|\in W')(\exists|y|\in W')(|x|R'_{12}|y|\ \&\ |y|R'_{12}|x|)$. Suppose $|x|\in W'$. Then there exists $y\in W$ such that $xR_{12}y\ \&\ yR_{12}x$. Then by the condition ($FR_{12}1$) of the filtration we obtain $|x|R'_{12}|y|\ \&\ |y|R'_{12}|x|$. ∎

Theorem 4.4. The logic L(ARROW) admits a filtration.

Proof. Let A be a formula and let Ψ be the smallest set of formulas containing A, closed under subformulas and satisfying the following condition

(*) If for some $i,j=1,2$ $[ij]A\in\Psi$ then for any $ij=1,2$ $[ij]A\in\Psi$.

It is easy to see that Ψ is finite and if n is the number of subformulas of A then $Card\Psi\leq 2^{4n}$. Then define W' and v' as in the definition of filtration. We define the relations R'_{ij} in W' as follows:

$|x|R'_{ij}|y|$ iff $(\forall[ij]A\in\Psi)(\forall k\in\{1,2\})(x\Vdash_{v}[ik]A\longleftrightarrow y\Vdash_{v}[jk]A)$.

First we shall show that the frame $\underline{W}'$is an a.f. The conditions (ρii) and (σij) follow directly from the definition of R'_{ij}. For the condition (τijk) suppose $|x|R'_{ij}|y|$ and $|y|R'_{jk}|z|$. To prove $|x|R'_{ik}|z|$ suppose $[ik]A\in\Psi$, $l\in\{1,2\}$ and for the direction ($\rightarrow$) suppose $x \Vdash_v [il]A$ and proceed to show that $z \Vdash_v [kl]A$. From $[ik]A \in \Psi$ we get $[ij]A, [jl]A \in \Psi$. Then $|x| R'_{ij} |y|$, $[ij]A \in \Psi$ and $x \Vdash_v [il]A$ imply $y \Vdash_v [jl]A$. This and $[jl]A \in \Psi$ and $|y| R'_{jk} |z|$ imply $z \Vdash_v [kl]A$.

The converse direction $(\leftarrow)$ can be proved in a similar way.

It remains to show that the conditions of filtration $(FR_{ij}1)$ and $(FR_{ij}2)$ are satisfied.

For the condition $(FR_{ij}1)$ suppose $xR_{ij}y$, $[ij]A \in \Psi$, $k \in \{1,2\}$ and for the direction $(\rightarrow)$ suppose $x \Vdash_v [ik]A$, $yR_{jk}z$ and proceed to show that $z \Vdash_v A$. From $xR_{ij}y$ and $yR_{jk}z$ we get $xR_{ik}z$ and since $x \Vdash_v [ik]A$ we get $z \Vdash_v A$. For the direction $(\leftarrow)$ suppose $y \Vdash_v [jk]A$, $xR_{ik}z$ and proceed to show that $z \Vdash_v A$. From $xR_{ij}y$ we get $yR_{ji}x$ and by $xR_{ik}z$ we get $yR_{jk}z$. From here and $y \Vdash_v [jk]A$ we obtain $z \Vdash_v A$. This ends the proof of $(FR_{ij}1)$.

For the condition $(FR_{ij}2)$ suppose $|x| R'_{ij} |y|$, $[ij]A \in \Psi$ and $x \Vdash_v [ij]A$. From here we obtain $y \Vdash_v [jj]A$ and since $yR_{jj}y$ we get $y \Vdash_v A$. This completes the proof of the theorem.■

Corollary 4.5.

(i) $BAL = L(ARROW) = L(ARROW_{fin}) = L(((nor)ARROW)_{fin})$.

(ii) BAL is a decidable logic.

Proof. (i) The first two equalities follow from corollary 3.4 and theorem 4.4. The last equality follows from theorem 2.7.

(ii) is a consequence of corollary 4.2 and corollary 3.4. ■

Theorem 4.6. Let $X \subseteq \{ser, ref, sym\}$. Then the logic $L((X)ARROW)$ admits a filtration.

Proof. Use the same filtration as in theorem 4.4 and apply lemma 4.3.■

Corollary 4.7. Let $X \subseteq \{ser, ref, sym\}$. Then:

(i) $B+X = L((X)ARROW) = L(((X)ARROW)_{fin})$

(ii) B+X is a decidable logic.

Theorem 4.8. The logic $L((total)ARROW)$ admits a filtration.

Proof. Use the same filtration as in theorem 4.4 and apply lemma 4.3.■

Corollary 4.9.

(i) $B+ref+sym+tr = L((pretotal)ARROW) = L((total)ARROW = L(((total)ARROW)_{fin}) = L(((pretotal)ARROW)_{fin})$

(ii) B+ref+sym+tr is a decidable logic.

5. An extension of BAL with a modality for equivalent arrows

We have seen that the condition of a normality is not modally definable. This means that the language $\mathcal{L}$ is not strong enough to tell us the difference between normal and non normal a.f. In this section we shall show that there exists a natural extension of the language $\mathcal{L}$ in which normality become modally definable.

Let $\underline{W}$ be an a.f. and for $x, y \in W$ define

$(\equiv)$ $x \equiv y$ iff $xR_{11}y$ & $xR_{22}y$.

Graphically x≡y: 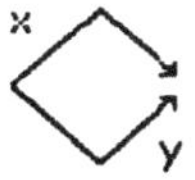

The relation ≡ is called an equivalence of two arrows.

By means of ≡ the normality condition is equivalent to the following one:

(Nor') $(\forall xy\in W)(x\equiv y \rightarrow x=y)$.

If we extend our language $\mathcal{L}$ with a new modality [≡], interpreted in a.f. with the relation ≡, then (Nor') is modally definable by the formula $p\rightarrow[\equiv]p$.

Let the extension of $\mathcal{L}$ with [≡] be denoted with $\mathcal{L}([\equiv])$.The general semantics of $\mathcal{L}([\equiv])$ is defined in the class of all relational structures /called also frames/ of the form $\underline{W}=(W, R_{11}, R_{22}, R_{12}, R_{21}, \equiv)$. The standard semantics of $\mathcal{L}([\equiv])$ is defined in the class of arrow frames with the relation ≡ defined by (≡).

We shall show that the condition (≡) is not modally definable. For that purpose we introduce the following nonstandard semantics of $\mathcal{L}([\equiv])$.

By a nonstandard ≡-arrow frame (≡-a.f.) we mean any system $\underline{W}=(W, R_{11}, R_{22}, R_{12}, R_{21}, \equiv)$ satisfying the following conditions for any $x,y,z\in W$ and $i,j,k\in\{1,2\}$:

(ρii), (σij) and (τijk),

($\equiv\rho$) $x\equiv x$,

($\equiv\sigma$) $x\equiv y \rightarrow y\equiv x$,

($\equiv\tau$) $x\equiv y \;\&\; y\equiv z \rightarrow x\equiv z$,

($\equiv\subseteq R_{ii}$) $x\equiv y \rightarrow xR_{ii}y$

The class of all nonstandard ≡-arrow frames is denoted by Nonstandard-≡-ARROW.

If a nonstandard ≡-a.f. satisfies the condition

($R_{11}\cap R_{22}\subseteq\equiv$) $xR_{11}y \;\&\; xR_{22}y \rightarrow x\equiv y$,

then it is called a standard ≡-a.f. It is easily seen that in any standard ≡-a.f. we have

$x\equiv y \longleftrightarrow xR_{11}y \;\&\; xR_{22}y$.

The class of all nonstandard and standard ≡-arrow frames are denoted respectively by Nonstandard-≡-ARROW and Standard-≡-ARROW.

All conditions from the definition of nonstandard ≡-a.f. are modally definable by the following formulas respectively:

(≡P) $[\equiv]A\rightarrow A$,

(≡Σ) $A\vee[\equiv]\neg[\equiv]A$,

(≡T) $[\equiv]A\rightarrow[\equiv][\equiv]A$,

($\subseteq_{ii}$) $[ii]A\rightarrow[\equiv]A$, $i=1,2$.

We shall show that the condition ($R_{11}\cap R_{22}\subseteq\equiv$) is not modally definable. For that purpose we shall proof first that L(Standard-≡-ARROW)=L(Nonstandard-≡-ARROW).

Lemma 5.1. Let $\underline{W}=(W, R_{11}, R_{22}, R_{12}, R_{21}, \equiv)$ be a nonstandard ≡-a.f. Then there exists a standard ≡-a.f. $\underline{W}'=(W', R'_{11}, R'_{22}, R'_{12}, R'_{21}, \equiv')$ and a copying from $\underline{W}$ to $\underline{W}'$ and if $\underline{W}$ is a finite a.f. then $\underline{W}'$ is a finite a.f. too.

Proof. Use the same construction as in the lemma 2.6 with the following modification. Let $\equiv(x)=\{y\in W/x\equiv y\}$. Since ≡ is an equivalence relation then $\equiv(x)=\equiv(y)$ implies $x\equiv y$. The definitions of R'_{ij} and ≡' are the following:

$f(x)R'_{ij}g(y)$ iff $xR_{ij}y$ & $(f+i.\equiv(x) = g+j.\equiv(y))$

$f(x)\equiv' g(y)$ iff $x\equiv y$ & $f=g$.

The details that this will do are left to the reader. ∎

Corollary 5.2. L(Standard-≡-ARROW)=L(Nonstandard-≡-ARROW).

Now the axiomatization of L(Nonstandard-≡-ARROW) is easy.

Denote by [≡]BAL the following axiomatic system:

Axioms and rules for [≡]BAL

(I) All axioms and rules of BAL,

(II) The following new axioms:

(≡P) $[\equiv]A\to A$,

(≡Σ) $A\vee[\equiv]\neg[\equiv]A$,

(≡T) $[\equiv]A\to[\equiv][\equiv]A$,

(⊆ii) $[ii]A\to[\equiv]A$, $i=1,2$.

Theorem 5.3. [≡]BAL is sound and complete in the class Nonstandard-≡-ARROW.

Proof - by the canonical construction. ∎

Corollary 5.4. [≡]BAL=L(Nonstandard-≡-ARROW)= L(Standard-[≡]-ARROW).

Theorem 5.5. (i) [≡]BAL= L(Nonstandard-≡-ARROW)= $L((\text{Nonstandard-}\equiv\text{-ARROW})_{fin})$

(ii) [≡]BAL is a decidable logic.

Proof. Apply the filtration technic with the following modification: the definition of the relations R'_{ij} is the same as in theorem 4.4., the definition of ≡' is the following

$|x|\equiv'|y|$ iff $(\forall[\equiv]A\in\Psi)(x\Vdash_v[\equiv]A \longleftrightarrow y\Vdash_v[\equiv]A)$ &

$|x|R'_{11}|y|$ & $|x|R'_{22}|y|$.

The details that this definition of filtration will do is left to the reader. ∎

6. Extensions of BAL with propositional constant Loop

We say that an arrow x forms a loop if $xR_{12}x$. Graphically

Let $\underline{W}$ be an a.f. We let $Loop_{\underline{W}}=\{x\in W/xR_{12}x\}$.

Lemma 6.1. In the language $\mathcal{L}$ Loop is not expressible in a sense that there is no a formula A in $\mathcal{L}$ such that for any a.f. $\underline{W}$, valuation v and $x\in W$: $x\Vdash_v A$ iff $x\in Loop_{\underline{W}}$.

Proof. Let $W=\{a, b, c\}$, $R_{11}=R_{22}=\{(a,a), (b,b), (c,c)\}$, $R_{12}=R_{21}=\{(a,a), (b,c), (c,b)\}$. It is east to see that W with the relations R_{ij} is an a.f. Let v be a valuation in W such that for any $p\in VAR$ $v(p)=\emptyset$. Then by induction on the complexity of a formula one can see that for any formula A the set $v(A)=\{x\in W/x\Vdash_v A\}$ is ether W or ∅. Suppose now that there exists a formula A such that $x\Vdash_v A$ iff $x\in$Loop. Then $v(A)=\{a\}$ which contradicts the previous result. ∎

Let $\mathcal{L}$(Loop) be an extension of the language $\mathcal{L}$ by a new propositional constant Loop with the following standard semantics: for any a.f. $\underline{W}$, valuation v and $x\in W$: $x\Vdash_v$ loop iff $x\in Loop_{\underline{W}}$. Loop has also a nonstandard semantics which can be defined in the following way. By a nonstandard Loop arrow frame we mean any system $\underline{W}$=(W,

R_{11}, R_{22}, R_{12}, R_{21}, δ) such that (W, R_{11}, R_{22}, R_{12}, R_{21}) is an a.f. and δ /sometimes denoted by $\delta_{\underline{W}}$/ is a subset of W. Then the interpretation of Loop in a nonstandard Loop a.f. $\underline{W}$ is: $x \Vdash_{\overline{V}}$ Loop iff $x \in \delta_{\underline{W}}$. A nonstandard Loop a.f. $\underline{W}$ is called a standard one if the following two conditions are satisfied stating together that $Loop_{\underline{W}} = \delta_{\underline{W}}$:

(Loop 1) $(\forall x \in W)(x \in \delta_{\underline{W}} \rightarrow x \in Loop_{\underline{W}})$,

(Loop 2) $(\forall x \in W)(x \in Loop_{\underline{W}} \rightarrow x \in \delta_{\underline{W}})$.

The class of all nonstandard Loop arrow frames is denoted by NonstandardLoopARROW. Accordingly the class of all standard Loop a.f. is denoted by StandardLoopARROW. It can be easily shown that the condition (Loop 1) is modally definable in NonstandardLoopARROW by the following formula

(Loop) Loop→([12]A→A).

If a nonstandard Loop a.f. satisfies (Loop 1) we call it a general Loop a.f. The class of general Loop arrow frames is denoted by GeneralLoopARROW. We shall show that condition (Loop 2) is not modally definable in GeneralLoopARROW. For that purpose we shall use the copying construction, which for frames with δ contains an additional condition

(δ) For any $x \in W$ and $f \in I$: $x \in \delta$ iff $f(x) \in \delta'$.

The copying lemma for this version of copying is also true.

Lemma 6.2. Let $\underline{W}$=(W, R_{11}, R_{22}, R_{12}, R_{21}, δ) be a general Loop a.f. Then there exists a standard Loop a.f. $\underline{W}'$=(W', R'_{11}, R'_{22}, R'_{12}, R'_{21}, δ') and a copying I from $\underline{W}$ to $\underline{W}'$ and if $\underline{W}$ is a finite then $\underline{W}'$ is a finite frame too.

Proof. The construction of I and W' is the same as in the proof of lemma 2.6. To define R'_{ij} we first define the function

$$\delta(x)=\begin{cases}0 & \text{if } x\in\delta\\ 1 & \text{if } x\notin\delta\end{cases}$$

where 0 and 1 are considered as zero and unit of the Boolean ring. Then:

$f(x)R'_{ij}g(y)$ iff $xR_{ij}y$ & $(f+i.\delta(x)=g+j.\delta(y))$.

$\delta'=\{x' / \exists x\in\delta\ \exists f\in I\ f(x)=x'\}$

The proof that this is a copying and that $\underline{W}$ is an a.f. is the same as in lemma 2.6. Let us show that $\underline{W}'$ is a standard Loop a.f.

For the condition (Loop 1) suppose $x'\in\delta'$. Then $x'=f(x)$ for some $x\in\delta$ and $f\in I$. So we have $xR_{12}x$, $\delta(x)=0$ and hence $f+1.\delta(x)=f+2.\delta(x)$. This shows that $f(x)R'_{12}f(x)$, hence $x'R'_{12}x'$.

For the condition (Loop 2) suppose $x'R'_{12}x'$. Then for some $x\in W$ and $f\in I$ we have $f(x)=x'$ and $f(x)R'_{12}f(x)$. Then $xR_{12}x$ and $f+\delta(x)=f$, so $\delta(x)=0$, which yields that $x\in\delta$. Thus $x'\in\delta'$. ■

Lemma 6.2 implies the following

Theorem 6.3. L(LoopARROW)=L(GeneralLoopARROW).

Corollary 6.4. Condition (Loop 2) is not modally definable.

Now the axiomatization of L(LoopARROW) is easy: we axiomatize L(GeneralLoopARROW) adding to the axioms of BAL the axiom

(Loop) Loop→([12]A→A).

The obtained system is denoted by LoopBAL. Using the canonical construction one can prove the following

Theorem 6.5. LoopBAL is sound and complete in GeneralLoopARROW.

Corollary 6.6. LoopBAL=L(GeneralLoopARROW)=L(LoopARROW).

The constant Loop makes possible to distinguish the logics L(LoopARROW) and L((nor)LoopARROW). Namely we have

Lemma 6.7. Let $\varphi=A\wedge Loop\rightarrow[12](Loop\rightarrow A)$. Then:

(i) $\varphi\notin$L(LoopARROW),

(ii) $\varphi\in$L((nor)LoopARROW),

(iii) L(LoopARROW)≠L((nor)LoopARROW).

Proof – straightforward by the completeness theorem.■

The formula φ from lemma 6.7 modally defines in GeneralLoopARROW the following condition

(nor_0) $(\forall xy)(xR_{11}y \ \&\ x\in\delta\ \&\ y\in\delta \longrightarrow x=y)$

Let $\underline{W}$ be a general Loop a.f. We call $\underline{W}$ quasi-normal if it satisfies the condition Nor_0.

Lemma 6.8. Let $\underline{W}=(W, R_{11}, R_{22}, R_{12}, R_{21}, \delta)$ be a quasi-normal general Loop a.f. Then there exists a normal Loop a.f. $\underline{W}'=(W, R'_{11}, R'_{22}, R'_{12}, R'_{21}, \delta')$ and a copying I from $\underline{W}$ to $\underline{W}'$ and if $\underline{W}$ is finite then $\underline{W}'$ is finite too.

Proof. The construction of I, W' and R'_{ij} is the same as in lemma 6.2 with the following modification of the function $\delta(x)$:

$$\delta(x)=\begin{cases}0 & \text{if } x\in\delta\\ \{x\} & \text{if } x\notin\delta\end{cases}$$

The proof that $\underline{W}'$ is a standard Loop a.f. is the same as in lemma 6.2. Let us show the condition of normality. For, suppose $f(x)R'_{11}g(y)$ and $f(x)R'_{22}g(y)$. Then we have $xR_{11}y$ & $(f+\delta(x)=g+\delta(y))$ and $xR_{22}y$ & $(f=g)$. From here we get $\delta(x)=\delta(y)$.

Case 1: $\delta(x)=\emptyset$. Then $\delta(y)=\emptyset$ and hence $x,y\in\delta_{\underline{W}}$. By $xR_{11}y$ and $x,y\in\delta_{\underline{W}}$ we get by (nor_0) $x=y$ and by $f=g$ we obtain $f(x)=f(y)$.

Case 2: $\delta(x)\neq\emptyset$. Then $\delta(y)\neq\emptyset$ and hence $\{x\}=\{y\}$, so $x=y$ and consequently $f(x)=g(y)$. This proves the condition of normality. ■

From lemma 6.8 we obtain the following

Theorem 6.9. L((nor_0)GeneralLoopARROW)=L((nor)LoopARROW).

Let NorLoopBAL=LoopBAL+$A\wedge Loop\rightarrow[12](Loop\rightarrow A)$. Using the canonical method we can easily prove the following

Theorem 6.10. NorLoopBAL is sound and complete in the class (nor_0)GeneralLoopARROW.

Corollary 6.11. NorLoopBAL=L((nor)LoopARROW).

Lemma 6.12. The logics L(GeneralLoopARROW) and L((nor_0)GeneralLoopARROW) admit a filtration and are decidable.

Proof. For the L(generalLoopARROW) use the same filtration as for the logic L(ARROW) with the following definition for δ: $\delta'=\{|x|/x\in\delta\}$.

We have to show that the filtered frame satisfies the condition (Loop 1). Suppose $|x|\in\delta'$. Then $x\in\delta$ and by (Loop 1) we get $xR_{12}x$. Then by the properties of filtration we get $|x|R'_{12}|y|$.

For the logic L((nor_0)GeneralLoopARROW) we modify the definition of R'_{ij} as follows:

$|x|R'_{ij}|y|$ iff $(\forall[ij]A\in\Psi)(\forall k\in\{1,2\})(x\Vdash_V[ik]A \longleftrightarrow x\Vdash_V[jk]A)$ &

$(x, y \in \delta' \rightarrow |x| = |y|)$.

The proof that this definition works is left to the reader. ∎

Corollary 6.13. The logics LoopBAL and NorLoopBAL possess finite model property and are decidable.

The language $\mathcal{L}([\equiv],\text{Loop})$ is an extension of the language $\mathcal{L}([\equiv])$ with the constant Loop. The standard semantics of this language is a combination of the standard semantics of $\mathcal{L}([\equiv])$ and $\mathcal{L}(\text{Loop})$. This semantics is also modally undefinable. To axiomatize it we introduce a general semantics for $\mathcal{L}([\equiv],\text{Loop})$ as follows.

A frame $\underline{W}=(W, R_{11}, R_{22}, R_{12}, R_{21}, \equiv, \delta)$ is called a general Loop-≡ arrow frame if $(W, R_{11}, R_{22}, R_{12}, R_{21})$ is an a.f. and ≡ and δ satisfy the conditions on the left side in the next table:

$(\rho\equiv)$	$x \equiv x$,	$(\equiv P)$	$[\equiv]A \rightarrow A$,
$(\sigma\equiv)$	$x\equiv y \rightarrow y \equiv x$,	$(\equiv\Sigma)$	$A \vee [\equiv]\neg[\equiv]A$,
$(\tau\equiv)$	$x\equiv y \,\&\, y\equiv z \rightarrow x \equiv z$,	$(\equiv T)$	$[\equiv]A \rightarrow [\equiv][\equiv]A$,
$(\equiv\subseteq R_{ii})$	$x\equiv y \rightarrow xR_{ii}y$, $i=1,2$,	$(\subseteq ii)$	$[ii]A \rightarrow [\equiv]A$, $i=1,2$,
(Loop 1)	$xR_{12}y$ [illegible],	(Loop)	$\text{Loop} \rightarrow ([\equiv]A \rightarrow A)$,
$(\equiv\delta)$	$x\equiv y \,\&\, x\in\delta \rightarrow y \in \delta$,	(≡Loop)	$\text{Loop} \rightarrow [\equiv]\text{Loop}$,
$(\equiv R_{11}\delta)$	$xR_{11}y \,\&\, x\in\delta \,\&\, y\in\delta \rightarrow x\equiv y$	(≡11Loop)	$\text{Loop} \wedge [\equiv]A \rightarrow [11](\text{Loop} \rightarrow A)$.

If in addition $\underline{W}$ satisfies the condition $(R_{11} \cap R_{22} \subseteq \equiv)$ and (Loop2) it is called standard Loop-≡ arrow frame. The classes of all general Loop-≡ arrow frames and standard Loop-≡-arrow frames are denoted by GeneralLoop-≡-ARROW and StandardLoop-≡-ARROW respectively.

All conditions from the left side of the above table are modally definable by the corresponding formulas from the right side.

We axiomatize the logic $\mathcal{L}([\equiv],\text{Loop})$ by adding all these formulas as axioms to the logic BAL The obtained system is /denoted by [≡]LoopBAL.

Theorem 6.14. The logic [≡]LoopBAL is sound and complete in the class GeneralLoop-≡-ARROW.

Proof - by the canonical construction. ∎

Lemma 6.15. Let $\underline{W}=(W, R_{11}, R_{22}, R_{12}, R_{21}, \equiv, \delta)$ be a general Loop-≡ a.f. Then there exist a standard Loop-≡ a.f. $\underline{W}'=(W', R'_{11}, R'_{22}, R'_{12}, R'_{21}, \equiv', \delta')$ and a copying I from $\underline{W}$ to $\underline{W}'$.

Proof. The set W', I, δ' and R'_{ij} are defined as in lemma 6.8 with the following modification of the function $\delta(x)$:

$$\delta(x) = \begin{cases} 0 & \text{if } x \in \delta \\ \equiv(x) & \text{if } x \notin \delta \end{cases}$$

The relation ≡' is defined as in lemma 5.1. The proof that this construction works is left to the reader. ∎

Corollary 6.16. [≡]LoopBAL=L(GeneralLoop-≡-ARROW)= L(StandardLoop-≡-ARROW).

7. Further perspectives

A. Extensions of BAL with additional connectives.

Sections 5 and 6 can be considered as examples of possible extensions of the language $\mathcal{L}$ with operators having their standard semantics in terms of arrow frames. There are many possibilities of such extensions, depending of what kind of relations between arrows we want to describe in a modal setting. The main scheme is the following: to each n+1-ary relation $R(x_0,x_1,\ldots,x_n)$ to introduce an n-place modal box operation $[R](A_1,\ldots,A_n)$ with the following semantics, coming from the representation theory of Boolean algebras with operators ([J&T 51], see also [VAK 91]):

$x_0 \Vdash_v [R](A_1,\ldots,A_n)$ iff

$(\forall x_1,\ldots,x_n \in W)(R(x_0,x_1,\ldots,x_n) \rightarrow x_1 \Vdash_v A_1$ or...or $x \Vdash_v A_n)$

The dual operator $\langle R\rangle(A_1,\ldots,A_n)$ is defined by $\neg[R](\neg A_1,\ldots,\neg A_n)$.

In the following we list some natural relations between arrows, which are candidates for a modal study:

$Path_n(x_1,\ldots,x_n)$ iff $x_1R_{21}x_2$ & $x_2R_{21}x_3$ & &$x_{n-1}R_{21}x_n$, $n\geq 2$

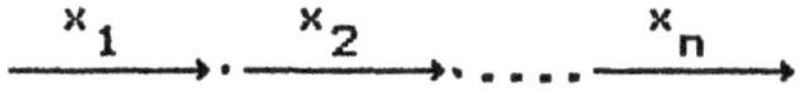

$Path_\infty(x_1,x_2,x_3,\ldots.)$ iff $(\forall n)Path(x_1,\ldots,x_n)$

$Loop_n(x_1,x_2,\ldots,x_n)$ iff $Path_{n+1}(x_1,\ldots,x_n,x_1)$

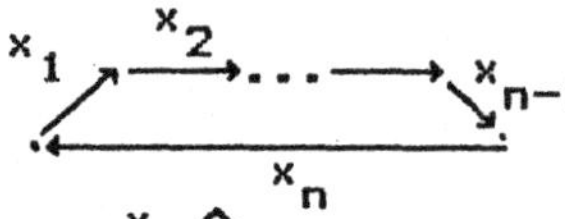

Converse: xSy iff $Loop_2(x,y)$

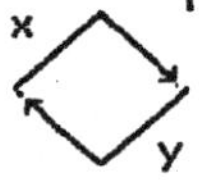

$Trapezium_n(x_1,\ldots,x_n,y)$ iff $Path_n(x_1,\ldots,x_n)$ & $x_1R_{11}y$ & $x_nR_{22}y$

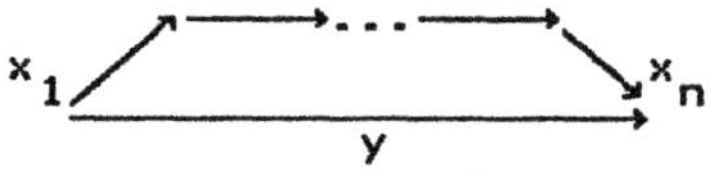

Triangle(x,y,z) iff $Trapezium_2(x,y,z)$

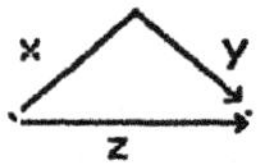

Connectedness: Con(x,y) iff

$\exists n\geq 2\exists x_1\ldots x_n$: $x=x_1$ & $x_n=y$ & $Path_n(x_1,\ldots,x_n)$

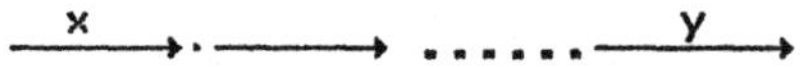

Double side connectedness: Dcon(x,y) iff Con(x,y) & Con(y,x)

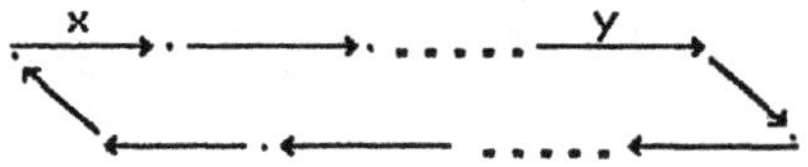

The relations $Path_n$, $Path_\infty$, $Loop_n$ can be used to define also semantics for suitable propositional constants:

$Path_n$, $Path_\infty$, $Loop_n$, Loop as follows.

$x_1 \Vdash_v Path_n$ iff $(\exists x_2,\ldots x_n \in W)Path_n(x_1,x_2,\ldots,x_n)$,

$x_1 \Vdash_v Path_\infty$ iff $(\exists x_2,x_3,\ldots)Path_\infty(x_1,x_2,x_3,\ldots)$,

$x_1 \Vdash_v Loop_n$ iff $(\exists x_2,\ldots,x_n)Loop_n(x_1,x_2,\ldots,x_n)$

$x \Vdash_v Loop$ iff $\exists n$ $x \Vdash_v Loop_n$.

These considerations motivate the following general problem: develop a modal theory /axiomatization, definability, (un)decidability/ of some extensions of BAL with modal operations corresponding to the above defined relations in arrow structures.

For example, the extension of BAL with the modal operations

A●B=<Triangle>(A,B), A^{-1}=[Converse]A and the propositional constant Id=Loop is a natural generalization of the modal logic of binary relations ([BEN 89], [VEN 89], [VEN 91]). This logic has a closed connection with various versions of representable relativized relational algebras ([KRA 89], [MA 82], [NÉM 91]).

B. Arrow semantics of Lambek Calculus and its generalizations.
Let A/B and A\B are "duals" of A●B with the following semantics:
x‖———B/A iff (∀yz∈W)(Triangle(x,y,z) & y‖———A → z‖———B),
  v   v   v
y‖———A\B iff (∀xz∈W)(Triangle(x,y,z) & x‖———A → z‖———B)
  v   v   v

The modal connectives A●B, A\B, and A/B can be considered as the operations in the Lambek Calculus. Mikulás [Mik 92] proves a completeness theorem for the Lambek Calculus with respect to a relational semantics of the above type over transitive normal arrow frames /this is an equivalent reformulation of Mikulás result in "arrow" terminology/. Roorda [Roor 91] and [Roor 91a] studies a modal version of Lambek Calculus extended with classical Boolean operations. So it is natural to study an extension of BAL with the above dyadic modal operations, which will give another intuition for the operations in the Lambek Calculus.

C. Arrow logics and point logics over arrow systems.
With each arrow structure S=(Ar, Po, 1, 2) we can associate thefollowing two relational systems: the arrow frame (Ar, R_{11}, R_{22}, R_{12}, R_{21}) and the point frame (Po, ρ). The first system is used as a semantic base of the logic BAL and the later can be used as a semantic base of an ordinary modal language with a modal operator □. So each class Σ of arrow systems determines a class of arrow frames Ar(Σ) and a class Po(Σ) of point frames. A general question, which arises is the problem of comparative study of the corresponding logics L(Ar(Σ)) and L(Po(Σ)). A kind of a correspondence between first order properties of Po(Σ) and Ar(Σ) was shown in section 1.

D. Many-dimensional generalizations of arrow systems.
The introduced in this paper arrow structures can be generalized to the so called n-dimensional arrow structures ([VAK 91a]) in the following way. Let S=(Ar, Po, 1,...n) be a two sorted algebraic system. S is called an n-dimensional arrow structure if for any i=1,...,n, i is a function from Ar to Po satisfying the axiom:
  (∀A∈Po)(∃i 1≤i≤n)(∃x∈Ar)(i(x)=A)
The arrows in an n-dimensional a.s. looks like as follows:

  1(x) 2(x) 3(x)........n(x)
  ●——————●——————●————————————→●

A natural example of n-dimensional a.s. is the set of all n-tuples of a given n-ary relation. Among the logics based on n-dimensional arrow frames are some generalizations of the so called cylindric modal Logics ([VEN 89], [VEN 91]). These logics have also a very closed connection with some versions of representable relativized cylindric algebras [NÉM 91].

Acknowledgments are due to Johan Van Benthem for calling my attention to arrows and arrow logics and to Hajnal Andréka and Istwán Németi for many stimulating discussions and pointing the connection of "arrow" approach to some problems of algebraic logic.

References

[BEN 86] VAN BENTHEM J.F.A.K., Modal Logic and Classical Logic, Bibliopolis, Napoli, 1986.

[BEN 89] VAN BENTHEM J.F.A.K. Modal Logic and Relational Algebra, manuscript, May 1989, to appear in the proceedingsof Malcev Conference on Algebra, Novosibirsk, 199?.

[BEN 90] VAN BENTHEM J.F.A.K. Private letter, June 1990.

[H&C 84] HUGHES G.E. & M.J.CRESSWELL, A companion to Modal Logic, Methuen, London, 1984.

[J&T 51] JONSSON B., TARSKI A. Boolean algebras with operators. Americ. J. Math., Part I: 73 891-993; Part II: 74,127-162, 1951.

[KRA 89] KRAMER R.L. Relativized Relational Algebras, manuskript, April 1989, to appear in the proc. of the Algebraic Logic Conference, Budapest 1988.

[MA 82] MADDUX R. D. Some varieties containing relational algebras, Trans. Amer. Math. Soc. Vol 272(1982), 501- 526.

[MIK 92] MIKULAÃS Sz., The completeness of the Lambek Calculus with respect to relational semantics, ITLI Prepublications, University of Amsterdam, 1992.

[NÊM 91] NÊMETI I. Algebraizations of Quantifier Logics, an introductory overview, manuscript, June1991, to appear in Studia Logica.

[ROOR 91] ROORDA D. Dyadic Modalities and Lambek Calculus, in Colloquium on Modal Logic 1991, ed. M. de Rijke, Amsterdam 1991.

[ROOR 91a] ROORDA D. Resource Logics, PhD thesis, Fac. Math. and Comp. Sc., University of Amsterdam, Amsterdam 1991.

[SEG 71] SEGERBERG K. An Essay in Classical Modal Logic, Filosofiska Studier 13, Uppsala, 1971.

[VAK 90] VAKARELOV D. Arrow logics, Manuscript, September 199?

[VAK 91] VAKARELOV D. Rough Polyadic Modal Logics, Journal of Applied Non-Classical Logics, v. 1, 1(1991), 9-35.

[VAK 91a] VAKARELOV D. Modal Logics for Reasoning about Arrows: Arrow Logics, in the proc. of 9-th International Congress of Logic Methodology and Philosophy of Sciences, Section 5 - Philosophical Logic, August 7-14, 1991, Uppsala.

[VAK 92] VAKARELOV D. Arrow logics with cylindric operators, abstract of a paper submeted to the 1992 European Summer Meeting of the ASL.

[VEN 89] VENEMA Y. Two-dimensional Modal Logic for Relational Algebras and Temporal Logic of Intervals, ITLI-prepublication series LP-89-03, University of Amsterdam, Amsterdam 1989.

[VEN 91] VENEMA Y. Many-dimensional Modal Logic, PhD thesis, September 1991, Fac. Math. and Comp. Sc., University of Amsterdam. To appear.

Knowledge without Modality: A Simplified Framework for Chronological Ignorance

Craig MacNish

Department of Computer Science
University of York
York YO1 5DD, UK
craig@minster.york.ac.uk

Abstract. Shoham's logic of chronological ignorance (CI) provides a formal approach for representing and reasoning about causal relationships. The utility of the formalism comes from imposing a partial order on Kripke interpretations for theories in a modal logic of *temporal knowledge*. In this paper we show that the use of modal logic in this context is unnecessary and that the same result can be achieved using classical logic. While semantically simpler, the classical approach is somewhat unwieldy. To overcome this problem we suggest an alternative underlying language, called *asserted logic*. The logic allows the convenient representation found in Shoham's approach without sacrificing the truth functional semantics and proof theory of classical logic. We demonstrate the utility of the logic by providing an equivalent nonmodal CI framework for causal theories.

1 Introduction

The logic of *chronological ignorance* (CI) [15, 16] was proposed by Shoham as a solution to the qualification and extended prediction problems [17] and as a formal account of causal reasoning in AI. For a broader discussion on the use of CI the reader is referred to [8].

CI is a nonmonotonic version of Shoham's modal logic of *temporal knowledge* (TK) [15]. The "knowledge" part comes from an interpretation of the modal operator $\Box$ as asserting that its argument is *known* to be true, and its dual $\Diamond$ as asserting that its argument is *possibly* true. The use of modal logic in this context follows earlier attempts to incorporate concepts such as consistency into an object language [11, 12]. The idea is to sanction weak inferences of the form $\Diamond\alpha$ (read "α is possible") when an assertion α is consistent with the knowledge base—that is when $\neg\alpha$ cannot be proven. This general principle is followed in CI by giving preference to interpretations in which the assertions which are *known* are minimised. Thus if p is a proposition and $\Box\neg p$ cannot be proven, precedence is given to interpretations in which $\neg\Box\neg p$ (or $\Diamond p$) is true.

Modal logic has been adopted for this task with little supporting motivation, and it is worth considering whether it is the appropriate choice of language. The shift from classical to modal logic sacrifices truth functionality [14]—for example, $\Box\alpha$ and $\Box\beta$ may have different truth values even though α and β have the same truth value—and complicates the semantics, axiomatisation and proof theory of the logic. Interpretations for modal logic require a family of structures or *possible worlds* related by an accessibility relation [6, 7]. Different accessibility relations correspond to different modal systems and axiomatisations, and classical proof theory is no longer adequate (see for example [18]). These complexities are further compounded when the logic is extended to incorporate such features as continuous time and strong negation [10], making it difficult to provide a sound proof theory.

In return for these sacrifices modal logic allows a great deal of freedom in the construction of formulas. The scope of the modal operator can include atomic formulas, compound formulas, and formulas which themselves include the modal operator. One test of the appropriateness of modal logic is whether these compound and nested forms have practical applications. For example, while it seems natural to interpret $\Box p \vee \Box q$ as asserting that p is known or q is known, it is less clear why we would wish to be able to use $\Box(p \vee q)$ (in each possible world p or q is true, but neither p nor q need be known) or $\Diamond\Diamond p$ (p is possibly possibly true) in this context.

This issue is taken further by Galton [3] who argues that from a representational point of view Shoham's use of $\Box$ does not resemble a modal operator, since the formulas $\Box\alpha$ and α have different statuses in the logic. Galton suggests that a more accurate reading of Shoham's operator would be "There is reason to believe that...".

2 The "Confidence" Spectrum

The one respect in which Shoham's operator does resemble a modal operator, according to Galton, is that it does not commute with negation. That is, $\Box\neg\alpha$ is in general not equivalent to $\neg\Box\alpha$. Another way of saying this is that $\Box\alpha \vee \Box\neg\alpha$ is not tautologically true. That is, the logic can be considered to escape the law of excluded middle at what might be called the *epistemic* level—it is not necessary for a formula to be either known to be true, or known to be false. The result is that the modal operator and negation connective can be used to obtain the spectrum of interpretations shown in Table 1, which we call a *confidence spectrum*.

Table 1. The confidence spectrum for modal logic.

Formal Interpretation			Informal
α	$\Box\alpha$	$\Box\neg\alpha$	Interpretation
true in all worlds	t	f	known to be true
true in some world	–	f	possibly true
true in some worlds, false in others	f	f	nothing known
false in some world	f	–	possibly false
false in all worlds	f	t	known to be false

With this spectrum available, a minimisation strategy is no longer forced to commit assertions to one extreme or the other, but can make use of the "middle ground". This is exactly what is done by the CI minimisation strategy: if an assertion is not forced to take the values "known to be true" or "known to be false" then it is assigned the value "nothing known".[1] In fact this is the *only* benefit of modality in chronological ignorance.

If we only require knowledge of atomic formulas, as is the case in CI, then this spectrum can be achieved using classical logic. To illustrate this point consider a classical propositional language. A simple language of knowledge, which we will call K_P, can be constructed by replacing each proposition p by the two propositions formed by prefixing p with **known_** and **known_not_** (we will use *known_p* and *known_not_p* as meta-variables representing the resulting propositions). An assignment of t to *known_p* (respectively *known_not_p*) is intended to mean that the original proposition p is known to be true (respectively false). To complete the formalism we add the formula

$$\neg(known\_p \wedge known\_not\_p) \tag{1}$$

for each pair of propositions to ensure that p cannot be known to be true and false simultaneously. The resulting confidence spectrum, which is shown in Table 2, is identical to the spectrum for modal logic. In this case, however, the formal interpretation is simply a classical truth valuation. (Note that we have also modified the informal interpretation for consistency with Galton [3]).

Table 2. The confidence spectrum for K_P.

Formal Interpretation		Informal
known_p	*known_not_p*	Interpretation
t	f	reason to believe
–	f	no reason to disbelieve
f	f	no reason to believe or disbelieve
f	–	no reason to believe
f	t	reason to disbelieve

Chronological ignorance can be redefined using a temporal version of the classical logic K_P, although the representation is somewhat unwieldy. The intended predicate symbols must be altered and formulas in the form of (1) must be explicitly added for all atomic formulas. What we would like instead is a language which has the convenient representation found in Shoham's TK without sacrificing the truth functional semantics and proof theory of K_P. Our solution is a language called asserted logic (AL) [9]. Asserted logic replaces the '**known_**' part of K_P formulas by a truth functional operator and the '**not_**' part by a strong negation connective. The collection of formulas of the form (1) can then be replaced by an axiom schema which is incorporated in the logic.

[1] Although Shoham informally interprets $\Diamond$ to mean possibly true and $\Diamond\neg$ to mean possibly false, chronological minimisation uses only the middle value "nothing known". In [8] we argue that this approach is counter-intuitive and suggest more appropriate strategies which use the full spectrum of values.

3 Asserted Logic

In this section we give an overview of asserted logic. For a more detailed treatment the reader is referred to [8, 9].

3.1 Syntax of AL

Let L be a classical first-order language. The symbols of AL are the symbols of L, a (3-valued negation) connective '$-$' and the operator **T**. The latter is called Bochvar's *assertion operator* after [5].

A new construction, called a *base formula*, is defined as follows:

1. If p is an atomic formula of L, then p is a base formula of AL.
2. If x is a base formula of AL, then $-x$ is a base formula of AL.

If p is an atomic formula of L, then p is called an *atomic base formula* of AL and p and $-p$ are called *base literals*. We will refer to base formulas with no free variables as *base sentences*, and to variable free atomic base formulas as *propositions*.

The atomic well-formed formulas (wffs) of AL are strings $\mathbf{T}x$ where x is a base formula. All other wffs are constructed from the atomic formulas in the usual way. A *theory* is a set of well-formed formulas. It is important to note that base formulas are not well-formed—they only appear in a theory within the scope of the assertion operator (hence the name *asserted logic*). Thus $\mathbf{T}p$ and $\mathbf{T}{-p} \rightarrow \neg\mathbf{T}q$ (where p and q are atomic base formulas) are examples of wffs while p, $\neg p \wedge \mathbf{T}p$, $\mathbf{T}(p \rightarrow q)$ and $\mathbf{TT}p$ are not wffs.

3.2 Propositional Semantics of AL

Since Shoham's causal theories are propositional we only describe the propositional semantics of AL in this paper. A first-order semantics for AL is proposed in [8], and similar arguments can be applied to the first-order extensions to causal theories recently discussed by Bell [2] and Nakata [13].

Base formulas of AL can be assigned any one of three truth values, t (true), f (false) or u (unknown). The strong negation connective '$-$' is defined according to the standard interpretation of 3-valued negation:

y	$-y$
t	f
u	u
f	t

The assertion operator **T** maps the three truth values onto the two traditional values t and f according to the following truth table:

y	$\mathbf{T}y$
t	t
u	f
f	f

Enclosing the three valued base formulas within the scope of the assertion operator allows the "two valued" connectives to adopt their usual meanings.

The formal semantics is given by the following definition.

Definition 1. *A* truth valuation *on* AL *is a mapping σ assigning to each base formula x a value x^σ from the set $\{t, f, u\}$ and to each wff α a value α^σ from the set $\{t, f\}$, such that for all base formulas y and wffs β and γ*

1. $(-y)^\sigma = t$ *iff* $y^\sigma = f$, *and* $(-y)^\sigma = f$ *iff* $y^\sigma = t$,
2. $(\mathbf{T}y)^\sigma = t$ *iff* $y^\sigma = t$,
3. $(\neg\beta)^\sigma = t$ *iff* $\beta^\sigma = f$,
4. $(\beta \rightarrow \gamma)^\sigma = t$ *iff* $\beta^\sigma = f$ *or* $\gamma^\sigma = t$.

The semantics differs from classical propositional semantics only by the addition of the first two conditions, which determine the meaning of the 3-valued negation connective and the assertion operator. Satisfaction and tautological consequence are defined in the usual way.

3.3 The Propositional Calculus in AL

The axioms of AL are those of the classical propositional calculus plus instances of the axiom schemata

$$\neg(\mathbf{T}x \wedge \mathbf{T}{-x}) \tag{2}$$

and

$$\mathbf{T}{-}{-}x \leftrightarrow \mathbf{T}x \ . \tag{3}$$

Axiom (2) ensures that $\mathbf{T}x$ and $\mathbf{T}{-x}$ cannot both be deduced from a consistent theory. Axiom (3) accounts for the nesting of strong negation connectives.

In [8] we show that this modified propositional calculus is sound and complete (that is, $\Phi \models \alpha$ if and only if $\Phi \vdash \alpha$). Note that since base formulas do not appear outside the scope of the assertion operator we do not require rules such as necessitation ($\vdash x \,/ \vdash \Box x$) and axioms such as reflexivity ($\Box x \rightarrow x$) which are associated with modal logics.

3.4 Epistemological Interpretation

The assertion operator **T**, like the modal operator $\Box$, can be interpreted as asserting that its argument is *known* to be true or, using Galton's interpretation, that *there is reason to believe* that its argument is true. This leads to the confidence spectrum for AL shown in Table 3. The logic escapes the law of excluded middle at the "epistemic" level since $\mathbf{T}x \vee \mathbf{T}{-x}$ is not tautologically true.[2]

[2] Note that adding $\mathbf{T}x \vee \mathbf{T}{-x}$ as an axiom schema reverts the logic to a classical 2-valued system.

Table 3. The confidence spectrum for asserted logic.

Formal Interpretation			Informal
x	$\mathbf{T}x$	$\mathbf{T}{-}x$	Interpretation
t	t	f	reason to believe
t or u	–	f	no reason to disbelieve
u	f	f	no reason to believe or disbelieve
u or f	f	–	no reason to believe
f	f	t	reason to disbelieve

For notational convenience we define **P** (possibly), **F** (known to be false) and **U** (unknown) as follows:

$$\mathbf{P}x =_{\mathsf{def}} \neg\mathbf{T}{-}x$$
$$\mathbf{F}x =_{\mathsf{def}} \mathbf{T}{-}x$$
$$\mathbf{U}x =_{\mathsf{def}} \neg(\mathbf{T}x \vee \mathbf{T}{-}x)$$

The five levels in the confidence spectrum therefore correspond to assignments of t to $\mathbf{T}x$, $\mathbf{P}x$, $\mathbf{U}x$, $\mathbf{P}{-}x$ and $\mathbf{F}x$ respectively. Note that **P**, **F** and **U** are not additional operators (although it may sometimes be convenient to refer to them as such) but meta-language symbols with the substitutions defined above.

While **T** exhibits epistemological characteristics often associated with the modal operator $\Box$, it is important to remember that **T** is simply a truth functional operator. In fact, any propositional AL theory can be mapped directly onto a $\mathrm{K_P}$ theory. Classical "off the shelf" theorem provers can therefore be used to test the validity of propositional AL theorems.

4 Chronological Ignorance in AL

The translation of CI to asserted logic is straightforward. A temporal aspect is given to base formulas, which in turn permits an appropriate partial order to be imposed on truth valuations. In this section we redefine CI for the propositional case (without temporal variables) which provides the basis for a comparison of causal theories.

4.1 Temporal AL

Shoham [15, 16] forms a temporal logic from a classical one by reifying propositions with pairs of temporal arguments denoting intervals. Bacchus et al. [1] argue that this approach is overly restrictive and sacrifices classical proof theory unnecessarily. They show that a two-sorted logic called BTK is more expressive and subsumes Shoham's temporal logic.

Since causal theories are propositional our choice here is not particularly important. However we will take advantage of the less restrictive syntax of BTK and allow single temporal arguments. One minor difference is that we switch the order of temporal and non-temporal terms in BTK to conform with the bulk of the literature on state based reasoning (for example [4]).

BTK is a standard many-sorted logic with two disjoint sorts, for nontemporal and temporal objects. Each predicate or function symbol has arity (m, n) where m and n are natural numbers—the first m arguments are nontemporal while the last n are temporal. The sort of a function is determined by the sort that the function returns. *Temporal asserted logic* (TAL) is formed in the same way as AL except that the underlying classical language L is replaced by the two-sorted language BTK.

In the remainder of this paper we will follow the convention that the temporal constant symbols are drawn from the set of integers $\mathbf{Z}$ and that all predicate symbols have arity $(m, 1)$ or $(m, 2)$. The former can be interpreted as associating an assertion with a particular time point or a state, while the latter associates an assertion with an interval. Thus $\texttt{loaded(gun,1)}$ indicates that a gun is loaded at time 1, while $\texttt{red(house17,2,12)}$ indicates that a particular house is red in the interval from 2 to 12.

The *latest time point* (*ltp*) of a propositional sentence is the greater (with respect to the normal interpretation of the integers) of the temporal constant symbols appearing in the sentence. We will follow the convention that the latest time point in an atomic base sentence is always written as the last argument. Thus the *ltp* of a base sentence $p(\ldots, t_1, t_2)$ is t_2.

4.2 Nonmonotonicity

TAL is made nonmonotonic by imposing a partial order on truth valuations and choosing a least element in the partial order which satisfies the theory in question. The partial order gives preference to truth valuations in which the base formulas that are *known* occur as *late* as possible (hence the name chronological ignorance). Our partial order is simpler than Shoham's [15, Def. 3.4] since the truth functionality of TAL allows us to specify the truth values of base formulas directly.

Definition 2. *A truth valuation σ_2 is* chronologically more ignorant *than a truth valuation σ_1 (written $\sigma_2 \prec_c \sigma_1$) if there exists a time t_0 such that*

1. *for any base sentence x whose ltp $\leq t_0$, if $x^{\sigma_1} = u$ then $x^{\sigma_2} = u$, and*
2. *there exists some base sentence y whose ltp is t_0 such that $y^{\sigma_2} = u$ but $y^{\sigma_1} \neq u$.*

Definition 3. *σ is a* chronologically maximally ignorant *(cmi) truth valuation for a theory Φ if $\sigma \models \Phi$ and there is no $\sigma' \prec_c \sigma$ such that $\sigma' \models \Phi$.*

According to Definition 3, preferred truth valuations assign the value u to base formulas wherever possible, choosing base formulas with smaller *ltp*'s where necessary. If there is a choice between base formulas with the same *ltp*'s then there will generally not be a unique cmi truth valuation. Causal theories, which are considered in the following section, are designed so that this problem does not arise.

5 Causal Theories

Causal theories are intended to capture the notion of physical causality—that what is true prior to some point in time determines what is true after that time. Shoham's causal theories [15, Defs. 4.1 and 4.2] can be redefined in AL as follows.

Definition 4. *A* causal sentence *is a sentence of the form*

$$\bigwedge_{i=1}^{m} \mathbf{T}x_i \;\wedge\; \bigwedge_{j=1}^{n} \mathbf{P}y_j \;\rightarrow\; \mathbf{T}z, \qquad m, n \in \mathbb{N} \tag{4}$$

where x_i, y_j $(i, j > 0)$ and z are base literals such that

1. *$ltp(x_i) < ltp(z)$ for $i = 1, \ldots, m$, and*
2. *$ltp(y_j) < ltp(z)$ for $j = 1, \ldots, n$.*

If $m = 0$ or $n = 0$ the corresponding (empty) conjunction is identically true. A causal sentence is called a boundary condition *if $m = 0$, and a* causal rule *otherwise.*

Definition 5. *A* causal theory *Φ is a set of causal sentences such that*

1. *there is a time point t_0 such that for all boundary conditions $\beta \in \Phi$, $t_0 < ltp(\beta)$,*
2. *there do not exist sentences $\alpha_1 \rightarrow \mathbf{T}z$ and $\alpha_2 \rightarrow \mathbf{T}{-z}$ in Φ such that $\{\alpha_1, \alpha_2\}$ is consistent, and*
3. *there is no x such that $\mathbf{P}x$ and $\mathbf{P}{-x}$ both appear on the left-hand side of sentences in Φ.*[3]

Causal sentences are thus implications from knowledge prior to some point in time to knowledge at that time. The partial order on interpretations gives a sense of unidirectionality to the implication connective which is appropriate for predictive reasoning. More importantly from a technical point of view, however, the constraints on causal theories guarantee that the partial order on interpretations has a unique least element satisfying any given theory. We now prove this to be the case.

5.1 CMI Truth Valuations

The utility of Shoham's causal theories rests largely on the "unique model theorem" [15, Thm. 4.4] which says that the same base sentences are *known* in all cmi interpretations of a causal theory. A stronger result follows in our formalism; namely, that there is exactly one cmi truth valuation for any causal theory.

Theorem 6 (Unique cmi truth valuation). *Let Φ be a causal theory and Σ be the set of all truth valuations on* TAL. *Reduce Σ to Σ_∞ as follows:*

- *Let t_0 be a time point such that for all boundary conditions $\beta \in \Phi$, $t_0 < ltp(\beta)$. Let Σ_0 be the set of truth valuations $\sigma \in \Sigma$ in which $x^\sigma = u$ for all atomic base sentences x whose $ltp \leq t_0$.*

[3] The third condition is not strictly necessary since we do not require Shoham's *soundness conditions* [15, Def. 5.1] and there is therefore a consistent interpretation of $\{\mathbf{P}x, \mathbf{P}{-x}\}$ in which x is assigned the value u. The soundness conditions are discussed further in [8]. The sentence is included here for compatibility with [15].

– *For all $i > 0$, $i \in \mathbb{N}$, let $t_i = t_{i-1} + 1$ and Σ_i be the set of truth valuations σ such that $\sigma \in \Sigma_{i-1}$ and*

$$x^\sigma = \begin{cases} t & \alpha \rightarrow \mathbf{T}x \in \Phi \text{ and } \sigma \models \alpha & (5a)\\ f & \alpha \rightarrow \mathbf{T}{-}x \in \Phi \text{ and } \sigma \models \alpha & (5b)\\ u & \text{otherwise} & (5c)\end{cases}$$

for all atomic base sentences x whose ltp is t_i. Let $\Sigma_\infty = \bigcap_{i=0}^{\infty} \Sigma_i$.

Then

1. *Σ_∞ contains a single truth valuation,*
2. *if $\sigma \in \Sigma_\infty$ then $\sigma \models \Phi$, and*
3. *σ is a cmi truth valuation for Φ if and only if $\sigma \in \Sigma_\infty$.*

That is, Φ is consistent and has a unique cmi truth valuation.

Proof. [1] The second condition of Definition 5 ensures that (5a) and (5b) cannot be simultaneously satisfied. The construction therefore defines a mapping from each atomic base formula onto a unique truth value in $\{t, f, u\}$. The mapping extends according to Definition 1 to a unique truth valuation.

[2] Let $\sigma \in \Sigma_\infty$. Any sentence $\phi \in \Phi$ whose $ltp \leq t_0$ must be a causal rule of the form $\mathbf{T}x_1 \wedge \ldots \rightarrow \mathbf{T}z$ where $ltp(x_1) < t_0$. Therefore $x_1^\sigma = u$, $(\mathbf{T}x_1)^\sigma = f$ and $\sigma \models \phi$. Any sentences whose $ltp > t_0$ must also be satisfied since it is of the form $\alpha \rightarrow \mathbf{T}z$ and from (5a)–(5c), if $\alpha^\sigma = t$ then $(\mathbf{T}z)^\sigma = t$.

[3] ($\Leftarrow$) Assume $\sigma \in \Sigma_\infty$ is not a cmi truth valuation for Φ. From [2] $\sigma \models \Phi$ so there must be some $\sigma' \prec_c \sigma$ such that $\sigma' \models \Phi$. Let t_e be the earliest time point at which σ and σ' differ. Then there is some atomic base sentence x with ltp t_e such that $x^\sigma \neq u$ and $x^{\sigma'} \neq x^\sigma$. From the construction there must be some sentence $\alpha \rightarrow \mathbf{T}x$ or $\alpha \rightarrow \mathbf{T}{-}x$ such that $\alpha^\sigma = t$. But $ltp(\alpha) < t_e$ so $\alpha^{\sigma'} = \alpha^\sigma = t$. Clearly σ' does not satisfy this sentence and hence $\sigma' \not\models \Phi$ — contradiction.

($\Rightarrow$) Assume σ is a cmi truth valuation for Φ but $\sigma \notin \Sigma_\infty$. From [1] and [2] there exists $\sigma' \in \Sigma_\infty$ such that $\sigma' \models \Phi$. Since $\sigma \notin \Sigma_\infty$ either

1. $x^\sigma \neq u$ for some atomic base sentence x whose $ltp \leq t_0$, or
2. there exists i such that $\sigma \in \Sigma_{i-1}$ but $\sigma \notin \Sigma_i$, in which case either
   (a) $\alpha \rightarrow \mathbf{T}x \in \Phi$ for some atomic base sentence x whose $ltp = t_i$ and $\sigma \models \alpha$, but $x^\sigma \neq t$,
   (b) $\alpha \rightarrow \mathbf{T}{-}x \in \Phi$ for some atomic base sentence x whose $ltp = t_i$ and $\sigma \models \alpha$, but $x^\sigma \neq f$, or
   (c) there is no base sentence x whose $ltp = t_i$ such that $\alpha \rightarrow \mathbf{T}x \in \Phi$ and $\sigma \models \alpha$, but $x^\sigma \neq u$.

In the case of 1 or 2(c), $\sigma' \prec_c \sigma$, and in the case of 2(a) or 2(b), $\sigma \not\models \Phi$. Therefore σ is not a cmi truth valuation for Φ — contradiction. □

From the construction it is clear that if Φ is a finite causal theory, then the number of atomic base sentences which are assigned a value other than u by the cmi truth valuation for Φ is no greater than the number of sentences in Φ. These atomic base sentences are found by the following algorithm, which simply assigns the truth values determined by (5a)–(5b) .

Algorithm 7. *Let Φ be a finite causal theory with cmi truth valuation σ.*

Step 1 *Let T be a list of all the sentences in Φ and S be a list of all the atomic base sentences appearing in T.*
Step 2 *Sort the sentences in T and S by ltp.*
Step 3 *If T is empty then halt. The atomic base sentences assigned t by σ are labeled t in S, and the atomic base sentences assigned f by σ are labeled f in S.*
Step 4 *Remove the first element $\bigwedge_{i=1}^{m} \mathbf{T}x_i \wedge \bigwedge_{j=1}^{n} \mathbf{P}y_j \rightarrow \mathbf{T}z$ from T. If one of the following conditions holds:*
1. *for some i, x_i is a positive base literal and x_i is not labeled t in S,*
2. *for some i, x_i is a negative base literal $-x_i'$ and x_i' is not labeled f in S,*
3. *for some j, y_j is a positive base literal and y_j is labeled f in S,*
4. *for some j, y_j is a negative base literal $-y_j'$ and y_j' is labeled t in S,*

then go to Step 3. Otherwise label the atomic part of z either t or f in S depending on whether z is a positive or negative base literal, and go to Step 3.

5.2 Equivalence of the Formalisms

Other than syntactic variations, Algorithm 7 is precisely the algorithm provided by Shoham [15, pp300-301]. This correspondence is formalised by the following corollary. We make use of a syntactic mapping *TK* from TAL base sentences to TK base sentences defined by:

$$\begin{aligned} TK(p(\ldots,t)) &=_{\mathrm{def}} \mathrm{TRUE}(t,t,p'(\ldots)); \\ &\qquad p \text{ has arity } (m,1),\ \ p' \text{ has arity } m \\ TK(p(\ldots,t_1,t_2)) &=_{\mathrm{def}} \mathrm{TRUE}(t_1,t_2,p'(\ldots)); \\ &\qquad p \text{ has arity } (m,2),\ \ p' \text{ has arity } m \\ TK(-x) &=_{\mathrm{def}} \neg TK(x) \end{aligned}$$

Corollary 8. *Let Φ be a finite causal theory with cmi truth valuation σ. Let*

$$\Phi' = \left\{ \bigwedge_{i=1}^{m} \Box TK(x_i) \wedge \bigwedge_{j=1}^{n} \Diamond TK(y_j) \supset \Box TK(z) \;\middle|\; \bigwedge_{i=1}^{m} \mathbf{T}x_i \wedge \bigwedge_{j=1}^{n} \mathbf{P}y_j \rightarrow \mathbf{T}z \ \in \Phi \right\}.$$

Then for any base literal w, $\mathbf{T}w^{\sigma} = t$ if and only if $TK(w)$ is known *in all cmi models of the TK theory Φ'.*

In other words, the formalisms are equivalent in terms of the beliefs which they sanction.

6 Conclusion

The preference order on models which forms the basis of CI does not require the use of modal logic. The same results can be achieved using classical logic, or more conveniently, using asserted logic. We have demonstrated the latter approach by providing a nonmodal account of CI for causal theories. In doing so we have simplified the semantics of CI and the proof of the "unique model" theorem for causal theories.

In [8] we show that this simplification to a truth-functional language permits a proof theory for chronological ignorance based on classical deduction.

On a broader level this work calls into question the ready acceptance of modal logic for representing epistemic concepts such as knowledge and possibility in automated reasoning systems. Since the inclusion of modality complicates the semantics and proof theory of a logic, its use should be justified—for example by the need to nest modal operators or apply them to compound formulas. If these complex forms are not needed then the shift to modal logic may be unnecessary.

References

1. Bacchus, F., Tenenberg, J. and Koomen, J. A. A Non-Reified Temporal Logic. In *KR'89: Proc. 1st International Conference on Principles of Knowledge Representation and Reasoning*, pp. 2–10, Toronto, Canada, 1989. Morgan Kaufmann, San Mateo, USA.
2. Bell, J. Extended Causal Theories. *Artificial Intelligence*, **48** pp. 211–224, 1991.
3. Galton, A. A Critique of Yoav Shoham's Theory of Causal Reasoning. In *Proc. AAAI-91*, Anaheim, USA, 1991.
4. Green, C. Theorem-Proving by Resolution as a Basis for Question-Answering Systems. In Meltzer, B. and Michie, D., editors, *Machine Intelligence 4*, pp. 183–205. Edinburgh University Press, Edinburgh, UK, 1969.
5. Haack, S. *Philosophy of Logics.* Cambridge University Press, Cambridge, UK, 1978.
6. Hughes, G. and Cresswell, M. *An Introduction to Modal Logic.* Methuen, London, UK, 1968.
7. Kripke, S. Semantical Considerations on Modal Logic. *Acta Philosophica Fennica*, **16**, 1963.
8. MacNish, C. *Nonmonotonic Inference Systems for Modelling Dynamic Processes.* PhD thesis, Technical Report CUED/F-INFENG/TR.102, Department of Engineering, University of Cambridge, Cambridge, UK, 1992.
9. MacNish, C. and Fallside, F. Asserted 3-valued Logic for Default Reasoning. In *AI'90: Proc. 4th Australian Joint Conference on Artificial Intelligence*, pp. 65–80, Perth, Australia, 1990. World Scientific Publishing, Singapore.
10. MacNish, C. and Fallside, F. Temporal Reasoning: A Solution for Multiple Agent Collision Avoidance. In *Proc. 1990 IEEE International Conference on Robotics and Automation*, pp. 494–499, Cincinnati, USA, 1990.
11. McDermott, D. Non-Monotonic Logic II: Non-Monotonic Modal Theories. *J. ACM*, **29** pp. 33–57, 1982.
12. McDermott, D. and Doyle, J. Non-Monotonic Logic I. *Artificial Intelligence*, **13** pp. 41–72, 1980.
13. Nakata, K. *Implementing Non-monotonic Temporal Logic.* MSc thesis, Department of Artificial Intelligence, University of Edinburgh, Edinburgh, UK, 1991.
14. Quine, W. V. Three Grades of Modal Involvement. In *The Ways of Paradox*, pp. 158–176. Harvard University Press, USA, 1976.
15. Shoham, Y. Chronological Ignorance: Experiments in Nonmonotonic Temporal Reasoning. *Artificial Intelligence*, **36** pp. 279–331, 1988.
16. Shoham, Y. *Reasoning About Change: Time and Causation from the Standpoint of Artificial Intelligence.* MIT Press, London, UK, 1988.
17. Shoham, Y. and McDermott, D. Problems in Formal Temporal Reasoning. *Artificial Intelligence*, **36** pp. 49–61, 1988.
18. Wallen, L. A. *Automated Proof Search in Non-classical Logics: Efficient Matrix Proof Methods for Modal and Intuitionistic Logics.* MIT Press, Boston, USA, 1990.

Design Complete Sequential Calculus for Continuous Fixpoint Temporal Logic

Regimantas Pliuškevičius

Institute of Mathematics and Informatics,
Akademijos 4, Vilnius 2600, LITHUANIA
email: logica @ ma-mii.lt.su Phone +7(0122)359511

Abstract. We present the method for design complete sequential calculus for continuous fixpoint temporal logic. The method relies on reduction of infinitary calculus for temporal logic under consideration to finitary one. As result of the reduction we can prove completeness and obtain the cut-free sequential calculus for temporal logic under consideration. The reduction present some way for searching so called invariant formulas.
Key words and phrases: fixpoint temporal logic, sequential calculi, infinitary rules of inference, completeness.

Introduction

Temporal logics become a very important tool for solving of various problems of artificial intelligence, cognitive science and theoretical computer science (see e.g.[4]). Discrete linear temporal logic is successfully used for specifying and verifying safety and liveness properties of concurrent systems. Unfortunately it has been shown that traditional temporal logic (with temporal operators $\bigcirc$ (next), $\Box$ (always) and $\mathcal{W}$ (unless)) is not expressive enough (e.g. in [11] it was demonstrated that in traditional temporal logic it is not possible to express the fact that a proposition had to be true on every even moment). For this reason the extension of traditional temporal logic with fixed point operators μ (minimal fixed point operator) and ν (maximal fixed point operator) has been recently advocated in the literature (see e.g. [1, 2, 10]). It is expected that temporal logic extended with fixed points operators will be the most useful logic for describing sets of execution behaviors.

There are two different approaches in the area of various applications of temporal logic: semantic and axiomatic. First one is related with the so called model checking method which is successfully applied when concurrent computing system can be described with a finite state space. But in the case of an infinite state space the methods of analysis and development of concurrent computing systems require more intimate knowledge about proofs methods and effective properties of calculus in various temporal logics.

In [7, 8, 9] it was proposed the method of investigation of a finitary calculus for temporal logic by means of infinitary ones. As result of reduction of infinitary calculus for temporal logic to finitary one we can prove completeness and obtain the cut-free sequential calculus for temporal logic under consideration. It must be stressed that the reduction presents some way for searching invariant formulas (e.g. for the temporal logic with $\circ$ (next) and $\Box$ (always) any invariant formula R is such are that $\models R \supset \circ R$). The finding of invariant formulas is the key element in the finitary derivation in temporal logics. Here we apply this method to investigate temporal logic with fixed point operators μ and ν. It is assumed that the fixpoint temporal logic satisfies monotonicity and continuity conditions (see sections 1, 2 and [2, 5, 6]).

Continuous fixpoint temporal logic is strictly weaker (see e.g. [5]) in expressive power than full (monotone) fixpoint temporal logic. Neither finiteness nor well-foundness is expressible in the continuous fixpoint temporal logic. Nevertheless, the continuous fixpoint temporal logic is itself strictly stronger than temporal logic with usual temporal operators, e.g. temporal operators $\Box$ (always), $\Diamond$ (sometimes), $\mathcal{W}$ (unless) and U (until) can be defined by the help of fixed point operators μ and $\nu : \Box A = \nu X(A \wedge \circ X)$; $\Diamond A = \mu X(A \vee \circ X)$; $AWB = \nu X(B \vee (A \wedge \circ X))$; $AUB = \mu X(B \vee (A \wedge \circ X))$.

1 Description of the infinitary calculus $G_{L\omega}$

The formulas are determined from constants T, F and propositional variables $X, Y, Z, X_1, Y_1, Z_1, \ldots$ with the help of logical symbols $\wedge, \vee, \neg$ and fixed point operators ν, μ. The formulas of the form $A \odot B$ (where $\odot \in \{\wedge, \vee\}$) and $\neg A$ are constructed as usual. The formulas $QXA(X)$ (where $Q \in \{\nu, \mu\}$) must satisfy the following monotonicity condition: all free occurrences of "temporal" variable X (simple: variable) in $A(X)$ fall under an even number of negation (X enters in $A(X)$ monotonically). Instead of formulas of the type $\circ A$ we shall consider the formulas with indices (denoted by A^i and certifying the truth value of A in the i-th moment of time) defined as follows:
1) $(Tr)^k := Tr$, where $Tr \in \{T, F\}$;
2) $(X^i)^k := X^{i+k}$, where i is zero (which is identified with an empty word) or any natural number, k is any natural number;
3) $(A \odot B)^k := A^k \odot B^k, \odot \in \{\wedge, \vee\}$;
4) $(\sigma A)^k := \sigma A^k, \sigma \in \{\neg, \nu X, \mu X\}$.

Expression $A \equiv B$ will mean the formula $(\neg A \vee B) \wedge (\neg B \vee A)$. A sequent is an expression of the form $\Gamma \rightarrow \Delta$, where Γ, Δ are arbitrary finite sets (not sequences or multisets) of formulas.

The infinitary calculus $G_{L\omega}$ is defined by the following postulates.
Axioms: $\Gamma, A \rightarrow \Delta, A$; $\Gamma \rightarrow \Delta, T$; $\Gamma, F \rightarrow \Delta$.
Rules of inference:
1) rules for maximal and minimal fixed points:

$$\frac{A(\nu XA(X)),\ \Gamma \rightarrow \Delta}{\nu XA(X),\ \Gamma \rightarrow \Delta}\ (\nu \rightarrow) \qquad \frac{\{\Gamma \rightarrow \Delta, A_{k+1}(T)\}_{k\in\omega}}{\Gamma \rightarrow \Delta,\ \nu XA(X)}\ (\rightarrow \nu_\omega)$$

$$\frac{\Gamma \rightarrow \Delta,\ A(\mu XA(X))}{\Gamma \rightarrow \Delta,\ \mu XA(X)}\ (\rightarrow \mu) \qquad \frac{\{A_{k+1}(F),\ \Gamma \rightarrow \Delta\}_{k\in\omega}}{\mu XA(X),\ \Gamma \rightarrow \Delta}\ (\mu_\omega \rightarrow)$$

where $A_1(X) = A(X)$; $A_{k+1}(X) = A(A_k(X))$;

2) logical rules of inference consist of traditional invertible rules of inference for $\wedge, \vee, \neg$;
3) structural rules: from the definition of a sequent it follows that $G_{L\omega}$ implicitly contains the structural rules contraction and exchange.

Derivations in the calculus $G_{L\omega}$ are built in an usual way (for the calculi with ω-rule), i.e., in the form of an infinite tree (with finite branches); the height of a derivation D is an ordinal (defined in a traditional way), denoted by $O(D)$.

Lemma 1.1. *(a)* $G_{L\omega} \vdash\rightarrow (\neg \nu X A(X) \equiv \mu X \neg A(\neg X))$; $G_{L\omega} \vdash\rightarrow (\neg \mu X A(X) \equiv \nu X \neg A(\neg X))$. *(b)* $G_{L\omega} \vdash\rightarrow (\nu X(B \vee (X \wedge C)) \equiv \nu X(B \vee C))$; $G_{L\omega} \vdash\rightarrow (\nu X(X \vee B) \equiv T)$; $G_{L\omega} \vdash\rightarrow (\mu X(B \wedge (X \vee C)) \equiv \mu X(B \wedge C))$; $G_{L\omega} \vdash\rightarrow (\mu X(X \wedge B) \equiv F)$.

Proof: applying the rules of inferences of $G_{L\omega}$.

It is convenient to consider a formulas in some normal forms (see e.g. [1]). A formula is positive if it is not of the form $\neg A$. A formula is in positive normal form if all its subformulas, with the possible exception of free propositional variables, are positive. Thus original monotonicity condition for positive normal form $QXA(X)$ ($Q \in \{\mu, \nu\}$) transforms to the following one: variable X in $A(X)$ does not enter in the scope of action of $\neg$.

Lemma 1.2. *Every formula can be transformed to positive normal form.*

Proof: using the original monotonicity condition, traditional logical laws (when the negation symbols are pushed through) and Lemma 1.1 (a).

A formula is in quarded normal form if all occurrences of bound variable X, excluding in $QX (Q \in \{\mu, \nu\})$ are such that $X = Y^i (i \geqslant 1)$.

Lemma 1.3. *Every formula can be transformed to quarded normal form.*

Proof: using Lemma 1.1 (b).

Example 1.1. Let $A = \nu X(U \wedge X \wedge \mu Y(V \vee Y \vee (X \wedge Y^1)))$, then using Lemma 1.1 (b) we get $G_{L\omega} \vdash\rightarrow (A \equiv \nu X((U \wedge V) \vee (U \wedge \mu Y(V^1 \vee (X^1 \wedge Y^2)))))$.

By Lemmas 1.2, 1.3 we can consider only formulas in positive and quarded normal form.

Lemma 1.4. *Let B be a formula, $\Delta(X)$ be a set of formulas and the variable X occurs in Δ monotonically, then (1)* $G_{L\omega} \vdash \Gamma, \Delta(T) \rightarrow \Theta \Rightarrow G_{L\omega} \vdash \Gamma, \Delta(B) \rightarrow \Theta$; *(2)* $G_{L\omega} \vdash \Gamma \rightarrow \Delta(F), \Theta \Rightarrow G_{L\omega} \vdash \Gamma \rightarrow \Delta(B), \Theta$.

Proof: by induction on $O(D)$, where D is the given derivation.

Lemma 1.5. *In $G_{L\omega}$ the following rules of inference:*

$$\frac{A(A_k(T)),\ \Gamma \rightarrow \Delta}{\nu X A(X),\ \Gamma \rightarrow \Delta}\ (\nu_k \rightarrow) \qquad \frac{\Gamma \rightarrow \Delta,\ A(A_k(F))}{\Gamma \rightarrow \Delta,\ \mu X A(X)}\ (\rightarrow \mu_k)\quad (k \in \omega)$$

are admissible.

Proof: using Lemma 1.4.

A derivation in the calculus I will be called atomic, if all axioms have the form $\Gamma, E \rightarrow \Delta, E$, where $E = X^i (i \geqslant 0)$.

Lemma 1.6. *An arbitrary derivation in $G_{L\omega}$ can be transformed to an atomic one.*

Proof. The case when $A = \neg X$ is evident. If $A = (M \odot N)$ (where $\odot \in \{\wedge, \vee\}$) then complexity of A is reduced with the help of rules for $\wedge, \vee$. Let $A = QXA(X)$ (where $Q \in \{\nu, \mu\}$), then complexity of A is reduced with the help of Lemma 1.5 and the rules $(\rightarrow \nu_\omega)$, $(\mu_\omega \rightarrow)$.

Lemma 1.7 (invertibility of rules of inference of $G_{L\omega}$). *Let (i) be a rule of inference of $G_{L\omega}$; S be the conclusion of (i), S_1 be any premise of (i). Then $I \vdash S \Rightarrow I \vdash S_1$.*

Proof: by the height of S using Lemma 1.6.

Lemma 1.8. *In $G_{L\omega}$ the following rule of inference:*

$$\frac{\Gamma \rightarrow \Delta}{\Gamma^1 \rightarrow \Delta^1}\ (+1)$$

is admissible.

Proof: by induction on $O(D)$, where D is a derivation of the premise of (+1).

Lemma 1.9. *Let $A_1, \dots . A_n, B_1, \dots, B_n$ be formulas, $D(X_1, \dots, X_n)$ be a formula and variables $X_1, \dots, X_n$ $(n \geqslant 1)$ enter in D monotonically, then in $G_{L\omega}$ the following rule of inference:*

$$\frac{A_1 \rightarrow B_1; \dots; \quad A_n \rightarrow B_n}{\Gamma, D(A_1, \dots, A_n) \rightarrow \Delta, D(B_1, \dots, B_n)}\ (M)$$

is admissible.

Proof: by induction on complexity of D using Lemmas 1.5, 1.8

2 Completeness and admissibility of cut in $G_{L\omega}$

At first we shall establish completeness of $G_{L\omega}$. The admissibility of cut in $G_{L\omega}$ is a corollary of the completeness theorem as formulated below.

A formula of temporal logic under consideration is interpreted in models M (see e.g. [2]):

$$M = <\alpha, i> = \omega \rightarrow 2^{A_G} \times \omega,$$

where ω is a set of all natural numbers, α provides an interpretation of propositional variables from the alphabet A_G of propositional symbols and assigns the truth values to them in $i \in \omega$ moment of time. The concept "A is valid in α, i" (in symbols $\alpha, i \models A$) is defined as follows:

1) $\alpha, i \models X^l \iff \alpha, i + l \models X \quad (l > 0)$
2) $\alpha, i \models \nu X A(X) \iff \forall n \in \omega(\alpha, i \models A_{n+1}(T))$
3) $\alpha, i \models \mu X A(X) \iff \forall n \in \omega(\alpha, i \models A_{n+1}(F))$.

In points 2) and 3) it is necessary that A would be $\wedge\vee$-continuous, i.e., $\models A(\underset{n<\omega}{\odot} B_n) = \models \underset{n<\omega}{\odot} A(B_n)$ $(\odot \in \{\wedge, \vee\})$.

Other cases are defined as in traditional logic. Using this concept we can define (in a traditional way) the concept of universally valid sequent .

Theorem 2.1 (soundness of $G_{L\omega}$). *If $G_{L\omega} \vdash S$, then S is universally valid.*

Proof: by induction on $O(D)$ using $\wedge\vee$-continuity.

To prove completeness of $G_{L\omega}$ let us introduce the "symmetric" calculus $G_{L\omega k}$, obtained from $G_{L\omega}$ by replacing $(\nu \to)$, $(\to \mu)$ by $(\nu_k \to)$, $(\to \mu_k)$ respectively (see section 1).

Lemma 2.1. *Let S be a sequent, then either there is a derivation in $G_{L\omega k}$ or $\exists M \not\models S$.*

Proof: by constructing the reduction tree for S (see e.g. [3]) and using the semantics of the temporal logic under consideration.

Theorem 2.2 (completeness and admissibility of cut in $G_{L\omega k}$, $G_{L\omega}$). *Let $I \in \{G_{L\omega k}, G_{L\omega}\}$ and a sequent S be universally valid, then $I \vdash S$.*

Proof: follows from Lemmas 2.1, 1.5.

Lemma 2.2. *(a). $G_{L\omega} \vdash \nu X A(X) \to \nu X(A(X) \vee B(X))$; (b) $G_{L\omega} \vdash \mu X(A(X) \wedge B(X)) \to \mu X A(X)$.*

Proof: using Lemma 1.5 and $\vee\wedge$-continuity.

3 Description of "invariant" calculus $\mathcal{R}$

The main rules in finitary calculus G_L are the following analytic cut-type rules:

$$\frac{\tilde{\nu}(\Omega),\ \Gamma \to \Delta,\ R; \quad R \to A(R)}{\tilde{\nu}(\Omega),\ \Gamma \to \Delta,\ \nu X A(X)}\ (\to \nu_1)$$

$$\frac{A(R) \to R; \quad R,\ \Gamma \to \Delta,\ \tilde{\mu}(\Omega)}{\tilde{\mu} X A(X),\ \Gamma \to \Delta,\ \mu(\Omega)}\ (\mu_1 \to),$$

where the invariant formula R in $(\to \nu_1)$, $(\mu_1 \to)$ is some formula constructed from formulas from $\tilde{Q}(\Omega)$ $(Q \in \{\nu, \mu\})$; $\tilde{Q}(\Omega) = QX_1A_1(X_1), \ldots, QX_nA_n(X_n)$.

Remark 3.1. The rules of inference $(\to \nu_1)$, $(\mu_1 \to)$ correspond the "Q-fixpoint induction" axioms $(Q \in \{\mu, \nu\})$: $R \wedge \nu X((\neg R \vee A(R)) \wedge X^1) \to \nu X A(X)$ and $\mu X A(X) \to R \vee \mu X((A(R) \wedge \neg R) \vee X^1)$.

The aim of the calculus $\mathcal{R}$ is to provide some way of constructing invariant formula R in $(\to \nu_1)$, $(\mu_1 \to)$. At first let us describe the rules of inference of $\mathcal{R}$. Let us introduce the "marked" forms of $(\nu \to)$, $(\to \mu)$:

$$\frac{A^{\nu}(\nu X A(X)),\ \Gamma \to \Delta}{\nu X A(X),\ \Gamma \to \Delta}\ (\nu' \to) \qquad \frac{\Gamma \to \Delta,\ A^{\mu}(\mu X A(X))}{\Gamma \to \Delta,\ \mu X A(X)}\ (\to \mu').$$

Formulas of the form $A^Q(Q \in \{\mu, \nu\})$ will be called Q-formulas. It is assumed that $(A \odot B)^Q := A^Q \odot B^Q$ $(Q \in \{\wedge, \vee\})$; $(QXA(X))^{Q_1} := QXA^{Q_1}(X)$,

$Q_1, Q \in \{\mu, \nu\}$; $(P^{Qk})^{Q_1} := P^{Qk} (k \in \omega)$, P is free or bound propositional variable. Remind that we consider formulas in positive and quarded normal forms.

Let us introduce the following rules of inference

$$\frac{\Gamma \to \Delta,\ A(\nu X A^{\nu}(X))}{\Gamma \to \Delta, \nu X A(X)} (\to \nu_2') \qquad \frac{A(\mu X A^{\mu}(X)),\ \Gamma \to \Delta}{\mu X A(X),\ \Gamma \to \Delta} (\mu_2' \to)$$

Lemma 3.1. *The rules of inference $(\to \nu_2'), (\mu_2' \to)$ are invertible in $G_{L\omega}$.*

Proof: using cut and its admissibility in $G_{L\omega}$.

Now we shall introduce the canonical form of a sequent S under consideration. The sequent S is in the canonical form and called primary if $S = \Sigma_1, \Pi_{11}^1, \Pi_{12}^{\nu 1}, \Pi_{13}^{\mu 1}, \widetilde{\mu}(\Omega_1^{\nu 1}), \widetilde{\mu}(\Theta_1^{\mu 1}), \widetilde{\nu}(\Delta_1^{\nu 1}) \to \Sigma_2, \Pi_{21}^1, \Pi_{22}^{\nu 1} \Pi_{23}^{\mu 1}, \widetilde{\mu}(\Theta_2^{\mu 1}), \widetilde{\nu}(\Delta_2^{\nu 1}), \widetilde{\nu}(\Omega_2^{\mu 1})$, where $\Sigma_i = \varnothing$ $(i = 1,2)$ or consists of formulas which do not contain ν, μ and indices (such formulas will be called logical ones); $\Pi_{i1}^1 = \varnothing$ $(i = 1,2)$ or consists of atomic formulas with indices; $\Pi_{ij}^{Q1} = \varnothing$ $(i = 1,2;\ j = 2,3)$ or consists of Q-atomic formulas $(Q \in \{\mu, \nu\})$ with indices, $\Omega_i^{Q1}, \Theta_i^{\mu 1}, \Delta_i^{\nu 1} = \varnothing$ $(i = 1,2)$ or consists of arbitrary Q-formulas with indices; besides $k \geqslant 2$, where k is the minimal index in $\Omega_i^{Q_1}, \Theta_i^{\mu 1}, \Delta_i^{\nu 1}$; if $k \geqslant 1$ the such sequent S will be called quasiprimary.

Let us define the notion of the reduction $R\{i\}$ in a calculus J of a sequent S to the sequents $S_1, \ldots, S_n$, where $\{i\}$ is the set of rules of inference invertible in J. $R\{i\}$ is a tree of sequents; the lowest sequent of $R\{i\}$ is the sequent S; every sequent in $R\{i\}$ except S is an upper sequent of the rule of inference $(k) \in \{i\}$ whose lower sequent is also in $R\{i\}$, i.e. $R\{i\}$ consists of the bottom-up applications of the rules of inference from $\{i\}$. The topmost sequents of $R\{i\}$ are the sequents $S_1, \ldots, S_n$ and any axiom of J. The notation $S\{i\} \Rightarrow \{S_1, \ldots, S_n\}$ will mean that it is possible to construct the reduction $R\{i\}$ of the sequent S to the sequents $S_1, \ldots, S_n$. From the definition of $R\{i\}$ we have that if $S\{i\} \Rightarrow \{S_1, \ldots, S_n\}$, then $J \vdash S \Rightarrow J \vdash S_j (1 \leqslant j \leqslant n)$. Let $G_{L\omega} \vdash S$ and let $S\{i\} \Rightarrow \{S_1, \ldots, S_n\}$, where $S_j (1 \leqslant j \leqslant n)$ is a primary (quasiprimary) sequent, $\{i\}$ consists of bottom-up applications of $(\nu' \to), (\to \mu'), (\to \nu_2'), (\mu_2' \to)$ and the rules of inference of $G_{L\omega}$ except $(\to \nu_\omega), (\mu_\omega \to), (\nu \to), (\to \mu)$ (consist of bottom-up applications of the rules of inference of $G_{L\omega}$ except of $(\to \nu_\omega), (\mu_\omega \to)$, $(\nu' \to), (\to \mu')$, $(\to \mu_2'), (\nu_2' \to)$, respectively); $\{S_1, \ldots, S_n\}$ will be called primary (quasiprimary) expansion of S and will be denoted by $Pr(S)$ (by $QPr(S)$, respectively).

Lemma 3.2. *Let $G_{L\omega} \vdash S$ then it is possible to construct $Pr(S)$ and $QPr(S)$.*

Proof: using the invertibility of the rules of inference of $G_{L\omega}$ and of $(\mu_2' \to), (\to \nu_2')$ in $G_{L\omega}$.

Let us consider the following examples. At first glance they (and others, which are implied by ones) are technical and difficult to understand. But since they correspond the Q-fixpoint induction axioms $(Q \in \{\mu, \nu\})$, they are natural in principal; besides they allow to understand the mechanism of reduction of derivations in infinitary calculus $G_{L\omega}$ to the derivations in finitary calculus G_L.

Example 3.1. (1). Let $S = P \wedge \nu X A_1(X^1) \to \nu X A(X^1)$, where $A_1(X^1) = (\neg P \vee P^1) \wedge X^1$; $A(X^1) = P \wedge X^1$. It is easy to verify that the sequent S corresponds to ν-fixpoint induction axiom (or in $\Box$-notation: $\Box$-induction axiom $P \wedge$

$\Box(\neg P \vee P^1) \to \Box P)$; also $G_{L\omega} \vdash S$. Using Lemma 3.2 we have $Pr(S) = S_1^{pr} = P,\ P^{\nu 1}, \nu X A_1^{\nu 1}(X^1) \to \nu X A^{\nu 1}(X^1)$; $QPr(S) = S' = P, \nu X A_1(X^1) \to \nu X A(X^1)$ and $G_{L\omega} \vdash S^{pr}; G_{L\omega} \vdash S'$.

(2). Let $S = P \wedge \nu X A(X^1) \to \nu Y \nu U B(Y^1, U^1)$, where $A(X^1) = (\neg P \vee \nu Z C(Z^1)) \wedge X^1$; $C(Z^1) = (P^1 \wedge Z^1) \wedge P_1$; $B(Y^1, U^1) = (Y^1 \wedge U^1) \wedge P_1$. It is easy to verify that the sequent S corresponds to ν-fixpoint induction axiom; also $G_{L\omega} \vdash S$. Using Lemma 3.2 we have $Pr\,(S) = \{S_1^{pr};\ S_2^{pr}\}$, where $S_1^{pr} = \Gamma \to \nu Y \nu U B^{\nu 1}(Y^1, U^1)$; $S_2 = \Gamma \to \nu U(\nu Y B^{\nu 2}(Y^1, U^1) \wedge U^{\nu 2} \wedge P_1^{\nu 1})$; $\Gamma = P, P^{\nu 1}, \nu Z C^{\nu 1}(Z^1), P_1^{\nu}, \nu X A^{\nu 1}(X^1)$; $QPr\,(S) = S' = P,\ \nu X A(X^1) \to \nu Y \nu U B(Y^1, U^1)$ and $G_{L\omega} \vdash S_i^{pr}$ $(i = 1, 2)$; $G_{L\omega} \vdash S'$.

(3). Let $S = P \wedge \nu X A(X^1) \to \nu Y \mu U(Y^1 \vee U^1)$, where $A(X^1) = (\neg P \vee \mu Z(P^1 \vee Z^1)) \wedge X^1$. It is easy to verify that S corresponds to ν-fixpoint induction axiom, also $G_{L\omega} \vdash S$ and $Pr(S) = \{S_1^{pr};\ S_2^{pr}\}$, where $S_1^{pr} = P, P^{\nu_1}, \nu X A^{\nu 1}(X^1) \to \Delta^1$; $S_2^{pr} = P, \mu Z(P^{\nu 2} \vee Z^{\nu 2}), \nu X A^{\nu 1}(X^1) \to \Delta^1$; $\Delta^1 = \nu Y \mu U(Y^{\nu 2} \vee U^{\nu 2}), \mu U(\nu Y \mu U(Y^{\nu 2} \vee U^{\nu 2}) \vee U^{\nu 2})$; $QPr(S) = S' = P, \nu X A(X^1) \to \nu Y \mu U(Y^1 \vee U^1)$ and $G_{L\omega} \vdash S_i^{pr}$ $(i = 1, 2)$; $G_{L\omega} \vdash S'$.

(4). Let $S = \mu Y \mu U B(Y^1, U^1) \to P, \mu X A(X^1)$, where $A(X^1) = (\mu Z C(Z^1) \wedge \neg P) \vee X^1$; $C(Z^1) = (P^1 \vee Z^1) \vee P_1$; $B(Y^1, U^1) = (Y^1 \vee U^1) \vee P_1$. It is easy to verify that S corresponds to μ-fixpoint induction axiom; also $G_{L\omega} \vdash S$ and $Pr(S) = \{S_1^{pr};\ S_2^{pr}\}$, where $S_1^{pr} = \mu Y \mu U B^{\mu 1}(Y^1, U^1) \to \Delta$; $S_2^{pr} = \mu U(\mu Y \mu U B^{\mu 2}(Y^1, U^1) \vee U^{\mu 2} \vee P_1^{\mu 1}) \to \Delta$; $\Delta = P, P^{\mu 1}, P_1^{\mu}, \mu Z C^{\mu 1}(Z^1), \mu X A^{\mu 1}(X^1)$; $QPr(S) = S$; $G_{L\omega} \vdash S_i^{pr}$ $(i = 1, 2)$.

(5). Let $S = \mu Y \nu U(Y^1 \wedge U^1) \to P, \mu X A(X^1)$, where $A(X^1) = (\nu Z(P^1 \wedge Z^1) \wedge \neg P) \vee X^1$. It is easy to verify that S corresponds to μ-fixpoint induction axiom; also $G_{L\omega} \vdash S$ and $Pr(S) = \{S_1^{pr};\ S_2^{pr}\}$, where $S_1^{pr} = \Gamma \to P, P^{\nu 1}, \mu X A^{\nu 1}(X^1)$; $S_2^{pr} = \Gamma \to P, \nu Z(P^{\mu 2} \wedge Z^{\mu 2}), \mu X A^{\mu 1}(X^1)$; $\Gamma = \mu Y \nu U(Y^{\mu 2} \wedge U^2), \nu U(\mu Y \nu U(Y^{\mu 2} \wedge U^{\mu 2}) \wedge U^{\mu 2})$ and $G_{L\omega} \vdash S_i^{pr}$ $(i = 1, 2)$.

Let us introduce the following rules of inference $\frac{S}{S_i^*}(A)$; $\frac{S_i^*}{S}(\bar{A})$, where S is a primary sequent, i.e. $S = \Sigma_1, \Pi_{11}^1, \Pi_{12}^{\nu_1}, \Pi_{13}^{\mu_1}, \tilde{\mu}(\Omega_1^{\nu_1}), \tilde{\mu}(Q_1^{\mu_1}), \tilde{\nu}(\Delta_1^{\nu_1}) \to \Sigma_2, \Pi_{21}^1, \Pi_{22}^{\nu_1}, \Pi_{23}^{\mu_1}, \tilde{\mu}(Q_2^{\mu_1}), \tilde{\nu}(\Delta_2^{\nu_1}), \tilde{\nu}(\Omega_2^{\mu_1})$; S_i^* has the following shape $(i \in \{1, 2, 3, 4\})$: $S_1^* = \Sigma_1 \to \Sigma_1$; $S_2^* = \Pi_{11} \to \Pi_{12}$; $S_3^* = \Pi_{12}^{\nu}, \tilde{\mu}(\Omega_1^{\nu}), \tilde{\nu}(\Delta_1^{\nu}) \to \Pi_{22}^{\nu}, \tilde{\nu}(\Delta_2^{\nu}), \tilde{\nu}(\Omega_2^{\mu})$; $S_4^* = \Pi_{13}^{\mu}, \tilde{\mu}(\Omega_1^{\nu}), \tilde{\mu}(Q_1^{\mu}) \to \Pi_{23}^{\mu}, \tilde{\mu}(Q_2^{\mu}), \tilde{\nu}(\nabla_2^{\mu})$. We shall omit the superscript $Q \in \{\mu, \nu\}$ in S, S_i^* if it will not bring to ambiguity. The sequents S_3^*, S_4^* in $(\bar{A})$ will be called temporal upper sequents.

Example 3.2. (1) Let $S = S_1^{pr}$ from Example 3.1.(2), then we can verify that $G_{L\omega} \vdash S_3^* = P, \nu Z C(Z^1), \nu X A(X^1) \to \nu Y \nu U B(Y^1, U^1)$.

(2) Let $S = S_1^{pr}$ from Example 3.1.(3), then we can verify that $G_{L\omega} \vdash S_3^* = P, \nu X A(X^1) \to \nu Y \mu U(Y^1 \vee U^1)$.

(3) Let $S = S_2^{pr}$ from Example 3.1.(3), then $S_{L\omega} \vdash S_3^* = \mu Z(P^1 \vee Z^1), \nu X A(X^1) \to \nu Y \mu U(Y^1 \vee U^1)$.

(4) Let $S = S_1^{pr}$ from Example 3.2.(5), then we can verify that $G_{L\omega} \vdash S_4^* = \mu Y \nu U(Y^1 \wedge U^1) \to P, \mu X A(X^1)$.

(5) Let $S = S_2^{pr}$ from Example 3.2.(5), then we can verify that $G_{L\omega} \vdash S_4^* = \mu Y \nu U(Y^1 \wedge U^1) \to \nu Z(P^1 \wedge Z^1), \mu X A(X^1)$.

Lemma 3.2 (alternation property). *The rule of inference* (A) *is admissible in* $G_{L\omega}$, *i.e., the rule of inference* $(\bar{A})$ *is invertible in* $G_{L\omega}$.

Proof: by induction on $< n, O(D) >$, where n is the number of positive (negative) occurrences of ν (μ, respectively), D is the given atomic derivation of S in $G_{L\omega}$.

Let $S_i \in Pr(S)(1 \leqslant i \leqslant n)$, let $S'_j(1 \leqslant j \leqslant m \leqslant n)$ be the temporal upper sequent of the application of the rule of inference $(\bar{A})$ whose lower sequent is the sequent S_i, then the set $\{S'_1, \ldots, S'_m\}$ will be called a temporal expansion of S and will be denoted by $T(S)$.

Lemma 3.3. *Let $G_{L\omega} \vdash S$ then it is possible to construct $T(S)$.*

Proof: follows from Lemmas 3.1, 3.2.

Now we can define the rules of inference of the calculus $\mathcal{R}$ consisting of: (1) the rules of inference of $G_{L\omega}$, except $(\rightarrow \nu_\omega), (\mu_\omega \rightarrow), (\nu \rightarrow), (\rightarrow \mu)$; (2) the rules of inference $(\nu' \rightarrow), (\rightarrow \mu'), (\rightarrow \nu'_2), (\mu'_2 \rightarrow)$, $(\bar{A})$. A derivation in $\mathcal{R}$ is constructed using bottom-up applications of rules of inference of $\mathcal{R}$. Let us describe the notion of closed sequent which will play a role of axiom of $\mathcal{R}$. Let D will be a derivation tree in $\mathcal{R}$. Let $S^* \in D$, then S^* will be called saturated if $\exists S' \in D$ and S' is below than S^* and $S^* = S'$. We shall say that a sequent S^* is absorbed by some sequent S' if S^* can be obtained from S' with the help of the structural rule of inference $(\mathcal{W})$. A sequent $S^* \in D$ will be called closed if either $S^* = \Gamma, A \rightarrow \Delta, A$ or saturated or absorbed by a saturated sequent which is in D. A derivation tree D in $\mathcal{R}$ will be called closed if all vertex in D are closed sequent. The notation $\mathcal{R} \vdash S$ will be mean that it is possible to construct closed derivation in $\mathcal{R}$.

Let us describe the natural tactics of constructing a closed derivation in $\mathcal{R}$. At first let us introduce the notion of resolvent of a sequent S (which will be denoted by *Re* (S). The construction of *Re* (S) consists of three stages. At the first one *Pr* (S) is constructed; let $S_i \in Pr\,(S)$ $(i = 1, \ldots n)$, then at the second stage the rule of inference $(\bar{A})$ is bottom-up applied to the each sequent S_i. Let $\{S^*_{i_1}, \ldots, S^*_{i_m}\}$ $(m \leqslant n)$ be the set consisting from temporal upper sequents of the bottom-up application of $(\bar{A})$. At the third stage $QPr\,(S^*_{ij})$ $(j = 1, \ldots, m)$ is constructed. As the final result we have $Re\,(S) = \{QPr\,(S^*_{i1}), \ldots, QPr\,(S^*_{ij})\}$.

Let us introduce the notion of n-th $(n \geqslant 0)$ resolvent of a sequent S (denoted by $Re^n(S)$). $Re^0(S) = S$ and $Re^1(S) = Re\,(S)$. Let $S_i \in Re^k(S)$ and S_i is not closed, $Re^{k+1} = \bigcup_i Re\,(S_i)$. If $\forall i$ S_i is closed then $Re^{k+1} = \varnothing$. If all the vertexes of $Re^n(S)$ are closed then such $Re^n(S)$ will be called closed and will be denoted by $CRe^n(S)$.

Lemma 3.4. *If $\mathcal{R} \vdash S$, then $\exists CRe^n(S)$.*

Proof: follows from the definition of $CRe^n(S)$.

A sequent S will be called singular if S contains positive (or negative) occurrences of ν (μ, respectively) and does not contain a negative (a positive) occurrences of ν (μ, respectively). A sequent S will be called ordinary if S contains both positive and negative occurrences of ν (and/or μ). A sequent S will be called simple if S does not contain any positive (negative) occurrences of ν (of μ, respectively).

Let S be a quasiprimary ordinary sequent. Let us define the notion of subformula and invariant subformula of an arbitrary formula from sequent S. These notions are obtained from the notion of subformula of an arbitrary formula in the propositional logic by adding the following two points: (1) the subformulas (invariant subformulas) of A^1 are the subformulas (invariant subformulas) of A and A^1 itself;

(2) the subformulas (invariant subformulas) of $QXB(X)$ ($Q \in \{\nu, \mu\}$) are subformulas of B, subformulas of $B(QXB(X))$ and the formula $QXB(X)$ itself (invariant subformulas of B and the formula $QXB(X)$ itself). Therefore the set of subformulas of the formula A containing Q is infinite contrarily to the set of invariant subformulas of A which is finite.

Lemma 3.5. *Let $G_{L\omega} \vdash S$ and S be an ordinary quasiprimary sequent. Then it is possible to construct $CRe^n(S)$, besides all saturated vertexes of $CRe^n(S)$ are either (1) the sequents of the form $S_i = \Pi_i, \tilde{\nu}(\Gamma_1) \to \Delta_i, \tilde{\nu}(\Gamma_2)$ (which will be called ν-type sequents), where $\Gamma_1, \Gamma_2 \neq \varnothing$; Π_i, Δ_i are invariant subformulas of Γ_1, besides Π_i consists of either atomic formulas or formulas of the form $\mu X A(X^1)$, Δ_i consists of atomic formulas, or (2) the sequents of the form $S_i = \Pi_i, \tilde{\mu}(\Gamma_1) \to \Delta_i, \tilde{\mu}(\Gamma_2)$ (which will be called μ-type sequents), where $\Gamma_1, \Gamma_2 \neq \varnothing$; Π_i, Δ_i are invariant subformulas of Γ_2, besides Π_i consists of atomic formulas; Δ_i consists of either atomic formulas or formulas of the form $\nu X A(X^1)$.*

Proof: using the finiteness of the set of invariant subformulas of formulas from S and the construction of $CRe^n(S)$.

The number of positive (negative) occurrences of ν (μ, respectively) entering in a ν-type (μ-type, respectively) saturated vertex S_i of $CRe^n(S)$ will be called inductive number of S_i and will be denoted by Ind (S_i). Let $S_1, \ldots, S_n$ be all saturated vertexes of $CRe^n(S)$, then max $(\text{Ind}(S_1), \ldots, \text{Ind}(S_n))$ will be called inductive number of $CRe^n(S)$ and will be denoted by $\text{Ind}(CRe^n(S))$. Let S_i be a saturated vertex of $CRe^n(S)$, if $\text{Ind}(S_i) = \text{Ind}(CRe^n(S))$, then S_i will be called inductive saturated vertex; if $\text{Ind}(S_i) < \text{Ind}\,(CRe^n(S))$, then S_i will be called simple saturated vertex.

Example 3.3. (1) Let S be the sequent from Example 3.1 (2). Applying to S the algorithm for finding $CRe^n(S)$ we get $CRe^n(S) = \{S_1, S_2\}$, where S_1, S_2 are ν-type saturated sequents and $S_1 = P, \nu Z\ C(Z^1),\ \nu X A(X^1) \to \nu Y \nu U B(Y^1, U^1)$; $\text{Ind}(S_1) = 2$; $S_2 = P, \nu Z C(Z^1),\ \nu X A(X^1) \to \nu U(\nu Y \nu U B^1(Y^1, U^1) \wedge U^1 \wedge P_1)$ $\text{Ind}(S_2) = 3$. Therefore S_1 is simple saturated vertex, S_2 is inductive one.
(2). Let S is the sequent from Example 3.1 (3), then $CRe^n(S) = \{S_1, S_2\}$, where S_1, S_2 are ν-type saturated sequents and $S_1 = P, \nu X A(X^1) \to \nu Y \mu U(Y^1 \vee U^1)$; $S_2 = P, \mu Z(P^1 \vee Z^1), \nu X A(X^1) \to \nu Y \mu U(Y^1 \vee U^1)$, both S_1, S_2 are inductive saturated vertexes.
(3). Let S is the sequent from Example 3.1 (4), then $CRe^n(S) = \{S_1, S_2\}$, where S_1, S_2 are μ-type saturated vertexes and $S_1 = \mu Y \nu U\ B(Y^1, U^1) \to P, \mu Z C(Z^1)$, $\mu X A(X^1)$; $S_2 = \mu U(\mu Y \mu U B^1(Y^1, U^1) \vee U^1 \vee P_1) \to P, \mu Z C(Z^1), \mu X A(X^1)$; S_1 is simple saturated vertex, S_2 is inductive one.
(4) Let S is the sequent from Example 3.1 (5), then $CRe^n(S) = \{S_1, S_2\}$, where S_1, S_2 are μ-type saturated vertexes and $S_1 = \mu Y \nu U(Y^1 \wedge U^1) \to P, \mu X A(X^1)$; $S_2 = \mu Y \nu U(Y^1 \wedge U^1) \to \nu Z(P^1 \wedge Z^1), \mu X A(X^1)$; both S_1, S_2 are inductive saturated vertexes.

Lemma 3.6 (main property of saturated vertexes of $CRe^n(S)$). *Let S' be a saturated vertex of $CRe^n(S)$, let $Re(S') = \{S_1, \ldots, S_n\}$. Then $\forall_i (1 \leqslant i \leqslant n)$ S_i is either saturated or absorbed by a saturated vertex of $CRe^n(S)$.*

Proof: follows from the construction of $CRe^n(S)$ and the finiteness of the set of vertexes of $CRe^n(S)$.

Example 3.4. (1). Let S, S_1, S_2 be the same as in Example 3.3 (1), then $Re(S_1) = \{S_1\}$; $Re(S_2) = \{S_1, S_2\}$.
(2). Let S, S_1, S_2 be the same as in Example 3.3. (2), then $Re(S_i) = \{S_1, S_2\}$, $(i = 1, 2)$.
(3). Let S, S_1, S_2 be the same as in Example 3.3. (3), then $Re(S_1) = \{S_1, S_2\}$; $Re(S_2) = \{S_1, S_2\}$.
(4). Let S, S_1, S_2 be the same as in Example 3.3. (4), then $Re(S_1) = \{S_1, S_2\}$; $Re(S_2) = \{S_1, S_2\}$.

Lemma 3.7. *Let $G_{L\omega} \vdash S$ and $S_i = \Pi_i, \widetilde{\nu}(\Gamma_1) \to \Delta_i, \widetilde{\nu}(\Gamma_2)$ ($S_i = \Pi_i, \widetilde{\mu}(\Gamma_1) \to \Delta_i, \widetilde{\mu}(\Gamma_2)$) $(1 \leqslant i \leqslant n)$ be any ν-type (μ-type) inductive saturated vertex of $CRe^n(S)$, then either (1) $\forall i(1 \leqslant i \leqslant n)$ there exists the formula $\nu X A(X^1) \in \nu(\Gamma_2)$ (the formula $\mu X A(X^1) \in \mu(\Gamma_1)$) such that $G_{L\omega} \vdash S'_i = \Pi_i, \widetilde{\nu}(\Gamma_1) \to \Delta_i, \nu X A(X^1)$ ($G_{L\omega} \vdash S'_i = \Pi_i, \mu X A(X^1) \to \Delta_i, \widetilde{\mu}(\Gamma_2)$), or (2) $G_{L\omega} \vdash \widetilde{\nu}(\Gamma_1) \to \widetilde{\nu}(\Gamma_2)$ ($G_{L\omega} \vdash \widetilde{\mu}(\Gamma_1) \to \widetilde{\mu}(\Gamma_2)$, respectively).*

Proof: using Lemmas 3.5, 3.6.

Remark 3.2. The set of sequents $\{S'_1 = \Pi_1, \widetilde{\nu}(\Gamma_1) \to \Delta_1, \nu X A(X^1); \ldots, S'_n = \Pi_n, \widetilde{\nu}(\Gamma_1) \to \Delta_n, \nu X A(X^1)\}$ ($\{S'_1 = \Pi_1, \mu X A(X^1) \to \Delta_1, \widetilde{\mu}(\Gamma_2); \ldots, S'_n = \Pi_n, \mu X A(X^1) \to \Delta_n, \widetilde{\mu}(\Gamma)\}$ corresponding ν-type (μ-type) inductive saturated vertexes will be called ν-type (μ-type) invariant set of the sequent S) and will be denoted by $\mathrm{INV}_\nu(S)$ ($\mathrm{INV}_\mu(S)$, respectively). The formula $R = \vee(\Pi_i^\wedge \wedge \neg\Delta_i^\vee) \wedge \widetilde{\nu}(\Gamma_1)^\wedge$ ($R = \wedge(\Delta_i^\vee \vee \neg\Pi_i^\wedge) \vee \widetilde{\mu}(\Gamma_2)^\vee$) will be called ν-type (μ-type, respectively) invariant formula; $\Gamma^\wedge(\Gamma^\vee)$ means conjunction (disjunction, respectively) of the formulas from Γ.

Remark 3.3. Let us consider any two applications of $(\to \nu_\omega)$ ($(\mu_\omega \to)$, respectively) in the given atomic derivation D of S. These applications will be called different if their principal formulas are either different (formulas of the type $A^n(Tr), A^m(Tr)$, where $n, m \geqslant 1$, $Tr \in \{T, F\}$, are considered as coincident) or coincide but have different descendants. Therefore, the number of different applications of $(\to \nu_\omega)$ (of $(\mu_\omega \to)$) in D is finite and let $Q[D]$ ($Q \in \{\nu, \mu\}$) means this number.

4 Description of finitary calculus G_L

The postulates of finitary calculus G_L are the same as in $G_{L\omega}$ except the $(\to \nu_\omega)$, $(\mu_\omega \to)$ which are replaced by following ones:

$$\frac{\widetilde{\nu}(\Omega), \Gamma \to \Delta, R; \quad R \to A(R)}{\widetilde{\nu}(\Omega), \Gamma \to \Delta, \nu X A(X)} \; (\to \nu_1)$$

$$\frac{\Gamma \to \Delta, \; A(\nu X A(X))}{\Gamma \to \Delta, \; \nu X A(X)} \; (\to \nu_2)$$

$$\frac{A(R) \to R; \quad R, \Gamma \to \Delta, \widetilde{\mu}(\nabla)}{\mu X A(X), \Gamma \to \Delta, \widetilde{\mu}(\nabla)} \; (\mu_1 \to)$$

$$\frac{A(\mu X A(X)), \Gamma \to \Delta}{\mu X A(X), \Gamma \to \Delta}\ (\mu_2 \to)$$

$$\frac{S_1;\ \ldots;\ S_n}{S}\ (\to \nu_3) \qquad \frac{S_1';\ \ldots;\ S_m'}{S'}\ (\mu_3 \to)$$

$$\frac{\Gamma \to \Delta}{\Pi, \Gamma \to \Delta, \Theta}\ (\mathcal{W}) \qquad \frac{\Gamma \to \Delta}{\Gamma^1 \to \Delta^1}\ (+1),$$

where $\widetilde{Q}(\Sigma) = QX_1A_1(X_1), \ldots, QX_nA_n(X_n)$, if $\Sigma = A_1(X_1), \ldots, A_n(X_n), Q \in \{\mu, \nu\}$; $\Omega, \nabla \neq \varnothing$ and $\Gamma, \Delta \neq \varnothing$ in the rules of inference $(\to \nu_1)$, $(\mu_1 \to)$, besides the invariant formula R in $(\to \nu_1)$, $(\mu_1 \to)$ satisfies the following condition: $\widetilde{\nu}(\Omega),\ \Gamma \to \Delta,\ \nu X A(X) \in INV(S)$; $\mathcal{R} \vdash S$ and R is ν-type invariant formula (see Remark 3.2) ($\mu X A(X),\ \Gamma \to \Delta,\ \widetilde{\mu}(\nabla) \in INV(S)$; $\mathcal{R} \vdash S$ and R is μ-type invariant formula (see Remark 3.2)). The rules of inference $(\to \nu_3)$, $(\mu_3 \to)$ satisfy the following condition: $S = \widetilde{\nu}(\Pi) \to \widetilde{\nu}(\Delta)$; $\widetilde{\nu}(\Delta) = \nu Y_1 B_1(Y_1), \ldots, \nu Y_n B_n(Y_n)$; $S_1 = \widetilde{\nu}(\Pi) \to B_1(T), \nu Y_2 B_2(Y_2), \ldots, \nu Y_n B_n(Y_n)$; $S_n = \widetilde{\nu}(\Pi) \to \nu Y_1 B_1(Y_1), \ldots, \nu Y_{n-1} B_{n-1}(Y_{n-1}), B_n(T)$; $S' = \widetilde{\mu}(\Theta) \to \widetilde{\mu}(\Pi)$; $\widetilde{\mu}(\Theta) = \mu X_1 A_1(X_1), \ldots, \mu A_m(X_m)$; $S_1' = A_1(F), \mu X_2 A_2(X_2), \ldots, \mu X_m A_m(X_m) \to \widetilde{\mu}(\Pi); \ldots;$ $S_m' = \mu X_1 A_1(X_1), \ldots, \mu X_{m-1} A_{m-1}(X_{m-1}), A_m(F) \to \widetilde{\mu}(\Pi)$.

Remark 4.1. The rules of inference $(+1)$, $(\mathcal{W})$ are included in G_L for the sake of simplicity.

5 Reduction of G_L + cut to $G_{L\omega}$ + cut

Now we shall prove that G_L+ cut $\vdash S \Rightarrow G_{L\omega}$+ cut $\vdash S$, where I+ cut is the calculus obtained from I by adding cut.

Lemma 5.1. *In $G_{L\omega}$ the structural rule of inference $(\mathcal{W})$ is admissible.*

Proof: by induction on $O(D)$.

Lemma 5.2. *In $G_{L\omega}$+ cut the following rule of inference:*

$$\frac{\Gamma \to \Delta, R;\quad R \to A(R)}{\Gamma \to \Delta, \nu X A(X)}\ (\to \nu_1^*)$$

is derivable.

Proof. Let us denote the left (right) premise of $(\to \nu_1^*)$ by (1) (by (2), respectively). It is easily to prove that from (2) we can get $R \to A(T)$ (3). Using Lemma 1.9, induction on k and cut from (1), (2), (3) we get $G_{L\omega}$+ cut $\vdash \{\Gamma \to \Delta, A_{k+1}(T)\}_{k \in \omega}$; using $(\to \nu_\omega)$ we get $G_{L\omega}$+ cut $\vdash \Gamma \to \Delta, \nu X A(X)$.

Lemma 5.3. *In $G_{L\omega}$+ cut the following rule of inference:*

$$\frac{A(R) \to R;\quad R, \Gamma \to \Delta}{\mu X A(X),\ \Gamma \to \Delta}\ (\mu_1^* \to)$$

is derivable.

Proof. Let us denote the right (left) premise of $(\mu_1^* \to)$ by (1) (by (2) respectively). It is easy to prove that from (2) we can get $A(F) \to R(3)$. Using Lemma 1.9, induction on k and cut, from (1), (2), (3) we get $G_{L\omega}$+ cut $\vdash \{A_{k+1}(F), \Gamma \to \Delta\}_{k\in\omega}$; using $(\mu_\omega \to)$ we get $G_{L\omega}$+ cut $\vdash \mu X A(X), \Gamma \to \Delta$.

Lemma 5.4. *In $G_{L\omega}$+ cut the rules of inference $(\to \nu_2)$, $(\mu_2 \to)$ is derivable.*

Proof: follows from Lemmas 5.2, 5.3 taking $R = A(QXA(X))$ $Q \in \{\nu, \mu\}$.

Lemma 5.5. *(a)* $\Gamma_{L\omega} \vdash \tilde{\nu}(\Theta) \to \Delta(T), \tilde{\nu}(\nabla) \Rightarrow G_{L\omega} \vdash \tilde{\nu}(\Theta) \to \tilde{\nu}(\Delta), \tilde{\nu}(\nabla)$; *(b)* $G_{L\omega} \vdash \Delta(F), \tilde{\mu}(\nabla) \to \tilde{\mu}(\Theta) \Rightarrow G_{L\omega} \vdash \mu(\Delta),\ \mu(\nabla) \to \tilde{\mu}(\Theta)$.

Proof: using induction on complexity of Δ and Lemma 2.2.

Lemma 5.6. *The rules of inference $(\to \nu_3), (\mu_3 \to)$ are admissible in $G_{L\omega}$.*

Proof: follows from Lemma 5.5.

Theorem 5.1. $G_L \vdash S \Rightarrow G_{L\omega}$+ *cut* $\vdash S$.

Proof: follows from Lemmas 1.8, 5.1, 5.2, 5.3, 5.4, 5.6.

6 Reduction of $G_{L\omega}$ to G_L

In this section using the tools of the section 3 we shall prove that $G_{L\omega} \vdash S \Rightarrow G_L \vdash S$.

Lemma 6.1. *Let $G_{L\omega} \vdash S$, let $R\{i\}$ be a reduction of S to the sequents $S_1, \ldots, S_n$, let $\{i\}$ consists of bottom-up applications of the rules of inference $(\to \nu_2), (\mu_2 \to), (\bar{A})$ and the rules of inference of $G_{L\omega}$, except $(\to \nu_\omega), (\mu_\omega \to)$; let $G_L \vdash S_j$ $(1 \leqslant j \leqslant n)$, then $G_L \vdash S$.*

Proof: the validity of Lemma follows from the fact that a bottom-up application of $(\bar{A})$ can be replaced by the applications of $(+1)$ and structural rule $(\mathcal{W})$.

Lemma 6.2. *Let $G_{L\omega} \vdash S$, where S is a quasiprimary ordinary sequent and $CRe^n(S)$ be such that all inductive saturated vertexes have the form $S_i = \Pi_i, \tilde{\nu}(\Gamma) \to \Delta_i, \nu X A(X)$ $(1 \leqslant i \leqslant n)$, then $G_L \vdash \mathcal{R} \to A(R)$, where $\mathcal{R}$ is ν-type invariant formula (see Remark 3.2).*

Proof. The proof of the Lemma is carried out by induction on (ν) $[D]$ (see Remark 3.3). Let $S_i \in INV_\nu(S)$, let $Re\,(S_i) = \{S_{i1}, \ldots, S_{im}\}$. Let us consider the reduction R_i of S_i to $S_{i1}, \ldots, S_{im}$. Then using induction hypothesis and Lemma 3.6 instead of reduction R_i we get $G_L \vdash \Pi_i, \tilde{\nu}(\Gamma) \to \Delta_i, A(R^1)$ $(1 \leqslant i \leqslant n)$, from which with the help of $(\wedge \to)$, $(\urcorner \to)$, $(\vee \to)$ we get $G_L \vdash \mathcal{R} \to A(R^1)$.

Example 6.1. (1). Let S, S_1, S_2 be the same as in Example 3.3 (1); S_1 is the simple saturated vertex, S_2 is inductive one, therefore $INV_\nu(S) = \{S_2 = P, \nu Z C(Z^1), \nu X A(X^1) \to \nu U(\nu Y \nu U B^1(Y^1, U^1) \wedge U^1 \wedge P_1)\}$ and $\mathcal{R} = P \wedge \nu Z C(Z^1) \wedge$

$\nu X A(X^1)$. Let us consider the reduction R_2 of S_2 to S_1, S_2:

$$
\dfrac{
D_1\left\{\begin{array}{l}
\dfrac{S_1}{\Gamma' \to \nu Y B^1(Y^1,U^1)}\,(\bar{A}) \\
\qquad\vdots \\
\Gamma \to \nu Y B^1(Y^1,U^1)
\end{array}\right.
\qquad
D_2\left\{\begin{array}{l}
\dfrac{\dfrac{S_2}{\Gamma' \to \nu U M^1(U^1)}\,(\bar{A}) \quad \Gamma_1', P_1 \to P_1}{\qquad\qquad}\,(\to \wedge) \\
\qquad\vdots \\
\Gamma \to \nu U M^1(U^1) \wedge P_1
\end{array}\right.
}{
\dfrac{\Gamma \to \nu Y \nu U B^1(Y^1,U^1) \wedge \nu U M^1(U^1) \wedge P_1}{\Gamma \to \nu U M(U^1)}\,(\to \nu_2)
}\,(\to \wedge),
$$

where $\Gamma = P, \nu Z C(Z^1), \nu X A(X^1)$; $\Gamma' = P, P^1, \nu Z C^1(Z^1), P_1, \nu X A^1(X^1)$; $\Gamma_1' = P, P^1, \nu Z C^1(Z^1), \nu X A^1(X^1)$; $M(U^1) = \nu Y \nu U(B^1(Y^1,U^1) \wedge U^1 \wedge P_1)$. Let us transform the part D_2 of R_2 to the following figure:

$$
D_2'\left\{\begin{array}{l}
\dfrac{\dfrac{S_2'}{\Gamma' \to R^1;}\,(\bar{A}) \qquad \Gamma_1', P_1 \to P_1}{\qquad\qquad}\,(\to \wedge) \\
\qquad\vdots \\
\Gamma \to R^1 \wedge P_1
\end{array}\right.
$$

where $S_2' = \Gamma \to R$. It is easy to verify that $G_L \vdash \Gamma \to R$. By induction hypothesis we get $G_L \vdash S$. Instead of $(\bar{A})$ in (D_1), D_2' we have $(\mathcal{W})$, $(+1)$. Therefore instead of D_1 and D_2 we get $G_L \vdash \Gamma \to \nu Y B^1(Y^1,U^1)$ and $G_L \vdash \Gamma \to R^1 \wedge P_1$. Applying $(\to \wedge)$, we get $G_L \vdash \Gamma \to M(R^1)$, applying $(\wedge \to)$ we get $G_L \vdash R \to M(R^1)$.

(2). Let S, S_1, S_2 be the same as Example 3.4 (2); both S_1, S_2 are inductive saturated vertexes, therefore $INV_\nu(S) = \{S_1 = P, \nu X A(X^1) \to \nu Y \mu U(Y^1 \vee U^1);\ S_2 = \mu Z(P^1 \vee Z^1), \nu X A(X^1) \to \nu Y \mu U(Y^1 \vee U^1)\}$ and $R = (P \vee \mu Z(P^1 \vee Z^1)) \wedge \nu X A(X^1)$. Let us consider the reduction R_1 of S_1 to S_1, S_2 :

$$
D\left\{\begin{array}{c}
\dfrac{S_1}{\Gamma_1 \to \nu Y M^1(Y^1),\ \mu U N^1(U^1)}\,(\bar{A}) \qquad \dfrac{S_2}{\Gamma_2 \to \nu Y M^1(Y^1),\ \mu U N^1(U^1)}\,(\bar{A}) \\
\diagdown \qquad\qquad \diagup \\
\dfrac{\dfrac{\dfrac{\Gamma \to \nu Y M^1(Y^1),\ \mu U N^1(U^1)}{\Gamma \to \nu Y M^1(Y^1) \vee \mu U N^1(U^1)}\,(\to \vee)}{\Gamma \to M(\mu Y M^1(Y^1))}\,(\to \mu)}{\Gamma \to \nu Y M(Y^1)}\,(\to \nu_2)
\end{array}\right.
$$

where $\Gamma = P, \nu X A(X^1), \Gamma_1 = P, P^1, \nu X A^1(X^1); \Gamma_2 = P, \mu Z(P^2 \vee Z^2)$; $M(Y^1) = \mu U(Y^1 \vee U^1), N(U^1) = \nu Y M(Y^1) \vee U^1$. Let us transform the part D of R_1 to the following figure D_1:

$$D_1 \left\{ \begin{array}{c} \dfrac{S_1'}{\Gamma_1 \to R^1,\ \mu U N^1(U^1)}\,(\bar{A}) \qquad \dfrac{S_2'}{\Gamma_2 \to R^1,\ \mu U N^1(U^1)}\,(\bar{A}) \\ \diagdown \qquad\qquad \diagup \\ \dfrac{\dfrac{\Gamma \to R^1,\ \mu U N^1(U^1)}{\Gamma \to R^1 \vee \mu U N^1(U^1)}\,(\to \vee)}{\Gamma \to M(R^1)}\,(\to \mu) \end{array} \right.$$

where $S_1' = P, \nu X A(X^1) \to R$; $S_2' = \mu Z(P^1 \vee Z^1), \nu X A(X^1) \to R$. It is easy to verify that $G_L \vdash S_1'$ and $G_L \vdash S_2'$. Instead of $(\bar{A})$ in D_1 we have $(W), (+1)$. Therefore instead of D_1 we get $G_L \vdash \Gamma \to M(R^1)$. Now let us consider the reduction R_2 of S_2 to the sequents S_1, S_2. Analogously as in the case of reduction R_1 we get $G_L \vdash \Gamma' \to M(R^1)$, where $\Gamma' = \mu Z(P^1 \vee Z^1),\ \nu X A(X^1)$. Applying $(\vee \to)$, $(\wedge \to)$ we get $G_L \vdash R \to M(R^1)$.

Lemma 6.3. *Let $G_{L\omega} \vdash S$, where S is a quasiprimary ordinary sequent and $CRe^n(S)$ be such that all inductive saturated vertexes have the form $S_i = \Pi_i, \mu X A(X^1) \to \Delta_i, \widetilde{\mu}(\Gamma_2)$ $(1 \leqslant i \leqslant n)$, then $G_L \vdash A(R) \to R$, where R μ-type invariant formula (see Remark 3.2).*

Proof. The proof of the Lemma is carried out by induction on (μ) $[D]$ (see Remark 3.2). Let $S_i \in INV_\mu(S)$, let $Re\,(S_i) = \{S_{i1}, \ldots, S_{im}\}$. Let us consider the reduction R_i of S_i to $S_{i1}, \ldots, S_{im}$. Then using induction hypothesis and Lemma 3.6 instead of reduction R_i we get $G_L \vdash \Pi_i, A(R^1) \to \Delta_i, R$ $(1 \leqslant i \leqslant n)$, from which with the help of $(\to \vee)$, $(\to \neg)$, $(\to \wedge)$ we get $G_L \vdash A(R^1) \to R$.

Example 6.2. (1). Let S, S_1, S_2 be the same as in Example 3.3 (3); S_1 is the simple saturated vertex, S_2 is inductive one, therefore $INV_\mu(S) = \{S_2 = \mu U M(U^1) \to P, \mu Z C(Z^1), \mu X A(X^1)\}$ $(M(U^1) = \mu Y \mu U B^1(Y^1, U^1) \vee U^1 \vee P_1)$ and $R = P \vee \mu Z C(Z^1) \vee \mu X A(X^1)$. Analogously as in Example 6.1.(1) we can transform the reduction R_2 of S_2 to S_1, S_2 into the derivation in G_L of the sequent $M(R^1) \to R$. (2). Let S, S_1, S_2 be the same as in Example 3.3 (4); both S_1, S_2 are inductive saturated vertexes, therefore $INV_\mu(S) = \{S_1, S_2\}$ and $R = (P \wedge \nu Z(P^1 \wedge Z^1)) \vee \mu X A(X^1)$. Analogously as in Example 6.1 (2) we can transform the reduction R_1 of S_1 to S_1, S_2 into the derivation in G_L of the sequent $\mu Y \nu U(R^1 \wedge U^1) \to P,\ \mu X A(X^1)$ and the reduction R_2 of S_2 to S_1, S_2 into the derivation in G_L of the sequent $\mu Y \nu U(R^1 \wedge U^1) \to \nu Z(P^1 \wedge Z^1),\ \mu X A(X^1)$. Applying $(\to \wedge)$, $(\to \vee)$ we get $G_L \vdash \mu Y \nu U(R^1 \wedge U^1) \to R$.

Lemma 6.4. *Let $G_{L\omega} \vdash S$, where S is a quasiprimary ordinary sequent and $CRe^n(S)$ be such that all inductive saturated vertexes have the form $S_i = \Pi_i, \widetilde{\nu}(\Gamma) \to \Delta_i, \nu X A(X)$ $(1 \leqslant i \leqslant n)$, where Π_i, Δ_i consist of invariant subformulas of Γ, then $G_L \vdash S$.*

Proof. The proof of the Lemma is carried out by induction on $(\nu)[D]$. As follows from Lemma 6.1, to prove the Lemma it is sufficient to prove that $G_L \vdash S_i$. With the help of $(\to \vee)$, $(\to \wedge)$, $(\to \urcorner)$ we have $G_L \vdash \Pi_i, \widetilde{\nu}(\Gamma) \to \Delta_i, \mathcal{R}$ (1) $(1 \leqslant i \leqslant n)$, where $\mathcal{R}$ is the same as in Lemma 6.2. By Lemma 6.2 we have $IN \vdash \mathcal{R} \to A(R)$ (2). Applying $(\to \nu_1)$ to (1), (2) we get $G_L \vdash S_i$ and therefore $G_L \vdash S$.

Lemma 6.5. *Let $G_{L\omega} \vdash S$, where S is a quasiprimary ordinary sequent and $CRe^n(S)$ be such that all inductive saturated vertexes have the form $S_i = \Pi_i, \mu X A(X) \to \Delta_i, \widetilde{\mu}(\Gamma)$ $(1 \leqslant i \leqslant n)$, where Π_i, Δ_i consist of invariant subformulas of Γ, then $G_L \vdash S$.*

Proof: analogously to Lemma 6.4.

Lemma 6.6. *Let $G_{L\omega} \vdash S$, where S is a quasiprimary ordinary sequent and $CRe^n(S)$ be such that all inductive saturated vertexes have the form $\widetilde{Q}(\Gamma) \to \widetilde{Q}(\Delta)(Q \in \{\mu, \nu\})$, then $G_L \vdash S$.*

Proof: by induction on $(Q)[D]$, using $(\to \nu_3)$, $(\mu_3 \to)$.

Lemma 6.7. *Let $G_{L\omega} \vdash S$, where S is ordinary sequent, then $G_L \vdash S$.*

Proof: using Lemmas 3.5, 3.6, 6.4, 6.5, 6.6.

Lemma 6.8. *Let $G_{L\omega} \vdash S$, where S is a singular sequent, then $G_L \vdash S$.*

Proof: by induction on $(Q)[D]$, $Q \in \{\mu, \nu\})$ using the rules of inference $(\to \nu_3)$, $(\mu_3 \to)$ and Lemma 6.1.

Theorem 6.1. *Let $G_{L\omega} \vdash S$, then $G_L \vdash S$.*

Proof. Let S be simple then $G_L \vdash S$. Let S be ordinary (singular), then the Theorem follows from Lemma 6.7 (Lemma 6.8, respectively).

Theorem 6.2. *If (i) is a rule of inference admissible in $G_{L\omega}$ then (i) is admissible in G_L.*

Proof: follows from Theorem 6.1.

Example 6.3. (1) Let $S_1 = \mu X(A(X,C) \vee A(X,B)) \to \mu Y A(Y,C) \vee \mu Z A(Z,B)$; $S_2 = \mu Y A(Y,C) \vee \mu Z A(Z,B) \to \mu X(A(X,C) \vee A(X,B))$; $S_3 = \mu X \mu Y A(X,Y) \to \mu Z A(Z,Z)$; $S_4 = \mu Z A(Z,Z) \to \mu X \mu Y A(X,Y)$, where X, Y, Z, C, B enters in A monotonically. Bottom-up applying to S_i $(i = 1,2,3,4)$ $(\mu_3 \to)$, $(\to \mu)$ and (M) (i.e. Lemma 1.9) we get $G_L \vdash S_i$ $(i = 1,2,3,4)$.
(2) Let $S = \mu X A(B(X)) \to A(\mu X B(A(X)))$, where X enters in A, B monotonically. Bottom-up applying $(\mu_1^* \to)$ (Lemma 5.3) with $R = A(\mu X B(A(X)))$ and after that (M), $(\to \mu)$ we get $G_L \vdash S$.
(3) Let $S = A(\mu X B(A(X))) \to \mu X A(B(X))$, where X enters in A, B monotonically. Bottom-up applying $(\to \mu)$ to S and after that (M), $(\mu_1^* \to)$ (with $R = B(\mu X(A(B(X)))))$, (M), $(\to \mu)$ we get $G_L \vdash S$.

Theorem 6.3. *(1) If S is universally valid, then $G_L \vdash S$; (2) $G_L + (\text{cut}) \vdash S \Rightarrow G_L \vdash S$.*

Proof: follows from Theorems 2.2, 6.1.

References

1. B.Baniegbal, H.Barringer: Temporal logic with fixed points. LNCS, Vol. 398, (1987), 62–74.
2. H.Barringer: The use of temporal logic in the compositional specification of concurrent systems. In A.Galton (ed.): Temporal logics and their applications. Academic Press, London, (1987), 53–90.
3. J.H.Gallier: Logic for computer Science: foundations of automatic theorem proving. Harper and Row, New York, (1986).
4. A.Galton: Temporal logic and computer science: an overview. In A.Galton (ed.): Temporal logics and their applications. Academic Pres, London, (1987), 1–52.
5. D.Park: Fixpoint induction and proofs of program properties. Machine Intelligence 5, Edinburgh University Press, (1969), 59–78.
6. D.Park: Finiteness is mu-ineffable. Theor. Comp. Sci, 3, (1976), 173–181.
7. R.Pliuškevičius: Investigation of finitary calculi for temporal logics by means of infinitary calculi. LNCS, Vol. 452, (1990), 464–469.
8. R.Pliuškevičius: Investigation of finitary calculus for a discrete linear time logic by means of finitary calculus. LNCS, Vol. 502, (1991), 504–528.
9. R.Pliuškevičius: Complete sequential calculi for the first order symmetrical linear temporal logic with UNTIL and SINCE. LNCS, Proc. of LFCS' 92 (in press).
10. M.Vardi: A temporal fixpoint calculus. Proc. of 15-th An. ACM Symp. on POPL (1988), 250–259.
11. P.Wolper: Temporal logic can be more expressive. Information and Control, 56, (1983), 72–99.

Logical Omniscience and Classical Logic*

Reinhard Muskens

Department of Linguistics, Tilburg University
P.O. Box 90153, 5000 LE Tilburg
rmuskens@kub.nl

Abstract. In all respectable logics a form of Leibniz's Law holds which says that logically equivalent expressions can be interchanged *salva veritate.* On the other hand, in ordinary language syntactically different expressions in general are not intersubstitutable in the scope of verbs of propositional attitude. It thus seems that logics of knowledge and belief should not be subject to Leibniz's Law. In the literature (e.g. Rantala [1982[a], 1982[b]], Wansing [1990]) we indeed find attempts to define epistemic logics in which the interchangeability principle fails. These systems are based on the notions of possible and impossible worlds: in possible worlds all logical connectives get their standard interpretation, but in impossible worlds the interpretation of the connectives is completely free. It is easily seen, however, that the resulting systems are no logics if we apply standard criteria of logicality. There is a simple way out that saves the idea: once it is accepted that the English words 'not', 'and', 'or', 'if', and the like are not to be treated as logical operations, we might as well be open about it and overtly treat them as non-logical constants. This allows us to retain classical logic. Possible worlds can be defined as those worlds in which 'not', 'and', 'or', 'if' etc. get the standard logical interpretation. On the basis of this idea a small fragment of English is provided with a very fine-grained semantics: no two syntactically different expressions get the same meaning. But on sentences that do not contain a propositional attitude verb there is a classical relation of logical consequence.

Keywords. Logical Omniscience, Propositional Attitudes, Impossible Worlds, Epistemic Logic.

1 Logical Omniscience

Let us call two expressions *synonymous* if and only if they may be interchanged in each sentence without altering the truth value of that sentence.[1] With the help of an argument by Benson Mates (Mates [1950]) it can be shown that synonymy is a very strong relation indeed. Consider, for example, the following two sentences.

(1) Everybody believes that whoever thinks that all Greeks are courageous thinks that all Greeks are courageous

(2) Everybody believes that whoever thinks that all Greeks are courageous thinks that all Hellenes are courageous

* I would like to thank Ed Keenan and Heinrich Wansing for comments and criticisms.

1 This essentially is Mates's [1950] formulation of the interchangeability principle. Note how close Mates's formulation is to Leibniz's:

Eadem sunt quorum unum potest substitui alteri salva veritate.

Some philosophers indeed believe that whoever thinks that all Greeks are courageous also thinks that all Hellenes are courageous.[2] But certainly not everyone agrees, and so (2) is false. We may assume, on the other hand, that (1) is true, and since (1) can be obtained from (2) by replacing 'Hellenes' by 'Greeks', the latter two words are not synonymous. By a similar procedure any pair of words that are normally declared synonyms can be shown not to be synonymous after all, if our definition of the term is accepted.

Worse, it seems that the relation of synonymy is even stronger than the relation of logical equivalence is. Sentences that are normally accepted to be logically equivalent need not be synonymous. Suppose[3] Jones wants to enter a building that has three doors, A, B, and C. The distances between any two of these doors are equal. Jones wants to get in as quickly as possible, without making detours and he knows that if A is locked B is not. Now, if our agent tries to open door B first and finds it locked there might be a moment of hesitation. The reasonable thing for Jones is to walk to A, since if B is locked A is not, but he may need some time to infer this. This contrasts with the case in which he tries A first, since if he cannot open this door he will walk to B without further ado. The point is that one may well fail to realize (momentarily) that a sentence is true, even when one knows the contrapositive to hold. For a moment (3) might be true while (4) is false.

(3) Jones knows that if A is locked B is not locked

(4) Jones knows that if B is locked A is not locked

It follows that the two embedded sentences are not synonymous, even though logically equivalent on the usual account.

All reasoning takes time. This means that

(5) Jones knows that φ

need not imply

(6) Jones knows that ψ

even if φ and ψ are logically equivalent. If the embedded sentences are syntactically distinct then, since Jones needs time to make the relevant inference, there will always be a moment at which (5) is true but (6) is still false.

Of course, the easier it is to deduce ψ from φ, the shorter this time span will be and the harder it is to imagine that Jones knows φ without realizing that ψ. We ascribe to Jones certain capacities for reasoning, even if we do not grant him logical omniscience. For instance, if Jones knows A and B, then it is hard to imagine that he would fail to know B and A as well. But even here there may be a split-second where the necessary calculation has yet to be made. Therefore 'Jones knows B and A' does not follow from 'Jones knows A and B'.

A real problem results if we try to formalise the logic of the verb *to know* and its like. All ordinary logics (including modal logics) allow logical equivalents to be interchanged, but

---

2 E.g. Putnam in Putnam [1954].

3 This example is adapted from Moore [1989].

epistemic contexts do not admit such replacements. It thus may seem that the logic of the propositional attitude verbs is very much out of the ordinary and that we must find a logic that does not support full interchangeability of equivalents if we want our theory of the propositional attitudes to fit the facts.

Systems that do not admit replacement of equivalents do in fact exist. For example Rantala [1982[a], 1982[b]], working out ideas of Montague [1970], Cresswell [1972] and Hintikka [1975], offers an 'impossible world semantics' for modal logic in which the interchangeability property fails. In the next section I'll criticise Rantala's system for not being a *logic* in the strict sense, but I think that its main underlying idea, the idea that we can use 'impossible' worlds to obtain a very fine grained notion of meaning, is important and useful. Despite appearances however, this idea is compatible with classical logic and in section 4 I shall show in some detail how we can use impossible worlds to treat the propositional attitudes without resorting to a non-standard logic. The logic that I shall use is the classical type theory of Church [1940]. Section 3 will be devoted to a short exposition of this logic for the convenience of those readers who are not already familiar with it.

2 Rantala Models for Modal Logic

The basic idea behind Rantala's interpretation of the language of propositional modal logic is to add a set of so-called impossible (or: non-normal) worlds to the usual Kripke frames. Equivalence can then be defined with respect to possible worlds only, but for interchangeability the impossible ones come into play as well. The net result will be that equivalents need not be interchangeable in the scope of the epistemic operator $\Box$.

Formally, a *Rantala model* for the language of propositional modal logic is a tuple $\langle W, W^*, R, V\rangle$ consisting of a non-empty set W of possible worlds, a set W^* of impossible worlds, a relation $R \subseteq (W \cup W^*)^2$, and a two-place valuation function V: FORM $\times$ $(W \cup W^*) \rightarrow \{0,1\}$ such that for all $w \in W$:

(i)	$V(\neg\varphi, w) = 1$	iff $V(\varphi, w) = 0$
(ii)	$V(\varphi \wedge \psi, w) = 1$	iff $V(\varphi, w) = 1$ and $V(\psi, w) = 1$
(iii)	$V(\Box\varphi, w) = 1$	iff $V(\varphi, w') = 1$ for all $w' \in W \cup W^*$ such that wRw'.

Note that the value of a complex formula in an impossible world can be completely arbitrary. The value of a complex formula does not depend on the values of its parts. Note also that Rantala models clearly generalize Kripke models: a Kripke model simply is a Rantala model with empty W^*.

A formula ψ is said to *follow from* a formula φ if $V(\varphi, w) = 1$ implies $V(\psi, w) = 1$ in each Rantala model $\langle W, W^*, R, V\rangle$ and $w \in W$. A formula φ is *valid* in a Rantala model if $V(\varphi, w) = 1$ in each $w \in W$. A formula φ is *valid* simpliciter if it is valid in each Rantala model. Formulae φ and ψ are *equivalent* iff φ follows from ψ and ψ follows from φ. Clearly, all propositional tautologies are valid, but Necessitation fails: validity of φ does not imply validity of $\Box\varphi$. Given the epistemic interpretation of $\Box$ this is as it should be: one may well fail to know that a sentence is true even if it is valid. Also the K schema fails: write $\varphi \rightarrow \psi$ for $\neg(\varphi \wedge \neg\psi)$, then $\Box(\varphi \rightarrow \psi) \rightarrow (\Box\varphi \rightarrow \Box\psi)$ is not valid. This is also as desired, since knowledge is not closed under modus ponens. The notion of validity just defined is indeed a minimal one: with the help of standard techniques it is easily shown that a sentence is valid if and only if it is a substitution instance of a propositional tautology.

Equivalent sentences need not be interchangeable in this system. For instance, p and $\neg\neg p$ are equivalent, but a model in which $\Box p$ and $\Box\neg\neg p$ are assigned different truth values in some possible world is easily constructed. The system thus meets the requirement discussed in the previous section.

Wansing [1990] shows that a number of formalisms that have arisen in AI research can in fact be subsumed under the impossible worlds approach. He proves in some detail that the knowledge and belief structures that were proposed in Levesque [1984], Vardi [1986], Fagin & Halpern [1988] and Van der Hoek & Meyer [1988] can be reduced to non-normal worlds models. Thijsse [1992] on the other hand proves that Rantala models can be reduced to the models of Fagin & Halpern [1988] (if the latter are slightly generalised) and thus, using Wansing's result, obtains equivalence between Rantala semantics and Fagin & Halpern's logic of general awareness.

The system is elegant enough, generalises Kripke's semantics for modal logic and subsumes other approaches to the logic of the propositional attitudes, then what are my qualms? Here is one. We have just defined the implication $\varphi \rightarrow \psi$ with the help of $\neg$ and $\wedge$. Alternatively, we could have introduced the arrow as a primitive, imposing the extra constraint that for all models $\langle W, W^*, R, V\rangle$ and for all $w \in W$:

(ii') $\quad V(\varphi \Rightarrow \psi, w) = 0 \quad$ iff $\; V(\varphi, w) = 1$ and $V(\psi, w) = 0$.

But the two methods lead to different results. For even while $\varphi \rightarrow \psi$ and $\varphi \Rightarrow \psi$ are equivalent, they are not interchangeable in all contexts: $\Box(\varphi \rightarrow \psi)$ and $\Box(\varphi \Rightarrow \psi)$ may have different truth values in some possible world. The addition of $\Rightarrow$ to the language and the addition of clause (ii') to the definition of a Rantala model really added to the logic's expressive power. Two Rantala models may validate exactly the same sentences of the original language, yet may differ on sentences of the new language.

This means that functional completeness fails for Rantala's system. Usually, when setting up a logic, we can contend ourselves with laying down truth conditions for some functionally complete set of connectives, e.g. for $\neg$ and $\wedge$. Adding connectives and letting them correspond to new truth functions usually does not increase expressive power since all truth functions are expressible with the help of $\neg$ and $\wedge$. But in the present case this is no longer so.

Why? The source of the trouble is that in Rantala models the interpretation of logical constants is not fixed. Even a reduplication of one of the logical constants would strengthen the system. Let us add a connective & to the system and impose a condition completely analogous to (ii), namely that for all $w \in W$:

(ii'') $\quad V(\varphi \,\&\, \psi, w) = 1 \quad$ iff $\; V(\varphi, w) = 1$ and $V(\psi, w) = 1$

The weird result is that $\Box(\varphi \wedge \psi)$ is *not* equivalent with $\Box(\varphi \,\&\, \psi)$ and that it is possible now to distinguish between models that were indistinguishable (i.e. validated the same sentences) before. The question arises: which is the real conjunction, $\wedge$ or &?

Is it possible to have a *logic* if the interpretations of the logical constants are allowed to vary with each model? This question can only be answered if some criterion of logicality is accepted. Such criteria have been developed within abstract model theory (see Barwise

[1974]), a branch of logic where theorems are proved of the form: "every logic that has such-and-such properties is so-and-so".[4] Rantala [1982b] notes that in fact his system does not meet the standards that are usually set here. There is a problem with *renaming*. In general the truth value of a formula should not change if we replace some non-logical constant in it by another constant which has the same semantic value. In Rantala models this fails, for example, it is easy to construct a Rantala model such that $V(p, w) = V(q, w)$ for all $w \in W \cup W^*$ but $V(\Box\neg p, w) \neq V(\Box\neg q, w)$ for some possible world $w \in W$. The value of $\Box\neg p$ thus may crucially depend on the particular name that we have chosen for the proposition that is denoted by p.

This may or may not be defensible, but, as I shall show below, the weird characteristics of Rantala's system are not essential to the idea of impossible world semantics. The idea can be formalised with the help of a system that meets all standards of logicality.

The basic intuition behind the introduction of impossible worlds is that, since we humans are finite and fallible, we fail to rule out worlds which would be ruled out by a perfect reasoner. What do such worlds look like? Well, for example, one of Jones' epistemic alternatives was the impossibility that 'if A is locked B is not' is true, but that 'if B is locked A is not' is false. For a short time, at least one impossible world in which the first sentence is true but the second is not was not ruled out by Jones' reasoning. But in such worlds the words 'not' and 'if' cannot get their usual Boolean interpretation, since this interpretation would simply force the sentences to be equivalent. We therefore end up with non-standard interpretations for the 'logical' words in English: 'and' cannot be intersection of sets of worlds, 'or' cannot be union, 'not' cannot be complementation and so on.

Rantala formalises this by treating the word 'and' as the connective $\wedge$ but by giving this last symbol a non-logical interpretation. This leads to a funny system. The obvious alternative is to keep the logic standard but to formalise the English word 'and' and its like as non-logical constants: once it is accepted that no *logical* operation strictly corresponds to the English word 'and', the most straightforward solution is be open about it and to formalise the word with the help of a non-logical constant.

Of course, some connection between the 'logical' words in English and the connectives that usually formalise them should remain intact. What connection? Even if we allow the interpretations of the 'logical' words to be completely arbitrary, there will be a subset of the set of all worlds where 'and' and its ilk behave standardly. These worlds where the logical words of English have their usual logical interpretation may be called the 'possible' or 'actualizable' ones. As we shall see below, the assumption that the actual world is actualizable leads to the desired relation of logical consequence.

But it is high time for a more precise formalisation. In the next section I give a short sketch of the classical logic that I want to use and in the last section I'll apply it to the propositional attitude verbs.

3 Classical Type Theory

Since we want to treat 'and', 'or', 'not', 'if', 'every' and 'some' as non-logical constants, we should use a logic that admits of non-logical constants for these types of expressions. Ordinary predicate logic will not do, but a logic that is admirably suited to the job is Church's

---

4 An example is Lindström's Theorem, which says that for no logic properly extending first-order predicate logic both the Compactness Theorem and the Löwenheim-Skolem Theorem hold.

[1940] formulation of Russell's Theory of Types (Russell [1908]). Since I expect that not all of my readers are familiar with this system, I'll give it a short exposition (and so readers who already know about the logic can skip this section). For a more extensive account one may consult the original papers (e.g. Church [1940], Henkin [1950, 1963]), Gallin [1975], the survey article Van Benthem & Doets [1983], or the text book Andrews [1986]. In Muskens [1989a, 1989b, 1989c] some variants of the logic are given, but I'll follow the standard set-up here.

In classical type theory each logical expression comes with a *type*. Types are either basic or complex. The type of truth values, here denoted with t, should be among the basic types, but there may be other basic types as well. In this paper, for example, we'll assume types for individuals (type e) and worlds (type s). Complex types are of the form $\alpha\beta$ [5] and an expression of type $\alpha\beta$ will denote a function which takes things of type α to things of type β. Formally we define:

DEFINITION 1 (Types). The set of *types* is the smallest set such that:

i. all basic types are types,
ii. if α and β are types, then $(\alpha\beta)$ is a type.

DEFINITION 2 (Frames). A *frame* is a set of non-empty sets $\{D_\alpha \mid \alpha$ is a type$\}$ such that $D_t = \{0, 1\}$ and $D_{\alpha\beta} \subseteq \{f \mid f: D_\alpha \rightarrow D_\beta\}$ for all complex types $\alpha\beta$.

The sets D_α will function as the *domains* of all things of type α. Note that we do not require domains $D_{\alpha\beta}$ to consist of *all* functions of the correct type, as this would make the logic essentially higher-order and non-axiomatisable. Let us assume for each type α the existence of denumerably infinite sets of variables and non-logical constants VAR_α and CON_α. From these we can build up terms with the help of lambda abstraction, application and the identity symbol.

DEFINITION 3 (Terms). Define, for each α, TERM_α, the set of *terms* of type α, by the following inductive definition:

i. $\text{CON}_\alpha \subseteq \text{TERM}_\alpha$;
   $\text{VAR}_\alpha \subseteq \text{TERM}_\alpha$;
ii. $A \in \text{TERM}_{\alpha\beta}, B \in \text{TERM}_\alpha \Rightarrow (AB) \in \text{TERM}_\beta$;
iii. $A \in \text{TERM}_\beta, x \in \text{VAR}_\alpha \Rightarrow \lambda x(A) \in \text{TERM}_{\alpha\beta}$;
iv. $A, B \in \text{TERM}_\alpha \Rightarrow (A = B) \in \text{TERM}_t$.

If $A \in \text{TERM}_\alpha$ we may indicate this by writing A_α. Terms of type t are called *formulae*. We obtain most of the usual logical signs by means of abbreviations.

DEFINITION 4 (Abbreviations).

T	abbreviates	$\lambda x_t(x_t) = \lambda x_t(x_t)$
$\forall x_\alpha\varphi$	abbreviates	$\lambda x_\alpha\varphi = \lambda x_\alpha \mathsf{T}$
$\bot$	abbreviates	$\forall x_t(x)$
$\neg\varphi$	abbreviates	$\varphi = \bot$
$\varphi \wedge \psi$	abbreviates	$\lambda X_{t(tt)}((X\mathsf{T})\mathsf{T}) = \lambda X_{t(tt)}((X\varphi)\psi)$

5 Sometimes denoted as $\alpha \rightarrow \beta$, sometimes as $\beta\alpha$.

The rest of the usual logical constants can be got in an obvious way.

In order to assign each term a value in a given frame, we must interpret all variables and non-logical constants in that frame. An *interpretation* function I for a frame $F = \{D_\alpha\}_\alpha$ is a function with the set of non-logical constants as its domain, such that $I(c) \in D_\alpha$ for each constant c of type α. Likewise, an *assignment* a for a frame $\{D_\alpha\}_\alpha$ is a function that has the set of all variables for its domain, such that $a(x) \in D_\alpha$ for each variable x of type α. If a is an assignment, then $a[d/x]$ is defined by $a[d/x](x) = d$ and $a[d/x](y) = a(y)$ for $y \neq x$.

A *very general model* is a tuple $\langle F, I\rangle$, consisting of a frame F and an interpretation function I for that frame. Given some very general model and an assignment, we can give each term a value.

DEFINITION 5 (Tarski Definition). The *value* $\|A\|^{M,a}$ of a term A on a very general model $M = \langle\{D_\alpha\}_\alpha, I\rangle$ under an assignment a for $\{D_\alpha\}_\alpha$ is defined as follows (To improve readability I write $\|A\|$ or $\|A\|^a$ for $\|A\|^{M,a}$):

i. $\|c\| = I(c)$ if c is a constant;
$\|x\| = a(x)$ if x is a variable;

ii. $\|A_{\alpha\beta}B_\alpha\| = \|A\|(\|B\|)$ if $\|B\| \in$ domain($\|A\|$)
$= \emptyset$ otherwise;

iii. $\|\lambda x_\alpha A\|^a =$ the function f with domain D_α such that $f(d) = \|A\|^{a[d/x]}$ for all $d \in D_\alpha$;

iv. $\|A = B\| = 1$ iff $\|A\| = \|B\|$.

We define a (*general*) *model* to be a very general model $M = \langle\{D_\alpha\}_\alpha, I\rangle$ such that $\|A_\alpha\|^{M,a} \in D_\alpha$ for every term A_α and we restrict our attention to general models. Note that in general models the second subclause of ii. does not apply (we needed it for the correctness of definition 5). The reader may verify that on general models the logical constants $\mathrm{T}, \forall, \bot, \neg$ and $\wedge$ get their usual (classical) interpretations.

The semantic notion of entailment is defined as follows.

DEFINITION 6 (Entailment). Let $\Gamma \cup \{\varphi\}$ be a set of formulae. Γ *entails* φ, $\Gamma \models \varphi$, if, for all models M and assignments a to M, $\|\psi\|^{M,a} = 1$ for all $\psi \in \Gamma$ implies $\|\varphi\|^{M,a} = 1$.

Henkin [1950] has proved that it is possible to axiomatise the logic. In fact, an elegant set of four axiom schemes and one derivation rule will do the job. For details see the literature mentioned above. For the present purposes it suffices to note that β-conversion and η-conversion hold and that we can reason with $=$ and the defined constants $\mathrm{T}, \forall, \bot, \neg$ and $\wedge$ as in (many-sorted) classical predicate logic with identity.

4 Classical Logic Without Logical Omniscience

Let us apply our logic to English.[6] Since we have decided to treat the 'logical' words as non-logical constants, we can now uniformly treat all words as such. Table 1 below gives a list of

6 The application of type logic to the formalisation of English discussed in this section beneficed greatly from Montague [1970ª, 1970ᵇ, 1973]. In fact we can think of it as a streamlined form of Montague semantics.

all constants that we shall use in this paper, most of them named in a way that makes it easy to see which words they are supposed to formalise (the others will not directly translate words of English; their use will become apparent below). The constants in the first column of the table have types as indicated in the second column.

NON-LOGICAL CONSTANTS	TYPE
`not`	$(st)(st)$
`and, or, if`	$(st)((st)(st))$
`every, a, some, no`	$(e(st))((e(st))(st))$
`is`	$((e(st))(st))(e(st))$
`hesperus, phosphorus, mary`	$(e(st))(st)$
`planet, man, woman, walk, talk`	$e(st)$
`believe, know`	$(st)(e(st))$
`i`	s
`h, p, m`	e
`B, K`	$e(s(st))$

Table 1

The idea behind the type assignment[7] is that the meaning of a sentence, a proposition, is a function that gives us a truth value in each world (and thus it is a function of type st), that the meaning of a predicate like `planet` is a function that gives a truth value if we feed it an individual and a world (type $e(st)$) and that an expression that expects an expression of type α should be of type $\alpha\beta$ if the result of combining it with such an expression should be of type β. So, for example, `not` is of type $(st)(st)$ since it expects a proposition in order to form another proposition with it; the name `mary` gives a sentence if it is followed by a predicate and may therefore be assigned type $(e(st))(st)$. Some easy calculation shows that the following are terms of type st.

(8) `(some woman)walk`
(9) `(no man)talk`
(10) `hesperus (is (a planet))`
(11) `(if((some woman)walk))((no man)talk)`
(12) `(if((some man)talk))((no woman)walk)`
(13) `mary(believe((if((some woman)walk))((no man)talk))`
(14) `mary(believe((if((some man)talk))((no woman)walk))`

Clearly, these terms bear a very close resemblance to the sentences of English that they formalise. For example, the structure of (13) is isomorphic or virtually isomorphic to the structure that most linguists would attach to the sentence 'Mary believes that if some woman is walking no man is talking'. But it should be kept in mind that these are terms of the logic and can be subject to logical manipulation.

We must of course make a connection between at least some of the non-logical constants that we have just introduced and the logical constants of the system. For example, (11) should be equivalent with (12), but as matters stand these two terms could denote two completely different (characteristic functions of) sets of worlds. Up to now, we have allowed the

---

7 Essentially this assignment was used in Montague [1970[b]] and Lewis [1974].

interpretations of the constants not, and, or, if and the like to be completely arbitrary, but it is not unreasonable to assume that at least in the actual world, which we denote with the constant i, these interpretations are standard. In order to ensure this we impose the following non-logical axioms.[8]

A1	$\forall p((\mathtt{not}p)\mathtt{i} \leftrightarrow \neg p\mathtt{i})$
A2	$\forall pq(((\mathtt{and}p)q)\mathtt{i} \leftrightarrow (p\mathtt{i} \wedge q\mathtt{i}))$
A3	$\forall pq(((\mathtt{or}p)q)\mathtt{i} \leftrightarrow (p\mathtt{i} \vee q\mathtt{i}))$
A4	$\forall pq(((\mathtt{if}p)q)\mathtt{i} \leftrightarrow (p\mathtt{i} \rightarrow q\mathtt{i}))$
A5	$\forall P_1P_2(((\mathtt{every}P_1)P_2)\mathtt{i} \leftrightarrow \forall x((P_1x)\mathtt{i} \rightarrow (P_2x)\mathtt{i}))$
A6	$\forall P_1P_2(((\mathtt{a}P_1)P_2)\mathtt{i} \leftrightarrow \exists x((P_1x)\mathtt{i} \wedge (P_2x)\mathtt{i}))$
A7	$\forall P_1P_2(((\mathtt{some}P_1)P_2)\mathtt{i} \leftrightarrow \exists x((P_1x)\mathtt{i} \wedge (P_2x)\mathtt{i}))$
A8	$\forall P_1P_2(((\mathtt{no}P_1)P_2)\mathtt{i} \leftrightarrow \neg\exists x((P_1x)\mathtt{i} \wedge (P_2x)\mathtt{i}))$
A9	$\forall Q\forall x(((\mathtt{is}Q)x)\mathtt{i} \leftrightarrow (Q\lambda y\lambda j(x = y))\mathtt{i})$
A10	$\forall P((\mathtt{hesperus}P)\mathtt{i} \leftrightarrow (P\mathtt{h})\mathtt{i})$
	$\forall P((\mathtt{phosphorus}P)\mathtt{i} \leftrightarrow (P\mathtt{p})\mathtt{i})$
	$\forall P((\mathtt{mary}P)\mathtt{i} \leftrightarrow (P\mathtt{m})\mathtt{i})$.

These axioms tell us that an expression not$p$ is true in the actual world i if and only if p is false in i, that (and$p$)q is true in i if and only if p and q are both true in i, and so on. Given these axioms many sentences get their usual truth value in the actual world. For example A7 tells us that ((some woman)walk)i is equivalent with $\exists x$((woman$x$)i $\wedge$ (walk$x$)i), A8 says that ((no man)talk)i is equivalent with $\neg\exists x$((man$x$)i $\wedge$ (talk$x$)i). Axiom A10 says that there is an individual h such that the quantifier hesperus holds of some predicate at i if and only if that predicate holds of h at i. We can use the axioms to see that the following terms are equivalent.

hesperus (is (a planet))i	
((is (a planet))h)i	(A10)
((a planet)$\lambda y\lambda j$(h = y))i	(A9)
$\exists x$((planet$x$)i $\wedge$ ($\lambda y\lambda j$(h = y)x)i)	(A6)
$\exists x$((planet$x$)i $\wedge$ h = x)	(β-reduction twice)
(planet h)i	(predicate logic)

Let Φ be the conjunction of our finite set of axioms and let $[k/\mathtt{i}]\Phi$ be the result of substituting the type s variable k for each occurrence of i in Φ. The st term $\lambda k[k/\mathtt{i}]\Phi$ denotes the set of those worlds in which not, and, or, if etc. have their standard logical meaning. We may view the term $\lambda k[k/\mathtt{i}]\Phi$ as formalising the predicate 'is logically possible' or 'is actualizable'. The axioms thus express that the actual world is logically possible or actualizable.

We define a notion of entailment on st terms with the help of the set of axioms AX = {A1,..., A12}(A11 and A12 will be given shortly). An argument is (weakly) valid if and

[8] Here and in the rest of the paper I shall let j and k be type s variables; x and y type e variables; (subscripted) P a variable of type $e(st)$; Q a type $(e(st))(st)$ variable; and p and q variables of type st. Variables are in *Times italic,* constants in Courier.

only if the conclusion is true in the actual world if all premises are true in the actual world, assuming that the actual world is actualizable.

DEFINITION 7 (Weak entailment). Let $\varphi_1, \ldots, \varphi_n$, ψ be terms of type *st*. We say that ψ *follows from* $\varphi_1, \ldots, \varphi_n$ if AX, $\varphi_1\mathtt{i}, \ldots, \varphi_n\mathtt{i} \models \psi\mathtt{i}$. Terms φ and ψ of type *st* are called *equivalent* if ψ follows from φ and φ follows from ψ.

That terms (11) and (12) are indeed equivalent in this sense can easily be seen now. The following terms are equivalent.

$$((\mathtt{if}((\mathtt{some\ woman})\mathtt{walk}))((\mathtt{no\ man})\mathtt{talk}))\mathtt{i}$$
$$((\mathtt{some\ woman})\mathtt{walk})\mathtt{i} \rightarrow ((\mathtt{no\ man})\mathtt{talk})\mathtt{i} \qquad \text{(A4)}$$
$$\exists x((\mathtt{woman}x)\mathtt{i} \wedge (\mathtt{walk}x)\mathtt{i}) \rightarrow ((\mathtt{no\ man})\mathtt{talk})\mathtt{i} \qquad \text{(A7)}$$
$$\exists x((\mathtt{woman}x)\mathtt{i} \wedge (\mathtt{walk}x)\mathtt{i}) \rightarrow \neg\exists x((\mathtt{man}x)\mathtt{i} \wedge (\mathtt{talk}x)\mathtt{i}) \qquad \text{(A8)}$$

In the same way we find that (12) applied to i is equivalent with

$$\exists x((\mathtt{man}x)\mathtt{i} \wedge (\mathtt{talk}x)\mathtt{i}) \rightarrow \neg\exists x((\mathtt{woman}x)\mathtt{i} \wedge (\mathtt{walk}x)\mathtt{i}),$$

and the equivalence of (11) and (12) follows with contraposition.

But if we try to apply a similar procedure to (13) and (14) the process quickly aborts. It is true that (13) applied to i with the help of A10 can be reduced to

$$((\mathtt{believe}((\mathtt{if}((\mathtt{some\ woman})\mathtt{walk}))((\mathtt{no\ man})\mathtt{talk})))\mathtt{m})\mathtt{i}$$

and that (14) applied to i can be reduced to

$$((\mathtt{believe}((\mathtt{if}((\mathtt{some\ man})\mathtt{talk}))((\mathtt{no\ woman})\mathtt{walk})))\mathtt{m})\mathtt{i}$$

but further reductions are not possible. In fact it is not difficult to find a model in which one of these formulae is true but the other is false. The reason is that it is not only the denotation in the actual world of the embedded terms (11) and (12) that matters now, but that their full meanings (i.e. denotations in all possible and impossible worlds) have to be taken into account. Not only their *Bedeutung* but also their *Sinn.* Since no two syntactically different terms have the same *Sinn,* no unwanted replacements are allowed.

Note that the solution does not commit us to a Hintikka style treatment of knowledge and belief. We have not assumed that belief is truth in all doxastic alternatives, knowledge truth in all epistemic alternatives. But we can, if we wish, make these assumptions by adopting the following two axioms.

$$\text{A11} \qquad \forall p \forall x(((\mathtt{believe}p)x)\mathtt{i} \leftrightarrow \forall j(((\mathtt{B}x)\mathtt{i})j \rightarrow pj))$$
$$\text{A12} \qquad \forall p \forall x(((\mathtt{know}p)x)\mathtt{i} \leftrightarrow \forall j(((\mathtt{K}x)\mathtt{i})j \rightarrow pj))$$

Here B and K are constants of type $e(s(st))$ that stand for the doxastic and epistemic alternative relations respectively. A term $((\mathtt{B}x)i)j$ can be read as: 'in world i, world j is a doxastic alternative of x' or 'in world i, world j is compatible with the beliefs of x'; $((\mathtt{K}x)i)j$

can be read as: 'in world i, world j is an epistemic alternative of x' or 'in world i, world j is compatible with the knowledge of x'.[9] If these axioms are accepted, we can reduce (13) to

$$\forall j(((\mathtt{Bm})\mathtt{i})j \rightarrow ((\mathtt{if}((\mathtt{some\ woman})\mathtt{walk}))((\mathtt{no\ man})\mathtt{talk}))j),$$

a formula that expresses that (11) holds in all Mary's doxastic alternatives. Clearly, no further reductions are possible and we can still find models such that (13) is true but (14) is false (in the actual world).

The mechanism helps us to solve some related puzzles as well. For example, (17) should not follow from (15) and (16) and it doesn't.

(15) hesperus (is phosphorus)
(16) (every man)(know(hesperus(is hesperus)))
(17) (every man)(know(phosphorus(is hesperus)))

Surely, (hesperus (is phosphorus))i reduces to h = p after a few steps, and h and p are thus interchangeable in all contexts if (15) is accepted, but (16), if it is applied to i, only reduces to

(18) $\forall x((\mathtt{man}\,x)\mathtt{i} \rightarrow \forall j(((\mathtt{K}x)\mathtt{i})j \rightarrow (\mathtt{hesperus\ (is\ hesperus)})j))$,

and (17) applied to i can only be reduced to

(19) $\forall x((\mathtt{man}\,x)\mathtt{i} \rightarrow \forall j(((\mathtt{K}x)\mathtt{i})j \rightarrow (\mathtt{phosphorus\ (is\ hesperus)})j))$.

Clearly, the premise h = p and (18) do not entail (19).

Terms (16) and (17) are the *de dicto* readings of the sentences 'Every man knows that Hesperus is Hesperus' and 'Every man knows that Phosphorus is Hesperus' respectively. Of course, we can also formalise *de re* readings, as is illustrated in (20) and (21). The reading that is formalised by (20) can be paraphrased as 'Of Hesperus, every man knows that it is Hesperus', while the other term can be paraphrased as 'Of Phosphorus, every man knows that it is Hesperus'. The reader may wish to verify that in this case the relevant entailment holds: (21) follows from (20) and (15).

(20) $\mathtt{hesperus}\lambda x((\mathtt{every\ man})(\mathtt{know}((\mathtt{is\ hesperus})x)))$

(21) $\mathtt{phosphorus}\lambda x((\mathtt{every\ man})(\mathtt{know}((\mathtt{is\ hesperus})x)))$

This possibility of quantifying-in, which the present theory shares with other semantic theories of the attitudes, distinguishes the approach from Quine's [1966] syntactic treatment. But

---

9 We may demand that B and K satisfy some axioms. The following seem a reasonable choice:
$\forall x((\mathtt{K}x)\mathtt{i})\mathtt{i}$
$\forall x\forall jk(((\mathtt{K}x)\mathtt{i})j \rightarrow (((\mathtt{K}x)j)k \leftrightarrow ((\mathtt{K}x)\mathtt{i})k))$
$\forall x\exists j((\mathtt{B}x)\mathtt{i})j$
$\forall x\forall jk(((\mathtt{B}x)\mathtt{i})j \rightarrow (((\mathtt{B}x)j)k \leftrightarrow ((\mathtt{B}x)\mathtt{i})k))$
$\forall x\forall j(((\mathtt{B}x)\mathtt{i})j \rightarrow ((\mathtt{K}x)\mathtt{i})j)$

our semantic theory is as fine-grained as any syntactic theory can be, for no two syntactically different expressions have the same meaning. The resemblance between the syntactic approach and ours is close: the syntactic theory treats the attitudes as relations between persons and syntactic expressions, we treat them as relations between persons and the meanings of those expressions. But since different expressions have different meanings, this boils down to much the same thing.

References

Andrews, P.B.: 1986, *An Introduction to Mathematical Logic and Type Theory: to Truth through Proof,* Academic Press, Orlando, Florida.

Barwise, J.: 1974, Axioms for Abstract Model Theory, *Annals of Mathematical Logic* 7, 221-265.

Benthem, J.F.A.K. Van, and Doets, K.: 1983, Higher-Order Logic, in Gabbay & Guenthner [1983] Vol I, 275-329.

Church, A.: 1940, A Formulation of the Simple Theory of Types, *The Journal of Symbolic Logic* 5, 56-68.

Cresswell, M.J.: 1972, Intensional Logics and Logical Truth, *Journal of Philosophical Logic* 1, 2-15.

Fagin, R. and Halpern J.Y.: 1988, Belief, Awareness and Limited Reasoning, *Artificial Intelligence* 34, 39-76.

Gabbay, D. and Guenthner, F. (eds.): 1983, *Handbook of Philosophical Logic*, Reidel, Dordrecht.

Gallin, D.: 1975, *Intensional and Higher-Order Modal Logic*, North-Holland, Amsterdam.

Henkin, L.: 1950, Completeness in the Theory of Types, *The Journal of Symbolic Logic* 15, 81-91.

Henkin, L.: 1963, A Theory of Propositional Types, *Fundamenta Mathematicae* 52, 323-344.

Hintikka, J.: 1975, Impossible Possible Worlds Vindicated, *Journal of Philosophical Logic* 4, 475-484.

Hoek, W. Van der, and Meyer, J.-J.: 1988, *Possible Logics for Belief,* Rapport IR-170, Vrije Universiteit, Amsterdam.

Levesque, H.J.: 1984, A Logic of Implicit and Explicit Belief, *Proceedings AAAI-84,* Austin, Texas, 198-202.

Lewis, D.: 1974, 'Tensions, in Munitz, M.K. and Unger, P.K. (eds.), *Semantics and Philosophy,* New York University Press, New York.

Mates, B.: 1950, Synonymity, reprinted in Linsky (ed.), *Semantics and the Philosophy of Language,* The University of Illinois Press, Urbana, 1952, 111-136.

Montague, R.: 1970, Universal Grammar, reprinted in Montague [1974], 222-246.

Montague, R.: 1973, The Proper Treatment of Quantification in Ordinary English, reprinted in Montague [1974], 247-270.

Montague, R.: 1974, *Formal Philosophy*, Yale University Press, New Haven.

Moore, R.C.: Propositional Attitudes and Russellian Propositions, in R. Bartsch, J.F.A.K. van Benthem and P. van Emde Boas (eds.), *Semantics and Contextual Expression, Proceedings of the Sixth Amsterdam Colloquium*, Foris, Dordrecht, 147-174.

Muskens, R.A.: 1989[a], A Relational Formulation of the Theory of Types, *Linguistics and Philosophy* 12, 325-346.

Muskens, R.A.: 1989[b], Going Partial in Montague Grammar, in R. Bartsch, J.F.A.K. van Benthem and P. van Emde Boas (eds.), *Semantics and Contextual Expression, Proceedings of the Sixth Amsterdam Colloquium*, Foris, Dordrecht, 175-220.

Muskens, R.A.: 1989[c], *Meaning and Partiality,* Dissertation, University of Amsterdam.

Putnam, H.: 1954, Synonymity and the Analysis of Belief Sentences, *Analysis* 14, 114-122.

Quine, W.V.O.: 1966, Quantifiers and Propositional Attitudes, in *The Ways of Paradox,* New York.

Rantala, V.: 1982[a], Impossible Worlds Semantics and Logical Omniscience, in I. Niiniluoto and E. Saarinen (eds.), *Intensional Logic: Theory and Applications,* Helsinki.

Rantala, V.: 1982[b], Quantified Modal Logic: Non-normal Worlds and Propositional Attitudes, *Studia Logica* 41, 41-65.

Russell, B.: 1908, Mathematical Logic as Based on the Theory of Types, *American Journal of Mathematics* 30, 222-262.

Thijsse, E.: 1992, *Partial Logic and Knowledge Representation,* Dissertation, Tilburg University.

Vardi, M.Y.: 1986, On Epistemic Logic and Logical Omniscience, in J.Y. Halpern (ed.), *Theoretical Aspects of Reasoning about Knowledge: Proceedings of the 1986 Conference,* Morgan Kaufmann, Los Altos, 293-305.

Wansing, H.: 1990, A General Possible Worlds Framework for Reasoning about Knowledge and Belief, *Studia Logica* 49, 523-539.

Weak Implication: Theory and Applications

Karen L. Kwast & Sieger van Denneheuvel
University of Amsterdam
Department of Mathematics & Computer Science,
Plantage Muidergracht 24, 1018 TV, Amsterdam.

Abstract. We study a generalization of the classical notion of implication, called *weak implication*. It extends unquantified predicate logic with a single level of *existential quantification*. We present a sound and complete set of deduction rules for weak implications.

The notion of weak implication was introduced for the sake of a formal specification of a *symbolic constraint solving* system. Other practical *applications* of can be found in the realm of relational database theory: query normalization and integrity constraints in the context of views.

1 Introduction

Weak equivalence is a generalization of the classical notion of equivalence. Two formulas are equivalent if they are satisfied by the same values; they are weakly equivalent if they are equivalent 'on' a set of variables X, that is, if their projections on X are equivalent. Consider for instance the non-equivalent constraints

$$x = y + 2,\ y = z + 3 \ \not\equiv\ x = z + 5 \tag{1}$$

If one wants to express x in terms of the known variable z, it makes sense to ask for the strongest condition on x and z that is entailed by the left-hand side. The actual value of y is irrelevant and becomes existentially quantified. Formally:

$$x = y + 2,\ y = z + 3 \ \equiv_{xz}\ x = z + 5 \tag{2}$$

A more interesting example derives from the context of constraint solving. The following set of constraints is underdetermined:

$$x + u = v,\ y + u = v,\ x + y = 6m,\ u < v \tag{3}$$

One can solve x and y in terms of m by means of the equivalent$_{mxy}$ constraints

$$x = 3 * m,\ y = 3 * m,\ m > 0 \tag{4}$$

Obviously, these 2 sets of constraints are not equivalent, but they express the same relation between x, y and m.

In this paper, we present the logical theory of weak implications. § 2 summarizes related research. Basic notions and definitions are provided in § 3. The system $\mathcal{WI}$ of derivation rules for weak implications is discussed in § 4. Applications are given in § 5: constraint solving, § 6: query normalization and § 7: integrity constraints.

2 Related Work

There are several areas of research that are related to the present subject. Obviously, the present system will be a subsystem of first order predicate logic, namely the logic of $\forall\exists$-sentences. This reduction in expressive power will be compensated for by an increase in transparency: the absence of (nested) quantifiers. Moreover, this fragment is known to be decidable.

Restrictions to the language are not uncommon: in logic programming (e.g. [1]), the language is restricted to unquantified predicate logic with equality and functions. *All* free variables are (implicitly and globally) universally quantified. Here we add in effect a second layer of quantification: (local) existential quantification.

In unification theory (see [7]), one studies the possibility to find a *most general unifier* Φ for a set of equations Γ in the context of an equational theory E. One can check Φ in terms of equivalence: $E \models \Gamma[\Phi] \equiv \text{TRUE}$, or, equivalently, $E \models \Gamma \equiv \Phi$. Here we study the more general issue: $\Gamma \equiv_X \Delta$ (in the context of E), where Φ may be part of Δ.

The present theory originates from yet another area: symbolic constraint solving. A good example is the system *Mathematica* ([9]). It allows one to declare a set of constraints, which can be simplified by the system. Moreover, one can ask for a symbolic solution, expressing some *wanted* variables in terms of the others (: the *known* variables). When the set of constraints (such as (3) above) is underdetermined, the system will fail.

Weak equivalence has been developed as a tool to describe and validate the **RL/1** symbolic constraint solving system, which allows for the specification of *intermediate* variables. This system is capable of producing a *conditional* solution: if the known variables satisfy some condition, then they define the wanted variables in the specified manner (cf. (4) above). We will illustrate this idea in § 5 by means of an example; for all details on **RL/1** see [2].

The system **RL/1** has much in common with **CLP** ([5]), where global constraints are combined with local conditions into rules and goals (: as in PROLOG). If a goal is satisfiable, a successful derivation will yield a set of *answer constraints* as *symbolic output*. In **CLP** the answer constraints are represented in *solved form*, which depends on the type of constraints. In **RL/1**, however, the answer consists of a *reduced constraint* and a *solution set*, that is, a symbolic solution expressing some variables in terms of others (see § 5). This has the advantage that the symbolic answer can be evaluated on a (relational) database, in accordance with the **RL** objective to integrate logic programming with database systems. On the other hand, it restricts the types of constraints that can be dealt with; **RL/1** does not cover recursive rules.

In § 5 we employ weak equivalence to give a formal specification of the **RL/1** constraint solver; its technical details and a comparison to other systems for constraint solving can be found in [2].

There are many other possible applications of the notion of weak implication. It can be applied to study the implication problem of any phenomenon that requires but a single level of existential quantification. In the present formalism, all quantifiers are removed and replaced by implicit quantification, by means of the implication variables. These variables connect the antecedent of the implication with the consequence and are (implicitly) universally quantified. All remaining variables are more or less *irrelevant*: they are (implicitly) existentially quantified, locally, that is, the scope of quantification is restricted to antecedent cq consequence.

The advantage of the weak implication formalism is that it corresponds with the algebraic projection operator: the projection set contains the relevant variables, all others can be ignored (: existentially quantified). As a consequence, weak implication statements are more intuitive and transparent, at least to people that are used to the relational algebra.

Weak implications can be applied to *normalize* terms of the relational algebra, high-lighting the natural boundaries of the projection normal form, $\Pi_X \sigma_\varphi(R \bowtie S)$, which corresponds with the basic **SQL** statement, `SELECT` X `FROM` R, S `WHERE` φ. Normalization theory does not *require* weak implications, of course, but its description can be simplified by means of weakly equivalent (sub)terms.

In a similar manner, weak implication can be applied to derive integrity constraints. There exist several formalisms to study the integrity implication problems, but none that includes views and equivalent term rewriting in a uniform manner.

The applications we give of the notion of weak implication are mainly illustrative. The results on term rewriting are freely adapted from previous work on term rewriting, by ourselves ([4]) as well as by others (e.g. [10]), only 'translated' into the weak implication formalism to illustrate the usefulness of the latter. The examples of derived integrity constraints have not yet been systematically developed; for the basics of database theory we refer to [8].

3 Definitions

Let $\mathcal{L}$ be a predicate language, consisting of constants CON $(a, b, c, \ldots)$, variables VAR $(x, y, z, \ldots)$, predicates $(R, S, T, \ldots)$ and functions $(f, g, \ldots)$. Terms $(s, t, \ldots)$ and formulas FORM $(\varphi, \psi, \ldots)$ are constructed as usual. $\mathcal{L}$ contains an identity symbol $=$ and the propositional constants TRUE and FALSE. $\alpha(\varphi)$ is the set of free variables of a formula φ. Attributes $\mathcal{A}$ and variables will be used indifferently; sets of attributes are denoted by X or $\mathbf{x}$.

A *solution* is an equation $x = t$ with $x \notin \alpha(t)$. A *solution set* (: $\Phi, \Psi, \ldots$) is a finite set of independent solutions, that is:

Definition 1 *A solution set Φ is a finite set $\{ x_1 = t_1, \ldots, x_n = t_n \}$, such that for all $i, j \leq n$: $x_i \neq x_j$ $(i \neq j)$ and $x_i \notin \alpha(t_j)$.*

The syntactic independence restriction on a solution set Φ can be reformulated in terms of its *head- and tail attributes*:

Definition 2 $\alpha_H(\Phi) := \{x_1, \ldots, x_n\}$ $\alpha_T(\Phi) := \alpha(t_1) \cup \ldots \cup \alpha(t_n)$.

The set $\Phi = \{x_1 = t_1, \ldots, x_n = t_n\}$ is a solution set iff the number of head attributes is n, that is, $\#(\alpha_H(\Phi)) = \#(\Phi)$, and $\alpha_H(\Phi) \cap \alpha_T(\Phi) = \emptyset$. Considered as a substitution, Φ is separated on its head attributes away from its tail attributes.

Let $\mathcal{M} = <\mathcal{D},\mathcal{I}>$ be a model for a language $\mathcal{L}$ and $\mathbf{H}$ the set of assignments $h : \text{VAR} \to \mathcal{D}$. $\mathcal{M} \models \varphi[h]$ is defined by the usual induction. Obviously, if $h =_{\alpha(\varphi)} h'$, that is, if $h(x) = h'(x)$ for all $x \in \alpha(\varphi)$, then $\mathcal{M} \models \varphi[h]$ iff $\mathcal{M} \models \varphi[h']$.

The notion of a valid implication is completely standard: $\varphi \models \psi$ iff every model and assignment that satisfies φ satisfies ψ as well. It is called *strong* here to discriminate it from its weaker generalization.

Weak implication is more general in the sense that it includes strong implication as a special case, but it is weaker in the sense that any weak implication is logically implied by its strong counterpart. Before we give the formal definition, an example will illustrate this point.

Example 1 *Some strong and weak implications.*
$x = y + 1 \models x > y \qquad x > y \not\models x = y + 1$
$x = y + 1 \models_{xy} x > y \qquad x > y \not\models_{xy} x = y + 1$
$x = y + 1 \models_{x} x > y \qquad x > y \models_{x} x = y + 1$

Weak implication is defined on model-level as well as in general:

Definition 3 *Weak implication*
1. $\mathcal{M} : \varphi \models_X \psi \;:=\; \forall h \in \mathbf{H} : \mathcal{M} \models \varphi[h] \Rightarrow \exists h' \in \mathbf{H} : h' =_X h \;\&\; \mathcal{M} \models \psi[h']$
2. $\varphi \models_X \psi \;:=\;$ *for all models* $\mathcal{M}$: $\mathcal{M} : \varphi \models_X \psi$

The relevance of the notion $\mathcal{M} : \varphi \models_X \psi$ can be explained by means of an example:

$$x = -1 \models_x x = y^2 \wedge x + 1 = 0$$

It depends on the model (: is $\sqrt{-1}$ defined?) whether or not there exists a satisfying assignment h'. As we do *not* want to fix the models from the outset, we *must* be able to conceive weak implications as a notion valid in a model $\mathcal{M}$ or in a class of models $\mathcal{C}$. Note that we do not define weak implication on assignment-level, that is, as a language connective. To see why, suppose we would define a connective $\supset_X$.

Definition 4 *Weak connective*
$\mathcal{M} \models \varphi \supset_X \psi[h] \;:=\; \mathcal{M} \models \varphi[h] \Rightarrow \exists h' \in \mathbf{H} : h' =_X h \;\&\; \mathcal{M} \models \psi[h']$

Given this definition we would derive at a system in which even transitivity is unvalid: $\varphi \supset_X \psi,\ \psi \supset_X \chi \not\models \varphi \supset_X \chi$.

Example 2 *Suppose* $h(x) = 3, h(y) = 5, h(z) = 2$, *then*
$\mathcal{M} \models x = 3 \supset_{xy} y = 5 * z\ [h]$
$\mathcal{M} \models y = 5 * z \supset_{xy} x = 11\ [h]$
$\mathcal{M} \not\models x = 3 \supset_{xy} x = 11\ [h]$

Note that this is not a counterexample under a higher level reading; the rule

$$\varphi \models_X \psi,\ \psi \models_X \chi \;\Rightarrow\; \varphi \models_X \chi$$

is valid, but most models will falsify the second assumption (: $y = 5 * z \models_{xy} x = 11$), thus avoiding the undesirable consequence (: $x = 3 \models_{xy} x = 11$).

Definition 5 $\varphi \equiv_X \psi \ := \ \varphi \models_X \psi \ \& \ \psi \models_X \varphi$

Before we turn to the logic of weak implications, we will try to demystify this notion by exploring its translation into quantified predicate formulas.

3.1 Clashing Variables

The major effect of indexing implication with a set of attributes $\mathbf{x}$ is that the remaining attributes (: $\mathbf{y}$ and $\mathbf{z}$ respectively) get existentially quantified. Hence:

Lemma 1 *Let* $\mathbf{y} = \alpha(\varphi) \setminus \mathbf{x}$ *and* $\mathbf{z} = \alpha(\psi) \setminus \mathbf{x}$, *then*
$\mathcal{M} : \varphi \models_{\mathbf{x}} \psi$ *iff* $\mathcal{M} \models \forall \mathbf{x}(\exists \mathbf{y}\, \varphi(\mathbf{xy}) \supset \exists \mathbf{z}\, \psi(\mathbf{xz}))$

Example 3 $m < x \models_x x = m + n \wedge n > 0$
corresponds with $\forall x(\exists m:\ m < x \supset \exists m, n:\ x = m + n \wedge n > 0)$

In general a weak implication may contain *clashing variables* (such as: m in the example above), that is, variables that appear on both sides of the implication sign but not in X. If that is the case, an implication must be *rectified* or at least *purified*, before it can be transformed into $\forall\exists$-normal form.

Definition 6 *An implication* $\varphi \models_X \psi$ *is rectified iff* $\alpha(\varphi) \subseteq X$.

Strong implication is a special case of weak implication: put $X = \mathcal{A}$, the set of *all* (relevant) attributes, that is: $\varphi \models \psi$ iff $\varphi \models_{\mathcal{A}} \psi$.

As a consequence, every strong implication is rectified. Moreover, every rectified *equivalence* is bound to be strong (cf. the rule irrelevance, see below). Hence we prefer a somewhat weaker property:

Definition 7 *An implication* $\varphi \models_X \psi$ *is purified iff* $\alpha(\varphi) \cap \alpha(\psi) \subseteq X$.

The absence of clashing variables guarantees that all variables that are shared by φ and ψ must be relevant (: in X).

Example 4 *Compare:*
$m < x \models_{xm} x = m + n \wedge n > 0$ *is rectified and purified,*
$m < x \models_{xm} x = k + n \wedge n > 0$ *is rectified and purified,*
$m < x \models_x x = k + n \wedge n > 0$ *is purified.*

Every implication can be transformed into an equivalent purified or rectified implication. On account of the confusing outlook of $x < y \models_x x > y$ it is advisable to purify an implication as soon as possible. Moreover, the corresponding predicate formula can be brought into $\forall\exists$-normal form, that is, a sentence with prefix of the format $\forall\exists$ and unquantified prenex.

Lemma 2 *Let* $\mathbf{y} = \alpha(\varphi) \setminus \mathbf{x}$, $\mathbf{z} = \alpha(\psi) \setminus \mathbf{x}$ *and* $\mathbf{y} \cap \mathbf{z} = \emptyset$ *(: purified), then:*
$\mathcal{M} : \varphi \models_{\mathbf{x}} \psi$ *iff* $\mathcal{M} \models \forall \mathbf{xy} \exists \mathbf{z}(\varphi(\mathbf{xy}) \supset \psi(\mathbf{xz}))$

(Note the change in quantifier for $\mathbf{y}$; cf. universally quantified Horn clauses.)

Example 5 *The weak implications of example 4 correspond with, respectively:*
$\forall x, m\ \exists n(m < x \supset x = m + n \wedge n > 0)$,
$\forall x, m\ \exists k, n(m < x \supset x = k + n \wedge n > 0)$,
$\forall x\ \exists m, k, n(m < x \supset x = k + n \wedge n > 0)$.

3.2 Substitution

In the sequel we will employ solution sets as *substitutions*. The notation $\varphi[\Phi]$ refers to the *result* of the substitution Φ on φ. Any solution set can be used as a substitution, replacing the head variables of Φ in φ by the corresponding tails. Note that the result is well-defined, on account of the well-formedness conditions on solution sets, which guarantee that the individual substitutions are independent.

Substitution has some well-known properties, which will be used here without further justification. In particular:

Lemma 3 *For all formulas φ, ψ and every solution set Φ:*
1. $\alpha_H(\Phi) \cap \alpha(\varphi[\Phi]) = \emptyset$.
2. If $\alpha(\varphi) \cap \alpha_H(\Phi) = \emptyset$, *then* $\varphi[\Phi] = \varphi$.
3. If $\Phi = \Phi_1 \cup \Phi_2$, *then* $\varphi[\Phi] = \varphi[\Phi_1][\Phi_2] = \varphi[\Phi_2][\Phi_1]$.
4. If $\Phi =_{\alpha(\varphi)} \Psi$, *then* $\varphi[\Phi] = \varphi[\Psi]$.
5. $\Phi[\Phi] \equiv \text{TRUE}$.

Any effective procedure that computes $\varphi[\Phi]$ from φ and Φ satisfies these properties. A trivial example of (3.5) is: $x = 8[x = 8]$ $(:= \; 8 = 8) \; \equiv \; \text{TRUE}$.

4 Rules for Weak Implication

The inference rules for strong implications can be generalized for weak implications. A list of them is given in figure 1: the system $\mathcal{WI}$ of derivation rules for weak implications. Figure 2 contains some derived rules and meta-rules.

All rules in figure 1, except for **abstraction**, are in fact strong, which means that these rules hold for strong implications as well as for arbitrary weak implications.

Some restrictions are placed on the meta-rules, since they are not sound unrestrictedly. This can be shown by means of examples; for instance, to explain the restriction on **union**:

Example 6 *Both* $x < 3 \models_x x < y \wedge y = 4$ *and* $x < 3 \models_x x < y \wedge y = 5$.
Still, $x < 3 \not\models_x x < y \wedge y = 4 \wedge y = 5$, *since* $(x < y \wedge y = 4 \wedge y = 5) \equiv \text{FALSE}$.

Similar examples can be given for all restrictions (see [6]).

Contraposition is de facto only valid for strong implications (: on account of the **strong** rule), but this is compensated for by the general validity of **absurdum**.

A rule worth noticing is **all cases**. Whereas **union** is restricted to compatible consequences, **all cases** is valid for all pairs of antecedents. In particular, putting $\varphi := \varphi \wedge \psi$ and $\phi := \varphi \wedge \neg\psi$, it implies the derived rule **either or**.

The **cut** rule is only applicable if the untrimmed implication is non-trivial. Another remarkable rule is **implication**. Its validity is restricted to antecedents with explicitly mentioned variables.

Theorem 4 *The rules and meta-rules of system $\mathcal{WI}$ (: figure 1) are all sound.*

Proof: Straightforward; see [6].
■

Name	Rule	Restriction
true	$\varphi \models_X \text{TRUE}$	
false	$\varphi \wedge \neg\varphi \models_X \text{FALSE}$	
weakening	$\varphi \wedge \psi \models_X \varphi \vee \chi$	
duality	$\neg\varphi \wedge \neg\psi \models_X \neg(\varphi \vee \psi)$	
substitution	$\varphi \wedge \Phi \models_X \varphi[\Phi]$	
generalization	$\varphi[\Phi] \wedge \Phi \models_X \varphi \wedge \Phi$	
abstraction	$\varphi[\Phi] \models_X \varphi \wedge \Phi$	$\alpha_H(\Phi) \cap X = \emptyset$
removal	$\varphi \models_{XY} \psi \Rightarrow \varphi \models_X \psi$	
irrelevance	$\varphi \models_X \psi \Rightarrow \varphi \models_{XY} \psi$	$Y \cap \alpha(\psi) \subseteq X$
transitivity	$\varphi \models_X \psi,\ \psi \models_X \chi \Rightarrow \varphi \models_X \chi$	
union	$\varphi \models_X \psi,\ \varphi \models_X \chi \Rightarrow \varphi \models_X \psi \wedge \chi$	$\alpha(\psi) \cap \alpha(\chi) \subseteq X$
all cases	$\varphi \models_X \psi,\ \phi \models_X \psi \Rightarrow \varphi \vee \phi \models_X \psi$	
contraposition	$\neg\varphi \models_X \neg\psi \Rightarrow \psi \models_X \varphi$	$\alpha(\psi) \subseteq X$

Figure 1: Rules and meta-rules of $\mathcal{WI}$.

Name	Rule	Restriction
tertium	$\text{TRUE} \models_X \varphi \vee \neg\varphi$	
absurdum	$\text{FALSE} \models_X \varphi$	
independence	$\text{TRUE} \models_\emptyset \Phi$	
Leibniz	$s = t \wedge \varphi(s) \models \varphi(t)$	
strong	$\varphi \models_X \psi \Rightarrow \varphi \models \psi$	$\alpha(\psi) \subseteq X$
instance	$\text{TRUE} \models_X \varphi(y) \Rightarrow \text{TRUE} \models_X \varphi(a)$	$y \in X$
instantiation	$\varphi \models_X \psi \Rightarrow \varphi[\Phi] \models_X \psi[\Phi]$	$\alpha(\Phi) \cap \alpha(\psi) \subseteq X$
MP	$\text{TRUE} \models_X \varphi,\ \varphi \models_X \psi \Rightarrow \text{TRUE} \models_X \psi$	
MT	$\text{TRUE} \models \neg\psi,\ \varphi \models_X \psi \Rightarrow \text{TRUE} \models \neg\varphi$	
augmentation	$\varphi \models_X \psi \Rightarrow \varphi \wedge \phi \models_X \psi \wedge \phi$	$\alpha(\phi) \cap \alpha(\psi) \subseteq X$
implication	$\phi \models_X \varphi \supset \psi \Rightarrow \phi \wedge \varphi \models_X \psi$	$\alpha(\varphi) \subseteq X$
deduction	$\phi \wedge \varphi \models_X \psi \Rightarrow \phi \models_X \varphi \supset \psi$	
either or	$\varphi \wedge \psi \models_X \chi,\ \varphi \wedge \neg\psi \models_X \chi \Rightarrow \varphi \models_X \chi$	
reduction	$\varphi \wedge \psi \models_X \neg\psi \Rightarrow \varphi \models_X \neg\psi$	
cut	$\varphi \wedge \psi \models_X \chi, \varphi \models_X \psi \Rightarrow \varphi \models_X \chi$	$\alpha(\varphi) \cap \alpha(\psi) \subseteq X$

Figure 2: Derived rules of the system $\mathcal{WI}$

Theorem 5 *All rules in figure 2 can be derived from $\mathcal{WI}$.*

Proof: We will only prove **instantiation**, by way of example.
To be proven: $\varphi \models_X \psi \;\Rightarrow\; \varphi[\Phi] \models_X \psi[\Phi]$, where $\alpha(\Phi) \cap \alpha(\psi) \subseteq X$.
Define $Y := \alpha(\varphi[\Phi] \wedge \psi[\Phi])$, which implies $Y \cap \alpha_H(\Phi) = \emptyset$.

1	given	$\varphi \models_X \psi$	
2	weakening	$\varphi \wedge \Phi \models_X \varphi$	
3	transitivity: 2,1	$\varphi \wedge \Phi \models_X \psi$	
4	weakening	$\varphi \wedge \Phi \models_X \Phi$	
5	union: 3,4	$\varphi \wedge \Phi \models_X \psi \wedge \Phi$	(: $\alpha(\Phi) \cap \alpha(\psi) \subseteq X$)
6	removal: 5	$\varphi \wedge \Phi \models_{X \cap Y} \psi \wedge \Phi$	
7	abstraction	$\varphi[\Phi] \models_Y \varphi \wedge \Phi$	(: $Y \cap \alpha_H(\Phi) = \emptyset$)
8	removal: 7	$\varphi[\Phi] \models_{X \cap Y} \varphi \wedge \Phi$	
9	transitivity: 8,6	$\varphi[\Phi] \models_{X \cap Y} \psi \wedge \Phi$	
10	substitution: 9	$\varphi[\Phi] \models_{X \cap Y} \psi[\Phi]$	
11	irrelevance: 10	$\varphi[\Phi] \models_X \psi[\Phi]$	(: $\alpha(\psi[\Phi]) \subseteq Y$)

For strong implications the restriction is trivial, so: $\varphi \models \psi \;\Rightarrow\; \varphi[\Phi] \models \psi[\Phi]$.
■

4.1 Propositional Logic

At first glance, $\mathcal{WI}$ can be partitioned into 3 parts: propositional rules (8), "quantification" rules to vary the implication variables (2) and rules dealing with identity in terms of solution sets (3). However, **abstraction** is in essence a quantification rule: the quantified analogon of $\varphi[x = c] \models_Y \varphi$ ($x \notin Y$) is the $\exists$-introduction rule $\varphi(c) \models \exists x\, \varphi(x)$.

The rules in system $\mathcal{WI}$ are still incomplete in the sense that basic properties of the propositional connectives are not listed, namely those that derive from basic set theory, such as $X = X \cup X$ and $X \cup Y = Y \cup X$. In particular, we need thinning and permutation rules (: $\vee$ and $\wedge$ are idempotent, symmetric and associative) and the law of double negation (: $\neg\neg\varphi \equiv \varphi$).

The rules in $\mathcal{WI}$ have been choosen to pinpoint where weak implications differ from strong ones. It should be obvious that all propositional valid inferences are derivable by the present system.

4.2 System $\mathcal{WI}$ is Complete

Let $\vdash_X$ be the smallest relation over unquantified formulas and sets of variables that satisfies all rules and meta-rules of the system $\mathcal{WI}$, listed in figure 1 and figure 2.

Lemma 4 expresses that all rules derivable from $\mathcal{WI}$ are sound:

$$\varphi \vdash_X \psi \text{ implies } \varphi \models_X \psi$$

The converse holds as well. This can be proven by means of a *Henkin construction*.

The definition of $\vdash_X$ is extended to sets of formulas $(\Gamma, \Delta, \ldots)$, in order to get a standard deduction system. Compactness guarantees that this does not affect the consequence relation.

A Henkin construction involves maximal consistent sets of formulas. Consistency is by definition strong; maximality will be relative to a set of variables X. FORM(X) is the set of formulas with free variables in X.

Definition 8 *Γ is X-maximal consistent iff*
1. Γ is consistent: $\Gamma \not\vdash$ FALSE.
2. for any $\varphi \in FORM(X)$ such that $\varphi \notin \Gamma$: $\Gamma \cup \varphi$ inconsistent.

Every maximal consistent set satisfies a *truth lemma.*

Lemma 6 *Let Δ be X-maximal consistent. For all $\varphi, \psi \in FORM(X)$:*
1. $\varphi \in \Delta$ or $\neg\varphi \in \Delta$
2. $\varphi \vee \psi \in \Delta$ iff $\varphi \in \Delta$ or $\psi \in \Delta$
3. $\varphi \wedge \psi \in \Delta$ iff $\varphi \in \Delta$ and $\psi \in \Delta$

Proof: Straightforward: since $\varphi, \psi \in$ FORM(X) all rules reduce to their standard strong format. ■
The existence of maximal X-consistent extensions will be the central issue in the completeness proof of $\mathcal{WI}$.

Theorem 7 *$\varphi \models_X \psi$ implies $\varphi \vdash_X \psi$.*

Proof: Outline; for further details see [6].
Suppose for some formulas φ_0, ψ_0 and some variables X_0 that $\varphi_0 \not\vdash_{X_0} \psi_0$.
We will construct a family of maximal consistent sets such that $\varphi_0 \in \Delta$ & $\psi_0 \notin \Delta$.
It can be assumed, without loss of generality, that $\alpha(\varphi_0) \subseteq X_0$.
The construction contains a number of steps:

1 Put $Y_0 := \alpha(\psi_0) \setminus X_0$,
so Y_0 is the set of existentially quantified variables in ψ_0.

2 Fix an enumeration $\{\theta_i\}_i$ of all equational solution sets $\theta : Y_0 \rightarrow$TERM(X_0),
where TERM(X_0) is the set of all object terms with variables in X_0.

3 Define $\Gamma := \varphi_0 \cup \{\neg\psi_0[\theta_i]\}_i$,
the extension of φ_0 with the set of all instantiations of $\neg\psi_0$.

Fact 1 Γ is consistent.
This can be proven formally in $\mathcal{WI}$, by deriving $\varphi_0 \vdash_{X_0} \psi_0$ from $\Gamma \vdash$ FALSE.

4 Embed Γ in a X_0-maximal consistent set Γ_*, by means of an enumeration $\{\xi_i\}_i$ of FORM(X_0), all formulas ξ with $\alpha(\xi) \subseteq X_0$.
1. $\Gamma_0 := \Gamma$
2. $\Gamma_{i+1} := \Gamma_i \cup \{\xi_i\}$, if this is consistent, $\Gamma_{i+1} := \Gamma_i$, otherwise.
3. $\Gamma_* := \bigcup_i \Gamma_i$

The restriction to FORM(X_0) entails that all implications are strong, so it is easy to prove the truth lemma (see lemma 6), restricted to formulas in FORM(X_0).

5 Define $\Delta^i := \Gamma_* \cup \{\theta_i\}$ for each θ_i in the enumeration.
Since already $\neg\psi_0[\theta_i] \in \Gamma_*$ this implies, by **generalization**, $\Delta^i \vdash \neg\psi_0$.

Fact 2 Every Δ^i is consistent.

6 Construct X_0Y_0-maximal extensions for each set Δ^i, as before, by means of an enumeration $\{\xi_j\}_j$ of FORM(X_0Y_0), all formulas ξ with $\alpha(\xi) \subseteq X_0Y_0$ (cf. **4**).

The truth lemma is extended to FORM(X_0Y_0).
The resulting family of X_0Y_0-maximal sets is a counterexample to $\varphi_0 \models_{X_0} \psi_0$:

7 Define a model $\mathcal{M} =< \mathcal{D}, \mathcal{I} >$ with $\mathcal{D}$ the set of equivalence classes generated by Γ_* and $\mathcal{I}$ induced by Γ_*, as follows:

1. $\mathcal{D} := \{\underline{t} \mid \alpha(t) \subseteq X_0\}$, where $\underline{t} := \{s | s = t \in \Gamma_*\}$
2. $\mathcal{I}(c) := \underline{c}$, for all constants c.
3. $\mathcal{I}(P) := \{< \underline{t_1} \dots \underline{t_n} > \mid P(t_1 \dots t_n) \in \Gamma_*\}$, for all predicates P.
4. $\mathcal{I}(f) : \mathcal{D}^m \rightarrow \mathcal{D} := \mathcal{I}(f)(\underline{t_1} \dots \underline{t_n}) = \underline{f(t_1 \dots t_n)}$, for all functions f.

Obviously, Leibniz' Law is essential to make this a meaningful definition (see below).

8 The relevant assignments $h_i : \mathcal{A} \rightarrow \mathcal{D}$ are defined relative to the larger sets Δ^i_*:
$h_i(y) = \underline{t}$, if $y \in Y_0$, $y = t \in \theta_i$,
$h_i(x) = \underline{x}$, if $x \in X_0$.

This model is well-defined, that is, it satisfies the following properties.
Fact 3 For this model $< \mathcal{D}, \mathcal{I} >$ and these assignments h_i:

1. $\mathcal{M} \models P(t_1 \dots t_n)[h_i]$ iff $P(t_1 \dots t_n) \in \Delta^i_*$, $t_k \in$ TERM(X_0Y_0)
2. $\mathcal{M} \models f(t_1 \dots t_n) = s[h_i]$ iff $f(t_1 \dots t_n) = s \in \Delta^i_*$
3. $\mathcal{M} \models \varphi[h_i]$ iff $\varphi \in \Delta^i_*$
4. $\mathcal{M} \models \xi[h_i]$ iff $\mathcal{M} \models \xi[h_j]$ for all $\xi \in \Gamma_*$, $\xi \in$ FORM(X_0)

9 $< \mathcal{D}, \mathcal{I} >$ is a well-defined countermodel for $\varphi_0 \models_{X_0} \psi_0$:
for all i: $\varphi_0 \in \Delta^i_*$, but $\psi_0 \notin \Delta^i_*$

As a consequence, $\mathcal{M} \models \varphi_0[h_i]$ and for all $h =_{X_0} h_i$, since $h = h_j$ for some j (!) $\mathcal{M} \not\models \psi_0[h]$, so $\mathcal{M} \not\models \varphi_0 \supset_{X_0} \psi_0$ and $\varphi_0 \not\models_{X_0} \psi_0$.

To summarize, we have constructed a countermodel to prove $\varphi_0 \not\models_{X_0} \psi_0$ for an arbitrary pair of formulas φ_0, ψ_0 and arbitrary set of variables X_0, on the assumption that $\varphi_0 \not\vdash_{X_0} \psi_0$. Therefore, if $\varphi_0 \models_{X_0} \psi_0$, then $\varphi_0 \vdash_{X_0} \psi_0$; the system $\mathcal{WI}$ is complete.
■

5 Constraint Solving

The original motivation to introduce the notion of weak equivalence was to validate the **RL/1** symbolic constraint solving system (cf. [2], [3]). In this paper we do not discuss the technical details of this constraint solver, which can be found in [2]. We will illustrate the aims of the solver system by means of an example and give a formal specification based on weak equivalence.

5.1 An Example

Consider a large set of user-defined constraints, such as the arithmetic laws dealing with ages, wages and commissions on sales. These laws can be expressed in terms of universal constraints with several parameters and variables. Not all queries make use of these laws in the same manner: some queries may compute taxes over net prices, others require gross prices, but a curious buyer may want to reconstruct rates from gross and net prices. In all these queries the same equations are involved:

```
TaxConstraints:
gross-price = net-price + taxes,
taxes = rates * net-price,
etcetera.
```

For some applications gross prices are listed in the database, other products, that are mainly sold to business customers, are stored with their net prices: the known variables may vary with the query as well. Note that in a spreadsheet, such as EXCEL, all constraints are 'directed' in that new fields are computed out of given ones. Arithmetic equations as used in the relational database language **SQL** are directed as well: one may define new attributes as a function of given ones.

Informally, we would like to extend a database system with facilities to store relevant arithmetic laws in *constraints*. Queries concerning these constraints and the related database tables should be formulated in a user-friendly manner, for instance by allowing unrestricted selections in the **SQL** `where`-clause, either explicitly or by means of imported modules:

```
SELECT  net-price
  FROM  Orders, Clients, Rates
 WHERE  client-name = 'Me'
        net-price > 1000
        TaxConstraints(*)
```

(The natural join equations are suppressed; `Rates` contains the tax-rates that may vary per product.) The constraints in `TaxConstraints` may include equations that are irrelevant to the present query, such as boundaries on commissions per salesman, as a percentage of his salary. In principle, each tuple in the join `Orders` ⋈ `Clients` ⋈ `Rates` can be used to instantiate the constraints, testing for satisfiability to decide inclusion (: inconsistency leads to rejection). This type of evaluation is rather inefficient, however, since for all accepted tuples the constraint solver must be invoked to compute the `net-price`.

In a more efficient system, all unrestricted queries such as the one given above are preprocessed by a symbolic constraint solver, yielding a restricted query to be evaluated on the database:

```
CREATE VIEW  Query (net-price)
  AS SELECT  gross-price / (rates + 1)
       FROM  Orders, Clients, Rates
      WHERE  client-name = 'Me'
             gross-price > 1000 * rates + 1000
```

The necessity to employ a view, in order to identify `net-price` as the result of `gross-price / (rates + 1)`, is of course an idiosyncrasy of the language **SQL**. However, it illustrates the difference between the *symbolic solution* and the actual answer. Note also the new restricted condition.

The symbolic answer is evaluated after the subsequent query:

```
SELECT  net-price
  FROM  Query
```

In this query only those tuples that will satisfy the constraints are selected to compute the `net-price` directly. The constraint solver has been invoked only once, to establish the symbolic solution and the remaining conditions, that is, the view definition.

More formally, the purpose of symbolic constraint solving is to determine whether or not a set of constraints is *separable* and if so, give an equivalent solution: the system should express the wanted variables W in terms of the known variables K, that have to satisfy some derived conditions. The input of a symbolic constraint solver is the unrestricted query $\Pi_X \sigma_\varphi(R)$. R is a database relation (or a view or join) to resolve known variables. The condition φ expresses all relevant knowledge; φ consists of equations and comparisons involving K, W and possibly some intermediate variables I, variables that are irrelevant to the present query, but that help to formulate the constraints in a user-friendly manner. The optimized output is a restricted database query of the format $\Pi_X(\kappa_\Phi(\sigma_\psi(R)))$, which can be evaluated directly on the database, without further intervention of the constraint solving system. The formal description and verification of the constraint solver can be formulated quite elegantly by means of the concept of weak equivalence.

5.2 Solvable Constraints

In general, a constraint solver is invoked to compute a relation $S(KW)$ from a relation $R(K)$ in accordance with the set of constraints φ. Let φ contain variables $\alpha(\varphi) = K \cup W \cup I$, that is, the disjoint union of known, wanted and intermediate variables. φ must be separated into a symbolic solution Φ and reduced condition ψ, but this will not be possible for all and arbitrary φ. The constraints φ are solvable in K and W if all variables in W can be expressed in terms of K; φ is reducible in K if the implicit condition φ poses on K can be expressed in terms of K alone.

- φ is **solvable** in K and W iff
  there exists a solution set Φ with $\alpha_H(\Phi) = W$ and $\alpha_T(\Phi) = K$, such that $\varphi \models_{KW} \Phi$.

- φ is **reducible** in K iff
  there exists a ψ with $\alpha(\psi) \subseteq K$, such that $\varphi \equiv_K \psi$.

- φ is **separable** in K and W iff
  there exists a solution set Φ with $\alpha_H(\Phi) = W$ and $\alpha_T(\Phi) = K$, and
  there exists a ψ with $\alpha(\psi) \subseteq K$, such that $\varphi \equiv_{KW} \Phi \wedge \psi$.

1	given: separable	$\varphi \equiv_{KW} \Phi \wedge \psi$
2	def: 1	$\varphi \models_{KW} \Phi \wedge \psi$
3	weakening: 2	$\varphi \models_{KW} \Phi$ (: solvable)
4	weakening: 2	$\varphi \models_{KW} \psi$
5	removal: 2	$\varphi \models_{K} \psi$
6	abstraction	$\psi \models_{K} \psi \wedge \Phi$
7	def:1	$\psi \wedge \Phi \models_{KW} \varphi$
8	transitivity	$\psi \models_{K} \varphi$
9	def: 5,8	$\varphi \equiv_{K} \psi$ (: reducible)

Figure 3: Separable implies solvable and reducible.

We say that φ is underdetermined on KW, if it is not solvable on K and W. (Note that in CLP the term *solvable* denotes satisfiability, $\varphi \equiv_{\emptyset}$ TRUE, that is, φ is not reducible to FALSE.

A good optimization strategy first tries to establish whether or not a constraint φ is separable. To be separable φ must be solvable as well as reducible:

Theorem 8 *φ is separable in K and W iff φ is reducible in K and φ is solvable in K and W.*

Proof: This can be proven formally in $\mathcal{WI}$, see figure 3 & 4.
■

Separability is a necessary requirement for efficient query evaluation. If the constraint φ is separable, then the constraint solver is only invoked once and the wanted variables can be computed by means of the symbolic solution Φ for all tuples that satisfy the reduced constraint ψ. In case φ is *not* separable, then only the more elaborate strategy remains of invoking the constraint solver for every tuple of known values, checking solvability for each individual tuple of known variables.

5.3 Symbolic Solutions

We would like to verify that the symbolic solution yields all and only correct answers, at least for separable constraints. Hence we must compare two *types* of constraint solvers, say $T1$ and $T2$. $T1$ corresponds with tuple-wise evaluation, and $T2$ with the more efficient evaluation strategy that employs symbolic solutions.

Definition 9 *Two types of constraint solvers.*

T1	*input:* φ, W; *output:* ψ, Φ.	**T2**	*input:* φ, K, W; *output:* ψ, Φ.
	1. $\varphi \equiv_W \psi \wedge \Phi$		1. $\varphi \equiv_{KW} \psi \wedge \Phi$
	2. $W \subseteq \alpha(\varphi)$		2. $KW \subseteq \alpha(\varphi)$, $K \cap W = \emptyset$
	3. $\psi \equiv$ TRUE *or* $\psi \equiv$ FALSE		3. $\alpha(\psi) \subseteq K$
	4. $\alpha_T(\Phi) = \emptyset$		4. $\alpha_T(\Phi) \subseteq K$
	5. $\alpha_H(\Phi) = W$		5. $\alpha_H(\Phi) = W$

Both solvers are partial in the sense that the output constraint ψ and the solution Φ are only generated if φ is not underdetermined on KW.

1	given: solvable	$\varphi \models_{KW} \Phi$
2	given: reducible	$\varphi \equiv_K \psi$
3	def: 2	$\varphi \models_K \psi$
4	irrelevance: 3	$\varphi \models_{KW} \psi$
5	union: 1,4	$\varphi \models_{KW} \psi \wedge \Phi$
6	def: 2	$\psi \models_K \varphi$
7	weakening: 6	$\psi \wedge \Phi \models_K \varphi$
8	augmentation: 1	$\varphi \models_{KW} \varphi \wedge \Phi$
9	transitivity: 7,8	$\psi \wedge \Phi \models_K \varphi \wedge \Phi$
10	substitution: 9	$\psi \wedge \Phi \models_K \varphi[\Phi]$
11	irrelevance: 10	$\psi \wedge \Phi \models_{KW} \varphi[\Phi]$
12	weakening	$\psi \wedge \Phi \models_{KW} \Phi$
13	union: 11,12	$\psi \wedge \Phi \models_{KW} \Phi \wedge \varphi[\Phi]$
14	generalization	$\Phi \wedge \varphi[\Phi] \models_{KW} \varphi \wedge \Phi$
15	transitivity: 13,14	$\psi \wedge \Phi \models_{KW} \varphi \wedge \Phi$
16	transitivity: 15	$\psi \wedge \Phi \models_{KW} \varphi$
17	def: 5, 16	$\varphi \equiv_{KW} \psi \wedge \Phi$ (: separable)

Figure 4: Solvable and reducible implies separable.

For a type $T1$ solver the solution Φ is a *tuple* which contains values for all wanted variables W. The constraint φ is solvable if $\psi \equiv$ TRUE, and inconsistent if $\psi \equiv$ FALSE. In the latter case a dummy Φ can be constructed to satisfy the specification. The type $T1$ solver is invoked for each tuple r in the relation R, with input W and $\varphi[r]$, the result of the substitution r on φ. Its output is a new tuple s (: over the attributes W) and an error-indicator ψ. In case $\psi \equiv$ TRUE, then $\varphi[r] \equiv_W s$, and the tuple $s \sqcup r$ is added to the answer relation S. If $\psi \equiv$ FALSE, then r is incompatible with φ, that is, $\varphi[r]$ is inconsistent, and no tuple is added to S.

A type $T2$ solver generalizes a $T1$ solver by the introduction of known variables. It generates a *symbolic* solution Φ and a *symbolic* solvability condition ψ. In case $\psi \equiv$ FALSE the query is rejected. Otherwise, solvability can be checked by simple evaluation of ψ for all tuples r in R. Those tuples r that pass the test can be extended to tuples $r \sqcup s$ on KW by calculating $s := \Phi[r]$. Therefore, the requested answer relation S can be computed with the following restricted database expression:

$$S \;:=\; \kappa_\Phi(\sigma_\psi(R)) \;=\; \{t \in S \mid \exists r \in R:\ t = r \sqcup s \;\&\; r \models \psi \;\&\; s = \Phi[r]\}$$

This is formalized in the following lemma.

Lemma 9 *If φ is separable, $\varphi \equiv_{KW} \psi \wedge \Phi$, and $\alpha(R) = K$, then for all $r \in R$ and all $s \in \mathcal{D}^W$:*

$$\varphi[r \sqcup s] \equiv_\emptyset \text{TRUE} \quad \textit{iff} \quad r \models \psi \;\&\; s = \Phi[r]$$

Proof:

1. Suppose $\varphi[r \sqcup s] \equiv_\emptyset$ TRUE.

Then $r \sqcup s \models_{KW} \varphi$, so, since $\varphi \equiv_{KW} \psi \wedge \Phi$, $r \sqcup s \models_{KW} \psi \wedge \Phi$, and, since $\alpha(\psi) \subseteq \alpha(r)$,

$r \models_{KW} \psi$, that is, $\psi[r] \equiv$ TRUE. Moreover, from $r \sqcup s \models_{KW} \Phi$, by instantiation, $(r \sqcup s)[r] \models_{KW} \Phi[r]$, that is, $s \models_{KW} \Phi[r]$. On account of the scopes involved this means that $s = \Phi[r]$.
2. Suppose that $\psi[r] \equiv_K$ TRUE and $s = \Phi[r]$.
Then $r \models \psi$ and $s \models \Phi[r]$, so, by mixed union $r \sqcup s \models \Phi[r] \wedge r \wedge \psi$, and, by generalization $r \sqcup s \models \Phi \wedge \psi$. Since $\varphi \equiv_{KW} \psi \wedge \Phi$, this implies $r \sqcup s \models_{KW} \varphi$, hence $\varphi[r \sqcup s] \equiv_{\emptyset}$ TRUE.
∎

As soon as W is determined by φ in terms of K the $T2$ solver must be preferred to the direct strategy that invokes $T1$ for each individual tuple. If the problem is underdetermined, however, it may turn out that some tuples are capable of being solved, as in the following example.

Example 7 $T2$: *input:* $a * x + b * y = c$, $K = \{a, b, c\}$, $W = \{x\}$; *output:* ? ? ? *For* r *such that* $r(b) = 0$ *the* $T1$ *solver gets input* $a * x + b * y = c[r]$, $\{x\}$, *yielding the answer* TRUE, $\{x = (c/a)[r]\}$. *Indeed, if* $r(a) * x = r(c)$, *then* $x = r(c)/r(a)$.

The $T1$ strategy is more complete in the sense that it can deal with inseparable constraints that happen to be separable after substitution for all tuples in R. Nevertheless, it may be assumed that the $T1$ solver is not invoked for all tuples of R on account of the general underdeterminedness of the constraint. Hence it is not unfair to change the constraint of example 7 for both solvers into the determined query

$$(a * x + b * y = c) \ \wedge \ (a = 0 \vee b = 0).$$

The $T2$ constraint solver that was described in this section has been implemented in the **RL/1** optimization strategy developed at the University of Amsterdam by S. van Denneheuvel (see [2]).

The $T1$ constraint solver corresponds with algorithms like *Gauss elimination* to solve sets of linear equations φ, where $W = \alpha(\varphi)$, and with the binary REDUCE procedure from *Mathematica.*

The system **CLP** ([5]) has a different strategy. A successful derivation (based on unification, as in PROLOG) expands a goal φ relative to the program P to yield a set of answer constraints. This corresponds with evaluating $\varphi[\theta]$ for a tuple θ (: atomic data) and does not require constraint solving. Then a type $T1$ solver is invoked to solve the remaining variables W from $\varphi[\theta]$ for each tuple θ. However, symbolic answers can be computed as well, by invocation of a type $T2$ solver with input φ, $K = \alpha(\varphi)$, $W = \emptyset$ and output ψ, $\Phi = \emptyset$ such that $\varphi \equiv_{\alpha(\varphi)} \psi$.

6 Query Normalization

Query optimization techniques appreciate standardized input, but, unfortunately, there is no general normal form for terms in the relational algebra. There are unconditional rewrite rules for a large fragment of the algebra though, leaving but a small set of cumbersome queries. In this section we will illustrate how weak equivalence can be used to normalize relational terms, yielding a normal form for almost all terms.

Relational terms are constructed from relation names by means of the operators Π (: projection), σ (: selection), $\cup$ (: union), $\bowtie$ (: join), $\setminus$ (: relational difference) and $[\,]$ (: renaming). Every term T is interpreted as a finite relation over its scope $\alpha(T)$, both defined by the usual induction (cf. [8] or [6]).

Definition 10 *Two relational terms R and S are equivalent, $R \equiv S$, if for every database $< \mathcal{D}, \mathcal{A}, \mathcal{I} >$: $\mathcal{I}(R) = \mathcal{I}(S)$.*

Example 8 $\sigma_\varphi \sigma_\psi(R) \equiv \sigma_{\varphi\wedge\psi}(R), \quad \sigma_\varphi(R) \setminus \sigma_\psi(S) \equiv \sigma_{\varphi\wedge\neg\psi}(R) \cup \sigma_\varphi(R \setminus S)$.

Renaming is essentially a syntactic operation, that is, $R[B/A]$ is a notational variant of R, renaming all attributes A in the constituent relations in R to B and actually performing the renaming in the projection- and selection sets. For basic relations $R[B/A]$ is a view over R, the same table with a new heading. Hence we may 'perform' all occurrences of renaming instantaneously on the terms themselves, leaving only 5 operations to be normalized.

Many relational equivalences are straightforward and well-known: cascades of projections and selections, set equivalences, union distribution, miscellaneous selection equivalences and the like (see e.g. [8] or [10]). Here only the projection equivalences will be mentioned.

Lemma 10 $R \bowtie \Pi_X(S) \equiv \Pi_{X\cup\alpha(R)}(R \bowtie S[Y/Z])$,
where $Y := (\alpha(R) \cap \alpha(S)) \setminus X$ and $Z \cap \alpha(T) = \emptyset$, for all basic relations T.

Proof: See [6], pg 52 ff.

■

The proof of this lemma is not very complicated, but it is tricky and a little messy. However, it can be formulated as a weak equivalence by means of a view Q for the projection subterm $\Pi_X(S)$ and a solution set Φ with 'new' variables as tails for the appropriate renaming of clashing variables:

Lemma 11 *If $Q :- \Pi_X(S)$, then $R \bowtie Q \equiv_{X\cup\alpha(R)} R \bowtie S[\Phi]$,*
where $\alpha_H(\Phi) := (\alpha(R) \cap \alpha(S)) \setminus X$ and $\alpha_T(\Phi) \cap \alpha(T) = \emptyset$, for all basic relations T.

Proof: Since $\alpha(Q) = X$, $Q \equiv_X S$. If $\alpha(R) \cap \alpha(S) \subseteq X$, then $R \bowtie Q \equiv_{X\cup\alpha(R)} R \bowtie S$. No need to give a proof; this is the augmentation rule! To avoid the condition, define a renaming solution set $\Phi := \{A_1 = B_1, \ldots\}$, with $\alpha_H(\Phi) = (\alpha(R) \cap \alpha(S)) \setminus X$ and $\alpha_T(\Phi) \cap \alpha(T) = \emptyset$, for all basic relations T. From abstraction we infer $Q \equiv_X S$ iff $Q \equiv_X S[\Phi]$. Moreover, $\alpha(R) \cap \alpha(S[\Phi]) \subseteq X$ by definition of Φ.

■

Example 9 *If $R(ABC)$ and $S(BCD)$, then: $R \bowtie \Pi_{CD}(S) \equiv \Pi_{ABCD}(R \bowtie S[E/B])$, in other words, if $Q \equiv_{CD} S$, then $R \bowtie Q \equiv_{ABCD} R \bowtie S[B = E]$.*

By a similar argument, we can pull projection over selection:

Lemma 12 *If $Q :- \Pi_X(R)$, then $\sigma_\varphi(Q) \equiv_{X\cup\alpha(\varphi)} \sigma_\varphi(R[\Phi])$,*
where $\alpha_H(\Phi) := (\alpha(R) \cap \alpha(\varphi)) \setminus X$ and $\alpha_T(\Phi) \cap \alpha(T) = \emptyset$, for all basic relations T.

The union operator interchanges freely with projection, though the resulting union need not be of compatible terms.

Constraint type	Example format
functional / primary key	$R(xyz), R(xy'z') \models y = y'$
inclusion / foreign key	$R(xy) \models_x S(xz)$
embedded multivalued	$R(xyzw), R(xy'z'w') \models_{xyzy'z'} R(xy'zw'')$
lossless join	$R(xyzw) \equiv_{xyzw} R(xyz'w'), R(xy'zw'), R(x'yzw)$

Figure 5: Examples of integrity constraints

Lemma 13 *If* $Q : - \Pi_X(R)$, *then* $Q \cup S \equiv_{X \cup \alpha(R)} R[\Phi] \cup S$
where $\alpha_H(\Phi) := (\alpha(R) \cap \alpha(S)) \setminus X$ *and* $\alpha_T(\Phi) \cap \alpha(T) = \emptyset$, *for all basic relations* T.

The difference operator interchanges with projection on the positive side:

Lemma 14 *If* $Q : - \Pi_X(R)$, *then* $Q \setminus S \equiv_{X \cup \alpha(R)} R[\Phi] \setminus S$,
where $\alpha_H(\Phi) := (\alpha(R) \cap \alpha(S)) \setminus X$ *and* $\alpha_T(\Phi) \cap \alpha(T) = \emptyset$, *for all basic relations* T.

Projection cannot be pulled over the negative side of a difference operator. This corresponds with the restriction on **contraposition**: from $Q \equiv_X S$ we cannot infer $\neg Q \equiv_X \neg S$, except when $\alpha(S) = X$ (in which case the projection would be trivial). Hence $R \setminus \Pi_X(S)$ cannot be expressed by a single equivalence statement (see [6]).

All positive occurrences of projection can be pulled to the outside, so if φ is a formula to express the projection-free subterm R, then $Q : - \Pi_X(R)$ is expressed by $Q(\mathbf{x}) \equiv_{\mathbf{x}} \varphi$. In this manner any relational term can be translated into a finite set of weak equivalences, one for every occurrence of the projection operator (cf. [8], pg 154 ff.). In particular, if all occurrences of projections in a relational term T are 'positive', then a single equivalence statement suffices to express T.

7 Integrity Constraints

There is yet another possible application of the notion of weak equivalence, namely to describe integrity constraints. Consider a set of functional- and inclusion dependencies, or a set of primary - and foreign key dependencies or even a set of embedded multivalued dependencies. If these constraints must be translated into predicate formulas one needs universal and existential quantifiers:

functional dependency $\forall x, y, y' : R(xy) \wedge R(xy') \supset y = y'$
inclusion dependency $\forall x, y : R(xy) \supset \exists z : S(xz)$

Note that identical variables are used to match attributes in R with corresponding attributes in S. To avoid quantification one can use weak implications, see figure 5.

The Armstrong axioms for functional dependencies and similar rules for other types of dependencies translate into valid implications that are derivable from $\mathcal{WI}$. To give an example of a more complicated rule, consider the mixed rule for functional and inclusion dependencies:

$$R[XY] \subseteq S[XY],\ R[XZ] \subseteq S[XZ],\ S: X \to Y \ \Rightarrow\ R[XYZ] \subseteq S[XYZ]$$

This rule translates into the following weak implication rule:

$$\left.\begin{array}{l} Rxyzu \models_{xy} Sxyzw, \\ Rxyzu \models_{xz} Sxyzw, \\ Sxyzw \wedge Sxy'z'w' \models y = y' \end{array}\right\} \Rightarrow Rxyzu \models_{xyz} Sxyzw$$

Given this notation for the mixed rule we can employ the (meta-) rules of weak implications to prove its correctness. The resulting proof is rigorous and simple and compares favourably with the usual informal argument in terms of R and S tuples.

Generally speaking, the formulation of integrity constraints by means of weak implication statements is hardly preferable to the standard notation. Still, it is very convenient for formal verification arguments, and can be used as a tool to integrate integrity with *query optimization* and *views*.

For instance, a *foreign key dependency* $\text{FK}(R, X, Q)$ may link the relation R to a previously defined view Q with *primary key* $\text{PK}(S) = X$. The expanded definition of Q contains a set of basic relations. One would like to infer the induced constraints on these basic relations from the constraints on Q.

Any view can be expressed by a finite set of weak equivalences, a single one if we ignore projection on the negative side of a difference (see § 6). Once views and integrity constraints are formulated in the same framework, it is possible to integrate the implication problems. We will only give a simple example to explain the basic idea: weak implications can be applied to constraints on views in order to induce constraints on basic relations:

Lemma 15 *Let* $\text{FK}(R, X, Q)$ *and* $Q :- S \bowtie T \cup S \bowtie U$.
Suppose X *is such that both* $X \subseteq \alpha(S)$ *and* $\alpha(T \bowtie U) \cap X = \emptyset$. *Then* $\text{FK}(R, X, S)$.

Proof: In terms of weak equivalences:

given : *foreign key*	$R(XY) \models_X Q(XZ)$
given : *view def.*	$Q(XZ) \equiv_{XZ} S(XYW) \wedge (T(YWZ) \vee U(YZ))$
weakening	$Q(XZ) \models_{XZ} S(XYW)$
removal	$Q(XZ) \models_X S(XYW)$
transitivity	$R(XY) \models_X S(XYW)$

This proves that $R[X] \subseteq S[X]$. By a similar argument one can show that $\text{PK}(S) = X$.
■

The integration works both ways: it can also be employed for query optimization purposes. In the following example the number of relations under a projection is reduced on account of the integrity constraint.

Lemma 16 *If* $X = \alpha(R) \cap \alpha(S)$ *and* $\text{FK}(R, X, S)$, *then* $R \bowtie S \equiv_X R$.

Proof: Let $Y = \alpha(R) \setminus X$ and $Z = \alpha(S) \setminus X$, then we need to prove:

$$R(XY) \models_X S(XZ) \Rightarrow R(XY), S(XZ) \equiv_X R(XY).$$

Straightforward, cf. [2] (§ Join optimization).
■

The examples in this section are not very profound, but they are only simple illustrations of a fundamental possibility: the application of the notion of weak equivalence on integrity constraints, views and query optimization. To do so in a systematic way remains as a future task.

8 Conclusion

We have given a sound and complete set of deduction rules for weak implications. This notion of implication generalizes unquantified predicate logic with a single level of ∀∃ quantification. This reduced class of predicate formulas is adequate for a large variety of subjects, ranging from formal specifications for constraint solving systems to implication problems in database theory. Since weak equivalence combines notational transparency with formal elegance, it offers a convenient semantic tool for knowledge representation.

References

[1] W.F. Clocksin & C.S. Mellish. *Programming in Prolog.* Springer-Verlag, 1981.

[2] S. van Denneheuvel. *Constraint solving on database systems. Design and implementation of the rule language* **RL/1**. Thesis. University of Amsterdam, 1991.

[3] S. v Denneheuvel & K.L. Kwast. *Weak equivalence for constraint solving.* In: Proceedings of IJCAI'91: Int. Joint Conference on A.I. Sydney. Morgan Kaufmann, 1991.

[4] S. v Denneheuvel, G.R. Renardel de Lavalette, E. Spaan & K.L. Kwast. *Query optimization using rewrite rules.* In: R. Book (ed.), Proceedings of RTA'91: Int. Conference on Rewriting Techniques and Applications. Como, LNCS488, 1991.

[5] J. Jaffar & S. Michaylov. *Methodology and Implementation of a CLP System.* In: J-L. Lassez (ed.), Proceedings of the 4th Int. Conference on Logic Programming, MIT Press, 1987.

[6] K.L. Kwast. *Unknown values in the relational database system.* Thesis. University of Amsterdam, 1992.

[7] J.H. Siekmann. *Unification Theory.* In: Journal of Symbolic Computation, Vol. 7, 1989.

[8] J.D. Ullman. *Principles of Data and Knowledge-Base Systems, Vol. I & II.* Computer Science Press, 1989.

[9] S. Wolfram. *Mathematica, a System for Doing Math by Computer.* Addison-Wesley, 1988.

[10] H.Z. Yang & P.A. Larson. *Query Transformation for PSJ-queries.* In: Proceedings of the 13th VLDB: Int. Conference on Very Large Databases, 1987.

Deriving Inference Rules for Terminological Logics*

Véronique Royer[1] and J. Joachim Quantz[2]

[1] Onera–Cert, 2 av. Edouard Belin, BP 4025, F-31055 Toulouse
[2] Technische Universität Berlin, KIT-BACK, FR 5–12, Franklinstr. 28/29, W-1000 Berlin 10

Abstract. Terminological Logics can be investigated under different perspectives. The aim of this paper is to provide the basis for a tighter combination of theoretical investigations with issues arising in the actual implementation of terminological representation systems. We propose to use inference rules, derived via the sequent calculus, as a new method for specifying terminological inference algorithms. This approach combines the advantages of the tableaux methods and the normalize-combine algorithms that have been predominant in terminological proof theory so far. We first show how a completeness proof for the inference rules of a relatively restricted terminological logic can be given. We then show how these inference rules can be used to construct normalize-compare algorithms and prove their completeness. Furthermore, these rules can be used in two ways for the characterization of terminological representation systems: first, the incompleteness of of systems can be documented by listing those rules that have not been implemented; second, the reasoning strategy can be described by spesifying which rules are applied forward and which backward.

1 Introduction

In the last ten years terminological logics have evolved as one of the central paradigms in knowledge representation. Research in this field focuses on theoretical investigations as well as on the development of system prototypes and applications. Terminological logics can therefore be viewed from very different perspectives: theoretical investigations are mostly interested in determining the complexity of a terminological logic; prototypical implementations aim to provide efficient representation systems that can be evaluated in applications; finally, application studies address the expressiveness of terminological logics and evaluate the usability of representation systems in practical application. Consequently, methods and interests of research in terminological logics differ considerably. The aim of this paper is to provide the basis for a tighter combination of theoretical investigations with issues arising in the actual implementation of terminological representation systems.

Terminological logics typically distinguish between *definitions*, *descriptions*, and *rules*. A definition has the form $t_n \doteq t$ and expresses that the name t_n is used as an

* This work was supported by the Commission of the European Communities and is part of the ESPRIT Project 5210 AIMS which involves the following participants: Datamont (I), ERIA (E), Non Standard Logics (F), Technische Universität Berlin (D), Deutsches Herzzentrum Berlin (D), Onera-Cert (F), Quinary (I), Universidad del Pais Vasco (E).

abbreviation for the term t. There are two types of terms in terminological logics, namely *concepts* (unary predicates) and *roles* (binary predicates). In a description an object is described as being an instance of a concept (o : c), or as being related to another object by a role ($\langle o_1, o_2 \rangle : r$). Rules have the form $c_1 \sqsubseteq c_2$ and stipulate that each instance of the concept c_1 is also an instance of the concept c_2. Note that from a theoretical point of view definitions are just a special case of rules and that $t_n \doteq t$ can be rewritten as $t \sqsubseteq t_n \wedge t_n \sqsubseteq t$. We will call such formulas terminological formulas, whereas TL-formulas in general encompass also descriptions.

Terminological logics can be distinguished with respect to the concept and role-forming operators they provide. In this paper we will illustrate our approach with the particular language $\mathcal{TL}_{\mathcal{SAN}}$ defined below. We choose this language because it is simple enough to be translated into First Order Logic without equality, but complex enough to involve implicit disjunction. We emphasize, however, that we are not so much interested in the results obtained for this particular logic as in the general method used to obtain these results. We are currently applying the method developed in this paper to the terminological logic of BACK V5 and intend to present the results in [18].

The following specifies the syntax $\mathcal{TL}_{\mathcal{SAN}}$. We use t_i to denote terms in general, c_i and r_i to denote concepts and roles respectively. The index n is used for names, the index p for primitive components:[3]

$$c \longrightarrow \top, \bot, c_n, c_p, c_1 \sqcap c_2, \exists r.c, \forall r.c, \neg\exists r.c$$

$$r \longrightarrow r_n, r_p, r_1 \sqcap r_2$$

$$\gamma \longrightarrow t_n \doteq t, t_2 \sqsubseteq t_1, o : c, \langle o_1, o_2 \rangle : r$$

For this logic a traditional model-theoretic semantics can be given where a model $\mathcal{M}$ is a pair of a domain D and an interpretation function $[\![\cdot]\!]^{\mathcal{I}}$. The interpretation function assigns to each object-name o injectively an element $[\![o]\!]^{\mathcal{I}}$ of D, to each concept c a subset $[\![c]\!]^{\mathcal{I}}$ of D, and to each role r a binary relation $[\![r]\!]^{\mathcal{I}}$ on D, respecting the following equations (we use $r(d)$ to denote $\{e : \langle d, e \rangle \in r\}$):

$$[\![\top]\!]^{\mathcal{I}} = D \tag{1}$$

$$[\![\bot]\!]^{\mathcal{I}} = \emptyset \tag{2}$$

$$[\![t_1 \sqcap t_2]\!]^{\mathcal{I}} = [\![t_1]\!]^{\mathcal{I}} \cap [\![t_2]\!]^{\mathcal{I}} \tag{3}$$

$$[\![\exists r.c]\!]^{\mathcal{I}} = \{d \in D : [\![r]\!]^{\mathcal{I}}(d) \cap [\![c]\!]^{\mathcal{I}} \neq \emptyset\} \tag{4}$$

$$[\![\forall r.c]\!]^{\mathcal{I}} = \{d \in D : [\![r]\!]^{\mathcal{I}}(d) \subseteq [\![c]\!]^{\mathcal{I}}\} \tag{5}$$

$$[\![\neg\exists r.c]\!]^{\mathcal{I}} = \{d \in D : [\![r]\!]^{\mathcal{I}}(d) \cap [\![c]\!]^{\mathcal{I}} = \emptyset\} \tag{6}$$

Satisfaction of formulas is then defined as follows:

$$\mathcal{M} \models t_n \doteq t \text{ iff } [\![t_n]\!]^{\mathcal{I}} = [\![t]\!]^{\mathcal{I}} \tag{7}$$

$$\mathcal{M} \models t_2 \sqsubseteq t_1 \text{ iff } [\![t_2]\!]^{\mathcal{I}} \subseteq [\![t_1]\!]^{\mathcal{I}} \tag{8}$$

$$\mathcal{M} \models o : c \text{ iff } [\![o]\!]^{\mathcal{I}} \in [\![c]\!]^{\mathcal{I}} \tag{9}$$

$$\mathcal{M} \models \langle o_1, o_2 \rangle : r \text{ iff } \langle [\![o_1]\!]^{\mathcal{I}}, [\![o_2]\!]^{\mathcal{I}} \rangle \in [\![r]\!]^{\mathcal{I}} \tag{10}$$

[3] See [13, p. 54ff] for the role of primitive components.

A structure $\mathcal{M}$ is a model of a formula γ iff $\mathcal{M} \models \gamma$; it is a model of a set of formulas Γ iff it is a model of every formula in Γ. A formula γ is entailed by a set of formulas Γ (written $\Gamma \models_{\mathrm{TL}} \gamma$) iff every structure which is a model of Γ is also a model of γ.

Theoretical research then focuses on the decidability of entailment and on the complexity of decision procedures for entailment. Most papers discuss algorithms that compute whether a terminological formula $c_2 \sqsubseteq c_1$ follows from a set of definitions. Since this formula is read as c_1 *subsumes* c_2 this specific form of entailment is sometimes referred to as the problem of concept subsumption. Though concept subsumption is probably the most important aspect of terminological logics, a system oriented point of view has to deal also with role subsumption and the entailment of assertional formulas.[4]

But even with respect to concept subsumption theoretical and implementation oriented research differ considerably. The most important terminological representation systems currently under construction, such as BACK, CLASSIC, or LOOM, contain inference algorithms that stand in the tradition of the normalize-compare paradigm.[5] In this paradigm concepts are first transformed into a canonical normal form. To determine subsumption between concepts these normal forms are then compared structurally. This format of presenting algorithms has a considerable drawback from a theoretical point of view. It seems that no systematic way can be specified to devise a complete normalize-compare algorithm for an arbitrary terminological logic.

For this reason, subsumption algorithms using a *tableaux method* have become more and more popular in the last years:[6] their general idea consists in reducing subsumption to unsatisfiability; the algorithm then tries to build a model and if it detects a clash it has proved subsumption. Clearly, the advantage of tableaux algorithms results from their theoretical properties: they can be specified in a systematic manner and it is comparatively easy to proof their completeness and their complexity. On the other hand, it is not so obvious how normal forms can be computed with these algorithms. We will therefore propose a new format for presenting TL-algorithms, which combines advantages of the two traditional approaches: the provability of completeness and the derivability of normal forms.

2 Inference Relations

In order to understand our perspective of terminological representation systems consider the application scenario of the AIMS project (Advanced Information Management System) which underlies the development of the BACK system. In a first step the terminology of a domain is modeled by entering definitions of concepts and roles relevant in the domain (this corresponds to the definition of a database schema). In a second step the concepts and roles defined in the terminology are used to enter descriptions of domain objects and to retrieve objects matching given descriptions. Given such a framework it is advantageous to precompute information that is implicitly given in order to speed up the query-answering in the retrieval phase. Thus in the current BACK implementation (cf. [17]) the subsumption relations between all concepts and roles of the terminology

---

[4] See [15] for a discussion of these aspects.

[5] Confer, for example, [13].

[6] Confer, for example, [5] and [6].

are computed and stored, as well as the most special description for each object. This way, most queries can be answered by simple table-look-ups and only in some cases reasoning is necessary to answer a query.[7] Note that from a theoretical point of view the caching of inferences computed by the algorithm is irrelevant, since it is only important whether the information is computable at all, or how complex it is to compute it (whether once or more often does not matter).

Classically the information implicitly given by a set of formulas is defined model-theoretically, that is by the closure of these formulas under the entailment relation ($\models_{TL}$). This is not satisfactory enough from the point of view of computation: this closure is better characterized *proof-theoretically*, by an *inference relation* $\vdash_{TL}$. This concept is seen as more and more fundamental in modern treatments of logic [2].

Such proof-theoretic considerations meet also the "least fixed point semantics" approach which is especially popular in the field of deductive databases. The semantics of deductive databases is defined by the closure of the extensional part (the factual information) under a so-called "immediate consequence operator" which formalizes the execution in forward chaining of the database rules [1, 14].

The task of specifying an inference relation $\vdash_{TL}$ for a terminological logic would thus consist in specifying a set of inferences rules, whose closure computes the set of all TL-formulas terminologically entailed by some given set of TL-formulas. Not surprisingly these rules resemble the rules used for normalization and comparison in the normalize-compare-style algorithms.

The attentive reader will have noted, however, that we face the same problem which is characteristic for the normalize-compare paradigm: how do we know that our set of rules is complete, i.e. that

$$\Gamma \models_{\mathrm{TL}} \gamma \text{ iff } \Gamma \vdash_{TL} \gamma \tag{11}$$

This is where the Sequent Calculus (SC) comes in handy: the Sequent Calculus is known to be sound and complete for First-Order Logic (FOL). We can specify a translation from TL-formulas into FOL-formulas[8] such that

$$\Gamma \models_{\mathrm{TL}} \gamma \text{ iff } \overline{\Gamma} \models_{\mathrm{FOL}} \overline{\gamma} \tag{12}$$

But because of the soundness and completeness of the Sequent Calculus we immediately get

$$\Gamma \models_{\mathrm{TL}} \gamma \text{ iff } \overline{\Gamma} \vdash_{SC} \overline{\gamma} \tag{13}$$

We will then look at the structure of the SC-proofs and thereby extract all inference rules that are necessary to operate directly on the TL-formulas. That is we will prove the soundness and completeness of the inference rules for terminological logics by proving the equivalence with a fragment of Sequent Calculus corresponding to the embedding of TL (more exactly, we prove completeness by showing that every proof in Sequent Calculus of $\overline{\gamma}$ can be rewritten into a derivation of γ using the inference rules).

[7] Kindermann compares implementation strategies for retrieval support in [11].

[8] For details see next section.

Remark. In [3] the Linear Sequent Calculus[9] was used to formalize a "tractable" notion of subsumption. Namely, the author defined subsumption to be linear sequent entailment: A$\sqsubseteq$B is *by definition* the sequent $\forall(x)A(x) \Rightarrow \exists x B(y)$ in Linear Sequent Calculus (a translation which is less than natural).

Here we do not enter into tractability consideration for Proof Theory, as people do with Linear Logic. We just exploit a natural translation of Terminological Logics into classical Sequent Calculus in order to specify *complete* inference systems for *classical* terminological entailment. Relying on other Sequent Calculi is quite possible but would characterize non-classical terminological entailments.

3 Translation of Terminological Formulas into First-Order Sequents

3.1 Classical Sequent Calculus

We assume that the reader is familiar with Gentzen's Sequent Calculi (e.g., [7, 9, 10]). The main advantages for using Sequent Calculi lies in their systematic and principled approach to Proof Theory.

Indeed, the proof process in Sequent Calculi is seen as a morphism of the set of formulas into the set of proofs. This gives rise in the classical first-order Sequent Calculus to the "subformula property": the notion of proof can be restricted to proofs which do not import formulas which are not subformulas of the proper axioms or the conclusion to be proved. Each logical connector is also axiomatized specifically by two rules (the so-called left- and right-rules) which can be used either as introduction or elimination rules (the rules are reversible). This gives rise to a top-down approach to proof, in a known by advance and closed universe of formulas, with the logical rules playing the role of elimination rules for the logical connectors, until axiom-sequents are finally reached.

In classical first-order Sequent Calculus the subformula property relies on the Cut Elimination theorem, we recall just below.

Proposition 1 Cut Elimination Theorem [9]. *Let S be a set of sequents (proper axioms) and s an individual sequent.* $S \vdash_{SC} s$ *iff there is a proof in* SC *of s whose leaves are either logical axioms or sequents obtained by substitution from sequents belonging to S, and where the Cut Rule is only applied with one premise being a proper axiom.*

We list the rules of the Sequent Calculus in Fig. 1, where "a" stands for a parameter (called "eigen-parameter" or "eigen-variable") which does not appear in the lower sequent; "t" stands for a term appearing in the lower sequent (thus we consider pure proofs [7]).

3.2 The Translation Function

We take the classical principles for the translation of TL-formulas into FOL-formulas [19]. Given the terminological language $\mathcal{TL}_{\mathcal{SAN}}$, the set of concept-names and primitive

[9] A sequent calculus for Linear Logic [8].

Operator	left-rule	right-rule
Weakening	$\frac{\Gamma\Rightarrow\Delta}{\Gamma,\alpha\Rightarrow\Delta}$	$\frac{\Gamma\Rightarrow\Delta}{\Gamma\Rightarrow\Delta,\alpha}$
Contraction	$\frac{\Gamma,\alpha,\alpha\Rightarrow\Delta}{\Gamma,\alpha\Rightarrow\Delta}$	$\frac{\Gamma\Rightarrow\alpha,\alpha,\Delta}{\Gamma\Rightarrow\alpha,\Delta}$
Axiom/Cut-Rule	$\Gamma,\alpha\Rightarrow\Delta,\alpha$	$\frac{\Gamma_1,\alpha\Rightarrow\Delta_1\,;\,\Gamma_2\Rightarrow\Delta_2,\alpha}{\Gamma_1,\Gamma_2\Rightarrow\Delta_1,\Delta_2}$
$\vee$	$\frac{\Gamma,\alpha\Rightarrow\Delta\,;\Gamma,\beta\Rightarrow\Delta}{\Gamma,\alpha,\beta\Rightarrow\Delta}$	$\frac{\Gamma\Rightarrow\Delta,\alpha\vee\beta}{\Gamma\Rightarrow\Delta,\alpha,\beta}$
$\wedge$	$\frac{\Gamma,\alpha,\beta\Rightarrow\Delta}{\Gamma,\alpha\wedge\beta\Rightarrow\Delta}$	$\frac{\Gamma\Rightarrow\Delta,\alpha\,,\Gamma\Rightarrow\Delta,\beta}{\Gamma\Rightarrow\Delta,\alpha\wedge\beta}$
$\rightarrow$	$\frac{\Gamma,\beta\Rightarrow\Delta\,;\Gamma\Rightarrow\alpha,\Delta}{\Gamma,\alpha\rightarrow\beta\Rightarrow\Delta}$	$\frac{\Gamma,\alpha\Rightarrow\Delta,\beta}{\Gamma\Rightarrow\Delta,\alpha\rightarrow\beta}$
$\neg$	$\frac{\Gamma,\alpha\Rightarrow\Delta}{\Gamma\Rightarrow\Delta,\neg\alpha}$	$\frac{\Gamma\Rightarrow\Delta,\alpha}{\Gamma,\neg\alpha\Rightarrow\Delta}$
$\forall$	$\frac{\Gamma,\forall xA(x),A(t)\Rightarrow\Delta}{\Gamma,\forall xA(x)\Rightarrow\Delta}$	$\frac{\Gamma\Rightarrow\Delta,A(a)}{\Gamma\Rightarrow\Delta,\forall xA(x)}$
$\exists$	$\frac{\Gamma,A(a)\Rightarrow\Delta}{\Gamma,\exists xA(x)\Rightarrow\Delta}$	$\frac{\Gamma\Rightarrow\Delta,\exists xA(x)}{\Gamma\Rightarrow\Delta,A(t),\exists xA(x)}$

Fig. 1. The rules of the Sequent Calculus.

concept components (resp. role-names and primitive role components) is bijectively mapped into a set of unary predicate symbols (resp. binary predicate symbols); object-names are bijectively mapped into constants. Let j be the corresponding bijection. We define the translation function $\overline{\gamma}$ as follows: a concept-name (or a primitive concept component) c is translated into $\lambda x C(x)$, where C denotes the unary predicate symbol representing c ($C = j(\mathsf{c})$); similarily, a role-name or primitive role component r is translated into $\lambda x \lambda y R(x, y)$, where R denotes the binary predicate symbol representing r ($R = j(\mathsf{r})$). We then define translation of arbitrary terms and terminological formulas:

$$\overline{\top} \stackrel{\text{def}}{=} \lambda x(True) \text{ (True is as special propositional letter)} \tag{14}$$

$$\overline{\bot} \stackrel{\text{def}}{=} \lambda x(False) \text{ (False is a special propositional letter)} \tag{15}$$

$$\overline{\mathsf{c}_1 \sqcap \mathsf{c}_2} \stackrel{\text{def}}{=} \lambda x(\overline{\mathsf{c}_1}(x) \wedge \overline{\mathsf{c}_2}(x)) \tag{16}$$

$$\overline{\forall \mathsf{r}.\mathsf{c}} \stackrel{\text{def}}{=} \lambda x(\forall y \overline{\mathsf{r}}(x, y) \rightarrow \overline{\mathsf{c}}(y)) \tag{17}$$

$$\overline{\exists \mathsf{r}.\mathsf{c}} \stackrel{\text{def}}{=} \lambda x(\exists y \overline{\mathsf{r}}(x, y) \wedge \overline{\mathsf{c}}(y)) \tag{18}$$

$$\overline{\neg\exists \mathsf{r}.\mathsf{c}} \stackrel{\text{def}}{=} \lambda x(\neg\exists y \overline{\mathsf{r}}(x, y) \wedge \overline{\mathsf{c}}(y)) \tag{19}$$

$$\overline{\mathsf{r}_1 \sqcap \mathsf{r}_2} \stackrel{\text{def}}{=} \lambda x \lambda y \overline{\mathsf{r}_1}(x, y) \wedge \overline{\mathsf{r}_2}(x, y) \tag{20}$$

$$\overline{\mathsf{c}_1 \sqsubseteq \mathsf{c}_2} \stackrel{\text{def}}{=} \forall x(\overline{\mathsf{c}_1}(x) \rightarrow \overline{\mathsf{c}_2}(x)) \tag{21}$$

$$\overline{\mathsf{r}_1 \sqsubseteq \mathsf{r}_2} \stackrel{\text{def}}{=} \forall x \forall y(\overline{\mathsf{r}_1}(x, y) \rightarrow \overline{\mathsf{r}_2}(x, y)) \tag{22}$$

$$\overline{\mathsf{o} : \mathsf{c}} \stackrel{\text{def}}{=} \overline{\mathsf{c}}(\overline{\mathsf{o}}) \tag{23}$$

$$\overline{\langle \mathsf{o}_1, \mathsf{o}_2 \rangle : \mathsf{r}} \stackrel{\text{def}}{=} \overline{\mathsf{r}}(\overline{\mathsf{o}_2})(\overline{\mathsf{o}_1}) \tag{24}$$

Note that, by beta-reduction, the lambda-abstractions are eliminated from the translation of terminological formulas. Thus terminological formulas are translated into pure FOL-formulas, which we will call *terminological clauses* in the following.[10]

Remark. Two remarks regarding notational conventions: we will use C_i and R_i to stand for the FOL-translations of c_i and r_i respectively; furthermore, we will occasionaly abbreviate the FOL-translation of concepts with quantificational structure as (**all** R C), (**some** R C), and (**no** R C). Thus, (**all** R C)(x) stands for $\forall y R(x, y) \rightarrow C(y)$.

Proposition 2 (easy proof omitted). *Let Γ be a set of* TL*-formulas. Let γ be a* TL*-formula. Let $\overline{\Gamma} = \{\,\overline{\gamma} : \gamma \in \Gamma\}$. Then, $\Gamma \models_{TL} \gamma$ iff $\overline{\Gamma} \models_{FOL} \overline{\gamma}$.*

The Sequent Calculus provides a sound and complete inference system for FOL. Thus we get the following proposition (easy proof omitted).

Proposition 3. *Let Γ be a set of* TL*-formulas. Let γ be a* TL*-formula. Let Δ and δ be respectively a set of sequents and a sequent defined as follows: $\Delta \stackrel{\text{def}}{=} \{\emptyset \Rightarrow \overline{\gamma} : \gamma \in \Gamma\} \cup \{False \Rightarrow \emptyset\, , \emptyset \Rightarrow True\}$, $\delta \stackrel{\text{def}}{=} \emptyset \Rightarrow \overline{\gamma}$. Then, $\Gamma \models_{TL} \gamma$ iff $\Delta \vdash_{SC} \delta$.*

3.3 Terminological Sequents

In this paragraph we give a syntactic characterization of the translation of terminological formulas into sequents (called *terminological sequents* in the following).

Let $\mathsf{c}_1 \sqsubseteq \mathsf{c}_2$ be a terminological formula, the above translation gives the sequent $\delta = \emptyset \Rightarrow \forall x(C_1(x) \rightarrow C_2(x))$. By using the $\forall$-right rule, δ is equivalent to a sequent of the form: $\emptyset \Rightarrow C_1(a) \rightarrow C_2(a)$, where "a" denotes a constant symbol which appears neither in C_1 nor in C_2. By using the $\rightarrow$-rule we get the simpler equivalent sequent: $C_1(a) \Rightarrow C_2(a)$. In the literature the constant symbol "a" is called the *eigen-parameter* of the universal formula [10]. Each application of $\forall$-right gives rise to a new different eigen-parameter.

Definition 4. A concept will be called **prime** iff it is a concept-name or a concept of the form (q r.c) for some quantor q.

As the language $\mathcal{TL}_{SAN}$ contains neither explicit negation nor disjunction, the propositional structure of the terminological sequents is purely implicative and conjunctive. By appropriate eliminations of the $\wedge$ connectors in the right-hand sides of sequents (by means of the $\wedge$-right rules) we can reduce the terminological sequents to sequents of the form: $C_1(a), C_2(a), ..., C_n(a) \Rightarrow C_{n+1}(a)$, where the C_i represent the first-order translations of prime concepts (instantiated by the same eigen-parameter "a").

[10] Note also that logical implication is used here for the translation of subsumption as well as for the translation of the all-quantor.

Let us notice that the terminological formulas $c \sqsubseteq \bot$ or $\top \sqsubseteq c$ can be translated directly into the sequents of the form $C(a) \Rightarrow \emptyset$ and $\emptyset \Rightarrow C(a)$ (by the axioms $\emptyset \Rightarrow True$ and $False \Rightarrow \emptyset$). We adopt this translation in the following.

When quantor-formulas appear in the sequent, the application of adequate quantification rules from SC results in the elimination of the quantors. In the following we refer to this process as the *unfolding* of quantor-formulas. The reverse process will be called *folding*.[11] For example, unfolding the all-formula (all R C) in the sequent $\Delta \Rightarrow (\textbf{all}\ R\ C)(a)$ gives the sequent $\Delta \Rightarrow R(a,b) \rightarrow C(b)$. Conversely, the sequent $R(a,b), C(a) \Rightarrow \Delta$ can be folded into $(\textbf{some}\ R\ C)(a) \Rightarrow \Delta$.

4 Terminological Sequent Calculi

4.1 Propositional Rules

The propositional terminological formulas are isomorphic to the propositional intuitionistic[12] conjunctive implicative fragment of SC. Then provability in the propositional fragment of $\mathcal{TL}_{\mathcal{SAN}}$ is provability by means of the $\wedge$-rules, the $\rightarrow$-rules and the Cut-Rule of SC (together with the structural rules), restricted to intuitionistic conjunctive implicative sequents.

We can now immediately derive terminological rules: from the axiom group of SC we get the *logical axiom* 25, from the $\wedge$-right rule of SC we obtain the *conjunction rules* 26, and from the Cut-Rule of SC we obtain the *Cumulative Transitivity* 27 (we use $\rightharpoonup$ for terminological inference rules):

$$\rightharpoonup c_1 \sqcap c_2 \sqsubseteq c_1 \tag{25}$$

$$c_1 \sqsubseteq c_2, c_1 \sqsubseteq c_3 \rightleftharpoons c_1 \sqsubseteq c_2 \sqcap c_3 \tag{26}$$

$$c_1 \sqsubseteq c_2, c_2 \sqcap c_3 \sqsubseteq c_4 \rightharpoonup c_1 \sqcap c_3 \sqsubseteq c_4 \tag{27}$$

Note that we have no negation rules and no significant structural rule. Left-contraction is assumed implicitly in this paper (we consider sequents as set of formulas instead of lists). Left-weakening gives 28 which can be derived by applying the Cut-Rule to an axiom and is thus redundant. The identity axioms 29 follow from the axioms os SC and implicit contraction. Simple transitivity (30) is just a special case of Cumulative Transitivity and therefore also redundant:

$$c_1 \sqsubseteq c_2 \rightharpoonup c_1 \sqcap c_3 \sqsubseteq c_2 \tag{28}$$

$$\rightharpoonup c \sqsubseteq c \tag{29}$$

$$c_1 \sqsubseteq c_2, c_2 \sqsubseteq c_3 \rightharpoonup c_1 \sqsubseteq c_3 \tag{30}$$

Note that all results presented for the propositional fragment carry over from concepts to roles. Finally we have the inference rules for the concepts $\top$ and $\bot$:

$$\rightharpoonup \bot \sqsubseteq c \tag{31}$$

$$\rightharpoonup c \sqsubseteq \top \tag{32}$$

$$c_1 \sqsubseteq c_2 \rightleftharpoons c_1 \sqcap \top \sqsubseteq c_2 \tag{33}$$

[11] The terminology comes from the jargon of rewrite systems.

[12] Intuitionistic sequents are sequents whose right-hand sides are restricted to one single formula.

4.2 Lifting Propositional Proofs into First-Order Proofs

In order to derive the proofs for quantor formulas we will consider "a lifting process", namely the process of "lifting" propositional proofs into first-order ones, by application of the quantification rules in the *introduction sense*. Before we show some technical details, we have to justify the completeness of the lifting process, i.e., that all the rules for the first-order terminological formulas (quantor formulas) can be obtained by "lifting" propositional rules.

Proposition 5. *The lifting process is complete for the rules relative to quantor formulas.*

Proof (Sketch). The proof relies on the normalisation property for the proofs in classical Sequent Calculus [7] which states that a sequent $\Gamma \Rightarrow \Delta$ is provable in first-order Sequent Calculus iff there is an "intermediate sequent" $\Gamma' \Rightarrow \Delta'$ such that:

1. $\Gamma' \Rightarrow \Delta'$ is provable *propositionally* (i.e., in the propositional fragment of SC).
2. $\Gamma \Rightarrow \Delta$ reduces to the sequent $\Gamma' \Rightarrow \Delta'$ by quantification rules (or structural rules different from weakening) only.

Remark. The property is based on a Permutability Lemma stating that any quantification rule can be permuted with a logical or structural rule under some conditions. One classical condition is that the sequents contain only prenex formulas, which is not the case with the terminological sequents. We rely here on a more general version of the normalization theorem for formulas in negation normal form NNF,[13] (cf. [7, Chap. 7]). The theorem also uses a more general version of the quantifier rules (quantifier rules for subformulas). To apply the NNF normalization theorem, we have to show that terminological sequents can be put into negation normal form. This is done by replacing every no-formula by an equivalent all-formula, and exchanging the quantifiers and negation as usual. The rewriting is an endomorphism in the set of terminological sequents.

In the general setting of provability from a set of assumptions (a set of proper axioms Δ), the arguments above still hold for the *cut-free parts of the proofs*, i.e., for the steps of the proofs preceding the applications of the Cut-Rule (which can be restricted with one proper axiom as premise).

Consequently, any proof (w.r.t. to some set of assumptions Δ) of a sequent in NNF can be supposed to be in normal form, that is containing, top-down from the root to the leaves:

- first-order steps consisting in applications of quantification rules (with eventual structural rules except from weakening),
- propositional steps consisting of application of logical rules (or structural rules), and
- Cut-Rule applications with one proper axiom of Δ and one sequent whose complexity (in terms of height of a normal proof) is less than the root sequent.

[13] A formula is in NNF iff the negation symbols appears only in front of atomic subformulas.

According to this normalization property, any first-order terminological sequent is provable iff it is provable by "lifting" propositional sequents.

The "lifting" process will consist in recovering quantor formulas from atomic concept and role formulas appearing in the terminological sequents. As no assumption can be formulated in Δ stating mutual logical properties of roles and concepts (unless the roles are encapsulated into quantor-concepts), we can state the following *Separation Lemma.*

Lemma 6. *Let σ be a propositional sequent of the form $\delta_r, \delta_c \Rightarrow \gamma_c, \gamma_r$ where δ_c, γ_c are formulas corresponding to the translation of concepts terms and δ_r, γ_r are formulas corresponding to the translation of role terms. Let Δ be a set of sequents representing terminological formulas (assumptions).*
Then $\Delta \vdash_{SC} \sigma$ iff $\Delta \vdash_{SC} \delta_c \Rightarrow \gamma_c$ or $\Delta \vdash_{SC} \delta_r \Rightarrow \delta_r$.

In the following, we show how to derive a complete set of consequence rules for the first-order fragment of our sample language $\mathcal{TL}_{SAN}$ by lifting propositional terminological formulas. Precisely, we will derive the rules for disjunctions from propositional formulas of the form $\mathsf{c}_1 \sqcap \mathsf{c}_2 \sqsubseteq \bot$, and the rules for quantors from propositional formulas $\mathsf{c}_1 \sqcap \mathsf{c}_2 \sqsubseteq \mathsf{c}$ and $\mathsf{c}_1 \sqsubseteq \mathsf{c}_2$.

4.3 Rules for Disjunctions

In the terminological language $\mathcal{TL}_{SAN}$ we have neither explicit disjunction nor negation. Thus we cannot express disjunctions, nor negations for propositional (quantor-free) concepts. But we have an implicit negation because of the no-quantor and consequently also an implicit disjunction, as the following equivalencies show:

$$\forall \mathsf{r}.\mathsf{c} \doteq \neg\exists \mathsf{r}.\neg \mathsf{c} \tag{34}$$

$$\top \doteq \exists \mathsf{r}.\mathsf{c} \sqcup \neg\exists \mathsf{r}.\mathsf{c} \tag{35}$$

In general the translation of such disjunctive concepts into SC gives a sequent of the form $\vdash_{SC} \emptyset \Rightarrow C_1(a), C_2(a)$. By the negation duality, this gives a pair of incompatible concepts, i.e., $\vdash_{SC} \neg C_1(a), \neg C_2(a) \Rightarrow \emptyset$. Thus the rules of disjunction give also, by duality, rules for incompatible concepts (as well as the admissible forms of implicit negation formulas in $\mathcal{TL}_{SAN}$).

According to the Lifting Lemma, the strategy for deriving the rules for disjunction is to examine the different ways to reduce an initial first-order sequent $C \Rightarrow C_1, C_2$ to a provable propositional sequent. The derivation is easy but rather long because it proceeds by a systematic case analysis of the possible quantor structure for C_1 and C_2. We just give some technical details of the derivation (the complete proof can be found in [18]).

Derivation Method. Let us suppose that we have to prove the terminological formula $\mathsf{c}_1 \sqsubseteq \mathsf{c}_2 \sqcup \mathsf{c}_3$, represented by the sequent $S = C_1(a) \Rightarrow C_2(a), C_3(a)$ for some eigenparameter "a". Let us suppose also that neither $C_1(a) \Rightarrow C_2(a)$, nor $C_1(a) \Rightarrow C_3(a)$, nor $C_1(a) \Rightarrow \emptyset$ nor $\emptyset \Rightarrow C_2(a), C_3(a)$ are provable (else everything becomes trivial).

We proceed by case analysis of the possible quantor structures of C_2 and C_3 in S. It is impossible to prove S if we suppose that both C_2 and C_3 come from quantor-free concepts, because the propositional fragment of $\mathcal{TL}_{\mathcal{SAN}}$ allows neither disjunction nor negation. One can also show that the case where C_2 is propositional and C_3 is a quantified formula is impossible (unless c_2 is $\bot$ or $\top$). We briefly sketch the analysis for the other cases.[14]

Case 1: C_2 and C_3 are both all-formulas. Assuming both are of the form (**all** R_i C_i) we have to prove the sequent $C_1 \Rightarrow \forall y R_1(a,y) \rightarrow C_4(y), \forall y R_2(a,y) \rightarrow C_5(y)$. By two successive applications of the $\forall$-right rule, i.e., creation of two new eigen-parameters, we get to prove: $C_1, R_1(a,b), R_2(a,c) \Rightarrow C_4(b), C_5(c)$.

The creation of two different parameters implies that either the proof is completely independent of C_4 and C_5 (hence independent from C_2 and C_3, which contradicts the initial assumption) or there is a separate proof for either $C_4(b)$ or $C_5(b)$ (which also contradicts the initial assumption). The same reasoning holds if C_2 and C_3 are quantified formulas involving no or all.

Case2: C_2 and C_3 are both some-formulas. Let us suppose that $C_2 = (\textbf{some}\ R_1\ C_4)(a)$ and $C_3 = (\textbf{some}\ R_2\ C_5)(a)$. In order to unfold C_2 and C_3 we need to instantiate by a term already in the sequent. The proof starts necessarily by the creation of a new eigen-parameter on the left-hand side (by $\exists$-left rule) which means that C_1 is a some-formula, say $(\textbf{some}\ R_3\ C_5)$. We then get

$$\begin{aligned}
\exists y(R_3(a,y) \wedge C(y)) &\Rightarrow \exists y(R_1(a,y) \wedge C_4(y)),\ \exists y(R_2(a,y) \wedge C_5(y)) \\
R_3(a,b),\ C(b) &\Rightarrow \exists y(R_1(a,y) \wedge C_4(y)),\ \exists y(R_2(a,y) \wedge C_5(y)) \\
R_3(a,b),\ C(b) &\Rightarrow R_1(a,b) \wedge C_4(b)),\ R_2(a,b) \wedge C_5(b))
\end{aligned}$$

Unfolding the conjunctions gives four sequents to prove, two of which have consequents "$R_1(a,b), C_4(b)$" and "$R_1(a,b), C_5(b)$", which are provable only by separation of the disjunction (Separation Lemma). By an easy case analysis we get:

$$\begin{aligned}
R_3(a,b) &\Rightarrow R_1(a,b) \\
R_3(a,b) &\Rightarrow R_2(a,b) \\
C(b) &\Rightarrow C_4(b), C_5(b)
\end{aligned}$$

This gives the monotonicity rule for the quantor some w.r.t. to disjunctions (37).

Case 3: C_2 is a some-formula, C_3 an all-formula. Let $C_2 = (\textbf{some}\ R_1\ C_4)(a)$ and $C_3 = (\textbf{all}\ R_2\ C_5)(a)$. To cope with the case where C_3 is a no-formula we may suppose in the proof that C_5 is possibly a negative formula ($C_5 = \neg C_6$). We get, by adequate eliminations of quantifications:

$$\begin{aligned}
C_1(a) &\Rightarrow \exists y(R_1(a,y) \wedge C_4(y)),\ R_2(a,b) \rightarrow C_5(b) \\
C_1(a) &\Rightarrow R_1(a,b) \wedge C_4(b),\ R_2(a,b) \rightarrow C_5(b) \\
C_1(a),\ R_2(a,b) &\Rightarrow R_1(a,b) \wedge C_4(b),\ C_5(b)
\end{aligned}$$

[14] In the following we omit the reference to "a".

By the $\wedge$-right rule, we get two sequents to prove:

$$C_1(a),\ R_2(a,b) \Rightarrow R_1(a,b),\ C_5(b)$$
$$C_1(a),\ R_2(a,b) \Rightarrow C_4(b),\ C_5(b)$$

The first sequent is provable (propositionally) only from an axiom sequent representing the terminological axiom $r_2 \sqsubseteq r_1$. In the second sequent there is an eigen-parameter "b" different from the eigen-parameter of $C_1(a)$. Trying to unfold a some-formula in C_1(a) does not work (a new eigen-parameter would be created). The only possibility is to unfold an all-formula (all r c) in C_1(a). Then we get :

$$C(b),\ R_2(a,b) \Rightarrow R_1(a,b),\ C_5(b)$$
$$C(b),\ R_2(a,b) \Rightarrow C_4(b),\ C_5(b)$$
$$R_2(a,b) \Rightarrow R(a,b),\ R_1(a,b),\ C_5(b)$$
$$R_2(a,b) \Rightarrow R(a,b),\ C_4(b),\ C_5(b)$$

By the Separation Lemma we obtain necessarily the conditions $C(b) \Rightarrow C_4(b), C_5(b)$ and $R_2(a,b) \Rightarrow R(a,b)$ and $R_2(a,b) \Rightarrow R(a,b)$. This gives Rule 39. As we allowed C_5 to be the translation of a negation concept, in case C_3 represents a no-formula, by applying the negation rule to C_5, we will get Rule 38.

Conclusion. We thus obtain the following inference rules for handling disjunction in $\mathcal{TL}_{SAN}$:

$$c_2 \sqsubseteq c_1, c_3 \sqsubseteq c_1 \rightleftharpoons c_2 \sqcup c_3 \sqsubseteq c_1 \tag{36}$$
$$c_1 \sqsubseteq c_2 \sqcup c_3, r_1 \sqsubseteq r_2, r_1 \sqsubseteq r_3 \rightharpoonup \exists r_1.c_1 \sqsubseteq \exists r_2.c_2 \sqcup \exists r_3.c_3 \tag{37}$$
$$c_2 \sqcap c_3 \sqsubseteq c_1, r_2 \sqsubseteq r_1, r_2 \sqsubseteq r_3 \rightharpoonup \forall r_3.c_3 \sqsubseteq \exists r_1.c_1 \sqcup \neg\exists r_2.c_2 \tag{38}$$
$$c_3 \sqsubseteq c_1 \sqcup c_2, r_2 \sqsubseteq r_1, r_2 \sqsubseteq r_3 \rightharpoonup \forall r_3.c_3 \sqsubseteq \exists r_1.c_1 \sqcup \forall r_2.c_2 \tag{39}$$

The need for these rules can be demonstrated by the following entailment, and the corresponding proof which is only possible by using the rules for disjunctions:

$$\exists r.c_1 \sqsubseteq c_2, \neg\exists r.c_1 \sqsubseteq c_2 \models_{TL} \top \sqsubseteq c_2 \tag{40}$$

$\exists r.c_1 \sqsubseteq c_2, \neg\exists r.c_1 \sqsubseteq c_2$	Axioms	(41)
$\exists r.c_1 \sqcup \neg\exists r.c_1 \sqsubseteq c_2$	from 41 and 39	(42)
$\forall r.\top \sqsubseteq \exists r.c_1 \sqcup \neg\exists r.c_1$	from 38	(43)
$\forall r.\top \sqsubseteq c_2$	from 42, 43, and 27	(44)
$\top \sqsubseteq c_2$	from 52 and 27	(45)

From the preceding analysis we can also derive the rules for mutually inconsistent concepts by exploiting the following equivalences:

$$C_1 \Rightarrow \neg C_2,\ \neg C_3 \text{ iff } C_1,\ C_2,\ C_3 \Rightarrow \emptyset$$
$$\neg\exists r.\neg c \doteq \forall r.c$$
$$\neg\neg\exists r.c \doteq \exists r.c$$

Inconsistent concepts will be written $c_1 \sqcap c_2 \sqsubseteq \bot$. The inference rules for inconsistent concepts are:

$$c_1 \sqsubseteq c_2, r_1 \sqsubseteq r_2 \rightleftharpoons \exists r_1.c_1 \sqcap \neg\exists r_2.c_2 \sqsubseteq \bot \quad (46)$$

$$c_1 \sqcap c_2 \sqsubseteq \bot, r_1 \sqsubseteq r_2 \rightleftharpoons \exists r_1.c_1 \sqcap \forall r_2.c_2 \sqsubseteq \bot \quad (47)$$

Exploiting systematically the negation duality, we can still derive the additional rules 48 and 49 (which follow easily from the precedent ones if we had an explicit negation rule - for more details see the complete paper [18]). In particular, through adequate instantiations by $\top$ or $\bot$, we get the equivalencies between $\forall r.c$ and $\neg\exists r.\neg c$.

$$c_3 \sqsubseteq c_1 \sqcup c_2, r_3 \sqsubseteq r_1, r_2 \sqsubseteq r_3 \rightharpoonup \forall r_3.c_3 \sqcap \neg\exists r_1.c_1 \sqsubseteq \forall r_2.c_2 \quad (48)$$

$$c_3 \sqsubseteq c_1 \sqcup c_2, r_3 \sqsubseteq r_2, r_3 \sqsubseteq r_1 \rightharpoonup \exists r_3.c_3 \sqcap \neg\exists r_2.c_2 \sqsubseteq \exists r_1.c_1 \quad (49)$$

$$c_1 \sqcap c_2 \sqsubseteq \bot, r_2 \sqsubseteq r_1 \rightharpoonup \forall r_1.c_1 \sqsubseteq \neg\exists r_2.c_2 \quad (50)$$

$$\top \sqsubseteq c_1 \sqcup c_2, r_2 \sqsubseteq r_1 \rightharpoonup \neg\exists r_1.c_1 \sqsubseteq \forall r_2.c_2 \quad (51)$$

4.4 Rules for the Quantors

Rules for $\top$ and $\bot$. We obtain the rules for the concepts $\top$ and $\bot$ by lifting trivial axioms with True or False. For instance, from True $\Rightarrow \emptyset$ we get immediately True $\Rightarrow$ (**all** r True)(a), which is a particular case of the rule 52 (we do not go into details here).

$$\top \sqsubseteq c \rightleftharpoons \top \sqsubseteq \forall r.c \quad (52)$$

$$c \sqsubseteq \bot \rightleftharpoons \exists r.c \sqsubseteq \bot \quad (53)$$

Remark. We could also get rules like $\rightharpoonup c \sqsubseteq \exists\bot.c$ if $\bot$ and $\top$ were introduced as particular role constructors, which is not the case in this paper. Similarly the above equivalences are no more true with inconsistent roles. For example, $r \sqsubseteq \bot$ makes $\forall r.c$ be $\top$ for every c.

Lifting Axioms $c_1 \sqsubseteq c_2$.

Proposition 7. *The following three terminological rules are sound and complete for lifting axioms of the form* $c_1 \sqsubseteq c_2$.[15]

$$c_1 \sqsubseteq c_2, r_1 \sqsubseteq r_2 \rightharpoonup \exists r_1.c_1 \sqsubseteq \exists r_2.c_2 \quad (54)$$

$$c_1 \sqsubseteq c_2, r_2 \sqsubseteq r_1 \rightharpoonup \forall r_1.c_1 \sqsubseteq \forall r_2.c_2 \quad (55)$$

$$c_1 \sqsubseteq c_2, r_1 \sqsubseteq r_2 \rightharpoonup \neg\exists r_2.c_2 \sqsubseteq \neg\exists r_1.c_1 \quad (56)$$

The proofs being all similar, we only give the proof for inference rule 54.

---

[15] These rules reflect the monotonicity properties of the quantifiers (cf. [16]).

Proof. Provability of $\Rightarrow \forall x C_1(x) \rightarrow C_2(x)$ is equivalent to provability of $C_1(b) \Rightarrow C_2(b)$.

The expansion rule allows to introduce a role term $R_1(a,b)$ in the left-hand side of the sequent. We get a proof of the sequent $R_1(a,b), C_1(b) \Rightarrow C_2(b)$, and equivalently $R_1(a,b) \wedge C_1(b) \Rightarrow C_2(b)$

The expression $R_1(a,b) \wedge C_1(b)$ can be folded into $(\mathbf{some}\ R_1\ C_1)(a)$ by the $\exists$-right rule, but with a parameter different from the parameter of C_2. This does not yield a proof for a terminological formula, as tail and head need the same parameter (cf. Sect. 3.3).

To lift the proof, we need to introduce role terms both on the left- and right-hand sides. For the right-hand side this requires necessarily the application of the $\wedge$-right if we want to fold a some-formula.

$$
\begin{array}{rl}
\vdash_{SC} C_1(b) \Rightarrow C_2(b) & \text{Axiom} \\
\vdash_{SC} R_1(a,b) \Rightarrow R_2(a,b) & \text{Axiom} \\
\vdash_{SC} R_1(a,b), C_1(b) \Rightarrow C_2(b) & \text{Expansion} \\
\vdash_{SC} R_1(a,b), C_1(b) \Rightarrow R_2(a,b) & \text{Expansion} \\
\vdash_{SC} R_1(a,b), C_1(b) \Rightarrow R_2(a,b) \wedge C_2(b) & \wedge\text{-right} \\
\vdash_{SC} R_1(a,b), C_1(b) \Rightarrow (\mathbf{some}\ R_2\ C_2)(a) & \exists\text{-right} \\
\vdash_{SC} (\mathbf{some}\ R_1\ C_1)(a) \Rightarrow (\mathbf{some}\ R_2\ C_2)(a) & \exists\text{-left} \\
\vdash_{SC} \Rightarrow (\mathbf{some}\ R_1\ C_1)(a) \rightarrow (\mathbf{some}\ R_2\ C_2)(a) & \\
\vdash_{SC} \Rightarrow \forall x(\mathbf{some}\ R_1\ C_1)(x) \rightarrow (\mathbf{some}\ R_2\ C_2)(x)) & \\
\vdash_{TL} \exists \mathsf{r}_1.\mathsf{c}_1 \sqsubseteq \exists \mathsf{r}_2.\mathsf{c}_2 &
\end{array}
$$

Note that the inference rules are reversible when inconsistent roles and concepts are excluded (see [18] for details).

Lifting Axioms $\mathsf{c}_1 \sqcap \mathsf{c}_2 \sqsubseteq \mathsf{c}_3$. We now lift conjunctive axioms of the form: $\mathsf{c}_1 \sqcap \mathsf{c}_2 \sqsubseteq \mathsf{c}_3$. So we start with the sequent $C_1(b), C_2(b) \Rightarrow C_3(b)$.

Folding induces necessarily a change of eigen-parameters from the initial sequent. The lifting process must concern simultaneously all the three concept-formulas C_1, C_2 and C_3, with the same eigen-parameter if we want to recover a terminological clause.

By lifting the axiom uniformly with all-formulas we get the *additivity rule* for the universal quantor:

$$\mathsf{c}_1 \sqcap \mathsf{c}_2 \sqsubseteq \mathsf{c}_3, \mathsf{r}_3 \sqsubseteq \mathsf{r}_1, \mathsf{r}_3 \sqsubseteq \mathsf{r}_2 \longrightarrow \forall \mathsf{r}_1.\mathsf{c}_1 \sqcap \forall \mathsf{r}_2.\mathsf{c}_2 \sqsubseteq \forall \mathsf{r}_3.\mathsf{c}_3 \tag{57}$$

It is easy to show that there is no uniform lifting of $\mathsf{c}_1 \sqcap \mathsf{c}_2 \sqsubseteq \mathsf{c}_3$ into some-formulas, because the folding of two expressions at the right-hand side of sequents will produce two different eigen-parameters. Similarly, there is no possibility to lift a left some-formula and a right all-formula.

By similar arguments, we can show that there is only one possibility to get a some-formula as right-hand term, by lifting an axiom $\mathsf{c}_1 \sqcap \mathsf{c}_2 \sqsubseteq \mathsf{c}_3$. This gives the following rule:

$$\mathsf{c}_1 \sqcap \mathsf{c}_2 \sqsubseteq \mathsf{c}_3, \mathsf{r}_2 \sqsubseteq \mathsf{r}_1, \mathsf{r}_2 \sqsubseteq \mathsf{r}_3 \longrightarrow \forall \mathsf{r}_1.\mathsf{c}_1 \sqcap \exists \mathsf{r}_2.\mathsf{c}_2 \sqsubseteq \exists \mathsf{r}_3.\mathsf{c}_3 \tag{58}$$

4.5 Descriptions

On the object level there are two types of formulas, called *descriptions*, namely $o_1 : c_1$ and $\langle o_1, o_2 \rangle : r_1$. We will now derive the inference rules for handling these descriptions.

$o_1 : c_1$. One way of deriving the sequent $\emptyset \Rightarrow C_1(o_1)$ is by application of the Cut-Rule. In order to apply the Cut-Rule we need to apply the $\rightarrow$-rule first, for which we need formulas containing an implication, i.e., the FOL-translations of either $c_1 \sqsubseteq c_2$ or of $o_2 : \forall r.c_1$. Since $o_2 : \forall r.c_1$ introduces the additional constraint $\langle o_2, o_1 \rangle : r$ we obtain the following inference rules:

$$o_1 : c_2, c_2 \sqsubseteq c_1 \rightharpoonup o_1 : c_1 \tag{59}$$
$$o_2 : \forall r.c_1, \langle o_2, o_1 \rangle : r \rightharpoonup o_1 : c_1 \tag{60}$$

The other way of deriving the sequent $\emptyset \Rightarrow C_1(o_1)$ is via the $\exists$-rule or the $\wedge$-rule which gives the following inference rules:

$$\langle o_1, o_2 \rangle : r_1, o_2 : c_1 \rightharpoonup o_1 : \exists r_1.c_1 \tag{61}$$
$$o_1 : c_1, o_1 : c_2 \rightharpoonup o_1 : c_1 \sqcap c_2 \tag{62}$$

$\langle o_1, o_2 \rangle : r_1$. To derive $\emptyset \Rightarrow R_1(o_1, o_2)$ the same considerations apply but there is no analogue to the $\rightarrow$ introduced by all-restrictions in concept terms and the $\exists$-rule is not applicable. Thus we only have:

$$\langle o_1, o_2 \rangle : r_2, r_2 \sqsubseteq r_1 \rightharpoonup \langle o_1, o_2 \rangle : r_1 \tag{63}$$
$$\langle o_1, o_2 \rangle : r_1, \langle o_1, o_2 \rangle : r_2 \rightharpoonup \langle o_1, o_2 \rangle : r_1 \sqcap r_2 \tag{64}$$

5 Normalize-Compare Algorithms

In this section we will indicate how the inference rules derived in the previous section can be used for the construction of normalize-compare algorithms. We begin with a complete list of the inference rules for $\mathcal{TL}_{\mathcal{SAN}}$ (with the exception of the rules for deriving descriptions):

$$\rightharpoonup t_1 \sqcap t_2 \sqsubseteq t_1 \tag{25}$$
$$t_1 \sqsubseteq t_2, t_1 \sqsubseteq t_3 \rightleftharpoons t_1 \sqsubseteq t_2 \sqcap t_3 \tag{26}$$
$$t_1 \sqsubseteq t_2, t_2 \sqcap t_3 \sqsubseteq t_4 \rightharpoonup t_1 \sqcap t_3 \sqsubseteq t_4 \tag{27}$$
$$\rightharpoonup \bot \sqsubseteq c \tag{31}$$
$$\rightharpoonup c \sqsubseteq \top \tag{32}$$
$$c_1 \sqsubseteq c_2 \rightleftharpoons c_1 \sqcap \top \sqsubseteq c_2 \tag{33}$$
$$c_2 \sqsubseteq c_1, c_3 \sqsubseteq c_1 \rightleftharpoons c_2 \sqcup c_3 \sqsubseteq c_1 \tag{36}$$
$$c_1 \sqsubseteq c_2 \sqcup c_3, r_1 \sqsubseteq r_2, r_1 \sqsubseteq r_3 \rightharpoonup \exists r_1.c_1 \sqsubseteq \exists r_2.c_2 \sqcup \exists r_3.c_3 \tag{37}$$
$$c_2 \sqcap c_3 \sqsubseteq c_1, r_2 \sqsubseteq r_1, r_2 \sqsubseteq r_3 \rightharpoonup \forall r_3.c_3 \sqsubseteq \exists r_1.c_1 \sqcup \neg\exists r_2.c_2 \tag{38}$$

$$c_3 \sqsubseteq c_1 \sqcup c_2, r_2 \sqsubseteq r_1, r_2 \sqsubseteq r_3 \rightharpoonup \forall r_3.c_3 \sqsubseteq \exists r_1.c_1 \sqcup \forall r_2.c_2 \tag{39}$$

$$c_1 \sqsubseteq c_2, r_1 \sqsubseteq r_2 \rightleftharpoons \exists r_1.c_1 \sqcap \neg\exists r_2.c_2 \sqsubseteq \bot \tag{46}$$

$$c_1 \sqcap c_2 \sqsubseteq \bot, r_1 \sqsubseteq r_2 \rightleftharpoons \exists r_1.c_1 \sqcap \forall r_2.c_2 \sqsubseteq \bot \tag{47}$$

$$c_3 \sqsubseteq c_1 \sqcup c_2, r_3 \sqsubseteq r_1, r_2 \sqsubseteq r_3 \rightharpoonup \forall r_3.c_3 \sqcap \neg\exists r_1.c_1 \sqsubseteq \forall r_2.c_2 \tag{48}$$

$$c_3 \sqsubseteq c_1 \sqcup c_2, r_3 \sqsubseteq r_2, r_3 \sqsubseteq r_1 \rightharpoonup \exists r_3.c_3 \sqcap \neg\exists r_2.c_2 \sqsubseteq \exists r_1.c_1 \tag{49}$$

$$c_1 \sqcap c_2 \sqsubseteq \bot, r_2 \sqsubseteq r_1 \rightharpoonup \forall r_1.c_1 \sqsubseteq \neg\exists r_2.c_2 \tag{50}$$

$$\top \sqsubseteq c_1 \sqcup c_2, r_2 \sqsubseteq r_1 \rightharpoonup \neg\exists r_1.c_1 \sqsubseteq \forall r_2.c_2 \tag{51}$$

$$\top \sqsubseteq c \rightleftharpoons \top \sqsubseteq \forall r.c \tag{52}$$

$$c \sqsubseteq \bot \rightleftharpoons \exists r.c \sqsubseteq \bot \tag{53}$$

$$c_1 \sqsubseteq c_2, r_1 \sqsubseteq r_2 \rightharpoonup \exists r_1.c_1 \sqsubseteq \exists r_2.c_2 \tag{54}$$

$$c_1 \sqsubseteq c_2, r_2 \sqsubseteq r_1 \rightharpoonup \forall r_1.c_1 \sqsubseteq \forall r_2.c_2 \tag{55}$$

$$c_1 \sqsubseteq c_2, r_1 \sqsubseteq r_2 \rightharpoonup \neg\exists r_2.c_2 \sqsubseteq \neg\exists r_1.c_1 \tag{56}$$

$$c_1 \sqcap c_2 \sqsubseteq c_3, r_3 \sqsubseteq r_1, r_3 \sqsubseteq r_2 \rightharpoonup \forall r_1.c_1 \sqcap \forall r_2.c_2 \sqsubseteq \forall r_3.c_3 \tag{57}$$

$$c_1 \sqcap c_2 \sqsubseteq c_3, r_2 \sqsubseteq r_1, r_2 \sqsubseteq r_3 \rightharpoonup \forall r_1.c_1 \sqcap \exists r_2.c_2 \sqsubseteq \exists r_3.c_3 \tag{58}$$

These inference rules give a proof theory for $\mathcal{TL}_{\mathcal{SAN}}$ but they do not constitute an algorithm on their own. To construct an algorithm we need to specify a control strategy which determines the order of application of inference rules and guarantees termination. (One advantage of tableaux methods is that such a control strategy can be easily specified.)

We stressed in Sect. 2 the importance of normal forms for TL-systems. The main idea is that the information explicitly specified by definitions, rules, and descriptions entails implicit information which is so important for the reasoning that it is worthwile to store it explictly instead of deriving it again and again. As an example, consider value restrictions, i.e. terms $\forall r.c$. In BACK V4 for each concept *the* value restriction at each relevant role is computed and stored [17]. This value restriction is used in comparing the concepts and also for forward propagation to role fillers.

In the following we will restrict ourselves to the problem of subsumption with respect to a set of cycle-free definitions. That is we will specify an algorithm SUB such that SUB(t_1,t_2) iff $\Gamma \models_{TL} t_2 \sqsubseteq t_1$, where Γ is a set of definitions, i.e., does not contain any rules. (We emphasize again that this is only one reasoning task among others in TL-systems, though an important one.) Due to this restriction we can ignore the inference rules 36 to 39 and the $\leftharpoondown$ direction for 46, 47, 52, and 53 (see [18] for details).

Our subsumption algorithm, which is similar to the one presented by Nebel in [13], will consist of two parts, namely a normalization part NORM and a comparison part COMP. NORM transforms a term t into a normal form n and COMP compares two normal forms. We will prove completeness of SUB by showing that the normal forms produced by NORM have a certain property and that COMP is complete for normal forms with these properties. These proofs will rely on the completeness of the inference rules listed above.

Definition 8. SUB takes two terms in set representation (see below) t_1 and t_2 and succeeds iff t_1 subsumes t_2:

```
SUB(t1,t2) :-
        NORM(t1,n1),
        NORM(t2,n2),
        COMP(n1,n2).
```

Note that we do not claim that our normal forms are optimal for TL-systems. Indeed we think that the determination of adequate normal forms is one of the most important topics in the development of TL-systems and has to rely on empirical studies (cf., for example [12]). All we want to show is that the completeness of normalize-compare algorithms can be proved via completeness of inference rules.

Before presenting NORM we will slightly change our notation for terms. As defined in the syntax a term is a conjunction of names, primitive components, and — for concepts — quantified restrictions. First, we will eliminate the names: for each name occurring in a term we substitute its definition until a term contains only primitive components and quantified restrictions. Note that this is done recursively, so that also terms occurring in the restrictions do not contain any names.[16] Next, we represent the conjunctions by a set-notation:

Definition 9. We define a function $\uparrow$ mapping concepts into sets as follows:

$$\uparrow t_p \stackrel{\text{def}}{=} \{t_p\} \tag{65}$$

$$\uparrow t_1 \sqcap t_2 \stackrel{\text{def}}{=} \uparrow t_1 \cup \uparrow t_2 \tag{66}$$

$$\uparrow \exists r.c \stackrel{\text{def}}{=} \{\exists \uparrow r. \uparrow c\} \tag{67}$$

$$\uparrow \forall r.c \stackrel{\text{def}}{=} \{\forall \uparrow r. \uparrow c\} \tag{68}$$

$$\uparrow \neg\exists r.c \stackrel{\text{def}}{=} \{\neg\exists \uparrow r. \uparrow c\} \tag{69}$$

Thus each term is now represented by a set. We will call the members of such a set atomic restrictions and the sets themselves normal forms.

Definition 10. The set of atomic restrictions $\mathcal{A}$ and the sets of normal forms $\mathcal{N}_C$ and $\mathcal{N}_R$ can be defined from the primitive components for concepts and roles, $\mathcal{P}_C$ and $\mathcal{P}_R$:

1. each $c_p \in \mathcal{P}_C$ is in $\mathcal{A}$;
2. $\top$ and $\bot$ are in $\mathcal{A}$;
3. if $n \in \mathcal{N}_C$ and $r \in \mathcal{N}_R$ then $\forall r.n$, $\exists r.n$, and $\neg\exists r.n$ are in $\mathcal{A}$;
4. if $a_1, \ldots, a_n \in \mathcal{A}$ then $\{a_1, \ldots, a_n\}$ is in $\mathcal{N}_C$.
5. if $r_1, \ldots, r_n \in \mathcal{P}_R$ then $\{r_1, \ldots, r_n\}$ is in $\mathcal{N}_R$.

In order to talk about subsumption between normal forms and atomic restrictions we need the inverse of $\uparrow$ which maps normal forms and atomic restrictions into concepts:

[16] This is called abstraction from term introductions in [13, p. 60].

Definition 11.

$$\downarrow\{a_1,\ldots,a_n\} \stackrel{\text{def}}{=} \downarrow a_1 \sqcap \ldots \sqcap \downarrow a_n \tag{70}$$

$$\downarrow\exists r.n \stackrel{\text{def}}{=} \exists \downarrow r. \downarrow n \tag{71}$$

$$\downarrow\forall r.n \stackrel{\text{def}}{=} \forall \downarrow r. \downarrow n \tag{72}$$

$$\downarrow\neg\exists r.n \stackrel{\text{def}}{=} \neg\exists \downarrow r. \downarrow n \tag{73}$$

Clearly, this switch between terms and sets is meaning preserving:

Lemma 12. *Let* t_1 *and* t_2 *be any terms.* $t_1 \sqsubseteq t_2$ *iff* $\Downarrow\Uparrow t_1 \sqsubseteq \Downarrow\Uparrow t_2$

We said above that NORM will map terms into normal forms with special properties:

Definition 13. We say that a normal form n is **vivid** iff for any atomic restriction a_1: $\downarrow n \sqsubseteq \downarrow a_1$ only if $\exists\, a_2 \in n$: $\downarrow a_2 \sqsubseteq \downarrow a_1$ and all normal forms occurring in n are vivid.

Clearly, vivid normal forms have the advantage that the subsumption test is somehow local: only simple atomic restrictions and no combinations have to be checked.

In order to construct a normalization algorithm that transforms terms into vivid normal forms we have to select the adequate inference rules, namely rules with the format

$$\gamma_1, \ldots, \gamma_n \rightharpoonup t_1 \sqcap t_2 \sqsubseteq t_3$$

The normalization strategy will be to add t_3 to the set representation of a term containing t_1 and t_2 provided that $\gamma_1, \ldots, \gamma_n$ hold. This strategy makes sense for 46 and 47, is redundant for 25, and poses problems for 27, 48, 49, 57, and 58. The problems arise because in these rules t_3 is not completly determined by t_1 and t_2.

We can replace, however, 57 and 58 by simpler rules:

$$r_1 \sqsubseteq r_2 \rightharpoonup \forall r_1.c_1 \sqcap \forall r_2.c_2 \sqsubseteq \forall r_1.c_1 \sqcap c_2 \tag{74}$$

$$r_1 \sqsubseteq r_2 \rightharpoonup \exists r_1.c_1 \sqcap \forall r_2.c_2 \sqsubseteq \exists r_1.c_1 \sqcap c_2 \tag{75}$$

We show the correctness of the first substitution by deriving $\forall r_2.c_1 \sqcap \forall r_3.c_2 \sqsubseteq \forall r_1.c_3$ from $c_1 \sqcap c_2 \sqsubseteq c_3$, $r_1 \sqsubseteq r_2$, $r_1 \sqsubseteq r_3$:

$$\forall r_1.c_1 \sqcap \forall r_1.c_2 \sqsubseteq \forall r_1.c_1 \sqcap c_2 \quad \text{Assumption and 74}$$

$$\forall r_1.c_1 \sqcap \forall r_1.c_2 \sqsubseteq \forall r.c_1 \sqcap c_2 \quad \text{Assumption and 27}$$

$$\forall r_2.c_1 \sqcap \forall r_3.c_2 \sqsubseteq \forall r_1.c_1 \sqcap c_2 \quad \text{Assumptions and 55}$$

The proofs for the other substitution is similar.

Similarily, we can replace the rules 48 and 49 by the following rules, since the only relevant disjunctive formula for $\mathcal{TL}_{SAN}$ without rules is $\top \sqsubseteq \exists r.c \sqcup \neg\exists r.c$:

$$\rightharpoonup \neg\exists r_1.\exists r_2.c_2 \doteq \forall r_1.\neg\exists r_2.c_2 \tag{76}$$

$$\rightharpoonup \neg\exists r_1.\neg\exists r_2.c_2 \doteq \forall r_1.\exists r_2.c_2 \tag{77}$$

$$r_1 \sqsubseteq r_2 \rightharpoonup \exists r_1.\top \sqcap \neg\exists r_2.\exists r_3.c_3 \sqsubseteq \exists r_1.\neg\exists r_3.c_3 \tag{78}$$

$$r_1 \sqsubseteq r_2 \rightharpoonup \exists r_1.\top \sqcap \neg\exists r_2.\neg\exists r_3.c_3 \sqsubseteq \exists r_1.\exists r_3.c_3 \tag{79}$$

Remark. These inference rules encode explicitly the information that $\exists r.c$ is the negation of $\neg\exists r.c$ and should be replaced by an explicit negation rule, when terminological rules are taken into account.

We thus get the following normalization algorithm:[17]

Definition 14. We define the algorithm NORM that maps a term t into a normal form n:

1. Set n to t.
2. If $\exists r_1.c_1 \in t$ and $\neg\exists r_2.c_2 \in t$ and SUB(c_2,c_1) and SUB(r_2,r_1) add $\perp$ to t (46).
3. If $\exists r_1.c_1 \in t$ and $\forall r_2.c_2 \in t$ and SUB($\perp$,$c_1 \cup c_2$) and SUB(r_2,r_1) add $\perp$ to t (47).
4. If $\forall r_1.c_1 \in t$ and $\forall r_2.c_2 \in t$ and SUB(r_2,r_1) add $\forall r_1.c_1 \cup c_2$ to t (74).
5. If $\exists r_1.c_1 \in t$ and $\forall r_2.c_2 \in t$ and SUB(r_2,r_1) add $\exists r_1.c_1 \cup c_2$ to t (75).
6. If $\forall r_1.c_1 \in t$ and $\neg\exists r_2.c_2 \in c_1$ add $\neg\exists r_1.\{\exists r_2.c_2\}$ to t (76).
7. If $\neg\exists r_1.c_1 \in t$ and $\exists r_2.c_2 \in c_1$ add $\forall r_1.\{\neg\exists r_2.c_2\}$ to t (76).
8. If $\forall r_1.c_1 \in t$ and $\exists r_2.c_2 \in c_1$ add $\neg\exists r_1.\{\neg\exists r_2.c_2\}$ to t (77).
9. If $\neg\exists r_1.c_1 \in t$ and $\neg\exists r_2.c_2 \in c_1$ add $\forall r_1.\{\exists r_2.c_2\}$ to t (77).
10. If $\exists r_1 c_1 \in t$ and $\neg\exists r_2.c_2 \in t$ and $\exists r_3.c_3 \in c_2$ and $r_1 \sqsubseteq r_2$ add $\exists r_1.\{\neg\exists r_3.c_3\}$ to t (78).
11. If $\exists r_1 c_1 \in t$ and $\neg\exists r_2.c_2 \in t$ and $\neg\exists r_3.c_3 \in c_2$ and $r_1 \sqsubseteq r_2$ add $\exists r_1.\{\exists r_3.c_3\}$ to t (79).
12. If $t \neq n$ go to 1.
13. Normalize the quantified restriction ($\exists r.c$, $\forall r.c$, or $\neg\exists r.c$) in t by normalizing the embedded roles and concepts.
14. Return the resulting t as n.

Proposition 15. NORM *yields for each term t a vivid normal form n.*

Proof (Sketch). Suppose NORM(t,n) and n is not a vivid normal form, i.e., there is an atomic restriction a_2 such that $\downarrow n \sqsubseteq \downarrow a_2$ but $\neg\exists a_1 \in n$ such that $\downarrow a_1 \sqsubseteq \downarrow a_2$. The proof for this subsumption must therefore contain one of the Rules 27, 46, 47, 48, 49, 57, or 58. But all these rules, except for 27, are applied in NORM. (The hard part is to show that the transitivity rule does not add any power to the other rules.) □

Definition 16. We define the algorithm COMP that takes two normal forms n_1 and n_2 and succeeds iff n_2 is more special than n_1.

---

[17] Note that we do not present the normalization algorithm as a rewrite-algorithm, but as adding information, which keeps the presentation simpler.

```
COMP([],_).
COMP([a1|n1],n2) :-
        MEMBER(a2,n2),
        COMP(a1,a2),
        COMP(n1,n2).
COMP(t,t).
COMP(⊤,_).
COMP(_,⊥).
COMP(_,∃r.n) :-
        COMP([⊥],n).
COMP(¬∃r._,∀r.n) :-
        COMP([⊥],n).
COMP(∀r.c1,_) :-
        COMP(c1,⊤).
COMP(∃r1.c1,∃r2.c2) :-
        COMP(r1,r2),
        COMP(c1,c2).
COMP(∀r1.c1,∀r2.c2) :-
        COMP(r2,r1),
        COMP(c1,c2).
COMP(¬∃r1.c1,¬∃r2.c2) :-
        COMP(r2,r1),
        COMP(c2,c1).
```

Proposition 17. COMP *is sound and complete for vivid normal forms*

We prove this by complete induction over the length of normal forms.

Definition 18. The function $|\cdot|$ assigns each atomic restriction a and each normal form n a natural number:

$$|t_p| \stackrel{\text{def}}{=} 1 \tag{80}$$

$$|\top| \stackrel{\text{def}}{=} 1 \tag{81}$$

$$|\bot| \stackrel{\text{def}}{=} 1 \tag{82}$$

$$|n| \stackrel{\text{def}}{=} \max(\{|a| : a \in n\}) \tag{83}$$

$$|\exists r.n| \stackrel{\text{def}}{=} 1 + \max(\{|r|, |n|\}) \tag{84}$$

$$|\forall r.n| \stackrel{\text{def}}{=} 1 + \max(\{|r|, |n|\}) \tag{85}$$

$$|\neg\exists r.n| \stackrel{\text{def}}{=} 1 + \max(\{|r|, |n|\}) \tag{86}$$

Proof. COMP is complete for vivid normal forms of length 1. (Trivial, the only subsumptions are $t_p \sqsubseteq t_p$, $c_p \sqsubseteq \top$, and $\bot \sqsubseteq c_p$.) Next, we prove that if COMP is complete for vivid normal forms of length n then it is complete for vivid normal forms of length $n+1$.

Suppose there are two vivid normal forms n_1, n_2 of length $n+1$ such that COMP(n_2,n_1) fails, but $\downarrow n_1 \sqsubseteq \downarrow n_2$. Now the only inference rule which has a conjunctive concept on the subsuming side is 26. This means that for all $a_i \in n_2$ $\downarrow n_1 \sqsubseteq \downarrow a_i$. And thus COMP must fail for one such a_i. Furthermore, n_1 is a vivid normal form which means that there is an $a_j \in n_2$ such that $\downarrow a_j \sqsubseteq \downarrow a_i$. Since $\vdash_{TL}$ is complete there is a proof for this subsumption and this proof must involve one of the Rules 32, 31, 53, 52, 54 55, or 56 since a_i and a_j are atomic restrictions. But COMP uses all these rules and the verification of the premises only requires comparison of normal forms of length n, for which COMP is complete by assumption. □

6 Conclusion

It has been shown that a set of inference rules can be specified for the restricted language $\mathcal{TL}_{\mathcal{SAN}}$. We have proved the completeness of this set by showing that every proof in SC of a translation of a TL-formula can be rewritten into a derivation of the TL-formula by the inference rules. This approach can be extended to more expressive languages. Due to the introduction of equality axioms by the translation of qualified number-restrictions, however, completeness proofs for languages containing these constructs will be a lot harder to achieve (cf. [18]).

From an implementation-oriented point of view the terminological inference rules can be used for several purposes. The most obvious is to check the completeness of a terminological representation system. Furthermore, incomplete representation systems can be compared by pointing out which rules have been implemented and which have not. This is a considerable step forward compared to the current method of just listing examples of incompleteness. Finally, implementation strategies will differ with respect to the application of the inference rules. In general, all rules can be applied forward or backward. In a sense, the rules used in the normalization part of a normalize-compare algorithm are applied forward, whereas the rules used in the compare part are applied backward. We have shown how the completeness of a particular normalize-compare algorithm can be proved via the completeness of the inference rules. It remains to be seen whether similar proofs are feasible for normalize-compare algorithms applying less rules in a forward manner, i.e. doing less normalization. Furthermore, these algorithms have to be extended to handle rules in addition to definitions.

References

1. K.Apt, H.Blair, A.Walker, "Toward a Theory of Declarative Knowledge", in *Workshop on Foundations of Deductive Databases*, 1986
2. A.Avron, "Simple Consequence Relations", *Information and Computation* **92**, 105–139, 1991
3. J.Castaing, "A New Formalisation of Subsumption in Frame-based Representation System", in *KR'91*, 78–88
4. C.L. Chang, R.C.T. Lee (eds), *Symbolic Logic and Mechanical Theorem Proving*, New York: Academic Press, 1973
5. F.M. Donini, M. Lenzerini, D. Nardi, W. Nutt, "The Complexity of Concept Languages", *KR'91*, 151–162

6. F.M. Donini, M. Lenzerini, D. Nardi, W. Nutt, "Tractable Concept Languages" *IJCAI'91*, 458–463
7. J.Gallier, *Logic for Computer Science; Foundations of Automatic Theorem Proving*, New York: Harper and Row, 1986
8. J.Y.Girard, "Linear Logic", *Theoretical Computer Science* **50**, 1–102, 1987
9. J.Y.Girard, Y.Laffont, P.Taylor, *Proofs and Types*, Cambridge: Cambridge University Press, 1989
10. G. Sundholm, "Systems of Deduction", in D.Gabbay, F.Guenthner (eds), *Handbook of Philosophical Logic, Vol. I*, Dordrecht: Reidel, 133–188, 1983
11. C. Kindermann, "Class Instances in a Terminological Framework–an Experience Report", in H. Marburger (ed.), *Proc. of GWAI-90*, Berlin: Springer, 48–57, 1990
12. R.M. MacGregor, D. Brill, "Recognition Algorithms for the Loom Classifier", to appear in *AAAI-92*
13. B. Nebel, *Reasoning and Revision in Hybrid Representation Systems*, Berlin: Springer, 1990
14. R.Kowalski, M.H.Van Emden, "The Semantics of Predicate Logic as a Programming Language", *Journal of ACM* 23(4), 1976
15. J. Quantz, *Modeling and Reasoning with Defined Roles in* BACK, KIT-Report 84, Technische Universität Berlin, 1990
16. J. Quantz, "How to Fit Generalized Quantifiers into Terminological Logics", to appear in *ECAI-92*
17. J. Quantz, C. Kindermann *Implementation of the* BACK *System Version 4*, KIT Report 78, Technische Universität Berlin, 1990
18. V. Royer, J. Quantz, "Deriving Inference Rules for Terminological Logics: a Rewriting Approach into Sequent Calculi". KIT Report in preparation
19. J. Schmolze, D. Israel, "KL-ONE: Semantics and Classification", in *Research in Knowledge Representation and Natural Language Understanding*, BBN Annual Report 5421, 27–39 1983

This article was processed using the LaTeX macro package with LLNCS style

Linear Proofs and Linear Logic

Bertram Fronhöfer

Institute of Informatics
Technical University Munich
Arcisstr. 21
Postfach 20 24 20
D-8000 Munich 2

Tel.: +49-89-2105-2031
Net: fronhoef@informatik.tu-muenchen.dbp.de
Fax: +49-89-526502

Abstract

In [3] a modification of the Connection Method [2] called Linear Proofs was introduced which constituted a new logical approach to plan generation. Inspired by this idea in [7] a similar approach based on Linear Logic was presented. The present paper analyses the relationship of these two approaches and shows to which extent they are equivalent and where they differ.

Keywords

Linear Proofs, Linear Logic, Logics for Planning

Introduction

Reasoning is an integral part of intelligent systems. In reality any such system has to cope with a changing world. This certainly is true for humans, but also for robots, for intelligent database management systems, for intelligent computer interfaces, and so forth. Hence reasoning must account for such changes in order to be truly useful in practice.

Classical logic as the major formalism for reasoning primarily is meant to deal with a static world in which no changes occur. Although there are ways to deal with changes in classical logic such as the situational calculus [8], or extensions like modal (or temporal) logic on the other hand, these have a number of drawbacks that have prevented them so far from becoming the obvious choice for the formalization of reasoning about changes. For instance, the situational calculus faces the well-known *frame problem* for which no satisfactory solution has been found.

In [3] a new approach—so-called *linear proofs*—to integrating actions into logic has been proposed. Inspired by this work other researchers reconstructed this approach on the basis of *linear logic* [7].

In the present paper we attempt to give translations between these approaches and discuss the common scope of both formalizations and their differences.

The plan of the paper is as follows:

In section 1 we survey the concept of linear proofs, while in section 2 we survey the reconstruction of linear proofs with Linear Logic as presented in [7]. Linear proofs are based on the concept of matrices, while Linear logic is based on sequent systems. In section 3 we examine general requirements for the comparison of matrices and sequent-derivations to be carried out and we define some fundamental concepts. Basically, the language to be used will be so-called Horn bundles—"Horn clauses" with conjunctive heads—of propositional logic. Since in sequent derivations various occurrences of the same sequent may occur, we introduce a general conception for distinguishing different occurrences of formulae, sequents, etc. (This is necessary to take into account occurrences in matrices as well.) In section 4 we define a generalized matrix representation of proofs which will later serve as a 'meta language'. The important point is that we base matrices on a multiset idea—sets of occurrences of elements instead of just sets of elements—which enables us to represent sequent derivations more faithfully in matrices by allowing to distinguish also different occurrences of columns (clauses) of a matrix as we are able to distinguish different occurrences of sequents in a sequent derivation. In section 5 we show that normal forms exist for these generalised "occurrence-based" matrices. Roughly spoken, we can transform matrices/proofs into a more 'tree-like' shape or alternatively such that extensive reuse of those subproofs is made of which occur various copies (thus reducing the size of proofs/matrices). "Linear matrix proofs" are the special case of matrices which satisfy both requirements. In section 6 we finally define translations between matrices and sequent derivations. In section 7 we point out some properties in matrices and sequent derivations which correspond with respect to these translations: to contraction steps in sequent derivations correspond in a matrix multiple involvements of literals in connections; to different occurrences of columns (clauses) in a matrix corresponds multiple usage of the corresponding sequents in a derivation. We show that "Linear matrix proofs" correspond to sequent derivations which are both contraction free and make no multiple use of unit sequents (i.e. sequents with empty left sides). On the bases of these results section 8 finally focuses on the comparison of linear proofs and the linear logic approach by means of considering both of them as special cases of our generalised matrices. *In subsection 8.3 we come to the conclusion that linear proofs are equally distant from both classical logic and linear logic, because they can be obtained from both of these logics by imposing basically the same requirement on proofs.*

1 Linear Proofs

A basic principle of the linear proofs approach is to view plan generation as the task of proving that a goal situation can be deduced from an initial situation and from the set of rules describing the actions. In order to present this approach, let us look at the following example of a plan generating proof, taken from [3], where a block b is moved from the top of a block a down on the table. It is specified in the following way:

The **initial situation** will be given by the formula

$$Sit(s) \wedge T(a) \wedge O(b,a) \wedge C(b) \wedge E \qquad (INIT)$$

which means, that we are in a situation s, denoted by $Sit(s)$, where a is on the table: $T(a)$, block b is on top of a : $O(b,a)$, the top of b is clear: $C(b)$ and the robot's hand is empty: E.

We need two **actions** to lift a block and to put a block down, which are specified by the following **rules**:

$$\forall w, x, y\, \exists l : Sit(w) \wedge O(x,y) \wedge C(x) \wedge E \longrightarrow Sit(l) \wedge H(x) \wedge C(y) \qquad (R_1)$$
$$\forall w', v\ \ \exists p : Sit(w') \wedge H(v) \longrightarrow Sit(p) \wedge T(v) \wedge E \qquad (R_2)$$

(R_1) can be read: for any situation w, where a block x is on a block y, the top of x is clear and the robot's hand is empty, exists a situation l in which the robot holds block x in his hand and the top of block y is clear.

(R_2) can be read: for any situation w' where the robot holds block v, exists a situation p in which v is on the table and the robot's hand is empty.

The **goal** is stated by the formula

$$\exists z : Sit(z) \wedge T(b) \wedge T(a) \wedge E \qquad (GOAL)$$

which says that we are asking for a situation z in which the blocks a and b are both on the table and the robot's hand is empty.

With our approach, a plan which brings about the desired situation will be extracted from a proof of the goal formula from the given situation and the known rules, i.e. we must prove the formula $INIT \wedge R_1 \wedge R_2 \longrightarrow GOAL$ or more explicitly

$$\begin{aligned} & Sit(s) \wedge T(a) \wedge O(b,a) \wedge C(b) \wedge E \\ & \wedge (\forall w, x, y\, \exists l : Sit(w) \wedge O(x,y) \wedge C(x) \wedge E \longrightarrow Sit(l) \wedge H(x) \wedge C(y)) \\ & \wedge (\forall w', v\ \ \exists p : Sit(w') \wedge H(v) \longrightarrow Sit(p) \wedge T(v) \wedge E) \\ & \longrightarrow \exists z : Sit(z) \wedge T(a) \wedge T(b) \wedge E \end{aligned}$$

A proof by means of the **Connection Method** is given by the following matrix. The set of bows represents the spanning set of connections which makes the matrix complementary[1].

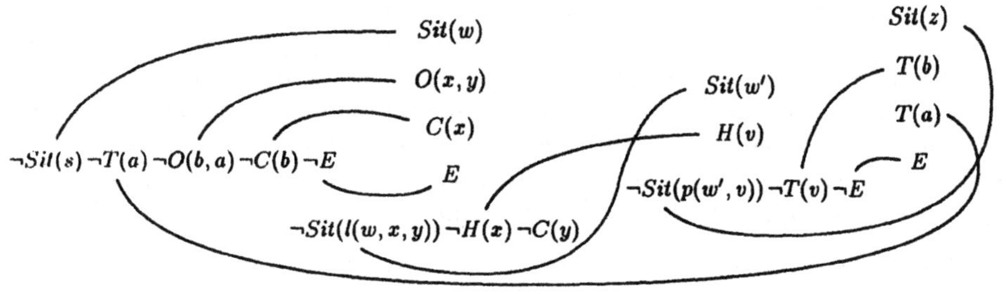

Explanations:

The horizontal row of atoms (or literals) on the left denotes the initial situation, the two towers in the middle denote the two applied rules—the left one picks up the block b and the right one puts the block b down on the table—and finally, the column of atoms on the right describes the goal situation.

In this proof the value of the variable z represents the plan we were looking for. It is the term $p(l(s,b,a),b))$ which is substantially built from the Skolem functions for the variables l and p in the rules. (We used the same symbol both for the variable and its corresponding Skolem function.) This term must be read: first lift block b from top of block a and then put block b down on the table. ■

[1] We will provide exact definitions of the concepts of the connection method in section 4, or consult [3] for a detailed presentation of the proof of this example.

Of course, we cannot expect every proof to yield a correct plan—since our atoms are time independent, they can be reused again and again, although they should no longer be "valid" after the execution of certain actions—but a notable feature of the proof above is that every instance of a literal is used at most once. Proofs of that kind were called **linear proofs** in [3] and it is claimed that this kind of linearity is the necessary restriction which must be imposed on proofs in order to generate correct plans.

2 Planning with Sequents

In this section we will recapitulate the approach presented in [7].

Since the issues which we will discuss in the following don't involve first-order features—we can assume our actions to be already completely instantiated—we will limit ourselves to propositional logic.

A planning problem is specified in [7] by two sets of axioms

- *current state axioms*: A set of "unit sequents" being of the form $\vdash A$, where A is a propositional variable.
- *transition axioms*: A set of sequents of the form $A_1, \ldots, A_n \vdash B_1 \otimes \ldots \otimes B_m$, where $A_1, \ldots, A_n, B_1, \ldots, B_m$ are propositional variables.

Differently from [7] we will use here ***goal sequents*** which are of the form $G_1, \ldots, G_k \vdash$ and the ***planning problem*** will be specified as the task to derive the empty sequent $\vdash$ from the current state axioms, the transition axioms and the goal sequent. (But see remark on page 5.)

There are several equivalent sequent systems which define classical logic. In absence of contraction and weakening—as in linear logic—nonequivalent systems result, which give rise to different groups of connectives. We will use the following rules of the so-called multiplicative group of the sequent system for linear logic: (The resulting derivational system will be denoted by S_L)

(axiom) $\quad \underline{A} \vdash \underline{A}$ for A atomic $\qquad$ *(cut)* $\quad \dfrac{\Gamma \vdash \underline{A} \qquad \Delta, \underline{A} \vdash B}{\Gamma, \Delta \vdash B}$

(exchange) $\quad \dfrac{\Gamma, \underline{A}, \underline{B}, \Delta \vdash C}{\Gamma, \underline{B}, \underline{A}, \Delta \vdash C}$ $\qquad$ $\mathcal{L}\text{-}\otimes$ $\quad \dfrac{\Gamma, \underline{A}, \underline{B} \vdash C}{\Gamma, \underline{A} \otimes \underline{B} \vdash C}$

Note that in [7] further rules are presented, for instance $\mathcal{R}\text{-}\otimes$ (right-introduction of $\otimes$), but we can restrict ourselves to derivations which make only so-called proper cuts, i.e. cuts with given axioms (see [7, 3.3, page 9]).

From the point of view of plan generation the absence of $\mathcal{R}\text{-}\otimes$ and the restriction to proper cuts sequentialises subplans which otherwise could be "executed" in parallel and precludes the composition of plans from subplans generated before. Since the applied actions and the resulting goal situation are nevertheless the same, we consider this limitation a negligible one for the present comparative purpose.

In addition we will make use of derivation rules which don't belong to linear logic, namely contraction and weakening. We denote by S_{LE} the derivational system resulting from adding contraction and weakening to S_L.

$$(weakening)\quad \frac{\Gamma \vdash B}{\Gamma, \underline{A} \vdash B} \qquad (contraction)\quad \frac{\Gamma, \underline{A}, \underline{A}, \Delta \vdash B}{\Gamma, \underline{A}, \Delta \vdash B}$$

Moreover, a meta predicate AU is introduced in [7] which keeps track of those facts from the initial situation which were used in the proof. Formally, it is defined:

- A current state axiom $\vdash F$ is a proof $\mathcal{D}$ of itself with $AU(\mathcal{D}) = \{F\}$.
- If $\mathcal{D}$ and $\mathcal{E}$ are proofs of sequents S and T respectively, such that $AU(\mathcal{D}) \cap AU(\mathcal{E}) = \emptyset$, then an application of the cut rule with premises S and T and conclusion S' yields a proof $\mathcal{D}'$ of S' with $AU(\mathcal{D}') = AU(\mathcal{D}) \cup AU(\mathcal{E})$.
- If $\mathcal{D}$ is a proof of a sequent S then the application of a $\mathcal{L}\text{-}\otimes$ rule or of the exchange rule with premise S and conclusion S' yields a proof $\mathcal{D}'$ of S' with $AU(\mathcal{D}') = AU(\mathcal{D})$. ∎

Remark:

Note that in [7] the presentation is more general: They treat the case of deriving an arbitrary sequent of the form

$$L_1, \ldots, L_n \vdash R_1 \otimes \ldots \otimes R_m$$

They show that a derivation $\mathcal{D}$ of such a sequent with $AU(\mathcal{D}) = \{F_1, \ldots, F_l\}$ can be transformed into a derivation $\mathcal{D}'$ of

$$F_1, \ldots, F_l, L_1, \ldots, L_n \vdash R_1 \otimes \ldots \otimes R_m$$

with $AU(\mathcal{D}') = \emptyset$. Such a derivation is considered "correct" and the derived sequent is called a *formal action* iff $F_1, \ldots, F_l, L_1, \ldots, L_n$ contains no duplicates and is a subset of the current state axioms. (Note that we identified a literal F with the unit sequent $\vdash F$.) This definition can be respecified as the task of constructing a derivation $\mathcal{D}$ of the empty sequent from the current state axioms, the transition axioms and the goal sequent $R_1, \ldots, R_m \vdash$ such that the set $AU(\mathcal{D})$ is a subset of the current state axioms. Thus we reduced the more general case of deriving arbitrary formal actions, being dealt with in [7], to the case of deriving empty sequents.

The requirement of empty intersection in case of $\mathcal{L}\text{-}\otimes$ and cut during the construction of the set $AU(\mathcal{D})$ guarantees that each current state axiom is used at most once in a derivation. ∎

Let us terminate this section with a presentation of example from section 1, page 3, as a sequent derivation:

We would have the current state axioms (we omitted the *Sit*-literals, since they were also missing in [7]):

$$\vdash T(a),\ \vdash ,O(b,a),\ \vdash C(b),\ \vdash E$$

and the (already instantiated) transition axioms:

$$O(b,a), C(b), E \vdash H(b) \otimes C(a)$$

$$H(b) \vdash T(b) \otimes E$$

and the goal sequent: $T(b), E, T(a) \vdash$

We get the following derivation:

$$\dfrac{O(b,a), C(b), E \vdash H(b) \otimes C(a) \qquad \dfrac{\dfrac{H(b) \vdash T(b) \otimes E \qquad \dfrac{T(b), E, T(a) \vdash}{T(b) \otimes E, T(a) \vdash}}{H(b), T(a) \vdash}}{H(b) \otimes C(a), T(a) \vdash}}{O(b,a), C(b), E, T(a) \vdash}$$

The empty sequent is now trivially derived through consecutive cuts with the current state axioms.

3 Preparatory Remarks

3.1 Occurrences

For comparing the two approaches presented in sections 1 and 2, we need a framework which is comprehensive enough to represent proofs of both approaches.

When speaking about derivations, proofs, formulae, literals, etc. we have to distinguish different occurrences of these objects. If matrices are defined as recursive sets (as in [2]) the distinction of different uses of a formula in a proof is not possible. For this reason we will define matrices here as recursive sets of *occurrences* of formulae in order to get a matrix notation which captures more structure of derivations. E.g. we want to express in the matrices whether a certain subproof has been used twice in a proof or whether a copy of it has been constructed. Moreover, when conjunctions and disjunctions of formulae are represented as sets, multiple occurrences of a formula are reduced to a single one. This is all right for classical logic where conjunction and disjunction are idempotent, but it is inconvenient for e.g. linear logic, where a non-idempotent conjunction and disjunction exist.

For these reasons we have to distinguish between different occurrences of the *objects* (matrices, formulae, derivations, etc.) which we want to deal with. *Formally*, this can be achieved by means of an infinite alphabet A of *occurrence symbols* and by labelling each literal with a symbol from A, i.e. we define a mapping ω from (a suitable subset of) A into the set of literals. This function can be represented as a set of pairs (L, s) consisting of a literal L and an occurrence symbol $s \in A$. When building our objects from the labelled literals obtained via ω, this mapping assures the intuitively clear requirement that no occurrence symbol may be used more than once in all the objects we deal with in the context of a proof, a definition, etc. Occurrences of objects will be defined as different iff the composing occurrences of literals are.

However, we will not make explicit use of this heavy notation, but will *informally* denote different occurrences of objects by different upper indices, e.g. O^1 and O^2 will mean that they are *copies* of each other, i.e. that there is a bijection π from the occurrence symbols in O^1 to those in O^2 and such that by replacing all occurrence symbols s in O^1 by $\pi(s)$ we obtain O^2. *Creating a copy* O^2 of an object O^1 means to create a bijection π from the occurrence symbols in O^1 to a new set of occurrence symbols and then to take an identical copy of O^1 in which we replace the occurrence symbols s in O^1 by $\pi(s)$. (Formally this corresponds to the composition of the mapping ω restricted to the occurrence symbols of O^1 and π, i.e. $\omega|_{O^2} := \omega|_{O^1} \circ \pi$, which extends the mapping ω to the set of occurrence symbols in O^2.) In the following we will also write O^1 to emphasize that we are referring to a particular occurrence of the object O. (Note that $\neg L^1$ is a particular occurrence of $\neg L$ and if we have L^1 and K^1 in the same object we just express that we refer to a particular occurrence of L and to a particular occurrence of K.) Finally, we will make use of two alphabets $1, 2, 3, \ldots$ and $\mathbf{1}, \mathbf{2}, \mathbf{3}, \ldots$ to refer to occurrences in matrices and in sequent derivations respectively.

3.2 Horn Bundles

Since the issues which we will discuss in the following don't involve first-order features, we will limit ourselves to propositional logic. We are just carrying out theoretical investigations where we can assume that all first order formulae are already suitably Skolemised and instantiated.

(In [7] only complete instantiations of transition axioms are considered. On the other hand, after having transformed formulae into matrices the only first-order features which are left are the uninstantiated variables. If we only consider complementary matrices and adopt the convention that the right side of an implication (or of a sequent) may not contain variables which are not already in the left side, we will have a ground instance.)

As we will not tackle here issues of disjunctive planning, the general shape of the actions in our planning problems—both in [3] and in [7]—is that a conjunction of facts is implied by another conjunction of facts, i.e.

$$A_1 \wedge \ldots \wedge A_n \longrightarrow B_1 \wedge \ldots \wedge B_m$$

resp. in the linear logic approach it is denoted by

$$A_1, \ldots, A_n \vdash B_1 \otimes \ldots \otimes B_m$$

(If the tail is empty, we speak about facts and if the head is empty, we are given a goal.) The difference between these formulae and Horn clauses lies in the conjunctive heads. Due to this similarity we will call such formulae ***Horn bundles***. In classical logic Horn bundles can be transformed into a conjunction (or a set) of Horn clauses (with identical tails). This means that Linear Proofs are basically Horn clause proofs. This is not possible with linear logic, where we have to distinguish the formula $H_1 \otimes H_2 \leftarrow T$ from the pair of formulas $H_1 \leftarrow T$ and $H_2 \leftarrow T$. This agrees with the intuition that all three formulae represent different actions.

It is well-known that Horn clause proofs can be displayed as proof trees by just copying multiply used subproofs. This result could be lifted to Horn bundles in a straightforward manner, e.g. by transforming Horn bundles into sets of Horn clauses. Apart from the impossibility to follow this way in linear logic, if we proceed this way we will get linear proofs only for trivial cases. E.g. after this transformation, the matrix for the planning example in section 1, page 3, would no longer be a linear proof.

Therefore, we adopt the convention that during the translation of formulae to matrices no use is made of distributive laws: Each Horn bundle $A_1 \wedge \ldots \wedge A_n \longrightarrow B_1 \wedge \ldots \wedge B_m$ yields exactly one column: $\{A_1, \ldots, A_n\}, \neg B_1, \ldots, \neg B_m$. Sets of Horn clauses yield matrices of nesting 2, while sets of Horn bundles yield matrices of nesting 3. In addition we will not exploit idempotency of classical conjunction and disjunction, i.e. if two of the A_i (or of the B_i resp.) are identical we get two different occurrences of literals. However, we assume that a tailless Horn bundle $H_1 \wedge \ldots \wedge H_n$ is translated into n unit columns $\{H_1\}, \ldots, \{H_n\}$.

4 Matrices

In this section we will define generalised matrices of Horn bundles and prove some basic properties about their structure.[2]

Definition 1 (complementary matrix)
A matrix *is a recursive set of* occurrences *of sets of ... of* occurrences *of literals.*[3]

A pair of literals (L_1, L_2) *is called* complementary *if there is a propositional variable* P *and* $L_1 = P$ *and* $L_2 = \neg P$. L_1 *will be called a* positive *literal, while* L_2 *will be called a* negative *literal.*

A connection *is a pair of occurrences of complementary literals* L^1 *and* K^1, *denoted by* (L^1, K^1) *or* (K^1, L^1). *(Note that different occurrences of the literals* L *and* K *form a different connection.)*

A path *through a matrix* $\mathcal{M} = \{C_1, \ldots, C_n\}$ *is a set of occurrences of literals which is obtained as the (necessarily disjoint) union of paths through* n *matrices* $\mathcal{M}_1, \ldots, \mathcal{M}_n$ *with* $\mathcal{M}_i \in C_i$. *If a* $\mathcal{M}_i$ *is just a literal* L, *then the respective path is* $\{L\}$.

A complementary path *is a path which contains a connection.*

A set of connections $\mathfrak{S}$ *is called* spanning *for a matrix* $\mathcal{M}$ *iff each path through* $\mathcal{M}$ *contains a connection from* $\mathfrak{S}$.

A complementary matrix *is a pair* $(\mathcal{M}, \mathfrak{S})$ *consisting of a matrix* $\mathcal{M}$ *and a spanning set of connections* $\mathfrak{S}$.

Matrices are a two dimensional representation of (negated) formulae in conjunctive $\wedge$- $\vee$ -form. For formulae in clausal form—and thus including Horn clauses—we get matrices of nesting 2: a set of occurrences of sets of occurrences of literals, which will be depicted as a *row* of *columns* of literals. (Matrices of nesting 2 will be said to be in *clausal normal form*.) For Horn bundles—remember that we don't transform them

[2] They are on the one hand quite familiar properties about Horn clauses which generalise without problems to Horn bundles, respectively on the other hand they are basic properties of matrices which also generalise easily to our "occurrence-based" matrices. Therefore we removed the proofs of the lemmata due to restrictions of space. For the same reason only proofs ideas are often given in the following sections. (The detailed proofs can be found in the technical report [4].)

[3] Speaking of a set of occurrences of elements shall mean that we are able to distinguish the different occurrences, i.e. that they (may) have different properties by which they can be identified. This is in contrast to multisets—generally defined as a mapping of a set into the natural numbers—where we just know that a certain element a occurs n times, but we don't have n distinguishable objects a which are only equivalent with respect to certain features.

to clausal normal form—we get matrices of nesting 3: a set of columns of occurrences of sets of occurrences of literals. (However, only the element of a column which corresponds to the conjunctive *head* of the Horn bundle is a set of (occurrences of) literals. These literals will be called *head literals*.) Columns which consist of a single literal are called *unit columns*.

We say that a set or an element $\mathcal{N}$ is *"in"* a set $\mathcal{M}$ iff there is a sequence of sets $\mathcal{N}_1, \ldots, \mathcal{N}_n$ such that $\mathcal{N} \in \mathcal{N}_1$, $\mathcal{N}_i \in \mathcal{N}_{i+1}$ and $\mathcal{N}_n \in \mathcal{M}$. I.e. "in" is the transitive closure of $\in$. We also say that $\mathcal{N}$ is *"contained"* in $\mathcal{M}$.

Definition 2
Given a matrix $\mathcal{M} = \{\mathcal{C}_1, \ldots, \mathcal{C}_n\}$ and a set of connections $\mathfrak{S}$ and given (possibly empty) subsets $\mathcal{D}_i \subset \mathcal{C}_i$ and $\mathcal{N} := \{\mathcal{D}_1, \ldots, \mathcal{D}_n\}$.

Then $\mathfrak{S}|_{\mathcal{N}}$, the restriction of $\mathfrak{S}$ to $\mathcal{N}$, *is the subset of $\mathfrak{S}$ which consists exactly of those connections (K^1, L^1) with K^1 and L^1 in $\mathcal{N}$.*

Analogously we can define for a path $\mathfrak{p}$ the restriction of $\mathfrak{p}$ to $\mathcal{N}$, *denoted as $\mathfrak{p}|_{\mathcal{N}}$, as the set of (occurrences of) literals from $\mathfrak{p}$ which are in $\mathcal{N}$.*

Definition 3
A complementary matrix $(\mathcal{M}, \mathfrak{S})$ is called minimal *iff the set $\mathfrak{S}$ is a minimal spanning set of connections for $\mathcal{M}$ and each column of $\mathcal{M}$ contains at least one literal which is involved in a connection from $\mathfrak{S}$.*

A complementary matrix $(\mathcal{M}, \mathfrak{S})$ is called connected *iff $\mathcal{M}$ is not the union of two nonempty submatrices $\mathcal{M}_1 \subset \mathcal{M}$ and $\mathcal{M}_2 \subset \mathcal{M}$, such that there is no connection $(K^1, L^1) \in \mathfrak{S}$ with K^1 in $\mathcal{M}_1$ and L^1 in $\mathcal{M}_2$.*

Lemma 1
A minimal complementary matrix $(\mathcal{M}, \mathfrak{S})$ is connected.

Lemma 2
If a complementary matrix $(\mathcal{M}, \mathfrak{S})$, $\mathcal{M} = \{\mathcal{C}_1, \ldots, \mathcal{C}_n\}$, has a partial path $\mathfrak{p}$ through the submatrix $\{\mathcal{C}_1, \ldots, \mathcal{C}_k\}$ $(0 < k < n)$, which contains no connection from $\mathfrak{S}$ and if there is no connection between literals in $\mathfrak{p}$ and literals in $\mathcal{C}_{k+1}, \ldots, \mathcal{C}_n$, then $(\mathcal{M}, \mathfrak{S})$ is not minimal.

Lemma 3
In a minimal complementary matrix of Horn bundles exists exactly one negative column, i.e. which is a set of occurrences of negative literals.

Lemma 4
Each negative literal in a minimal complementary matrix of Horn bundles is involved in exactly one connection.

Definition 4 (subproof)
Given a complementary matrix $(\mathcal{M}, \mathfrak{S})$.
A subproof for a negative literal $\neg F^1$ in $\mathcal{M}$ is a pair $(\mathcal{M}_s^1, \mathfrak{S}_s^1)$ with $\mathcal{M}_s^1 \subset \mathcal{M}$, $\mathfrak{S}_s^1 \subset \mathfrak{S}$ together with a literal F^1 in $\mathcal{M}_s^1$ and a connection $(F^1, \neg F^1) \in \mathfrak{S}$ such that

$$(\mathcal{M}_s^1 \cup \{\{\neg F^1\}\}, \mathfrak{S}_s^1 \cup \{(F^1, \neg F^1)\})$$

is a minimal complementary matrix.

Two subproofs $(\mathcal{M}_1^1, \mathfrak{S}_1^1)$ *and* $(\mathcal{M}_2^2, \mathfrak{S}_2^2)$ *are called* equivalent subproofs *if they are copies of each other. If the bijection on the occurrence symbols is the identity, then they are called* identical subproofs.

Lemma 5
For each negative literal $\neg F^1$ *in a complementary matrix* $(\mathcal{M}, \mathfrak{S})$ *of Horn bundles exists a subproof.*

Lemma 6
Given a minimal complementary matrix $(\mathcal{M}, \mathfrak{S})$.
For each negative literal $\neg F^1$ *in* $\mathcal{M}$ *exists a unique subproof* $(\mathcal{M}_s, \mathfrak{S}_s)$ *of the following structure:*
There is a column $C^1 = \{\{H_1^1, \dots, H_n^1\}, \neg T_1^1, \dots, \neg T_m^1\}$ *and a connection* $(\neg F^1, H_k^1) \in \mathfrak{S}$ *and subproofs* $(\mathcal{M}_1, \mathfrak{S}_1), \dots, (\mathcal{M}_m, \mathfrak{S}_m)$ *of the* $\neg T_1^1, \dots, \neg T_m^1$ *respectively together with connections* $(\neg T_i^1, T_i^1)$ *with* T_i^1 *in* $\mathcal{M}_i$.
Then $\mathcal{M}_s = \bigcup_{i=1}^m \mathcal{M}_i \cup \{C\}$ *and* $\mathfrak{S}_s = \bigcup_{i=1}^m \mathfrak{S}_i \cup \bigcup_{i=1}^m \{(\neg T_i^1, T_i^1)\}$
In addition $\mathfrak{S}_s = \mathfrak{S}|_{\mathcal{M}_s}$ *and* $\mathfrak{S}_i = \mathfrak{S}|_{\mathcal{M}_i}$.

Definition 5
With the notation of lemma 6 above the column C^1 *is called the* top column *of the subproof* $(\mathcal{M}_s, \mathfrak{S}_s)$ *of* $\neg F^1$ *and the literal* F^1 *is called the* pivot literal.

5 Normal Forms of Matrices

It is well-known that various different proofs may exist for the same theorem. When proofs are represented as matrices some of the differences disappear—e.g. different orderings of derivation steps. However, other differences still remain and matrices can be required to fulfil certain "normalizing" conditions.

As already mentioned in section 3.1, Horn clause proofs can be displayed as proof trees by just copying multiply used subproofs. For Horn bundles we define below the somewhat similar concept of weakly linear. Furthermore, we take into account the possibility to make multiple use of subproofs via factoring (or less directly via lemmas) by introducing the concept of completely reduced.

Definition 6
A minimal complementary matrix $(\mathcal{M}, \mathfrak{S})$ *is called* weakly linear *iff no (occurrence of a) literal is involved in more than one connection of* $\mathfrak{S}$.

A minimal complementary matrix is called completely reduced *iff it does not contain two different equivalent subproofs.*

A minimal complementary matrix is called linear *iff it is both weakly linear and completely reduced.*

Lemma 7
For every minimal complementary matrix $(\mathcal{M}^1, \mathfrak{S}^1)$ *of Horn bundles exists a weakly linear (minimal complementary) matrix of copies of columns of* $\mathcal{M}^1$.

Proof idea:
Due to lemma 4 every negative literal is involved in exactly one connection. Therefore we only have to treat the case that a positive literal L^1 is involved in (at least) two

connections $(L^1, \neg L^2)$ and $(L^1, \neg L^3)$. Weak linearity can be achieved by adding a copy of the common subproof of $\neg L^2$ and $\neg L^3$ to the matrix and by exchanging one of the connections, say $(L^1, \neg L^3)$, through a a connection from $\neg L^3$ into the new submatrix.

Lemma 8
For every minimal complementary matrix $(\mathcal{M}, \mathfrak{S})$ of Horn bundles exists a completely reduced (minimal complementary) matrix $(\mathcal{M}_s, \mathfrak{S}_s)$ with $\mathcal{M}_s \subset \mathcal{M}$.

Proof idea:
When ever there are subproofs which are copies of each other, we can replace all connections into the top column of the first one by connections into the top column of the second one, and afterwards delete the first subproof. This process must be carried out iteratively starting with the smallest subproofs consisting of unit columns.

We are are primarily interested in matrices which are both weakly linear and completely reduced, because these proofs correspond to the "real" linear ones. The combination of these two conditions, however, can still be achieved by weaker requirements.

Definition 7
A complementary matrix is called basically reduced *if it does not contain two unit columns which are copies of each other.*

Theorem 9
A weakly linear (minimal complementary) matrix of Horn bundles, which is basically reduced is already completely reduced.

Proof idea:
The proof goes by showing that in weakly linear matrices (different) copies of subproofs entail the existence of different copies of unit columns.

Theorem 10
A minimal complementary matrix is linear *iff it is both weakly linear and basically reduced.*

Proof: Immediately from theorem 9 and the definition of linear matrix. ■

6 Translations

Now let's look at sequent systems.
In order to compare matrices and sequent derivations, we have to determine what corresponds in sequent derivations to the concepts of weakly linear and completely/basically reduced. This can be found out best by defining translations from matrices into sequent derivations and vice versa.

We first give the definition of these translations and prove afterwards the necessary lemmata which guarantee their existence and their correctness.

Definition 8 (translation of matrices)
For a minimal complementary matrix $(\mathcal{M}, \mathfrak{S})$ this translation is defined as follows:

We start with translating the negative column (see lemma 3)
$C_1 := \{\neg A_1^1, \ldots, \neg A_n^1\}$ *in $\mathcal{M}$ to the sequent* $A_1^1, \ldots, A_n^1 \vdash$.

Next we create for each connection $(\neg A_i^1, A_i^2) \in \mathfrak{S}$ *a* link $[A_i^1, A_i^2]$ *and the connection* $(\neg A_i^1, A_i^2)$ *is* marked *somehow as having "created a link".*

More generally we assume that we have already translated the columns $\mathcal{C}_1, \ldots, \mathcal{C}_f$ *(with* $\mathcal{M} := \{\mathcal{C}_1, \ldots, \mathcal{C}_g\}$ *and* $f < g$*) and that*

$$T_1^1, \ldots, T_m^1 \vdash$$

is the last derived sequent of the derivation under construction. In addition we assume for each literal T_i^1 *a link* $[T_i^1, T_i^2]$ *to a literal* T_i^2 *in the matrix.*

If two literals T_k^1 *and* T_l^1 *with* $(k \neq l)$ *in the sequent are linked to the same literal* T_h^2 *in the matrix, we apply a contraction to the sequent which means a deletion of* T_l^1*. (The double line below means a series of exchanges which must precede the contraction.)*

$$\frac{T_1^1, \ldots, T_k^1, \ldots, T_l^1, \ldots, T_m^1 \vdash}{T_1^3, \ldots, T_k^3, \ldots, T_m^3 \vdash}$$

For the two links $[T_k^1, T_h^2]$ *and* $[T_l^1, T_h^2]$ *a new link is created* $[T_k^3, T_h^2]$*, while the other links* $[T_i^1, T_i^2]$ $(i \neq k, l)$ *are "passed on" creating thus the new links* $[T_i^3, T_i^2]$.

After the execution of all possible contractions, we can now assume a last derived sequent

$$T_1^1, \ldots, T_m^1 \vdash$$

in our derivation such that each of the literals $T_1^1, \ldots, T_m^1$ *is linked to a different (occurrence of a) literal in the matrix.*

(*) *Next we select a column*

$$\{\{G_1^4, \ldots, G_n^4\}, \neg S_1^4, \ldots, \neg S_{n'}^4\}$$

from the matrix, which has no unmarked connections starting from its head literals, i.e. for all connections in $\mathfrak{S}$ *involving literals from* $G_1^4, \ldots, G_n^4$ *we have already created links to the sequent derivation.*

(That such a column always exists is proved in the lemma 12 below.)

Let's assume that we have links $[T_1^1, G_1^4], \ldots, [T_g^1, G_g^4]$ $(g \leq n)$. *If* $g \neq n$ *we weaken iteratively the sequent*

$$T_1^1, \ldots, T_m^1 \vdash$$

to

$$G_{g+1}^2, \ldots, G_n^2, T_1^2, \ldots, T_m^2 \vdash$$

After applications of the exchange rule, we introduce iteratively $\otimes$ *operators which yields*

$$T_1^3 \otimes \ldots \otimes T_g^3 \otimes G_{g+1}^3 \otimes \ldots \otimes G_n^3, T_{g+1}^3, \ldots, T_m^3 \vdash$$

During these weakenings and $\otimes$ *introductions we create links between the new occurrences of the* T_i *and the respective* G_i *till we finally get links* $[T_i^3, G_i^4]$.

The last sequent is now "cut" with the sequent

$$S_1^4, \ldots, S_{n'}^4 \vdash G_1^4 \otimes \ldots \otimes G_n^4$$

and we obtain

$$S_1^5, \ldots, S_{n'}^5, T_{g+1}^5, \ldots, T_m^5 \vdash$$

Next we create links $[S_i^5, S_i^6]$ *for each connection* $(\neg S_i^4, S_i^6)$ *in* $\mathfrak{S}$ *and mark these connections as having created a link. For the links from* T_i^3 $(i = g{+}1, \ldots, m)$ *we create links from* T_i^5.

Lemma 11

The just defined translation of a minimal complementary matrix terminates and results in a sequent derivation of the empty sequent.

Proof: Trivial

Lemma 12

At point (*) *in the definition of the translation above always exists a column without unmarked connections from the head literals.*

Proof: After having already translated the columns $\mathcal{C}_1, \ldots, \mathcal{C}_f$ (with $\mathcal{M} := \{\mathcal{C}_1, \ldots, \mathcal{C}_g\}$ and $f < g$), we obtained

$$T_1^1, \ldots, T_m^1 \vdash$$

as the last derived sequent of the derivation under construction. In addition we assume for each literal T_i^1 a unique link $[T_i^1, T_i^2]$ to a literal T_i^2 in the matrix.

Assuming that the literals T_i^2 are pivot literals of subproofs $(\mathcal{M}_i, \mathfrak{S}_i)$ we have that the *remaining matrix* is $\mathcal{R} := \bigcup_{i=1}^m \mathcal{M}_i = \{\mathcal{C}_{f+1}, \ldots, \mathcal{C}_g\}$

and the set of unmarked connections is $\bigcup_{i=1}^m \mathfrak{S}_i = \mathfrak{S}|_{\mathcal{R}}$

This is rather evident for the initial case: $f = 1$.

In the iterative case we assume a top column (without unmarked connections)[4], say $\mathcal{C}_{f+1} := \{\{G_1, \ldots, G_n\}, \neg S_1, \ldots, \neg S_{n'}\}$, from one of the subproofs, say $(\mathcal{M}_1, \mathfrak{S}_1)$. Due to lemma 6 exist subproofs $(\mathcal{M}'_1, \mathfrak{S}'_1), \ldots, (\mathcal{M}'_{n'}, \mathfrak{S}'_{n'})$ of the $\neg S_1, \ldots, \neg S_{n'}$ respectively together with connections $(\neg S_i, S_i)$ with S_i in $\mathcal{M}'_i$.

We get $\mathcal{R} := \bigcup_{i=1}^m \mathcal{M}_i = \bigcup_{i=2}^m \mathcal{M}_i \cup \bigcup_{i=1}^{n'} \mathcal{M}'_i \cup \mathcal{C}_{f+1}$

i.e. the new remaining matrix $\mathcal{R}' := \bigcup_{i=2}^m \mathcal{M}_i \cup \bigcup_{i=1}^{n'} \mathcal{M}'_i = \{\mathcal{C}_{f+2}, \ldots, \mathcal{C}_g\}$

For the set of connections we get

$\mathfrak{S}|_{\mathcal{R}} = \bigcup_{i=1}^m \mathfrak{S}_i = \bigcup_{i=2}^m \mathfrak{S}_i \cup \bigcup_{i=1}^{n'} \mathfrak{S}'_i \cup \bigcup_{i=1}^{n'} \{(\neg S_i, S_i)\}$

and without the newly marked connections, i.e. the set of unmarked connections:

$\bigcup_{i=2}^m \mathfrak{S}_i \cup \bigcup_{i=1}^{n'} \mathfrak{S}'_i = \bigcup_{i=2}^m \mathfrak{S}|_{\mathcal{M}_i} \cup \bigcup_{i=1}^{n'} \mathfrak{S}|_{\mathcal{M}'_i} = \mathfrak{S}|_{\mathcal{R}'}$

If now a head literal in the top column of one of the subproofs $\mathcal{M}_i$ is involved in an unmarked connection, then this connection must be in one of the $\mathfrak{S}_i$. This means that the head literals of the top column of a maximal element among the $\mathcal{M}_i$ must not be involved in an unmarked connection and we have a column to select for being translated next. ■

Remark:

According to the way we created the links, each (occurrence of a) literal in the sequent derivation is linked to at most one (occurrence of a) literal in the matrix. ■

For the inverse translation we assume derivations with Horn bundles, which consist of applications of cut, weakening, contraction and left-introduction of $\otimes$.

---

[4] Since the negative goal column has no head literals (which consequently cannot be involved in any connection), it can be considered a column without unmarked connections.

Definition 9 (translation of sequent derivations)
We can define this translation as follows:

From lemma 13 we know that in the sequent derivation there must be a terminal node of the form $A_1^1, \ldots, A_n^1 \vdash$ *. (The node corresponding to the query.) This sequent is translated into the column* $\{\neg A_1^1, \ldots, \neg A_n^1\}$ *and we create links* $\langle A_i^1, \neg A_i^1 \rangle$*.*

Assuming now to have arrived during the translation process at a last sequent $T_1^1, \ldots, T_m^1 \vdash$ *of the derivation being translated with links* $\langle T_i^1, \neg T_i^2 \rangle$ *into the matrix for some of the* T_i^1*.*

If the next step is an exchange, *a* weakening *or a* left-introduction of $\otimes$, *we just pass on the links. In case of a* contraction *of, say* T_1^1 *and* T_2^1

$$\frac{T_1^1, T_2^1, \ldots, T_m^1 \vdash}{T_2^2, \ldots, T_m^2 \vdash}$$

We create for the two links $\langle T_1^1, \neg T_1^2 \rangle$ *and* $\langle T_2^1, \neg T_2^2 \rangle$ *the two new links* $\langle T_2^2, \neg T_1^2 \rangle$ *and* $\langle T_2^2. \neg T_2^2 \rangle\rangle$*. All other links are passed on.*

If the next step is a cut *with another sequent*

$$S_1^4, \ldots, S_h^4 \vdash G_1^4 \otimes \ldots \otimes G_n^4$$

and assuming now a last sequent $T_1^3 \otimes \ldots \otimes T_n^3, \ldots, T_m^3 \vdash$ *of the derivation being translated with*

$$T_1^3 \otimes \ldots \otimes T_n^3$$

equal to $G_1^4 \otimes \ldots \otimes G_n^4$ *we add the column* $\{\{G_1^4, \ldots, G_n^4\}, \neg S_1^4, \ldots, \neg S_h^4\}$ *to the matrix, create links* $\langle S_i^4, \neg S_i^4 \rangle$ $(i = 1, \ldots, h)$ *and for each link* $\langle T_i^3, \neg T_i^2 \rangle$ *with* $i \leq n$ *we create a connection* $(G_i^4, \neg T_i^2)$*.*

Since we assume a sequent derivation with only proper cuts, we have that the sequent

$$S_1^4, \ldots, S_h^4 \vdash G_1^4 \otimes \ldots \otimes G_n^4$$

is a terminal node of the sequent derivation tree.

Lemma 13
In every sequent derivation of Horn bundles resulting in the empty sequent exists at least one terminal node of the form $A_1^1, \ldots, A_n^1 \vdash$ *.*

Proof: In order to derive the empty sequent from Horn bundles

$$A_1, \ldots, A_n \vdash B_1 \otimes \ldots \otimes B_m$$

we need rules which reduce the number of literals in the left and right part of sequents. The exchange rule and the left introduction of $\otimes$ don't reduce and weakening increases. Only the cut rule has "reductive power"

On the left side we can cut and reduce with sequents

$$C_1, \ldots, C_{n'} \vdash D_1 \otimes \ldots \otimes D_{m'}$$

such that $n' + (n - m') < n$ and finally we need sequents with empty left side: $\vdash F$.

Analogously, to get an empty right side we must finally have sequents which add nothing, i.e. being of the form $\Gamma \vdash$. ∎

Lemma 14
The just defined translation of a sequent derivation of the empty sequent terminates and results in a minimal complementary matrix $\mathcal{M}$*.*

Proof: The negative literals in the matrix which are linked to the literals in the sequent at which we just arrived during the translation process are exactly the 'frontends' of those paths through the matrix which are not yet complementary due to connections in the submatrix constructed so far.

This is immediately clear for the initial sequent $A_1^1, \ldots, A_n^1 \vdash$. Iteratively it becomes clear if we look at the translation of cuts, where some paths—those for which we create connections $(G_i^4, \neg T_i^2)$—are made complementary, while the other continuations through $\neg S_i^4$ remain 'open' and are memorized in the links $\langle S_i^4, \neg S_i^4 \rangle$ (Note that links to the frontends of open paths never get lost during the translation of weakening, left-introduction of $\otimes$ and contraction.)

From the way how the matrix is constructed during the translation process follows as well that the matrix $\mathcal{M}$ is connected. For assume $\mathcal{M} = \mathcal{M}_1 \cup \mathcal{M}_2$ then one of the submatrices, say $\mathcal{M}_1$, must contain the negative column—there can be only one, because it is impossible to cut two negative columns. Then following the construction process of the matrix we would reach all columns of $\mathcal{M}$ while still remaining in $\mathcal{M}_1$. Consequently, $\mathcal{M}_2$ is empty. Since we only added columns 'by need' via a connection, we have no columns in $\mathcal{M}$ which are completely unconnected. During the translation process we created exactly one connection for each negative literal, consequently none of these connections is superfluous. This means that the matrix $\mathcal{M}$ is minimal ∎

Definition 10 (translation of sequent sub-derivations)
The translation of sequent derivations to matrices can also be restricted to subderivations starting at a single literal on the left side of a sequent.

We take an arbitrary sequent $T_1^1, \ldots, F^1, \ldots, T_m^1 \vdash$ containing a literal F and we start translating by first creating a matrix consisting of the single column $\{\neg F^1\}$. We also create a link $\langle F^1, \neg F^1 \rangle$.

From this starting point we now continue the translation process. In contrast to the translation of the whole derivation, we may now encounter derivation steps in which no literal with a link is involved. These steps will cause no change to the matrix and links are just propagated. The connected submatrix containing $\neg F^1$ is a subproof of $\neg F^1$.

Lemma 15
The translation process of a subderivation results in matrix $\mathcal{M}_s \cup \{\{\neg F^1\}\}$ together with a set of connections $\mathfrak{S}_s$ which contains a connection $(\neg F^1, F^1)$ such that $(\mathcal{M}_s, \mathfrak{S}_s \setminus \{(\neg F^1, F^1)\})$ is a subproof of $\neg F^1$.

Proof: Analogous to the proof of lemma 14. ∎

7 Correspondences

With the translations defined in section 6 we obtain analogues for the lemmata proved in section 5 for matrices.

Lemma 16
Weakly linear (minimal complementary) matrices correspond to contraction free sequent derivations.

Proof: During the translation of a matrix the introduction of a contraction

$$\frac{T_1^1, \ldots, T_k^1, \ldots, T_l^1, \ldots, T_m^1 \vdash}{T_1^3, \ldots, T_k^3, \ldots, T_m^3 \vdash}$$

in a sequent derivation presupposes two links $[T_k^1, T_h^3]$ and $[T_l^1, T_h^3]$ to the same literal T_h^3 in the matrix. Since each link is derived from a connection and different links from different connections, we must have had connections $(\neg T_k^1, T_h^3)$and $(\neg T_l^1, T_h^3)$ in the matrix as causes for the links above. This means that the literal T_h^3 is involved in two connections. Consequently, a sequent derivation which is not contraction free results from a matrix which is not weakly linear.

On the other hand, when translating a sequent derivation the creation of two connections $(K^1, \neg K^1)$ and $(K^1, \neg K^2)$ involving the same (occurrence of a) literal K^1, presupposes two links $\langle K^3, \neg K^1 \rangle$ and $\langle K^3, \neg K^2 \rangle$ from a literal K^3 in a sequent. Since the only possibility for the literal K^3 to get involved in two links is during a contraction, a matrix which is not weakly linear results form a sequent derivation which is not contraction free. ∎

We now get immediately:

Lemma 17

For each sequent derivation (of the empty sequent) with Horn bundles exists also a contraction free sequent derivation using instances of the same Horn bundles.

Proof: To obtain such a contraction free sequent derivation we translate the given sequent derivation into a matrix, transform the matrix into a weakly linear one (according to lemma 7) and by finally translating back into a now contraction free sequent derivation. ∎

The shape of a derivation starting at a certain literal F^1 in a sequent depends on the ordering of the derivation steps. Moreover, the derivation is loaded with the contexts around this literal, which are either just carried along, or may obfuscate the derivation starting at F^1 by intermingling it with other derivation steps. For this reason we define the concept of equivalent subderivation via the translation to matrices.

Definition 11

Two (occurrences of) literals F^1 and F^2 in the left parts of sequents in a sequent derivation have equivalent subproofs if subderivations starting at F^1 and F^2 have equivalent matrix proofs.

A sequent derivation $\mathcal{D}$ is completely reduced *iff there are no two different occurrences F^1 and F^2 which have equivalent subproofs in a matrix translation of the derivation $\mathcal{D}$.*

Lemma 18

For every sequent derivation (of the empty sequent) with Horn bundles exists a completely reduced derivation using occurrences of the same Horn bundles.

Proof: We translate the sequent derivation into a matrix, completely reduce it and translate back. ∎

Definition 12

A sequent derivation is called factually unique *if it contains at most one occurrence of each unit sequent.*

Lemma 19
A sequent derivation (of the empty sequent) with Horn bundles corresponds to a basically reduced matrix iff it is factually unique.

Proof: Follows immediately from the fact that during the translations of sequent derivations into matrices and vice versa a bijection is created between the sequents at the leaves of a sequent derivation and the columns of a matrix, which restricts to a bijection between unit sequents and unit columns. ■

Theorem 20
A contraction free sequent derivation (of the empty sequent) with Horn bundles which is factually unique, corresponds already to a linear minimal complementary matrix.

8 Comparison

In the preceding section we proved the following:

Theorem 21
Given a minimal complementary matrix $(\mathcal{M}, \mathfrak{S})$ *and a sequent derivation* $\mathcal{D}$ *in the system* S_{LE} *—both with Horn bundles—which are translations of each other, then* $(\mathcal{M}, \mathfrak{S})$ is weakly linear and basically reduced *iff* $\mathcal{D}$ is contraction free and factually unique.

We want to discuss next how this theorem relates to

- Linear Proofs in the sense of [3]
- the Linear Logic approach to planning in [7]

8.1 Linear Proofs

Linear Proofs (in the sense of [3]) differ from the presentation given so far in two points:

- While we defined matrices as recursive sets of occurrences of ... occurrences of literals, in the connection method matrices are defined as recursive sets of ... of literals. (Let us speak of *s-matrices* in case of the latter concept and of *o-matrices* in case of the matrices defined in definition 1, page 8.)
- In [3] the Horn bundles contained distinguished literals which kept track of the consecutive situations of a plan (and which were denoted by the predicate *Sit* in the example on page 3).

Basing everything on s-matrices has the disadvantage that certain applications which are typical examples of the use of linear logic cannot be done. For instance the plan to buy two packets of cigarettes (see [5]) each costing '*FF5*', would involve two occurrences of '*FF5*'. For this plan we get a weakly linear and basically reduced o-matrix, but no linear proof (based on s-matrices), because the two unit columns '*FF5*' would become one, but now with two connections.

However, we might ask whether a planning logic should be able to cope with those kinds of applications. It is true that traditional logic has been severely criticized as being not apt for planning, but even the badly reputed situational calculus has never been accused of having an unsatisfying semantics. On the other hand, aiming to cope with examples like the one above, which cannot be translated into e.g. situational calculus in a straightforward manner, cause a rupture with the 'semantical' tradition of planning.

Another consequence of s-matrices is that they are already implicitly completely reduced, because we cannot represent different occurrences of columns. This makes the condition of weak linearity a defining criterion for linear proofs and we have:

Theorem 22
A minimal complementary s-matrix is linear iff it is weakly linear.

Another consequence of s-matrices is that different non-equivalent subproofs for two occurrences of a literal $\neg F$ are precluded (Note that in s-matrices we may still have various occurrences of a literal, but only in different columns.). This restriction might prevent the generation of a plan for the towers of Hannoi problem, where the action of shifting a disk N from tower a to tower b would be applied various times. This means that the column corresponding to this action would be the top column of different subproofs—which results in different occurrences in o-matrices—but in s-matrices all of these different subproofs (except a single one) would be redundant and consequently would be cut off when creating a minimal complementary matrix.

This problem is remedied by the introduction of the situation literals, which by keeping track of the consecutive situations of a plan, assure that applications of actions at different points in a plan correspond to different instantiations of the action, and thus yielding different subproofs.

On the other hand, these situation literals have the drawback that they obligatorily impose a sequential order on the actions of which a plan is composed. (That our sequent derivations correspond also to sequential plans stems from the fact that we didn't allow right introductions of conjunctions, a restriction which we imposed just for keeping our translations simpler.)

8.2 Linear Logic

Due to the discarding of contraction linear logic disallows the multiple use of subderivations: If we derived sequents $\vdash F$ and $F, F, \Delta \vdash \Gamma$, we are not able to derive first $F, \Delta \vdash \Gamma$ by contraction and next to "cut" with $\vdash F$, but we have to "cut" immediately with $\vdash F$ yielding $F, \Delta \vdash \Gamma$ and then have to rederive $\vdash F$ and to cut again. This rederivation of $\vdash F$ is only possible if we are given a further set (or copies) of the sequents used in the first derivation of $\vdash F$. This rederivation, however, is avoided by recording the used facts in the AU predicate, which is required to contain no duplicates: *no fact must be contained there more often than at most once.* So theorem 21 specialises to:

Theorem 23
A Linear Logic derivation (of the empty sequent) with Horn bundles corresponds to a weakly linear minimal complementary o-matrix.

A Linear Logic derivation (of the empty sequent) with Horn bundles corresponds to a linear minimal complementary o-matrix, if it is factually unique.

If we look back at the translation of o-matrices into sequent derivations, we encounter various applications of weakening which became necessary whenever we wanted to translate columns which had unconnected literals in their head. (Such a columns is also found, for instance, in the example on page 3.) These unconnected literals tell us that the actions produced more facts than were consumed later by following actions, resp. more than were requested by the given goal. In other words, such unconnected literals would not appear if we had specified the goal sufficiently complete, i.e. the goal has to comprise all facts which are generated by the actions which compose the plan. However, this requirement is a tremendously difficult one from a practical point of view. It means that when specifying a planning problem we have to know already in advance the plan to be executed, because only after knowing this plan we will be able to tell exactly how the world will look like after its execution. Let us add, however, that this problem has not to be faced, if we are not concerned with goal-oriented planning, but with the problem of temporal projection, where the goal is not given, but is asked for. (It might be that the authors of [7] have this concept in mind when they speak about planning.)

An interesting extension of the linear logic approach to planning would be to specify the initial situation as a multiset of facts instead of requiring it to be a set. This would quite naturally allow to cope with the cigarettes example.

8.3 Some Final Subjective Remarks

We showed in section 5 that two alternative normal forms exist for minimal complementary o-matrices. For classical logic where conjunctions of formulae are usually identified with sets of formulae, the completely reduced normal form may seem to be a natural representation. This naturalness becomes at least obsolete if we go over to logics with non-idempotent connectives. However, even for classical logic this naturalness can be challenged from the point of view of proof construction as carried out by proof procedures: The proofs constructed by the algorithms given for the connection method can be represented quite naturally as o-matrices, and as long as no use of factorisation or lemmata is made, these o-matrices are weakly linear. This would suggest to consider weakly linear o-matrices as the normal concept which is to be implicitly assumed. (Note that this would also largely conform with the common representation of proofs in form of proof trees.) Adopting this point of view, we would urged us to impose the condition of being basically reduced as an additional constraint on the matrices. (Note the coincidence with common ways to proceed in theorem proving: e.g in current work on the implementation of a linear proof system it turned out that the main problem is not to obtain weakly linear proofs, but to assure basically reduced ones.)

Summarizing we can say: *Assuming implicitly that minimal complementary o-matrices are weakly linear, then a basically reduced matrix is already linear.*

How does this relate to linear logic? The question which should be answered is: What is a suitable consequence relation for linear logic? i.e. what shall it mean that a formula/sequent is a consequence of other formulas/sequents? When we look at the literature, several different consequence relations are found; e.g in [6] or two different consequence relations in [1].

The definition given in [7]—and which we called here factually unique derivation—differs from all of them. In particular, in none of the consequence relations referred to above, a *AU* predicate appears. (Note also that if the *AU* predicate were generally used, Horn clause reasoning would not be in linear logic.)

For this reason we suppose that the condition of factual uniqueness is an additional constraint to be imposed on linear logic consequence relation in order to enable planning.

Since factually unique and basically reduced coincide we are tempted to say that in order to do plan generation we have to impose on classical logic as well as on linear logic the same constraint.

References

[1] Arnon Avron. The Semantics and Proof Theorie of Linear Logic. *Theoretical Computer Science*, (57):161–184, 1988.

[2] W. Bibel. *Automated Theorem Proving*. Vieweg, 1982.

[3] W. Bibel. A Deductive Solution for Plan Generation. *New Generation Computing*, 6:115–132, 1986.

[4] B. Fronhöfer. Linear proofs and linear logic. Technical report, Technical University Munich, 1992.

[5] Jean-Yves Girard. La Logique Lineaire. *Pour la Science, Edition française de Scientific American*, pages 74–85, April 1990.

[6] Pat Lincoln, John Mitchel, Andre Scedrov, and Natarajan Shankar. Decision problems for propositional linear logic. Technical Report SRI-CSL-90-08, SRI International, 1990.

[7] M. Masseron, C. Tollu, and J. Vauzeilles. Generating Plans in Linear Logic. Technical Report 90-11, Université Paris Nord, Département de mathématiques et informatique, Avenue J.B.Clément, F-93430 Villetaneuse, December 1990.

[8] J. McCarthy and P. Hayes. Some philosophical problems from the standpoint of artificial intelligence. In B. Meltzer and D. Michie, editors, *Machine Intelligence 4*, pages 463–502. Edinburgh University Press, 1969.

Relevance and Revision

About Generalizing Syntax-based Belief Revision

Emil Weydert

emil@adler.philosophie.uni-stuttgart.de
IMS, University of Stuttgart,
Azenbergstrasse 12, W-D 7000 Stuttgart 1, Germany

Abstract. This paper proposes a syntax-based revision-procedure for belief sets structured by an arbitrary relevance pre-order and an elementary epistemic dependency relation and investigates the consequences for the Gaerdenforsian rationality postulates. In particular, it generalizes Nebel's epistemic relevance revision.

1 Introduction

Intelligent behaviour in a complex, dynamic world, where the available "knowledge" is mostly incomplete, speculative or incorrect, requires rational and feasible belief revision strategies telling cognitive agents how to deal with new and possibly conflicting informations. In the last years, there have been several attempts to give a normative or procedural (qualitative) formal account of this particularly important aspect of commonsense reasoning. A very influential proposal goes back to Alchourron, Gärdenfors and Makinson [85], who formulated a set of rationality postulates for belief revision and related epistemic transformations, based on the idea of minimal change, and investigated different formal realizations of these principles. In particular, Gärdenfors and Makinson [88] used epistemic entrenchment (a kind of preference) orderings as a tool for constructing revision functions.

Unfortunately, this theory and many others are suffering from several serious defects, notably their intuitive inadequacy, their unrealistic assumptions of complete (i.e. total) epistemic preference orderings and deductively closed belief sets, their neglect of more complex forms of (e.g. default, meta or control) knowledge and / or, last but not least, their silence concerning preference updating and iterated revisions, which is also related to their inability to deal with the famous Gärdenforsian impossibility theorem [Gärdenfors 88]. This is especially disturbing if we want to get a realistic formal model of communicating agents, where minds are finitary, belief changes frequent and epistemic contents sophisticated.

In recent times, different aspects of these problems have been addressed by different means. Within the realm of theory revision, some authors have turned towards weakened epistemic entrenchment relations [Lindström, Rabinowicz 91, Rott 92, Weydert 91, 92], which are no longer required to form linear hierarchies and may allow protected non-tautological beliefs, doxastic preference modalities or epistemic revision conditionals [Morreau 90, Rott 91, Weydert 91, 92], implicitly or explicitly linked to a nonmonotonic background logic and partly figuring as control knowledge within the belief set itself [Weydert 91, 92]. But we should also mention finitary model-based approaches (e.g. for updates reflecting changing world states) obeying Dalal's *Principle of Irrelevance of Syntax* (cf. [Katsuno, Mendelzon 91]), which are offering interesting partial answers as well.

However, if we intend to give not only an abstract but also a practical account of revision rationality, we have to look first of all at syntax-based approaches, where not only the meaning (models) but also the form (syntax) of the knowledge base and the input may affect the epistemic revision process. Here, the most prominent proposals are due to Nebel[91, 92] and Doyle[91].

Nebel considers (finitary or infinitary) prioritized belief bases, i.e. axiom sets for full (deductively closed) belief sets, structured by an arbitrary total pre-order with maximum called an epistemic relevance ordering. The idea is to preserve preferably the most relevant (whatever this means in practice) beliefs during revisions. For the rest, the standard methodology of Gärdenfors (inspired by Levi) is adopted to belief bases, which allows to meet all the less controversial rationality requirements.

Doyle on the other hand argues for revision processes guided by an economic rationality concept and rejects the minimal change dogm (≈ saving as many beliefs as is consistently possible), which may lead to the violation of some basic Gärdenforsian postulates and in particular to syntax-dependency. Whereas Nebel assigns priorities (degrees of relevance) to sentences in the belief base, Doyle assumes in his standard theory of rational belief revision a total pre-order of sets of sentences (whatever its decision theoretic origins), which provides additional flexibility.

However, for both accounts we have a mixture of too liberal as well as too restrictive features. For instance, complete preference relations positioning sets of beliefs are hardly realistic, and Doyle knows that. Furthermore, it is not very appealing to think of cognitive agents as ranking potential belief bases or sets of sentences in a purely top-down manner. A bottom-up approach starting with preferences structuring individual beliefs seems to be more natural, but Doyle doesn't give us concrete indications how this could be achieved. At this level of generality, not very much can be said. So, let's turn to Nebel's framework. His epistemic relevance orderings are total pre-orders on belief bases. This presupposes that cognitive agents should always have a clear idea about the relative relevance of different beliefs, which doesn't exactly correspond to our daily experience of epistemic partiality. Moreover, in the case of infinitary belief bases (which may become interesting when considering or simulating first-order languages), his revision method is only applicable if the relevance ordering satisfies additional, in fact undesirable constraints (e.g. no strictly ascending infinitary sequences of relevance degrees). Finally, this approach is also unable to express dependencies between beliefs affecting the Gärdenforsian paradigm.

It is the purpose of this paper to develop and investigate a generalization of epistemic relevance revision using arbitrary pre-orders, correspondingly revised contraction and revision procedures and a rudimentary notion of dependency. In the present context, we concentrate on single revision steps and don't address the problem of iterated revisions.

After the presentation of our epistemic state concept, we shall disclose the principles behind our revision strategy and introduce a partial order on belief subsets which will be derived from our doxastic relevance and dependency relations. This will allow us to define a belief contraction procedure based on a kind of filtered prioritized intersection and to apply the Levi-principle for realizing revisions. In the main part, we shall consider the implications for the Gärdenforsian postulates and, to conclude, we shall try to explain the differences between our and other accounts for belief revision.

2 Doxastic states

Before we can begin to think about a formal account for belief revision, we have to fix our minds about what we are going to revise and in which context this will be done. In other words, we need an appropriate abstract notion of belief and of doxastic (= epistemic) states or contexts figuring with the incoming information as arguments of the revision functions to be constructed. Our position here is a liberal representational one, allowing us to model (to a very decent extent) finite real-life cognitive agents as well as their idealized infinitary omniscient counterparts. In practice, this means identifying the basic belief structures with possibly infinitary, not necessarily inferentially closed sets of sentences w.r.t. some given language and logic. Note that this doesn't force us to accept every set of sentences as a reasonable belief base. For instance, we may want to restrict ourselves to finite or recursive belief sets or to impose some weak closure conditions (e.g. forming and resolving conjunctions). Nevertheless, because we would like to close up any set of sentences, we shall only deal with belief set classes closed under arbitrary intersections, which will guarantee the existence of smallest belief supersets. The language itself should be closed under the usual propositional connectives and, for the purpose of this paper, the logic should be sufficiently close to what we are used to. This means nontriviality (consistency), monotonicity, closure under classical tautological consequence (supraclassicality), finite conditionalization (i.e. the deduction theorem is satisfied for material implication) and compactness (which reflects the reality of cognitive agents). But these conditions exactly describe classical propositional consequence relativized to some language and theory (e.g. the theorems of the initial inference relation).

Definition 2.1 Let L be a language closed under the propositional connectives T, ¬, &, ∨, →, ↔, and ⊢ be the classical propositional consequence relation on L w.r.t. a given consistent set of axioms S (= *classical (monotonic) background logic*). A pair (IB, ⊢) will be called a *belief set structure* iff IB is a nonempty set of subsets of L, closed under arbitrary intersections and containing in particular L, and ⊢ is a classical background logic. IB's elements will be called *admissible belief sets*. (Note that our belief sets don't have to be closed under ⊢)

But of course, this will not be enough. Cognitive agents do not only have beliefs about the world, they do also handle higher-order epistemic commitments (i.e. attitudes towards beliefs concerning degrees of confidence, pragmatic relevance, justification links, ... etc). Depending on the expressive power and strength of the basic logic we adopt, they will or will not be represented within the belief set itself. However, they will always form a part of the epistemic state as a whole, even if we want or have to handle them separately (e.g. if there is no suitable logic at hand). Doing belief revision, these internal structures, valuations or hierarchies should allow us to evaluate the consequences of accepting or rejecting different beliefs. Now, we could imagine a notion of epistemic preference summing up all those individual aspects (evidential, economic, pragmatic, esthetic, ...) which determine our strength of belief in or our willingness to act according to a given sentence or system of sentences. Here, we shall not be (directly) concerned with the question of how to calculate such a relation but only with its structural characteristics. It seems plausible and practical to ask for a (partial) pre-order, not more, not less. Indeed, we noted above that completeness (i.e. totality) is unrealistic as a general constraint. Similarly, it seems difficult to avoid strictly ascending prioritization sequences (A physicist thinking about the mass of the universe might be absolutely convinced that the number of atoms is finite. Furthermore, he might believe with increasing confidence that it lies below 10^{90n} for growing n. But this requires ascending sequences of preference degrees. Note also that it wouldn't be an easy task for him to tell us the exact position of his beliefs about the identity of John F. Kennedy's murderer w.r.t. this context. In other words, we need unrestricted partial preference orderings). Because we are considering arbitrary belief sets and don't want to assume a priori any kind of omniscience, this minimal structural condition for what we shall call a doxastic relevance ordering will not be supplemented by any logical constraints (e.g. like those characterizing epistemic entrenchment).

In the past, epistemic preference relations sometimes had to carry an additional burden [Lindström, Rabinowicz 91, Weydert 91], namely to encode dependencies between beliefs, telling us for instance to reject a given belief as soon as we have to reject some other one (this may have profound consequences for the Gärdenforsian postulates). However, we would prefer not to confuse epistemic preference and epistemic dependency. Therefore, we shall introduce a rudimentary form of doxastic dependency as an independent second epistemic relation, which will give us additional possibilities to control the revision process in a natural way. In a sense, this relation plays a role similar to that played by logical consequence in the traditional epistemic entrenchment framework. Seeking for generality, we restrict ourselves again to reflexivity and transitivity. A very general notion of an epistemic state would now be the following one.

Definition 2.2 We call (B, ≤, «) a *(possible) doxastic state* w.r.t. the belief set structure (IB, |-) iff B is an admissible belief set in IB and ≤, « are pre-orders (i.e. reflexive, transitive relations) on B. ≤ is called a *doxastic relevance ordering* and « a *doxastic dependency relation* on B. Let IB" be the set of all possible doxastic states w.r.t. (IB, |-). IB' is called a *doxastic system* w.r.t. the belief set structure (IB, |-) iff IB' is a subset of IB" and every element of IB is instantiated by a doxastic state in IB'.

Given that our revision-procedure requires consistency-checks all the time anyway, we could have assumed consistency right from the beginning (for belief sets B≠L). But for the sake of generality and in view of other, still more realistic approaches, we shall accept the existence of a whole bunch of inconsistent doxastic states.

3 Doxastic relevance revision

Doxastic relevance revision is a generalization of Nebel's proposal for syntax-based revision [Nebel 92], which extends his prioritized revision strategy to deal with arbitrary relevance pre-orders (e.g. containing strictly ascending infinitary chains, incomparable elements or no maxima) and considers dependency-closed subsets of sentences. It is organized around at least three fundamental principles.

1. **Levi-principle.** Belief revision should be done following a two-step strategy :
   1. Reducing the initial belief set, taking care of the other principles, to neutralize any obstacle to the consistent addition of (the acceptable part of) the epistemic input.
   2. Adding what can and has to be added and closing up to get a new admissible belief set (and ideally to reach a new doxastic state).

2. **Symmetry.** Any choice between alternatives has to be justified by the given basic or derived preference orderings. Consequently, in the case of multiple, equally preferred extensions, we should be sceptical and consider their intersection (contrary to what Doyle says).

3. **Minimal change.** Accepted new information should always be integrated in the least damaging way. This means preserving as many of the doxastically most relevant beliefs as possible without violating dependency constraints.

Now, let (IB, |-) be an arbitrary belief set structure. We begin our partial formal implementation of the above principles by defining for any doxastic state (B, ≤, «) in IB" a relation << on the powerset 2^B, measuring the overall doxastic relevance of a given subset w.r.t. ≤ in accordance with our third postulate.

Definition 3.1 Let ≤ be a pre-order on B. Then we define for all X, Y ∈ 2^B,

$$X << Y \text{ iff } \forall x \in X\backslash Y\ \exists y \in Y\backslash X\ (x \leq y\ \&\ \forall x \in X\backslash Y\ \neg(y \leq x))$$

This means that in the difference set, all the elements of X have to be strictly (irreversibly) dominated by elements of Y (note the similarity with lexicographic orderings where we are also comparing the best positions in the disagreement domain). The reflexive version of Nebel's set ordering restricted to complete pre-orders without strictly ascending infinitary sequences (where his basic definition "X $<<_n$ Y iff among the (linearly ordered) relevance equivalence classes, there is a highest layer V such that

$V \cap X \neq V \cap Y$, and it satisfies $V \cap X \subset V \cap Y$" works) is subsumed by our definition, which seems to be the natural extension to deal with arbitrary pre-orders on belief sets. If X is a subset of Y, we immediately get $X \ll Y$, and in particular, we have $X \ll X$. In the general case of non-empty differences however, clear-cut relevance distinctions are required. Note that "preferring supersets" derives its justification from the minimal change principle. In fact, this constraint partly explains why our definition has received its present "lexicographic" form. In general, we can show that $\ll$ has very nice, partly expected, partly surprising, but rather useful properties.

Proposition 3.1 If $\leq$ is a pre-order on B and $\ll$ is defined on 2^B as in Def. 3.1, then $\ll$ is a partial order, i.e. reflexive, transitive and antisymmetric.

Proof : Reflexivity is obvious. Transitivity is a bit more tricky. Let X, Y, Y' be subsets of B with $X \ll Y$ and $Y \ll Y'$. Then, if $x \in X \backslash Y'$, we have to distinguish four cases:

1. $x \in Y$. Hence $x \in Y \backslash Y'$, and by $Y \ll Y'$ there is $y \in Y' \backslash Y$ with $x \leq y$ and $\forall u \in Y \backslash Y'$ $\neg(y \leq u)$ (*).

1.1 $y \in Y' \backslash X$. Suppose there is $x' \in X \backslash Y'$ with $y \leq x'$. Then $x' \in X \backslash Y$ by (*), and from $X \ll Y$, we get $x'' \in Y \backslash X$ with $x' \leq x''$ and $\forall u \in X \backslash Y$ $\neg(x'' \leq u)$. So $y \leq x''$, $\forall u \in Y \backslash Y'$ $\neg(x'' \leq u)$ and $x'' \in Y' \backslash X$, both by (*). In particular, $\forall u \in X \backslash Y'$ $\neg(x'' \leq u)$. So, either we don't have such an x', or we can switch from y to ($x \leq$) x" and we are done.

1.2 $y \notin Y' \backslash X$. Hence $y \in X \backslash Y$ and by $X \ll Y$ there is $y' \in Y \backslash X$ with $y \leq y'$ and $\forall u \in X \backslash Y$ $\neg(y' \leq u)$. But then, $y' \in Y' \backslash X$ by (*) and $x \leq y'$. From $y \leq y'$ follows $\forall u \in Y \backslash Y'$ $\neg(y' \leq u)$ and therefore $\forall u \in X \backslash Y'$ $\neg(y' \leq u)$. Hence y' does the job.

2. $x \notin Y$. Then $x \in X \backslash Y$ and there is $y \in Y \backslash X$ with $x \leq y$ and $\forall u \in X \backslash Y$ $\neg(y \leq u)$ (**).

2.1 $y \in Y'$. Suppose we have $u \in X \backslash Y'$ with $y \leq u$. By (**) $u \in Y \backslash Y'$. Then there is $u' \in Y' \backslash Y$ with $u \leq u'$ and $\forall v \in Y \backslash Y'$ $\neg(u' \leq v)$. Hence $y \leq u'$ and by (**) $u' \in Y' \backslash X$, but also $\forall v \in X \backslash Y'$ $\neg(u' \leq v)$. Thus, either such an u doesn't exist and y is right, or u' is as desired.

2.2 $y \notin Y'$. So we have $y' \in Y' \backslash Y$ with $y \leq y'$ and $\forall u \in Y \backslash Y'$ $\neg(y' \leq u)$. But with (**), this gives us $y' \in Y' \backslash X$ and $\forall u \in X \backslash Y'$ $\neg(y' \leq u)$. Hence y' is enough.

Thus, transitivity holds because we can always find a dominating element in $Y' \backslash X$. Now let's see what happens if $X \ll Y$ and $Y \ll X$. If $X \backslash Y$ is non-empty, there must be $x \in X \backslash Y$ and $y \in Y \backslash X$ with $x \leq y$ and $\forall u \in X \backslash Y$ $\neg(y \leq u)$. But this contradicts $Y \ll X$. Hence $X \backslash Y$ and similarly $Y \backslash X$ will have to be empty, that is X=Y.

Our doxastic relevance comparing tool at the set-level will now allow us to introduce two notions of relevance-based contraction. The basic idea behind the traditional minimal change approaches to belief revision, as proposed by Gärdenfors for theories and by Nebel for belief bases, has been the following: Given a belief set K and a new, conflicting information A, consider the "best" inclusion-maximal A-permitting (i.e. consistent with A) subsets of K, take the intersection of their closures, add A and close up. If closing up means deductive closure, this procedure (***a-revision***) is equivalent to a second one (***b-revision***), which consists in adding A directly to all the preferred subsets, closing up and intersecting afterwards. Note that belief sets revised according to the a-strategy will always be contained (if closure means intersection of the admissible supersets, which is of course monotonic) in their b-revised counterparts. In general, b-revision seems to be preferable, being in a sense as plausible but more productive than a-revision. Nevertheless, because in the following we shall be mainly concerned with logical closure, a-revision as the classical procedure will do the job.

When trying to define prioritized contraction / revision w.r.t. doxastic relevance, we are immediately confronted to a serious problem. Consider the following example:

Example 3.1 Let A, A', A_n be independent propositional variables ($<\, := $ "$\leq \& \neg \geq$").

$$B = \{A \vee A_1, A \vee (A_2 \& \neg A_1), \ldots, A \vee (A_n \& \neg A_{n-1} \& \ldots \& \neg A_1), \ldots, A, A', T\}$$
with $A < A' < A \vee A_1 < \ldots < A \vee (A_n \& \neg A_{n-1} \& \ldots \& \neg A_1) < \ldots < T$

Obviously, all the maximally consistent ¬A-permitting subsets of B will be of the form $\{A \vee (A_n \& \neg A_{n-1} \& \ldots \& \neg A_1), A', T\}$. On the other hand, the partial order << induced by ≤ will provide a strictly ascending sequence ($\{A', T\}$ <<) $\{A', A \vee A_1, T\}$ << $\{A', A \vee (A_2 \& \neg A_1), T\}$ << ... << $\{A', A \vee (A_n \& \neg A_{n-1} \& \ldots \& \neg A_1), T\}$ << ... without a maximal element. That is, there are no <<-maximal ¬A-permitting sub-bases. In this context, the usual contraction procedures would yield either an inconsistent set, namely L, or the input B, which is hardly better. Our intuitions, however, would favour an outcome like {T, A'}. Working only with theories also doesn't help. By putting all the proper logical consequences into the lowest relevance degree, we are left with the same result. Hence, the problem is a real one for infinitary relevance revision.

So, it will not be sufficient to consider only the <<-maxima, which may not exist, but we have to inspect the rest of the <<-structure as well to define appropriate procedures fitting the minimal change principle. One way to proceed would be to pick up those sentences which are supported by every <<-dense set of the subsets we are considering (S is called a *«-dense* subset of S' iff for all x in S', there is an y in S with x«y) and hence in particular by all the <<-maxima. In fact, we even have to go one step farther.

Definition 3.2 Let P be a set of subsets from U and « be a pre-order on P. Then the *filtered prioritized intersection* of the elements of P w.r.t. « is defined by «-ΔP : = $\{A \in U \mid \forall X \in P\, \exists Y \in P\, (X \ll Y \;\&\; \forall Y' \in P\, (Y \ll Y' \rightarrow \exists Y'' \in P\, (Y'' \supseteq Y' \;\&\; Y' \ll Y'' \;\&\; A \in Y'')))\}$

Note that the weakening $\exists Y'' \in P\, (Y'' \supseteq Y' \;\&\; Y' \ll Y'' \;\&\; \ldots)$ is required to ensure intuitively correct results in examples like the previous one. Otherwise, because (for all n>2) $\{A', A \vee (A_{n-1} \& \neg A_{n-2} \& \ldots \& \neg A_1), T\} \ll \{A \vee (A_n \& \neg A_{n-1} \& \ldots \& \neg A_1), T\}$ but $A' \notin \{A \vee (A_n \& \neg A_{n-1} \& \ldots), T\}$, A' would not belong to a naively defined filtered prioritized intersection. On the other hand, the strengthening subcondition $Y'' \supseteq Y'$ is needed to prevent oversized outcomes. Taking « to be the identity, we get the usual straightforward intersection notion. Also, if « is a complete pre-order on P without strictly ascending infinitary chains or, more generally, if «-maxima are «-dense, filtered prioritized becomes maxima intersection. Observe however that even when maxima are available, their standard intersection might contain much more elements than the filtered prioritized one. The following result will be rather useful.

Proposition 3.2 If P' = {X | X is inclusion-maximal in P} is inclusion-dense in P and $Y \supseteq X$ always implies $X \ll Y$, then we have $\ll\text{-}\Delta P = \{A \in U \mid \forall X \in P\, \exists Y \in P'\, (X \ll Y \;\&\; \forall Y' \in P'\, (Y \ll Y' \rightarrow A \in Y'))\}$

Proof : Suppose A∈ «-ΔP. Hence $\forall X \in P\, \exists Y \in P\, (X \ll Y \;\&\; \forall Y' \in P\, (Y \ll Y' \rightarrow$

$\exists Y''\in P\ (Y''\supseteq Y'\ \&\ A\in Y''))$. By our assumption, we can find $Y^\circ\in P'$ with $Y^\circ\supseteq Y$ and $Y\ll Y^\circ$. Because $X\ll Y$ implies $X\ll Y^\circ$, $\forall Y'\in P\ (Y\ll Y' \rightarrow \ldots)$ implies $\forall Y'\in P'\ (Y^\circ\ll Y' \rightarrow \ldots)$, and $Y'\in P'\ \&\ Y''\in P\ \&\ Y''\supseteq Y'$ requires $Y' = Y''$, we get effectively $\forall X\in P\ \exists Y\in P'\ (X\ll Y\ \&\ \forall Y'\in P'\ (Y\ll Y' \rightarrow A\in Y'))$.
Now, suppose $\forall X\in P\ \exists Y\in P'\ (X\ll Y\ \&\ \forall Y'\in P'\ (Y\ll Y' \rightarrow A\in Y'))$. Then we get immediately $\forall X\in P\ \exists Y\in P\ (X\ll Y\ \&\ \forall Y'\in P'\ (Y\ll Y' \rightarrow A\in Y'))$. But $\forall Y'\in P'\ (Y\ll Y' \rightarrow A\in Y'))$ and $\forall Y'\in P\ \exists Y''\in P'\ (Y''\supseteq Y')$ (by density) imply $\forall Y'\in P\ (Y\ll Y' \rightarrow \exists Y''\in P'\ (Y''\supseteq Y'\ \&\ Y\ll Y''\ \&\ A\in Y'')$, which gives us $A\in \ll\text{-}\Delta P$.

Now, guided by the Levi-principle, our search for symmetry and the minimal change conception, we are ready to introduce our basic contraction and revision notions. Let's begin with a simplified version which is free from dependency considerations.

Definition 3.3 Let $(\mathbb{B}, \vdash)$ be an arbitrary belief set structure and *cl* be the associated closure function with $cl(X) := \cap\{Y\in\mathbb{B} \mid Y\supseteq X\}$.
A function $- : \mathbb{B}'' \times L \to \mathbb{B}$ is called the *free doxastic relevance contraction function* w.r.t. $(\mathbb{B}, \vdash)$ iff for all doxastic states $K = (K^\circ, \leq, \text{«}) \in \mathbb{B}''$ and sentences $A\in L$, if $\ll$ is the partial set-order derived from $\leq$, $K\text{-}A = cl(\ll\text{-}\Delta\{X\in 2^{K^\circ} \mid X \nvdash A\})\cap K^\circ$.
A function $* : \mathbb{B}'' \times L \to \mathbb{B}$ is called the *free doxastic relevance revision function* w.r.t. $(\mathbb{B}, \vdash)$ iff for all doxastic states $K = (K^\circ, \leq, \text{«}) \in \mathbb{B}''$ and sentences $A\in L$, $K*A = cl((K\text{-}\neg A)\cup\{A\})$ if consistent, and otherwise $K*A = K^\circ$.
$(K\text{-}A := -(K, A)$ and $K*A := *(K, A))$

Note that we have already realized, at least implicitly and forced by our disapproval of unnatural ordering constraints, a departure from the AGM (relational) partial meet contraction / revision which depends on the existence of suitable maxima. In fact, our doxasic states provide yet another tool, which may lead us even farther away from the classical paradigm, namely a (primitive) dependency-structure. It allows us to connect different elements of our belief bases so as to relate their respective removals. This seems to be an interesting step towards a marriage of the coherentist and the foundationalist view of belief revision. In practice this means that we are considering only those subsets of a belief set underlying a given doxastic state which are closed under the associated dependency relation. But (inclusion-)maximal dependency-closed $\neg A$-permitting subsets don't need to be (inclusion-)maximal $\neg A$-permitting ones. So, the Gärdenforsian presuppositions may be violated, even in the finite case. The general definition for our dependency-depending epistemic transformations is as follows.

Definition 3.4 (Conditions as in Definition 3.3)
$- : \mathbb{B}'' \times L \to \mathbb{B}$ is called the *(general) doxastic relevance contraction function* and $* : \mathbb{B}'' \times L \to \mathbb{B}$ the *(general) doxastic relevance revision function* w.r.t. $(\mathbb{B}, \vdash)$ iff
$K\text{-}A := -(K, A) = cl(\ll\text{-}\Delta\{X\in 2^{K^\circ} \mid X \nvdash A \text{ and } X \text{ a } K^\circ,\text{«-cone}\})\cap K^\circ$
$K*A := *(K, A) = cl((K\text{-}\neg A)\cup\{A\})$ if consistent, and otherwise $K*A = K^\circ$.
X is called an Y,«- *cone* (w.r.t. $\vdash$) iff $cl_{\vdash}(X)\cap Y = X$ and $\forall A\in X(A\in Y\ \&\ \forall A'\in Y\ (A \text{«} A' \rightarrow A'\in X))$.

Free doxastic relevance revision obviously yields the same results than (general) doxastic relevance revision if « is the identity. Since $\vdash$ is compact, it easily follows from Zorn's lemme that any K°,« - cone X with $X \nvdash A$ can be extended to an inclusion-maximal one X', which of course will satisfy $X \ll X'$. Hence, we could apply

Prop. 3.2 and restrict ourselves in the previous definitions to inclusion-maximal such X's. Because we are dealing with pre-orders, dependency-constraints and a sceptical approach, taking intersections to combine equally preferred alternatives, our account will not be subsumed by Doyle's rational revision, which has nevertheless a less strict view on the permitted subsets issue. It will be interesting to see what this means for Gärdenfors' rationality postulates, some of which have already been criticized in the literature for different reasons anyway. Note that in a sense, our present, fragmentary notion of revision is nondeterministic, because it doesn't specify a new doxastic state.

Within our framework, prioritized base-revision can be implemented in at least two different ways. One possibility is to identify prioritized belief bases with the doxastic states (e.g. with finite or arbitrary sets of sentences as belief sets) of an appropriate doxastic system. Here, prioritized base-revision is seen as an instance of doxastic relevance revision and the revision process depends only on the internal base structure and the closure constraints adopted for belief sets, but not on external logical consequences. However, without sufficiently strong closure conditions (often affecting efficiency), contraction may become rather destructive (e.g. if the admissible belief sets are just the finite subsets of L closed under conjunction, and K is the doxastic state $(K^\circ, id_{K^\circ}, id_{K^\circ})$ with $K^\circ = \{A, B, A\&B, T\}$, then $K\text{-}(A\&B) = \{T\}$ instead of what we would normally expect, namely $\{A \vee B\}$). But there is another form of prioritized base-revision, closer to theory revision and handling arbitrary belief bases, which has been adopted by Nebel and others. The idea is to consider and manipulate instead the logical closures of subbases and to look at their base parts for relevance-based evaluations. In other words, a kind of base-controlled theory revision. Because intersections of finitely many finitely axiomatised theories can be finitely described by the disjunction of the axiom set conjunctions, this revision procedure nevertheless doesn't force us (a priori) to leave the realm of the finite (if we switch to and handle finite representations). A possible implementation of this methodology in our context would be the following one.

Definition 3.5 (Conditions as in Definition 3.3 with $cl_{\vdash}(X) \in \mathbb{B}$ for all $X \in 2^L$)
Doxastic base revision $*_b$ and *contraction* $-_b$:

$K\text{-}_bA := cl_{\vdash}(\ll''\text{-}\Delta\{cl_{\vdash}(X) \mid X \in 2^{K^\circ}, X \text{ a } K^\circ, \ll\text{-cone}, X \nvdash A\})$

$K*_bA := cl_{\vdash}((K\text{-}_b\neg A) \cup \{A\})$ if consistent, and otherwise $K*_bA = K^\circ$,

with $X \ll'' Y$ iff $X \cap K^\circ \ll Y \cap K^\circ$.

As Nebel [92] remarked, base revision is related to relevance-based revision. What you find in the knowledge base should a priori be more relevant than any consequence to be derived. We conjecture that Nebel's equivalence results concerning prioritized base and epistemic relevance theory revision can be extended to the present framework. In the following, we shall therefore concentrate on doxastic relevance revision for theories.

4 AGM-properties

The framework and in particular the postulates for belief revision and other epistemic transformations proposed by Gärdenfors and his colleagues during the eighties are and will be for some time to come a very valuable tool to discuss, compare, situate and handle different approaches to this subject. The paradigm itself and most of its principles have been criticized in the literature for various reasons, but they have provided a much clearer picture of the problems involved.

In the following, we shall give an evaluation of doxastic relevance contraction and revision w.r.t. the AGM-postulates. Because the latter have been conceived for belief sets closed under logical consequence, we shall restrict ourselves to *full* belief set models (IB, $\vdash$), where IB is the set of $\vdash$ - theories and cl is $\vdash$ - closure. Note that our doxastic transformations are always applied to what we call doxastic states. So, we have to read the principles accordingly, with K's of the form (K°, ≤, «), where K° is in IB.

A and A' will be sentences in L, and K+A, K°+A (Gärdenfors' expansion by A) will be interpreted as cl(K°∪{A}) (⊂ will denote non-strict inclusion).

Gärdenforsian contraction postulates

K-1 K-A is $\vdash$ - closed

K-2 K-A ⊂ K°

K-3 If A ∉ K°, then K-A = K°

K-4 If $\nvdash$ A, then A ∉ K-A

K-5 K° ⊂ (K-A)+A

K-6 If $\vdash$ A↔A', then K-A = K-A'

K-7 K-A ∩ K-A' ⊂ K-A&A'

K-8 If A ∉ K-A&A', then K-A&A' ⊂ K-A

Gärdenforsian revision postulates

K*1 K*A is $\vdash$ - closed

K*2 A ∈ K*A

K*3 K*A ⊂ K+A

K*4 If ¬A ∉ K°, then K+A ⊂ K*A

K*5 K*A = L implies $\vdash$ ¬A

K*6 If $\vdash$ A↔A', then K*A = K*A'

K*7 K*A&A' ⊂ (K*A)+A'

K*8 If ¬A' ∉ K*A, then (K*A)+A' ⊂ K*A&A'

The first six (seven / eight) of the contraction resp. revision postulates are or have been considered as basic (very / rather reasonable) and describe (relational / transitively relational) partial meet contraction and revision (i.e. intersecting (greatest w.r.t. a relation / a transitive relation on belief subsets) inclusion-maximal A-permitting belief subsets and adding A). In fact, these principles also characterize epistemic transformations defined by very different means, for instance epistemic entrenchment relations [Gärdenfors, Makinson 88] (which obey strong logical constraints and have a more semantic touch) or appropriate safe contraction functions [Alchourron, Makinson 85, Rott 91]. Of course, conditions backed by such representation results will be particularly attractive ones. Nevertheless, if you try to generalize the notion of minimal change and bring in some dependency concepts, seeking for a slightly more realistic account of belief revision, things will become different. To see what happens to the postulates when we switch to doxastic relevance procedures, we shall give an alternative formulation, closer to the traditional account.

Let K = (K°, ≤, «) be a doxastic state w.r.t. the full belief set model (IB, $\vdash$), and P be a set of K°-subsets s.t. P' : = {X | X is inclusion-maximal in P} is inclusion-dense in P. By Prop. 3.2, we have <<-ΔP = {A∈K° | ∀X∈P ∃Y∈P' (X<<Y & ∀Y'∈P' (Y<<Y' → A∈Y'))} = {A∈K° | ∃f∈P'^P ∀X∈P (X<<f(X) & ∀Y'∈P' (f(X)<<Y' → A∈Y'))} = {A∈K° | ∃f∈P'^P((∀X∈P (X<<f(X)) & ∀Y'∈P' ((∃X∈P f(X)<<Y') →

$A \in Y'))\} = \cup\{\cap U(f,P,<<) \cap K^\circ \mid f \in H(P,<<)\}$ where $U(f,P,<<) := \{Y' \in P' \mid \exists X \in P\ f(X) << Y'\}$ and $H(P,<<) := \{g \mid \forall X \in P\ (X << g(X))\ \&\ g \in P'^P\}$. Note that all the $U(f,P,<<)$ (for $f \in H(P,<<)$) are in $F(P,<<) := \{S \in 2^{P'} \mid S$ is $<<$-dense in P and $<<$-upwards closed in $P'\} \neq \emptyset$ (because of our assumption, $P' \in F(P,<<)$), and any set S in $F(P,<<)$ is of the form $U(f,P,<<)$ (set $f(X) \in \{Y \in S \mid X << Y\}$ for $X \in P \backslash S$ and $f(X) = X$ for $X \in S$). Hence $<<\text{-}\Delta P = \cup\{\cap S \cap K^\circ \mid S \in F(P,<<)\}$ if P' is dense in P w.r.t. inclusion. Note also that if $S, S' \in F(P,<<)$, $S \cap S' \in F(P,<<)$ as well, and that $S' \supseteq S$ implies $\cap S \cap K^\circ \supseteq \cap S' \cap K^\circ$. It follows that $(\{\cap S \cap K^\circ \mid S \in F(P,<<)\}, \subset)$ is directed because two elements $\cap S \cap K^\circ$, $\cap S' \cap K^\circ$ are always majorized by $\cap(S \cap S') \cap K^\circ$.

Let $P(A) := \{X \in 2^{K^\circ} \mid X$ a $K^\circ,\ll$-cone and $X \nvdash A\}$ and $I : L \to \cup\{F(P(A),<<) \mid A \in L\}$ be a choice function with $I(A) \in F(P(A),<<)$ (if $P(A) = \emptyset$, then $I(A) = \emptyset$ because $F(\emptyset,<<) = \{\emptyset\}$ and the converse also holds) and $I(A) = I(A')$ for logically equivalent A, A' (where $P(A) = P(A')$). Let $\mathrm{II}(K)$ be the set of all such I for a given $K = (K^\circ, \leq, \ll)$. For any I in $\mathrm{II}(K)$, we can define corresponding contraction- and revision-like functions $-_I$ and $*_I$ with $K -_I A = cl_{\vdash}(\cap I(A) \cap K^\circ) = \cap I(A) \cap K^\circ$ and $K *_I A = cl_{\vdash}(K -_I \neg A \cup \{A\})$ if consistent, otherwise K°.

Proposition 4.1 In general, $-_I$ contraction / $*_I$ revision functions only support K-1, 2, 3, 6 / K*1, 3, 4, 6. If $\ll = id_{K^\circ}$, we get in addition K-4, 5 and K*2, 5.

Proof : ($\approx$ stands for "$\leq \& \geq$", and in the finitary counterexamples, I is assumed to pick up the set of $<<$-maxima)

- K-1 Immediate.
- K-2 Immediate.
- K-3 If $A \notin K^\circ$, then $K^\circ \nvdash A$, K° is the $<<$-greatest $K^\circ,\ll$- cone and $P(A)' = \{K^\circ\}$, i.e. $K -_I A = K^\circ$.
- K-4 Not true. If $\nvdash A$ but $T \ll A$, $I(A) = \emptyset$ and $K -_I A = \cap I(A) \cap K^\circ = K^\circ$.
- K-5 Not valid. Consider $K = (cl_{\vdash}(\{A, B\}), \leq, \ll)$ with $A \vee \neg B, \neg A \vee B, A \vee B \ll A$. Then $K -_I A = cl_{\vdash}(\{T\})$, but $(K -_I A) + A = cl_{\vdash}(\{A\}) \neq cl_{\vdash}(\{A, B\})$ (A, B are assumed to be logically independent).
- K-6 Immediate.

K-7 and K-8 are usually wrong, because $I(A\&A')$ can be chosen independently (e.g. too large, making $\cap I(A\&A')$ too small) from $I(A)$ and $I(A')$.

- K*1 Immediate.
- K*2 Fails for instance if $\vdash \neg A$ or $T \ll \neg A$.
- K*3 Immediate.
- K*4 If $\neg A \notin K^\circ$, then $K -_I \neg A = K^\circ$ by K-3 and $K *_I A = cl_{\vdash}(K^\circ \cup \{A\}) = K + A$.
- K*5 Wrong, because $K *_I A = L$ can also happen if $K = L$ and $T \ll \neg A$.
- K*6 Immediate.

K*7 and K*8 are usually wrong (see K-7, 8).

If « = $id_{K°}$, inclusion-maximal A-permitting K°,«-cones are inclusion-maximal A-permitting K°-subsets. Hence, $-_I$ / $*_I$ becomes an instance of partial meet contraction / revision, which satisfies K-1 - 6 / K*1 - 6.

Fortunately, we have another result relativizing the failures of K-7, K*7 for some instances of I.

Proposition 4.2 For all K = (K°, ≤, «)∈ IB" and I∈ II(K), we can find an I'∈ II (K) with ∀A∈L(I(A) ⊇ I'(A)), s.t. $-_{I'}$ satisfies any given instance of K-7 involving K.

Proof : Given A, A'∈ L, define I'(X) = {X∈ I(A&A') | ∀Y∈ P(A&A')'(X<<Y → Y∈ I(A)∪I(A'))} for |- X↔(A&A'), as well as I'(Y) = I(Y) for all the other Y∈ L. Because it satisfies the density (majorize X∈ P(A&A') = P(A)∪P(A') successively in I(A) / I(A'), I(A&A'), (if possible) I(A') / I(A), I(A&A'), and use P(A)'∪P(A')' ⊇ P(A&A')') and the upwards closure (trivial) requirements, I' will belong to II(K). Given that I'(A)∪I'(A') = I(A)∪I(A') ⊇ I'(A&A') (X<<X), the instantiations of K-7 for A, A' will then directly follow from the definitions.

Now, we can use these results to analyze the behaviour of our doxastic transformations - and *. Because of {I(A) | I∈ II(K)} = F(P(A),<<), it follows from what we said before that K-A = $cl_{|-}$(<<-ΔP(A))∩K° = $cl_{|-}$(∪{∩I(A)∩K° | I∈ II(K)})∩K° = ∪{$cl_{|-}$(∩I(A)∩K°) | I∈ II(K)}) = ∪{K-$_I$A | I∈ II(K)}) (*) (note that ({∩I(A)∩K° | I∈ II(K)}, ⊂) is a directed system and for all A in L, if I(A) ⊇ I'(A), then K-$_{I'}$A ⊇ K-$_I$A). Similarly, we have K*A = $cl_{|-}$((K-¬A)∪{A}) = $cl_{|-}$(∪{K-$_I$A | I∈ II(K)}∪{A}) = $cl_{|-}$(∪{K-$_I$A∪ {A} | I∈ II(K)}) = ∪($cl_{|-}$(K-$_I$A∪ {A}) | I∈ II(K)}) = ∪{K*$_I$A | I∈ II(K)} (**). Our next theorem will be mainly a corollary of these considerations.

Theorem 4.3 Given a full belief set structure (IB, |-), the corresponding doxastic relevance contraction function - satisfies K-1, 2, 3, 6, 7 and the corresponding revision function * supports K*1, 3, 4, 6. For those doxastic states with « = $id_{K°}$ ("freeness"), we get in addition K-4, K-5, K*2, K*5 and K*7, but neither K-8 nor K*8.

Proof : First of all, note that the counterexamples in 4.1 will not be affected by our switch to - and * because for the chosen I, $-_I$ = - and $*_I$ = *. Given (*), (**), 4.1, 4.2 and the form of the postulates (often Horn-like), the mentioned principles are easily checked by inspection (using directedness for K-7) or derived (like K*7 for free states) from the standard AGM-representation theorems. Concerning K-8, K*8, we have the following (free) counterexamples.

K-8 is not valid. Consider e.g. K = ($cl_{|-}$({A, A'}), ≤, $id_{K°}$) with two relevance degrees, where (for logically independent A, A') A&A', A∨¬A',¬A∨A', A' < T, A, A↔A', A∨A'. Then it is easy to check that (A ∉) K-A&A' = $cl_{|-}$({A∨¬A'}) is not included in K-A = $cl_{|-}$({A'∨¬A}).

K*8 is not valid. Let $K = (cl_{\vdash}(\{\neg A\}), \leq, \ll)$ be a free doxastic state and < be given by *rest of* $K^{\circ} < A\&A' \rightarrow \neg A'' \approx A\&A' \rightarrow A'' \approx A \rightarrow A'\&A''$. Now, $\neg A' \notin K*A$ and $A'' \in (K*A)+A'$, but $A'' \notin K*A\&A'$.

Whether K*7 is valid in general is for the moment still an open problem. But we think that it probably fails (note the role of K-5 for the AGM-theorems).

5 Comments and conclusions

Now, there are several reasons for this friendly but sensible departure from the Gaerdenforsian normative framework.

First of all, we have to mention the intrusion of doxastic dependency, which allows a more direct control of the revision process. Earlier approaches (cf. [Lindström, Rabinowicz 91, Weydert 91]), which saw the need for such an instrument on the borderline of coherentism and foundationalism (cf. [Gärdenfors 88]), tried to attribute this role to the epistemic preference relation itself. Our intuitions, however, don't support such a confusing identification of preference / relevance and dependency. That's why we have preferred, at least for the moment, to work with two different pre-orders. Further investigations should determine the exact relationship between these concepts and / or lead to some useful constraints. We can model unconditional protection of non-tautological A's (by T«A), but also conditional protection (by B«A, i.e. contingent sentences on the left), which explains the failures of K-4, K*2, K*5 and K-5. So, we don't force our cognitive agents to stay passive and welcome every inherently consistent incoming information. In particular, we get rid of the often criticized recovery postulate K-5, which in a sense is antithetical to the existence of epistemic dependencies. A less desirable consequence would be the possible failure of K*7, which could be partly due to missing logical constraints for our dependency notion.

Another important change is induced by our switch from well-behaved (e.g. well-founded), logically constrained (e.g. basic preference and entrenchment relations) and layered (epistemic relevance / entrenchment) structures to arbitrary pre-orders, which reflects much better the very general character of the priorities agents may use when reorganizing their beliefs. Note that in doxastic relevance contexts, contrary to what happens for weak epistemic entrenchment relations, it is the relevance-maximizing contraction procedure which is responsible for the violation of K-8 and K*8, not the use of complete pre-orders (cf. counterexamples).

As far as the satisfaction of the AGM-postulates is concerned, doxastic relevance revision is formally incomparable with (but intuitively stronger than) Doyle's rational revision [Doyle 91] (K-1, 2, 4) and (in non-free contexts) weaker than Nebel's epistemic relevance approach [Nebel 92] (K*1 - 7). But in general, our framework appears to be able to provide a slightly better mixture of flexibility, intuitive adequacy, expressive power and formal strength.

References

Alchourrón, C., Gärdenfors, P. and Makinson, D. : 1985, On the Logic of Theory Change : Partial Meet Contraction and Revision Functions, *Journal of Symbolic Logic* **50**, 510-530.

Alchourrón, C. and Makinson, D. : 1985, On the Logic of Theory Change : Safe Contraction, *Studia Logica* **44**, 405 - 422.

Doyle, J. : 1991, Rational Belief Revision (preliminary report). In *Principles of Knowledge Representation and Reasoning: Proceedings of the Second International*

Conference, J.A. Allen et al. (eds.), Morgan Kaufmann, CA, pp. 163-174.
Gärdenfors, P. : 1988, *Knowledge in Flux : Modeling the Dynamics of Epistemic States*, Bradford Books, MIT Press, Cambridge, Mass. .
Gärdenfors, P. and Makinson, D. : 1988, Revisions of Knowledge Systems using Epistemic Entrenchment. In *Theoretical Aspects of Reasoning about Knowledge*, M. Vardi (ed.), Morgan Kaufmann, Los Altos, pp. 83-95.
Katsuno, H. and Mendelzon, A.O. : 1991, Propositional Knowledge Base Revision and Minimal Change. In *Artificial Intelligence* **52**, pp. 263-294.
Lindström, S. and Rabinowicz, W. : 1991, Epistemic Entrenchment with Incomparabilities and Relational Belief Revision. In *The Logic of Theory Change*, A. Fuhrmann and M. Morreau (eds.), Springer LNCS 465, Berlin etc., pp. 93-126.
Morreau, M. : 1992, Epistemic Semantics for Counterfactuals. In *Journal of Philosophical Logic* **21**, 33-62.
Nebel, B. : 1991, Belief Revision and Default Reasoning: Syntax-Based Approaches. In *Principles of Knowledge Representation and Reasoning: Proceedings of the Second International Conference*, J.A. Allen et al. (eds.), Morgan Kaufmann, CA, pp. 417-428.
Nebel, B. : 1992, Syntax-Based Approaches to Belief Revision. In Belief Revision, P. Gärdenfors (ed.), Cambridge University Press, Cambridge, forthcoming.
Rott, H. : 1991, A Nonmonotonic Conditional Logic for Belief Revision I. In *The Logic of Theory Change*, A. Fuhrmann and M. Morreau (eds.), Springer LNCS 465, Berlin etc., pp. 135-183.
Rott, H. : 1992, Preferential Belief Change using Generalized Epistemic Entrenchment, *Journal of Logic, Language and Information* , to appear.
Weydert, E. : 1991, Doxastic Preference Logic, a New Look at Belief Revison. In *Logics in AI*, J. van Eijck (ed.), Springer LNAI 478, Berlin etc., pp. 526-543.
Weydert, E. : 1992, *Preferential Belief Revision.* Manuscript.

Modellings for Belief Change: Base Contraction, Multiple Contraction, and Epistemic Entrenchment (Preliminary Report)

Hans Rott

Fachgruppe Philosophie, Universität Konstanz, Postfach 5560, D–7750 Konstanz, Germany

Abstract. This paper draws a distinction between the set of explicit beliefs of a reasoner, the "belief base", and the beliefs that are merely implicit. We study syntax-based belief changes that are governed exclusively by the structure of the belief base. In answering the question whether this kind of belief change can be reconstructed with the help of something like an epistemic entrenchment relation in the sense of Gärdenfors and Makinson [8], we extract several candidate relations from a belief base. The answer to our question is negative, but an approximate solution is possible, and in some cases the agreement is even perfect. Two interpretations of the basic idea of epistemic entrenchment are offered. It is argued that epistemic entrenchment properly understood involves multiple belief changes, i.e., changes by sets of sentences. Since none of our central definitions presupposes the presence of propositional connectives in the object language, the notion of epistemic entrenchment becomes applicable to the style of knowledge representation realized in inheritance networks and truth maintenance systems.

1. Introduction

1.1. Representation of beliefs

Our model of belief will be a simple one. A *belief* is represented by a sentence in some (regimented) language. Research done in Artificial Intelligence has recently lead to a revival of the logic of belief. It was felt that a clear distinction should be drawn between the *explicit* and the *implicit* beliefs of a reasoner [14, 18]. The former ones are those that the reasoner would assent to if asked and for which he has some kind of independent warrant. The latter ones are those that follow, by some specified logic, from the set of explicit beliefs.

We distinguish a *belief base*, the set of explicit beliefs, from a *belief set*. A belief set is closed under logical consequences, it is a *theory* in the logician's sense. In general, we conceive of belief sets as generated by belief bases. Let us say that H is a *belief base for the belief set* K if and only if K is the set of all logical consequences of H, i.e., if $K = Cn(H)$.

We must make a decision what to count as a *belief state*. A belief state is that kind of thing, pre-theoretically understood, that is changed when we change our beliefs. As we cannot read off from a belief set K which beliefs in it are the explicit ones, a belief state cannot be just a belief set. Should we then say that a belief state is modelled by a belief base H? Of course, we then have no problem in generating the full belief set, provided we have fixed an appropriate logic Cn. So everything we could possibly want to know about the set of currently entertained beliefs can be answered if H is known. However, as we shall see, there is a dynamical problem with this conception. In the sort of changes we shall consider, we cannot satisfy two desiderata at the same time: the desideratum that the changed belief state can be characterized by a belief base, and the desideratum that this belief base contains the set of explicit beliefs after the change has been effected. This is an unpleasant state of affairs which we shall have to put up with in this paper. Giving an answer to our question, we say that a *belief state* is a pair $\langle H, K\rangle$ such that H is a belief base for K. However, the reader be warned that our change operations are not making belief states out of belief states in response to a certain input. We shall explain this in the next section.

Before doing that, let us delineate the object language and its logic. The logic of belief change, and especially the theory of epistemic entrenchment, has been discussed for a language with the expressiveness of propositional logic, including all its connectives $\neg$, $\wedge$, $\vee$, $\rightarrow$ and $\leftrightarrow$, as well as the truth and falsity constants $\top$ and $\bot$. In contrast to this, we aim at reducing the linguistic prerequisites. Our considerations are to apply also to systems using severely restricted languages, as encountered e.g. in inheritance nets or truth (reason) maintenance systems.

Correspondingly, the logic governing our language has to obey only structural rules. We require that it is *reflexive*, *monotonic*, *transitive*, and *compact*. We refer to our logic either as a consequence operator Cn or as an inference relation $\vdash$, with the usual understanding that $\phi \in Cn(H)$ iff $H \vdash \phi$. In the first notation our four requirements become

(R) $H \subseteq Cn(H)$

(M) If $H \subseteq H'$ then $Cn(H) \subseteq Cn(H')$

(T) $Cn(Cn(H)) \subseteq Cn(H)$

(C) If $\phi \in Cn(H)$ then $\phi \in Cn(H')$ for some finite subset H' of H

When we link our considerations to earlier work, we make use of connectives. Then the logic is further supposed to be supraclassical, i.e., what follows classically from a given premise set should follow from it in Cn. We also assume that Cn satisfies the deduction theorem.

1.2. Dynamics

A belief change occurs if a belief state is changed in order to accommodate it to a certain input. In the case we are going to deal with, the input comes in the form of (explicit) beliefs. In the research program initiated by Alchourrón, Gärdenfors and Makinson ([3]; for excellent surveys, see [6] and [16]), belief states are identified with belief sets, and inputs are single sentences. Still working in broadly the same research program, Fuhrmann [4, 5] and Hansson [9, 10, 11] offer modellings for two important generalizations. They investigate what happens when belief states are identified with belief bases (with belief sets as special cases) and when the input comes in sets of sentences (with singletons as special cases). In short, they generalize the theory of belief change to *base changes* and *multiple changes*.

It is clear from the very beginning that the idea of base change is indeed compelling. True, it is reasonable to say that what an agent really believes is the belief set K, including the full set of his implicit beliefs. But it is at least as reasonable to think of *belief change operations* as acting on the set of explicit beliefs alone. After all, merely implicit beliefs have a secondary status, they are derived from the explicit ones. And if some of the explicit beliefs they depend on should have to give way, so should they! This is a foundationalist picture of belief revision and contrasts with the coherentist picture predominant in the current theory of belief revision [7, 13]. We will endorse the philosophy of base change in this paper.

Again, it is a good idea to be ready for set-like inputs. But this issue does not seem to have the same philosophical force as base contraction. Philosophically, base change is an alternative to theory change, while multiple change is just an extension of singleton change. There seems to be no intimate connection between these two kinds of deviation from the original framework of Alchourrón, Gärdenfors and Makinson. However, we shall argue that multiple belief changes play a significant role in the analysis of base changes.

1.3. Three types of belief change

The simplest type of belief change is the addition of a new belief ϕ (or a set of beliefs) which is consistent with the old beliefs. In this case, we have no problem to identify the relevant operations. We can effect theory change through base change. Using the symbol '+', we define *consistent additions* as follows:

$$H + \phi = H \cup \{\phi\}$$
$$K + \phi = Cn(K \cup \{\phi\})$$

Notice that '+' has two different meanings here, depending on whether its first argument is supposed to be a belief base or a belief set. It is obvious how to generalize these definitions when the input comes in sets. However, the generalization will be far from obvious in the remaining cases, so we shall restrict ourselves to singleton inputs in the rest of this section.

The operation of accommodating a belief state to some input is considerably more difficult if the latter is inconsistent with the former. In this case, it is held that consistency should act as an integrity constraint for our belief system. For such *belief-contravening additions*, we shall adopt the following idea: In order to rationally include ϕ into the set H (or K) of your beliefs, first make H (or K) consistent with ϕ, i.e., recant the commitment to $\neg\phi$, and then add ϕ consistently to the resulting set. It is common to use the term '*revision*' to cover both consistent and belief-contravening additions, and to use the symbols '$*$' for revisions and '$\dot{-}$' for *contractions*. The above idea which is credited to Isaac Levi in the literature then becomes:

$$H * \phi = (H \dot{-} \neg\phi) + \phi = (H \dot{-} \neg\phi) \cup \{\phi\}$$
$$K * \phi = (K \dot{-} \neg\phi) + \phi = Cn((K \dot{-} \neg\phi) \cup \{\phi\})$$

This is the *Levi identity*, in its two versions for base and for theory change. One may think that the Levi identity is not of much help as long as we do not know how the contraction operation $\dot{-}$ behaves. This is right, but still it reduces the problem of finding suitable revision operations to the problem of finding suitable contraction operations. Philosophically, contraction appears to be the more fundamental operation. Like most authors in belief revision, we shall follow Levi's advice and study contractions in the following.

What is this fundamental interesting operation called 'contraction'? The contraction of a set of beliefs with respect to an input sentence ϕ is a subset of the original beliefs which does not logically imply ϕ. (In a sense, "input sentences" for contractions are rather "output sentences".) The concept of logical consequence is obviously relevant here. In case we start with a belief set K, we should end up with another belief set $K\dot{-}\phi$ which is logically closed again. In contrast to the case of additions, we do not want to stipulate that the contracted belief set $K\dot{-}\phi$ can always be identified with the set of logical consequences of a new belief base $H\dot{-}\phi$. We will explain why presently.

1.4. The basic idea of minimal change

When forced to perform a belief change, it seems rational to preserve as many of the prior beliefs as possible . Many writers have embraced such a condition of minimal change (minimum mutilation, maximal conservativity, informational economy) for many different purposes [17]. We will use the label 'minimal change approach' as a proper name for that account of belief revision which covers at least maxichoice, full, and partial meet contraction in the sense of Alchourrón, Gärdenfors and Makinson [3].

This is the *basic idea of minimal change:* In order to contract a belief base H (or a belief set K) with respect to ϕ, look at the *maximal* subsets of H (of K) which do not imply ϕ. Since every piece of information is valuable, no gratuitous loss of beliefs is tolerated. Accordingly, we may say that a set H_1 of beliefs is *better than* (or *preferred to*) a set H_2 (relative to the belief base H) if H_1 preserves more explicit beliefs than H_2, that is, if $H_2 \cap H$ is a proper subset of $H_1 \cap H$. If H_1 and H_2 are subsets of H, this of course reduces to $H_2 \subset H_1$. Generalizing a bit, we say that a set $\mathcal{H}_1$ of sets of sentences is *better than* (or *preferred to*) a set $\mathcal{H}_2$ of sets of sentences, in symbols $\mathcal{H}_2 \sqsubset \mathcal{H}_1$, if for every H_2 in $\mathcal{H}_2$ there is an H_1 in $\mathcal{H}_1$ with $H_2 \cap H \subset H_1 \cap H$.

2. Base contraction and multiple contraction

In the following, the term 'base contraction' is not to be taken literally. *What* is changed is the theory $K = Cn(H)$ generated by a base H. But *how* the theory is changed depends on the way it is axiomatized, on the form of H. For instance, while $H = \{p, q\}$ and $H' = \{p \wedge q\}$ generate the same theory K, we expect that $K\dot{-}p$ contains q if K is axiomatized by H, but that q is lost if K is axiomatized by H'. In the latter case, q is inseperable from p. Throughout this paper, we assume that syntactical information (the structure of explicit beliefs) is the *sole* mechanism controlling belief change.

The minimal change approach is afflicted with a decisive difficulty. In general, there is more than one solution to the minimal change problem, i.e., more than one maximal set of beliefs which does not imply ϕ. Following Alchourrón and Makinson [1], we let $H\bot\phi$ denote the set of all maximal subsets of H which fail to imply ϕ. The point is that there is usually more than just one member in $H\bot\phi$. What then to do? Assuming that only the syntactical information provided by the base governs a theory's dynamical behaviour, we adopt an egalitarians's point of view. All elements of $H\bot\phi$ are to be treated equally.[1]

The *bold* or *credulous* option is *maxichoice* base change: In order to eliminate ϕ from K, choose one element of $H\bot\phi$ at random, and close under Cn.

DEFINITION 1. *Let H be a base for H and γ be a (single-valued) choice function which selects, for every nonempty set $H\bot\phi$, an arbitrary element H' of $H\bot\phi$. Then the* maxichoice base contraction *over K determined by H and γ is given by*

$$\psi \in K\dot{-}\phi \quad \textit{iff} \quad \nvdash \phi \textit{ and } \gamma(H\bot\phi) \vdash \psi, \textit{ or } \vdash \phi \textit{ and } \psi \in K.$$

Being maximally conservative, maxichoice contraction comes as close to the idea of minimal change as possible. However, if we do not have any information to govern the choice of some particular element of $H\bot\phi$, there is no guarantee that γ selects "the right" one. Believers do not play dice. The arbitrariness of maxichoice contractions is avoided by the next model for belief revision.

The *skeptical* option is *full meet* base change: In order to eliminate ϕ from K, take all the elements of $H\bot\phi$, then close each under Cn, and finally take the intersection.[2]

DEFINITION 2. *For any base H for K, the* full meet base contraction *over K determined by H is given by*

$$\psi \in K\dot{-}\phi \quad \textit{iff} \quad \nvdash \phi \textit{ and } H' \vdash \psi \textit{ for every } H' \in H\bot\phi, \textit{ or } \vdash \phi \textit{ and } \psi \in K.$$

Full meet contractions depart from the idea of minimal change, because the intersection of a set of maximal non-implying subsets is not itself a maximal non-implying subset. However, the symmetrical consideration of each element of $H\bot\phi$ is required by our decision to let in no other information than is encoded in the structure of the explicit beliefs. Opting for full meet contraction thus means adhering to the equality of rights of the members in $H\bot\phi$.

Alchourrón and Makinson [1] have shown that both maxichoice and full meet contraction make good sense only if H is a non-theory. So let us emphasize right at the beginning that it is indeed essential for the following constructions that we have at our disposal a differentiation between explicit and implicit beliefs. This is not only a distinction which is desirable intuitively, but also a technical prerequisite.

It would not be quite right to characterize our proposals as "theory change through base change" [5]. We do not want to stipulate that $K\dot{-}\phi = Cn(H\dot{-}\phi)$

[1] This should not be confused with the idea that all elements of H are equally well *entrenched*. In general they are not, according to Definition 10 below.

[2] Essentially the same method is applied by Veltman [24] and Kratzer [12] for the analysis of counterfactuals, and by Poole [20] for nonmonotonic reasoning.

for some appropriate $H \dot{-} \phi$. Let us illustrate why. Consider $H = \{p, q\}$ and retract $p \wedge q$ from $K = Cn(H)$. $H \bot (p \wedge q) = \{\{p\},\{q\}\}$, so under $K \dot{-} \phi = Cn(H \dot{-} \phi)$, maxichoice would give us either $K \dot{-} (p \wedge q) = Cn(\{p\})$ or $K \dot{-} (p \wedge q) = Cn(\{q\})$, while full meet would give us $K \dot{-} (p \wedge q) = Cn(\emptyset)$. Neither of these solutions seems satisfactory. Intuitively, $K \dot{-} (p \wedge q) = Cn(\{p \vee q\})$ would be good. Assuming that one of p and q may be false, we should still cling to the belief that the other one is true. But $H' = \{p \vee q\}$ is no base which can be constructed naturally from H—it certainly does not record any explicit belief. So we give up the aim of getting $\langle K \dot{-} \phi, H \dot{-} \phi \rangle$ from $\langle K, H \rangle$ and stay content with the more modest aim of getting $K \dot{-} \phi$ from K with the help of the belief base H. That is, H is relevant, and indeed all that is relevant, for the construction of $K \dot{-} \phi$ from K, but H will not get revised itself. Pictorially, instead of the desirable transition $\langle K, H \rangle \overset{\phi}{\longmapsto} \langle K \dot{-} \phi, H \dot{-} \phi \rangle$ we will study the transition $K \overset{\phi, H}{\longmapsto} K \dot{-} \phi$. There will be no suggestion as to the contents of $H \dot{-} \phi$.

When inputs come in sets, we are presented with two different kinds of contraction. The aim of a *pick contraction* is to give up at least one element of a set S, while the aim of a *bunch contraction* is to give up every element of a set S, both times with minimal mutilation of the original belief state.[3] In conformity with the basic idea of minimal change, we again focus on maximal non-implying subsets of H.

Let $H \bot \langle S \rangle$ be the set of all maximal subsets of H which do not imply every element of S, and $H \bot [S]$ the set of all maximal subsets of H which do not imply any element of S. Clearly, $H \bot \langle \{\phi\} \rangle = H \bot [\{\phi\}] = H \bot \phi$.

The concepts of maxichoice and full meet base contraction can be generalized naturally to cover pick and bunch contractions as well. As the case of maxichoice contractions is entirely analogous, we restrict ourselves to full meet contractions. Borrowing Fuhrmann's [4] symbols, we introduce

DEFINITION 3. *For any base H for K, the* pick *and* bunch *versions of* multiple full meet base contraction *over K are defined as follows:*

$$\psi \in K \dot{-} \langle S \rangle \ \textit{ iff } \ H' \vdash \psi \textit{ for every } H' \in H \bot \langle S \rangle \neq \emptyset, \textit{ or } H \bot \langle S \rangle = \emptyset \textit{ and } \psi \in K.$$
$$\psi \in K \dot{-} [S] \ \textit{ iff } \ H' \vdash \psi \textit{ for every } H' \in H \bot [S] \neq \emptyset, \textit{ or } H \bot [S] = \emptyset \textit{ and } \psi \in K.$$

From now on, we will drop curly brackets within pointed and square brackets, so $K \dot{-} [\{\phi, \psi\}]$ will simplify to $K \dot{-} [\phi, \psi]$, and $H \bot \langle \{\phi, \psi\} \rangle$ to $H \bot \langle \phi, \psi \rangle$, etc.

3. Epistemic entrenchment

The concept of epistemic entrenchment has turned out to be a natural and fruitful instrument for the analysis of belief change [6, 7, 8, 15, 21, 22, 23]. 'Epistemic entrenchment' is just another word for comparative retractability. Intuitively, $\phi < \psi$ means that it is easier to discard ϕ than to discard ψ. We may call this *the basic*

[3] André Fuhrmann [4] was probably the first to study pick and bunch contractions. He called them choice and meet contractions. For danger of confusion with maxi*choice* and full *meet* contraction, we introduce new names.

idea of epistemic entrenchment. Below we shall offer two interpretations of this idea in order to make it more precise.

Technically, epistemic entrenchment relations are known to have a number of characteristic properties. The basic postulates are

(EE1) $\top \not< \top$ (Non-Triviality)

(EE2$^\uparrow$) if $\phi < \psi$ and $\psi \vdash \chi$, then $\phi < \chi$ (Continuing Up)

(EE2$^\downarrow$) if $\phi < \psi$ and $\chi \vdash \phi$, then $\chi < \psi$ (Continuing Down)

(EE3$^\uparrow$) if $\phi < \psi$ and $\phi < \chi$, then $\phi < \psi \wedge \chi$ (Conjunction Up)

(EE3$^\downarrow$) if $\phi \wedge \psi < \psi$, then $\phi < \psi$. (Conjunction Down)

There is an equivalent and more economical set of postulates which does not mention any connective of the object language. First, we replace Non-Triviality by irreflexivity. Second, we note that postulates (EE2$^\uparrow$) and (EE3$^\uparrow$) taken together are equivalent to (EE$^\uparrow$), while (EE2$^\downarrow$) and (EE3$^\downarrow$) taken together are equivalent to (EE$^\downarrow$) (see [23]):

(EE$^\uparrow$) if $\phi < \psi$ for every ψ in a non-empty set S and $S \vdash \chi$, then $\phi < \chi$

(EE$^\downarrow$) if $\phi < \psi$ and $\{\psi, \chi\} \vdash \phi$ then $\chi < \psi$.

The set of basic postulates may be supplemented by the following ones.

(EE4) if $H \not\vdash \bot$, then: $\bot < \phi$ iff $H \vdash \phi$ (Minimality)

(EE5) if $\not\vdash \phi$, then $\phi < \top$ (Maximality)

(EE6) if $\phi < \psi$, then $\phi < \chi$ or $\chi < \psi$ (Virtual Connectivity)

Again purely structural formulations of (EE4) and (EE5) are possible by substituting 'there is a ψ such that $\psi < \phi$ (such that $\phi < \psi$)' for '$\bot < \phi$' (for '$\phi < \top$'). For the motivation and discussion of all these postulates, see Gärdenfors and Makinson [8] and Rott [23]. Epistemic entrenchment relations are required to satisfy (EE1) – (EE3$^\downarrow$) in [23], and in addition (EE4) – (EE6) in [8]. (In fact, Gärdenfors and Makinson work with the non-strict relation $\leq$ which can be defined from the strict relation $<$ by taking the converse complement.)

Given a relation of epistemic entrenchment, how can we get a contraction function from it? For the principal case, where $\phi \in K$ and $\phi < \top$, the standard definition [8, 23] is

DEFINITION 4. *For any relation $<$ of epistemic entrenchment, the* large EE-contraction *with respect to $<$ is given by*

$$\psi \in K \dot{-} \phi \ \textit{ iff } \ \psi \in K, \textit{ and } \phi < \phi \vee \psi \textit{ or} \vdash \phi$$

The presence of the disjunction $\phi \vee \psi$ here is somewhat mysterious (to say the least). An alternative idea was ventilated in Rott [21]:

DEFINITION 5. *For any relation* $<$ *of epistemic entrenchment, the* small EE-contraction *with respect to* $<$ *is given by*

$$\psi \in K \dot{-} \phi \;\; \textit{iff} \;\; \psi \in K, \textit{ and } \phi < \psi \textit{ or} \vdash \phi$$

Both Definition 4 and Definition 5 make sure that $K \dot{-} \phi$ is a theory and that the contraction function $\dot{-}$ satisfies a number of rationality postulates. Large EE-contractions, but not small EE-contractions, satisfy the so-called postulate of *recovery:* $K \subseteq (K \dot{-} \phi) + \phi$.

It follows from ($EE2^{\uparrow}$) that $K \dot{-} \phi$ according to Definition 5 is a subset of $K \dot{-} \phi$ according to Definition 4—whence the names. Lindström and Rabinowicz [15, Section 5] argue convincingly that given an epistemic entrenchment relation $<$, any reasonable contraction of K with respect to ϕ should result in a belief set which includes the small and is included in the large EE-contraction.

The basic idea of epistemic entrenchment is still very vague and ought to be made more precise. The *first* or *competitive interpretation* of it suggests to determine the relative ease of retracting a sentence by looking at the fate of ϕ and ψ in a direct competition between ϕ and ψ. It reconstructs epistemic entrenchment from observed contraction behaviour [8, 23]:

DEFINITION 6. *For any contraction function* $\dot{-}$ *over* K, *the* epistemic entrenchment relation revealed by $\dot{-}$ *is given by*

$$\phi < \psi \;\; \textit{iff} \;\; \psi \in K \dot{-} (\phi \wedge \psi) \textit{ and } \phi \notin K \dot{-} (\phi \wedge \psi).$$

Definition 6 yields extremely nice results for large EE-contraction functions over a theory K. If the contraction function $\dot{-}$ satisfies certain rationality postulates, then $<$ as obtained by Definition 6 is a relation of epistemic entrenchment from which we can recover $\dot{-}$ with the help of Definition 4. And conversely, if $<$ is a relation of epistemic entrenchment, then $\dot{-}$ as obtained by Definition 4 satisfies certain rationality postulates and permits a reconstruction of $<$ with the help of Definition 6. Details can be found in Gärdenfors and Makinson [8] and Rott [23].

In [23], I emphatically adopt the idea that $K \dot{-} (\phi \wedge \psi)$ is to be interpreted as a *multiple* contraction, viz. the pick contraction with respect to $\{\phi, \psi\}$. Contracting K with respect to $\phi \wedge \psi$, I argued, is exactly the same as retracting at least one of ϕ and ψ. In symbols, $K \dot{-} (\phi \wedge \psi) = K \dot{-} \langle \phi, \psi \rangle$. The motivation of Definition 6 is then clear: if you have to give up either ϕ or ψ, and you give up ϕ and keep ψ, then ψ has been more entrenched than ϕ.

I do not see any intuitive reason for supposing that the identity $K \dot{-} (\phi \wedge \psi) = K \dot{-} \langle \phi, \psi \rangle$ is inadequate in some applications. Still it is good to be prepared for this possibility. Another motive for modifying Definition 6 is that we want to avoid explicit mentioning of particular connectives, in order to make the epistemic entrenchment approach applicable to restricted languages as encountered for instance in semantic networks. We take the motivation of Definition 6 seriously and suggest the following improvement:

DEFINITION 7. *For any contraction function* $\dot{-}$ *over* K, *the* epistemic entrenchment relation revealed by $\dot{-}$ *is given by*

$$\phi < \psi \;\; \textit{iff} \;\; \psi \in K\dot{-}\langle\phi,\psi\rangle \;\textit{and}\; \phi \notin K\dot{-}\langle\phi,\psi\rangle.$$

This interpretation of the basic idea of epistemic entrenchment builds on the concept of pick contraction.

4. Full meet base contraction as extended epistemic entrenchment contraction

In order to be able to deal with full meet base contractions in terms of epistemic entrenchment, we extend the basic idea of epistemic entrenchment to sets of sentences. From now on, '$\langle S\rangle \ll \langle T\rangle$' is intended to mean that it is easier to discard some element of S than to discard some element of T. '$[S] \ll [T]$' is intended to mean that it is easier to discard all elements of S than to discard all elements of T. We shall speak of *extended epistemic entrenchment* in the sequel, with the two types *pick* and *bunch entrenchment*.

Let us try to extend the competitive interpretation of epistemic entrenchment accordingly. For pick entrenchment, this is easy. The obvious suggestion is

$$\langle S\rangle \ll \langle T\rangle \;\text{ iff }\; T \subseteq K\dot{-}\langle S\cup T\rangle \text{ and } S \not\subseteq K\dot{-}\langle S\cup T\rangle.$$

But for bunch entrenchment, there is no sensible condition which can be formalized with the present means.

So we propose another understanding of—possibly extended—epistemic entrenchment. The *second* or *minimal change interpretation* of the basic idea of epistemic entrenchment builds on the basic idea of minimal change. It reads 'is easier' as 'does not require as great an informational loss as' or 'sacrifices fewer explicit beliefs than'. Formally, preference is identified with the proper subset relation '$\subset$'. Let us define the following version of extended epistemic entrenchment:

DEFINITION 8. *For any base H for K, the relation $\ll$ of* bunch entrenchment *generated by H is defined by*

$[S] \ll [T]$ *iff for every $H' \subseteq H$ such that $Cn(H') \cap T = \emptyset$ there is an H'' such that $H' \subset H'' \subseteq H$ and $Cn(H'') \cap S = \emptyset$, and $\not\vdash \psi$ for every ψ in S.*

The following equivalent formulation is sometimes more convenient:

OBSERVATION 1. *Let H be a base for K, and $\ll$ be the bunch entrenchment generated by H. Then $[S] \ll [T]$ iff for every $H' \in H\bot[T]$ there is an $H'' \in H\bot[S]$ such that $H' \subset H''$, and $H\bot[S] \neq \emptyset$, i.e., iff $H\bot[T] \sqsubset H\bot[S]$, and $H\bot[S] \neq \emptyset$.*

(The proofs of the observations are given in the full paper.) We spare the reader the analogous definition of pick entrenchment, and we do not want to enter into a discussion of the properties of $\ll$. We now observe that full meet base contractions allow an elegant characterization in terms of bunch entrenchment.

OBSERVATION 2. *Let H be a base for K, $\dot{-}$ be the full meet base contraction determined by H, and $\ll$ be the relation of bunch entrenchment generated by H. Then $\psi \in K\dot{-}\phi$ iff $[\phi] \ll [\phi,\psi]$.*

We can directly represent full meet base contractions as extended EE-contractions in the following sense.

DEFINITION 9. *For any relation* $\ll$ *of bunch entrenchment, the* EEE-contraction *with respect to* $\ll$ *is given by*

$$\psi \in K \dot{-} \phi \text{ iff } \psi \in K, \text{ and } [\phi] \ll [\phi, \psi] \text{ or } \vdash \phi$$

Like Definition 5, this definition is connective-free.

5. Epistemic entrenchment generated by belief bases

Given a base H for K, we now try to find a more familiar, i.e., non-extended, relation of epistemic entrenchment without beforehand committing ourselves to a certain contraction method. We again exploit the basic idea of epistemic entrenchment. As it happens, the competitive and the minimal change interpretation of it can be unified in the present case. For the definition, we employ the latter one:

DEFINITION 10. *For any base* H *for* K, *the* relation of epistemic entrenchment generated by H *is given by*

$$\phi < \psi \text{ iff } \nvdash \phi \text{ and for every } H' \subseteq H \text{ such that } H' \nvdash \psi \text{ there is an } H'' \text{ such that } H' \subset H'' \subseteq H \text{ and } H'' \nvdash \phi.$$

This is a singleton version of Definition 8. Clearly, it is a *negative* interpretation of epistemic entrenchment, focussing on the ways to *discard* a belief. It is intuitively well-motivated. Roughly, ψ is more entrenched than ϕ iff for every way of discarding ψ there is a better way of discarding ϕ. As a special case of Observation 1, we take down

OBSERVATION 3. *Let* H *be a base for* K, *and* $<$ *be generated by* H. *Then* $\phi < \psi$ *iff* $\nvdash \phi$ *and for every* $H' \in H \bot \psi$ *there is an* $H'' \in H \bot \phi$ *such that* $H' \subset H''$, *i.e., iff* $\nvdash \phi$ *and* $H \bot \psi \sqsubset H \bot \phi$.

So ψ is more entrenched than ϕ if for every "best" way of discarding ψ there is a still better "best" way of discarding ϕ. More exactly, in terms of maxichoice contraction functions, if for every γ there is a γ' such that the maxichoice base contraction of K with respect to ϕ determined by γ' properly includes the maxichoice base contraction of K with respect to ψ determined by γ.

In the following, we trace some of the implications of this definition. First, we verify that Definition 10 generates a relation of epistemic entrenchment in the generalized sense of Rott [23], but not in the standard sense of Gärdenfors and Makinson [8]. Then we show that for full meet base contractions the two interpretations of the basic idea of epistemic entrenchment coincide. In the next section, we show that full meet base contractions, which are EEE-contractions characterized by Definitions 8 and 9, can be interpolated by means of small and large EE-contractions based on the epistemic entrenchment relation generated by the base. Although the approximation cannot in general be strengthened to an identity, sometimes a perfect agreement can be attained.

It is easy to verify that the relation $<$ defined in Definition 10 has the following properties:

Observation 4. *For every belief base H, the relation $<$ generated by H satisfies (EE1) - (EE3$^{\downarrow}$) and (EE4) - (EE5), but it does not satisfy (EE6).*

That is, Definition 10 does not yield an epistemic entrenchment relation in the sense of Gärdenfors and Makinson [8], but it does yield an epistemic entrenchment relation in the less demanding sense of Rott [23].

If pick contractions are formalized as in Definition 3, Definition 10 turns out to be equivalent with the first interpretation of the basic idea of epistemic entrenchment as formalized in Definition 7. In the present context, the competitive and the minimal change interpretations of epistemic entrenchment coincide.

Observation 5 (Coincidence Lemma). *Let H be a base for K, let $<$ be the epistemic entrenchment relation generated by H and $<'$ be the epistemic entrenchment relation revealed by the pick version of multiple full meet base contraction. Then $\phi < \psi$ iff $\phi <' \psi$.*

6. Full meet base contractions as approximated by epistemic entrenchment contractions

Suppose that the epistemic entrenchment relation $<$ is generated by the belief base H for K. We wonder about the relation between full meet base contractions generated by H (or alternatively, by Definitions 8 and 9) on the one hand and the large and small EE-contractions based on $<$ on the other. Is it possible to get what we got by extended epistemic entrenchment above with the help of singleton epistemic entrenchment? As for large EE-contractions, the answer must be negative, because they are known to satisfy the recovery postulate, which base contractions notoriously do not.

It turns out that singleton epistemic entrenchment is insufficient in general, but an approximation by upper and lower bounds is possible. The entrenchment relation determined by a belief base H with the help of Definition 10 allows us to follow the above-mentioned recommendation of Lindström and Rabinowicz. We can interpolate full meet base changes according to Definition 2, i.e., EEE-changes according to Definitions 8 and 9, by large and small EE-changes according to Definitions 4 and 5.

Observation 6 (Interpolation Lemma). *Let H be a base for K and $<$ be the entrenchment relation generated by H. Furthermore, let $\dot{-}$ be the full meet base contraction function determined by H, let $\dot{-}_1$ be the small and $\dot{-}_2$ be the large EE-contraction with respect to $<$. Then*

$$K \dot{-}_1 \phi \;\subseteq\; K \dot{-} \phi \;\subseteq\; K \dot{-}_2 \phi \,.$$

The converse inclusions are not valid.

In a couple of cases, the correspondence between full meet base contraction and epistemic entrenchment contraction is perfect, if the latter is to mean large EE-contraction based on the relation $<$ generated by the belief base H. There are at least three ways of equivalence. We list them in increasing importance.

Theories. The first case is when the base H is already a belief set, i.e., when $H = K$. However, this case is of limited relevance. The epistemic entrenchment relation generated by K is nearly empty if K is a theory, because then $K' \in K\bot\psi$ and $K'' \in K\bot\phi$ imply $K' \not\subseteq K''$, unless K'' equals K. So in this case, $\phi < \psi$ according to Definition 10 can hold only if either $\psi \in K$ and $\phi \notin K$, or $\vdash \psi$ and $\nvdash \phi$. This corresponds to a well-known trivialization result of Alchourrón and Makinson [1, Observation 2.1] for full meet contractions of theories.

Nebel's blown-up contractions. The second case in point is when full meet base contraction is supplemented with a mechanism to enforce the recovery postulate. This is basically the suggestion of Nebel [19]:

DEFINITION 11. *For any base H for K, the* blown-up contraction $\dot{-}$ *is given by*

$\psi \in K\dot{-}\phi$ *iff* $(K\dot{-}_1 \phi) \cup \{\phi\rightarrow\chi : \chi \in H\} \vdash \psi$, *where* $K\dot{-}_1 \phi$ *is the full meet base contraction determined by* H.

The set $Rec = \{\phi\rightarrow\chi : \chi \in H\}$ is a recovery ticket which allows one to "undo" a base contraction with respect to ϕ. It is easy to check that on Definition 11, $K = (K\dot{-}\phi) + \phi$, for every ϕ in K. But since clearly $\neg\phi$ implies every element of *Rec*, and *Rec* in turn implies $\phi\rightarrow\psi$ for every ψ in K, we find that $\psi \in K\dot{-}\phi$ according to Definition 11 iff $\phi \vee \psi \in K\dot{-}\phi$ according to Definition 2. It is not difficult to see that this is equivalent to saying the ϕ is in the large EE-contraction of K based on the epistemic entrenchment relation generated by H.

Revisions based on the Levi identity. Thirdly, the correspondence is perfect if a contraction is only an intermediate for a revision constructed with the help of the Levi identity. Since, by Levi and the deduction theorem, ψ is in $K * \phi$ iff $\phi\rightarrow\psi$ is in $K\dot{-}\neg\phi$, we have to check $K\dot{-}\neg\phi$ only for sentences of the form $\phi\rightarrow\psi$. But clearly, for every EE-relation $<$, $\neg\phi < (\phi\rightarrow\psi)$ is equivalent to $\neg\phi < \neg\phi\vee(\phi\rightarrow\psi)$, so Definitions 4 and 5 are equivalent for sentences of the form $\phi\rightarrow\psi$ in $K\dot{-}\neg\phi$. Hence, by the Interpolation Lemma, either form of EE-revision is identical with full meet base revision.

7. Conclusion

The aim of this report has been to provide an illustration for the versatility of the concept of epistemic entrenchment, to apply epistemic entrenchment to belief states ("bases") which are not supposed to be logically closed, and to further the intuitive understanding of epistemic entrenchment and its relation to multiple contraction.

Our starting point has been a fixed belief base H generating a belief set K. Our concern is "syntax-based" belief change, or belief change determined by belief bases, and we assume that the structure of H is the sole information governing the changes of K. We have given a reformulation of full meet base contractions as extended EE-contractions: Definition 2 is equivalent to the combination of Definitions 8 and 9. This representation depends on an extension of epistemic entrenchment to sets of sentences ("bunch entrenchment"). We elaborated on the basic idea of epistemic entrenchment as comparative retractability by giving it two different readings. The usual "competitive" interpretation was distinguished from what we called the "minimal change interpretation" of the phrase 'ψ is harder to discard than ϕ'.

We proposed a method of extracting an epistemic entrenchment relation $<$ from a belief base H. Discovering that Definition 10 is equivalent to the combination of Definitions 3 and 7, we observed a confluence of the two interpretations of epistemic entrenchment (the "Coincidence Lemma"). It was demonstrated that upper and lower bounds of full meet base contractions can be specified in the form of large and small EE-contractions based on the relation $<$ generated by the belief base (the "Interpolation Lemma").

Since the publication of Gärdenfors's *Knowledge in Flux*, relations of epistemic entrenchment have been known to be interdefinable with belief contractions. For theory change by singletons, the following transitions are standard in the literature:

$$\phi < \psi \text{ iff } \psi \in K \dot{-} (\phi \wedge \psi) \text{ and } \phi \notin K \dot{-} (\phi \wedge \psi)$$

$$\psi \in K \dot{-} \phi \text{ iff } \psi \in K, \text{ and } \phi < \phi \vee \psi \text{ or } \vdash \phi$$

To my mind, there is no denying that these bridge principles are the pivotal points of an illuminating and well-developed theory of belief change [6, 8, 21, 22, 23]. However, the occurrences of '$\wedge$' and '$\vee$' are somewhat mysterious. This is why I suggest a more transparent way to think of the interdefinability between epistemic entrenchment and belief change.

$$\phi < \psi \text{ iff } \psi \in K \dot{-} \langle \phi, \psi \rangle \text{ and } \phi \notin K \dot{-} \langle \phi, \psi \rangle \qquad \text{(Definition 7)}$$

> *Full meet base specialization:* if $\dot{-}$ is the full meet base contraction determined by H, then, by Observations 3 and 5, $\phi < \psi$ is definable by $H \bot \psi \sqsubset H \bot \phi$
>
> *Singleton reformulation:* $\psi \in K \dot{-} (\phi \wedge \psi)$ and $\phi \notin K \dot{-} (\phi \wedge \psi)$

$$\psi \in K \dot{-} \phi \text{ iff } \psi \in K, \text{ and } [\phi] \ll [\phi, \psi] \text{ or } \vdash \phi \qquad \text{(Definition 9)}$$

> *Full meet base specialization:* if $\ll$ is the bunch entrenchment generated by H, then, by Observation 2, full meet base contraction coincides with EEE-contraction, and $\psi \in K \dot{-} \phi$ is definable by $H \bot [\phi, \psi] \sqsubset H \bot [\phi]$
>
> *Singleton interpolation:* in so far as $\phi < \psi$ implies $[\phi] \ll [\phi, \psi]$, and this in turn implies $\phi < \phi \vee \psi$, large and small EE-contractions can serve as upper and lower bounds of EEE-contractions

Our deviation from the standard account is clear. We invoke sets with two elements as arguments for contraction operations and entrenchment relations. More specifically, we replace, in the direction from belief change to epistemic entrenchment, the contractions with respect to conjunctions by pick contractions, and in the direction from epistemic entrenchment to belief change, the entrenchments of disjunctions by bunch entrenchments.

What is the reward for this exercise? First and foremost, we get a better understanding of the relevant interrelations. They sometimes happen to reduce to the standard definitions. But what is really meant by the latter is, I submit, precisely what is made explicit by the new definitions. In one direction, I should think there is virtually no difference: $K \dot{-} (\phi \wedge \psi)$ seems to be intuitively identifiable with $K \dot{-} \langle \phi, \psi \rangle$.

In the other direction, however, it is only the restricted context of theory change by singletons that makes our new definition reduce to the old one: $[\phi] \ll [\phi, \psi]$ may—and must!—then be identified with $\phi < \phi \vee \psi$.

Secondly, we manage without reference to any particular connective of the object language. Thus the theory of epistemic entrenchment becomes applicable to systems using a severely restricted language. For instance, we can speak of the entrenchment of the nodes in inheritance nets or reason maintenance systems (also called "truth maintenance systems"). There ought to be a corresponding connective-free formulation of the so-called Gärdenfors postulates for contraction operations. The obvious suggestion is to replace occurrences of '$K \dot{-} (\phi \wedge \psi)$' by '$K \dot{-} \langle \phi, \psi \rangle$'. The elimination of connectives, however, works only for belief contractions. Belief revisions constructed according to the Levi identity make use of negations, and there does not seem to be a straightforward way to avoid this.[4]

At last, we should like to give two warnings. The connective-free formulation of epistemic entrenchment relations and theory contractions is only a by-product of this paper, slightly improving on the presentation in [23]. It is not necessary for the analysis of syntax-based belief change which turns essentially on the syntactical structure of the items in a belief base. There is no immediate transfer of insights from belief base update to updates in inheritance networks or reason maintenance systems ("RMSs") with their unstructured "nodes". It may be expedient for some purposes to identify RMS "justifications" with Horn clauses. But this certainly does not suffice for nonmonotonic systems. Our Cn is supposed to be monotonic.

Multiple contraction and extended epistemic entrenchment have been found to be an appropriate means for analyzing base contraction. However—this is the second warning—, the conepts of multiple contraction and extended epistemic entrenchment themselves, cut loose from the special context of maxichoice and full meet base contraction, are still very much in need of a thoroughgoing analysis. This is evidently beyond the scope of the present paper.

References

[1] Alchourrón, Carlos, and David Makinson (1982): "On the logic of theory change: Contraction functions and their associated revision functions", *Theoria* **48**, 14–37.

[2] Alchourrón, Carlos, and David Makinson (1985): "On the logic of theory change: Safe contraction", *Studia Logica* **44**, 405–422.

[3] Alchourrón, Carlos, Peter Gärdenfors and David Makinson (1985): "On the logic of theory change: Partial meet contraction and revision functions", *Journal of Symbolic Logic* **50**, 510–530.

[4] Fuhrmann, André (1988): *Relevant Logics, Modal Logics, and Theory Change*, PhD thesis, Australian National University, Canberra.

[5] Fuhrmann, André (1991): "Theory contraction through base contraction", *Journal of Philosophical Logic* **20**, 175–203.

[6] Gärdenfors, Peter (1988): *Knowledge in Flux: Modeling the Dynamics of Epistemic States*, Bradford Books, MIT Press, Cambridge, Mass.

[4] Thanks to an anonymous referee for pointing this out.

[7] Gärdenfors, Peter (1990): "The dynamics of belief systems: Foundations vs. coherence theories", *Revue Internationale de Philosophie* **44**, 24–46.

[8] Gärdenfors, Peter, and David Makinson (1988): "Revisions of knowledge systems using epistemic entrenchment", in *Proceedings of the Second Conference on Theoretical Aspects of Reasoning about Knowledge*, Moshe Vardi ed., Los Altos, CA: Morgan Kaufmann, pp. 83–95.

[9] Hansson, Sven Ove (1989): "New operators for theory change", *Theoria* **55**, 114–132.

[10] Hansson, Sven Ove (1991): *Belief base dynamics*, Doctoral Dissertation, Uppsala University.

[11] Hansson, Sven Ove (1992): "Reversing the Levi identity", *Journal of Philosophical Logic*, forthcoming.

[12] Kratzer, Angelika (1981): "Partition and revision: The semantics of counterfactuals", *Journal of Philosophical Logic* **10**, 201–216.

[13] Harman, Gilbert (1986): *Change in View*, Bradford Books, MIT Press, Cambrige, Mass.

[14] Levesque, Hector J. (1984): "A logic of implicit and explicit belief", *Proceedings 3rd National Conference on Artificial Intelligence*, Austin TX, pp. 198–202.

[15] Lindström, Sten, and Wlodzimierz Rabinowicz (1991): "Epistemic entrenchment with incomparabilities and relational belief revision", in André Fuhrmann and Michael Morreau eds., *The Logic of Theory Change*, Springer-Verlag, Lecture Notes in Artificial Intelligence 465, Berlin, pp. 93–126.

[16] Makinson, David (1985): "How to give it up: A survey of some formal aspects of the logic of theory change", *Synthese* **62**, 347–363.

[17] Makinson, David (1992): "The five faces of minimal change", unpublished manuscript.

[18] McArthur, Gregory L. (1988): "Reasoning about knowledge and belief: A survey", *Computational Intelligence* **4**, 223–243.

[19] Nebel, Bernhard (1989): "A knowledge level analysis of belief revision", in Ronald Brachman, Hector Levesque and Raymond Reiter eds., *Principles of Knowledge Representation and Reasoning. Proceedings of the 1st International Conference*, Morgan Kaufmann, San Mateo, Ca., pp. 301–311.

[20] Poole, David (1988): "A logical framework for default reasoning", *Artificial Intelligence* **36**, 27–47.

[21] Rott, Hans (1991): "Two methods of constructing contractions and revisions of knowledge systems", *Journal of Philosophical Logic* **20**, 149–173.

[22] Rott, Hans (1992a): "On the logic of theory change: More maps between different kinds of contraction function", in Peter Gärdenfors ed., *Belief Revision*, Cambridge University Press, Cambridge, pp. 122–141.

[23] Rott, Hans (1992b): "Preferential belief change using generalized epistemic entrenchment", *Journal of Logic, Language and Information* **1**, 45–78.

[24] Veltman, F. (1976): "Prejudices, presuppositions and the theory of counterfactuals", in Jeroen Groenendijk and Martin Stokhof eds., *Amsterdam Papers of Formal Grammar*, Vol. I, Centrale Interfaculteit, Universiteit Amsterdam, pp. 248–281.

A Framework for Default Logics*

Christine Froidevaux and Jérôme Mengin**

Laboratoire de Recherche en Informatique
Bat. 490 - Université Paris Sud
91405 ORSAY CEDEX
France

Abstract. We present a general framework for default logics, which encompasses most of the existing variants of default logic. It allows us to compare those variants from a knowledge representation point of view. We then exploit this framework to generalize Brewka's work on cumulative default logic, and to give an operational definition of extensions. Proof theoretical and semantical aspects are investigated.

1 Introduction

Default logic is one of the most famous formalisms for representing general rules that have exceptions, and reasoning with them. Its original formulation, by means of fixed points of some operator [Rei80], is very natural, but hides some undesirable effects, due in particular to the interaction between the defaults, which is not well handled. In order to correct some of these undesirable effects, several authors give other formulations of default logic. But, as these variants are still expressed in terms of fixed points of some operators, the interaction between the defaults, and thus the kind of reasoning that these variants really express remain unclear.

We present in Sect. 1 another approach to default logic, which is inspired by the work by Bonté and Levy [BL89]. It is based on the usual language of default logic. But instead of using a fixed-point construction, we build *regular* and *saturated* paths of reasoning, that lead to what one might consider as valid and complete views of the world, given some default theory under consideration. In Sect. 2, we show the link with the usual fixed-point approach to default logic. In Sect. 3 we give some applications of our approach to default logic: we study cumulativity in default logic, generalizing Brewka's work [Bre91]; we provide an operational definition for the extensions of a default theory, and investigate proof theoretical and semantical aspects, in the spirit of Lukaszewicz [Luk88].

Due to the lack of space, we omit the proofs of some simple results, and give a sketch of proof for the others. All complete proofs can be found in [FM92].

2 The Framework

2.1 General Principles

Nonmonotonic rules are represented in default logic by special inference rules, called defaults, of the form $\frac{a(x):b(x)}{c(x)}$, where $a(x)$, $b(x)$ and $c(x)$ are formulas of some logical

* This work was partially supported by the Esprit - BRA project DRUMS

** Work by the second author was partially performed during his stay at the Queen Mary and Westfield College, London, United Kingdom

language $\mathcal{L}$, and x is their vector of free variables. $a(x)$ is called the *prerequisite* of the default, $b(x)$ its *justification*[3] and $c(x)$ its *consequent*. It can be read for an individual t: if $a(t)$ is true and if $b(t)$ is consistent, then infer $c(t)$. A default is closed if its prerequisite, justification and consequent contain no free variable. A closed default theory is a pair $\Delta = (W, D)$, where W is a set of closed formulas of $\mathcal{L}$, and D a set of closed defaults. We consider, in the rest of the paper, closed default theories only.

W contains the facts that are certain about a situation (Etherington [Eth88] calls them "hard facts"), whereas D contains some general rules about a domain, rules that might have exceptions. These rules, together with the usual inference rules and axioms of some logic L, are used to make deductions from W, whenever it is coherent to do so. Roughly, the difficulty is to formalize when it is coherent to use some defaults. Most variants of default logic define some operator that associates, to a set of formulas E, the set of formulas that can be deduced from E with the defaults that are coherent with respect to E: the extensions of a default theory are defined as the fixed points of such an operator.

In this paper, we prefer to consider that W is extended by applying some default rules, according to the following general principles:

P1 A default can be applied only if it is *active*, i.e. if its prerequisite can be proved (from W using first order logic and maybe other default rules that can be applied too)[4].

P2 A set of defaults that are applied together in order to build an extension must remain *regular*: this regularity condition varies from one variant of default logic to another, but in most of them it requires the resulting extension to be consistent. The regularity condition also specifies in which sense the justifications of the defaults used to build an extension have to be consistent. If the application of several defaults together leads to an irregular set of defaults, then several "smaller" extensions are built instead, by separately applying several regular sets of defaults.

P3 An extension is *saturated*: any default which is applicable to an extension must be applied to it. The applicability condition is the other varying element (with the regularity) from one variant of default logic to another one (note that the first principle only gives a necessary condition to the applicability of a default).

P4 An extension is *deductively closed* under first order logic provability.

Let us now introduce some notations that we will use in the rest of this paper. Given a default d, we denote its prerequisite by $\mathrm{Pre}(d)$, its justification by $\mathrm{Jus}(d)$, and its consequent by $\mathrm{Cons}(d)$, i.e. $d = \frac{\mathrm{Pre}(d):\mathrm{Jus}(d)}{\mathrm{Cons}(d)}$. Similarly, if D is a set of defaults, we denote by $\mathrm{Pre}(D)$ the set $\{\mathrm{Pre}(d), d \in D\}$, by $\mathrm{Jus}(D)$ the set $\{\mathrm{Jus}(d), d \in D\}$, and by $\mathrm{Cons}(D)$ the set $\{\mathrm{Cons}(d), d \in D\}$. The symbol $\vdash$ denotes the consequence relation of some logic L. We suppose that this consequence relation is monotonic, reflexive, and compact. Given a set of formulas, we denote by $\mathrm{Th}(E)$ the set of its logical consequences, $\mathrm{Th}(E) = \{f, E \vdash f\}$. Lastly we suppose that W is consistent (most authors agree to say that when W is inconsistent, the theory has only one extension which is inconsistent, that is, the set $\mathcal{L}$ of all formulas).

[3] For the sake of clarity, we consider in this paper that the justification of a default consists of one formula only. However, we briefly describe in Sect. 2.6 how most results can be generalized to defaults that have multiple justifications.

[4] All the variants of default logic that we study in this paper agree on this principle. See e.g. [ZM89], [Wil90] for variants of default logic that do not agree on this principle.

2.2 Grounded Sets of Defaults

The first principle, which gives the notion of a "directed path of reasoning", is formalized in Schwind [Sch90] with "grounded sets of defaults", and in Bonté and Levy [BL89] with "universes". We give below a definition of a grounded set of defaults which is equivalent to that of [Sch90].

Definition 1. Let $\Delta = (W, D)$ be a default theory, a **set of defaults grounded in** W (or, when there is no possible confusion, a grounded set of defaults) is a subset U of D such that $U = \bigcup_{n \in \mathbb{N}} U_n$, where the sequence $(U_n)_{n \in \mathbb{N}}$ is defined by:

- $U_0 = \emptyset$,
- $U_{n+1} = \{d \in U, W \cup \mathrm{Cons}(U_n) \vdash \mathrm{Pre}(d)\}$, for all n in $\mathbb{N}$.

A default d is **active** with respect to a set of defaults U grounded in W if and only if $W \cup \mathrm{Cons}(U) \vdash \mathrm{Pre}(d)$.

A set of defaults is grounded in W if the prerequisites of all its elements can be proved from W and the consequents of other defaults in the same set, and without there being any cycle in those proofs. Notice that if U is grounded in W, then a default d is active with respect to U if and only if $U \cup \{d\}$ is grounded in W, if and only if there exists $n \geq 0$, such that $W \cup \mathrm{Cons}(U_n) \vdash \mathrm{Pre}(d)$.

We discuss in subsequent sections the formalization of the notions of regularity and applicability (Principles 2 and 3). However, we can already formalize the notion of saturation, whatever the notion of applicability may be.

Definition 2. Let $\Delta = (W, D)$ be a default theory. A set $U \subseteq D$ of defaults, grounded in W, is **saturated** if all the defaults of D which are applicable to U are in U.

In this paper, we will focus on different grounded sets of defaults U characterized by various properties. But, whatever these properties may be, an extension is obtained as the set of theorems of $W \cup \mathrm{Cons}(U)$ (in the sense of the logic L).

Definition 3. Let $\Delta = (W, D)$ be a default theory. A set E of formulas is **generated by** a set of defaults U, grounded in W, if $E = \mathrm{Th}(W \cup \mathrm{Cons}(U))$. Moreover, E is an **extension** of Δ if E is generated by a regular and saturated set of defaults grounded in W.

2.3 Regularity

The weak regularity is the notion which underlies the first definitions of extensions, i.e. Reiter's [Rei80] and Lukaszewicz' [Luk88]. Like all notions of regularity that we will study, the weak regularity of a grounded set of defaults requires the set of theorems that it generates to be consistent. The following example illustrates another required condition:

Example 1. Let $\Delta = (W, D)$, with $W = \{a\}$, $D = \{d_1, d_2\}$, $d_1 = \frac{a:b}{c}$, $d_2 = \frac{a:\neg c}{d}$. Here the prerequisite of the two defaults, a, is in W and $W \cup \mathrm{Cons}(\{d_1, d_2\})$ is consistent. But if the default d_1 is used to generate an extension then, the default d_2 cannot be applied in the same extension, because its justification, $\neg c$, is not consistent with the extension.

The crucial point in the example above is that a grounded set of defaults can be considered valid only if for any default d in U, the justification of d is consistent with W and the consequents of the defaults in U. The intuitive meaning of a default $\frac{a:b}{c}$ is in this case: "if a is known, and if b is consistent with what is known, then infer c". As Delgrande and Jackson say [DJ91], "the justifications are individually consistent with the extension". Levy has formalized this with the following definition:

Definition 4. [Lev91] Let $\Delta = (W, D)$ be a default theory, a set of defaults $U \subseteq D$ grounded in W is said to be **weakly regular**[5] if $W \cup \text{Cons}(U) \cup \{\text{Jus}(d)\}$ is consistent for all d in U.

The second notion of regularity interprets the justification of a default as an assumption, which is implicitly concluded whenever the default is applied, even if not explicitly concluded. In this sense, it is not possible to assume that a proposition is both true and false at the same time. More generally this notion of regularity requires that the justifications of the defaults in a grounded set of defaults are all together consistent with the generated set of theorems. It can be justified by the following, example produced by Poole:

Example 2 [Poo89]. Let $W = \{\text{broken(left-arm)} \vee \text{broken(right-arm)}\}$, $D = \{d_1, d_2\}$, with $d_1 = \frac{:\text{usable(left-arm)} \wedge \neg\text{broken(left-arm)}}{\text{usable(left-arm)}}$, $d_2 = \frac{:\text{usable(right-arm)} \wedge \neg\text{broken(right-arm)}}{\text{usable(right-arm)}}$. This theory has one weakly regular grounded set of defaults, $\{d_1, d_2\}$, which would generate an extension containing usable(left-arm) and usable(right-arm), although the justifications of the two defaults together are contradictory in this theory. It can seem counterintuitive to make both assumptions $\neg$broken(left-arm) and $\neg$broken(right-arm) when we know broken(left-arm)$\vee$ broken(right-arm).

Definition 5. Let $\Delta = (W, D)$ be a default theory, a set of defaults $U \subseteq D$ grounded in W is said to be **strongly regular** if $W \cup \text{Cons}(U) \cup \text{Jus}(U)$ is consistent.

Example 2 continued. The grounded set of defaults $\{d_1, d_2\}$ is not strongly regular. Instead, $\{d_1\}$ and $\{d_2\}$ are two strongly regular grounded sets of defaults. They generate two extensions containing usable(left-arm) and usable(right-arm) respectively.

With this new definition of regularity, the intuitive meaning of a default $\frac{a:b}{c}$ becomes " if a is known, and if b is consistent with what is known, then *assume* b and infer c", where "assume b" means that $\neg b$ is not consistent in this extension. Although b is not inferred, there is a kind of commitment to the assumption b, which blocks the use of any other default, whose justification would be $\neg b$.

Rychlik [Ryc91] considers the defaults as arguments in favor of their conclusions. For him, an agent needs only one argument in favor of a conclusion. Thus, in order to generate an extension, a grounded set of defaults should not contain any default, whose consequent is subsumed by the consequents of other defaults of this set. His definition of extensions is based on the definition of the subsumption of a default by a set of defaults:

Definition 6. [Ryc91] Let $\Delta = (W, D)$ be a default theory, a default d is said to be **weakly subsumed** by a subset D' of D if there exists a set of defaults $U \subseteq D'$ grounded in W such that

[5] Levy simply says "regular"

$d \notin U$, and
$\text{Cons}(d) \in \text{Th}(W \cup \text{Cons}(U))$.

Then d is said to be **subsumed** by $D' \subseteq D$ if there exists $D'' \subseteq D'$ such that d is weakly subsumed by D'' and no element of D'' is weakly subsumed by D'.

This naturally leads to the following notion of regularity:

Definition 7. Let $\Delta = (W, D)$ be a default theory. Suppose that some notion of regularity is given. A set of defaults $U \subseteq D$ grounded in W is **concisely regular** if it is regular and for any $d \in U$, d is not subsumed by U.

2.4 Applicability

We now study the notion of applicability of a default to a grounded set of defaults. We assume in this section that a notion of regularity of a grounded set of defaults is given. Both notions of weak and strong regularity give the same results for the examples of this section.

The simplest notion of applicability of a default to a grounded set of defaults supposes of course that the default is active with respect to the set of defaults, and requires the resulting grounded set of defaults to be still regular. This can be formalized with the following definition:

Definition 8. Let $\Delta = (W, D)$ be a default theory, and suppose that we have defined a notion of regularity for the grounded sets of defaults of Δ. A default d is **cautiously applicable** to a grounded set of defaults U if d is active with respect to U and if the resulting grounded set of defaults $U \cup \{d\}$ is regular. A grounded set of defaults saturated for this notion of applicability is **cautiously saturated.**

With this definition, starting from an empty set of defaults and by iteratively applying to it defaults which are cautiously applicable to it, one always obtains a regular grounded set of defaults[6]. Moreover, a cautiously saturated grounded set of defaults is a grounded set of defaults which is maximally regular (for set inclusion). This property allows for a very constructive way of building extensions, as shown in Sect. 4.2. As a consequence, any default theory such that W is consistent always has an extension in this sense (maybe generated by the empty set of defaults).

Another property of this notion of applicability is that it leads to **semi-monotonic** default logic in the following sense:

Proposition 1. *Let D and D' be two sets of defaults, with $D \subseteq D'$, and let W be a set of formulas. If U is a regular and cautiously saturated grounded set of defaults of (W, D) then there exists a regular and cautiously saturated grounded set of defaults U' of (W, D') such that $U \subseteq U'$.*

It is also worth mentioning that, when considering cautious applicability, a default theory is equivalent to its semi-normal form:

[6] Thus the name "cautious".

Proposition 2. *Let $\Delta = (W, D)$ be a default theory. Then the sets of formulas generated by the weakly regular (respectively strongly regular) and cautiously saturated subsets of D grounded in W are the sets of formulas generated by the weakly regular (respectively strongly regular) and cautiously saturated subsets of* $\{\frac{a:b\wedge c}{c}, \frac{a:b}{c} \in D\}$ *that are grounded in W.*

The definition of extensions given by Reiter [Rei80] involves another notion of applicability which we formalize as follows:

Definition 9. Let $\Delta = (W, D)$ be a default theory. A default d is **hazardously applicable** to a grounded set of defaults U if d is active with respect to U and if $W \cup \text{Cons}(U) \cup \{\text{Jus}(d)\}$ is consistent. A grounded set of defaults U such that all defaults hazardously applicable to U are in U is **hazardously saturated**[7].

With this notion of applicability, a default $\frac{a:b}{c}$ is not always equivalent to its semi-normal form $\frac{a:b\wedge c}{c}$. Consider for example the theory $(\emptyset, \{\frac{:a}{\neg a}\})$. This theory has no hazardously saturated and regular grounded set of defaults, though the corresponding semi-normal default theory $(\emptyset, \{\frac{:a\wedge\neg a}{\neg a}\})$ has one hazardously saturated and regular grounded set of defaults, namely $\emptyset$. This example also shows that a default theory does not always have an extension generated by a hazardously saturated universe. Furthermore, default logic is not, in general, with respect to hazardous applicability, semi-monotonic.

2.5 Normal Defaults

A default is *normal* if its consequent is logically equivalent to its justification. When a default theory contains normal defaults only, the weak regularity and the strong regularity are equivalent, as well as the cautious applicability and the hazardous applicability.

Proposition 3. *Let $\Delta = (W, D)$ be a normal default theory, and U a grounded set of defaults of Δ. Then U is weakly regular if and only if U is strongly regular, if and only if $W \cup \text{Cons}(U)$ is consistent. Moreover, a grounded subset U of D is cautiously saturated if and only if U is hazardously saturated, if and only if (for all $d \in D$, if d is active with respect to U and $U \cup \{d\}$ is regular then $d \in U$).*

2.6 Multiple Justifications

As we mentioned at the beginning of this section, the original form of a default is $d = \frac{a:b_1,\ldots,b_n}{c}$, where a, $b_1, \ldots, b_n$ and c are logical formulas. The defaults that we have considered until now are defaults whose set of justifications has a unique element. Let us now study how we can extend the various notions of regularity and applicability to the case of defaults having multiple justifications.

In Reiter's definition of extensions, the meaning of such a default $\frac{a:b_1,\ldots,b_n}{c}$ is: "if a is known, and if for all i, b_i is consistent with what is known, then infer c". This is strongly related to the notion of weak regularity. For example, the default theory $(\emptyset, \{\frac{:b,\neg b}{c}\})$ has two weakly regular grounded sets of defaults, $\{\frac{:b,\neg b}{c}\}$ and $\emptyset$. So we should rewrite the notion of weak regularity as follows:

[7] Hazardously saturated grounded sets of defaults are what Levy calls "complete universes"

A set of defaults U will be said to be *weakly regular* if for any default $\frac{a:b_1,\ldots,b_n}{c} \in U$, $W \cup \text{Cons}(U) \cup \{b_i\}$ is consistent for any $1 \leq i \leq n$.

But if we consider the notion of strong regularity, then no grounded set of defaults containing the default $\frac{:b,\neg b}{c}$ in any default theory should be strongly regular, because the conjunction of the justifications $b \wedge \neg b$ of this default is inconsistent. We can rewrite the definition of strong regularity for this more general form of default:

A grounded set of defaults U is *strongly regular* if $W \cup \text{Cons}(U) \cup \left(\bigcup_{d \in U} \{b_i \in \text{Jus}(d)\}\right)$ is consistent.

For this second notion it is equivalent to replace a default $\frac{a:b_1,\ldots,b_n}{c}$ by the one which has only one justification $\frac{a:b_1 \wedge \ldots \wedge b_n}{c}$ (this is not the case with Reiter's definition of extensions, see [Bes89]).

As the cautious applicability does not involve the justifications of the defaults excepted through the notion of regularity, it does not have to be modified for defaults having multiple justifications. The hazardous applicability can be expressed as follows:

Let $\Delta = (W,D)$ be a default theory, a default $d = \frac{a:b_1,\ldots,b_n}{c}$ is *hazardously applicable* to a grounded set of defaults U if d is active with respect to U and if it is true that $W \cup \text{Cons}(U) \cup \{b_i\}$ is consistent for $1 \leq i \leq n$.

3 A Classification of the Various Definitions of Default Logic

3.1 Equivalence Fixed Points - Regular and Saturated Grounded Sets of Defaults

The aim of this chapter is to show that Reiter's default logic and most of its variants fit in our framework, by exhibiting the regularity and saturation conditions that correspond to extensions in each of them. Most of these variants have been introduced as fixed points of some operator Γ which depends on a set D of defaults. This operator Γ is defined on some space of the form $\mathcal{E} = 2^{\mathcal{L}} \times 2^{\mathcal{H}}$, where $\mathcal{L}$ is the logical language, and $\mathcal{H}$ is the logical language too, or the set of defaults, depending on the variant of default logic considered. The extensions of a default theory are then defined as follows:

Definition 10. Let $\Delta = (W,D)$ be a default theory. A set E of formulas is an extension of Δ if there exists $H \subseteq \mathcal{H}$ such that (E,H) is a fixed point of the operator Γ, which associates (E',H') with (E,H), where E' and H' are the smallest sets that verify:

P1 $W \subseteq E'$,
P2 $\text{Th}(E') = E'$,
P3$_{(E,H)}$ $\forall d \in D$, if $\text{Pre}(d) \in E'$ and $Q((E,H),d)$ holds then $\text{Cons}(d) \in E'$ and $f(d) \in H'$,

where

$f(d)$ returns some information about d that must be stored in H', in order to take in account some interaction between the defaults.

Q is some predicate on $\mathcal{E} \times D$, that defines which defaults are used to construct $\Gamma(E,H)$.

Both the mapping f and the predicate Q depend on the variant of default logic considered.

When, as in the case of Reiter's definition, the operator is defined on the set of subsets of the logical language itself, we consider a predicate Q defined on $2^{\mathcal{L}}$, and there is no mapping f.

The next theorem shows the link between regular and saturated grounded sets of defaults, and the fixed points of the operator Γ above.

Theorem 4. *Let $\Delta = (W,D)$ be a default theory. A pair (E,H) is a fixed point of Γ if and only if there exists a subset U of D such that $E = \mathrm{Th}(W \cup \mathrm{Cons}(U))$ and $H = f(U)$ and for all d in D:*

$$d \in U$$
$$\Leftrightarrow$$
$$U \cup \{d\} \textit{ is grounded in } W \textit{ and } \mathbf{Q}((\mathrm{Th}(W \cup \mathrm{Cons}(U)), f(U)), d) \textit{ holds.}$$

Sketch of the proof. Define $\sigma(U) = (\mathrm{Th}(W \cup \mathrm{Cons}(U)), f(U))$. Clearly, σ is an increasing mapping from D on $\mathcal{E}$, i.e. if $U \subseteq U'$, and if $(E,H) = \sigma(U)$, $(E',H') = \sigma(U')$, then $E \subseteq E'$ (because the consequence relation $\vdash$ is reflexive), and $H \subseteq H'$. We also introduce the predicate $\mathbf{P}$ defined by $\mathbf{P}((E,H),(E',H'),d) \equiv (\mathrm{Pre}(d) \in E'$ and $\mathbf{Q}((E,H),d)$ holds). Define, for $(E,H) \in \mathcal{E}$:

$$\rho_{E,H}: \quad \begin{array}{c} \mathcal{E} \to 2^D \\ (E',H') \to \{d \in D, \mathbf{P}((E,H),(E',H'),d)\}, \end{array} \quad \text{and } \Theta_{E,H} = \bigcup_{n\geq 1}(\sigma\rho_{E,H})^n.$$

We show that for all $(E,H) \in \mathcal{E}$,

$$\Gamma(E,H) = \Theta_{E,H}\,\sigma(\emptyset) = \sigma(\bigcup_{n\geq 0} U_n), \qquad (1)$$

where $U_0 = \emptyset$ and for $n \geq 0$, $U_{n+1} = \{d \in D, \mathbf{P}((E,H),\sigma(U_n),d)\}$. The fixed points of Γ are then generated by the sets of defaults U such that $U = \bigcup_{n\geq 0} U_n$, where $U_0 = \emptyset$ and for $n \geq 0$, $U_{n+1} = \{d \in D, \mathbf{P}(\sigma(U),\sigma(U_n),d)\}$.The result follows.

□

The "if and only if" condition in the theorem above leads to a characterization of extensions in terms of regular and saturated grounded sets of defaults, as far as one implication expresses the regularity condition and the groundedness requirement (for all $d \in U$, $U \cup \{d\} = U$ is grounded and $\mathbf{Q}(\sigma(U),d)$ holds), while the other one expresses the saturation condition.

We show, in [FM92], that most definitions of extensions for default logic are instanciations of Definition 10. This enables us to find the regularity and applicability conditions on which these variants are based.

3.2 Lukaszewicz' Extensions

As an example, we show the application of the theorem of the previous section to Lukaszewicz' extensions, in order to find the regularity and saturation conditions to which they correspond.

Definition 11 [Luk88]. E is an **m-extension** of a default theory Δ, **with respect to a set of formulas** F, if $E = \Gamma_{L,1}(E,F)$ and $F = \Gamma_{L,2}(E,F)$ where $\Gamma_{L,1}(E,F)$ and $\Gamma_{L,2}(E,F)$ are the smallest sets of formulas satisfying:

P1 $W \subseteq \Gamma_{L,1}(E,F)$,
P2 $\Gamma_{L,1}(E,F) = \text{Th}(\Gamma_{L,1}(E,F))$,
L3 if $d \in D$, such that $\text{Pre}(d) \in \Gamma_{L,1}(E,F)$ and for all $x \in F \cup \{\text{Jus}(d)\}$, $E \cup \{x, \text{Cons}(d)\}$ is consistent, then $\text{Cons}(d) \in \Gamma_{L,1}(E,F)$ and $\text{Jus}(d) \in \Gamma_{L,2}(E,F)$.

Theorem 5. *Let $\Delta = (W,D)$ be a default theory, and let E be a set of formulas. Then E is an m-extension of Δ if and only if E is generated by a weakly regular and cautiously saturated grounded set of defaults.*

Although a slightly different theorem has already been proved in [SR91], we give here a proof which uses the result of the previous section.

Proof. The definition of Lukaszewicz is an instanciation of definition 10, with $\mathcal{E} = 2^{\mathcal{L}} \times 2^{\mathcal{L}}$, $f(d) = \text{Jus}(d)$, and $Q_L((E,F),d) \equiv \forall x \in F \cup \{\text{Jus}(d)\}, E \cup \{x, \text{Cons}(d)\}$ is consistent. From Theorem 4, the m-extensions of Δ are generated by the grounded sets of defaults U such that: for all $d \in D$, $d \in U$ if and only if (d is U-active and $\forall d' \in U \cup \{d\}, W \cup \text{Cons}(U \cup \{d\}) \cup \{\text{Jus}(d')\}$ is consistent). This is equivalent to: for all $d \in D$, $d \in U$ if and only if (d is U-active and $U \cup \{d\}$ is weakly regular). Hence the result.

□

3.3 A Classification of the Various Definitions of Default Logic

The application of Theorem 4 to the various definitions of default logic that can be found in the literature yields the classification shown in Table 1.

Because they are not given in terms of fixed points of some operator, the definitions of [DJ91] and [dTGCH90] do not exactly correspond to saturated grounded sets of defaults. However, when the set of defaults is *finite*, the J-extensions are generated by strongly regular and cautiously saturated grounded sets of defaults, while the C-extensions are the maximal sets (for set inclusion) generated by weakly regular and cautiously saturated grounded sets of defaults.

The notions of regularity and saturation that we give in Table 1 for Brewka's CDL-extensions and priority preserving CDL-extensions correspond to the extensions obtained when considering the variants introduced by Brewka without the labels that he associates to formulas, as shown in Theorem 7 of Sect. 4.1.

Table 1. Classification of the variants of default logic

	Regularity condition	**Applicability condition**
Reiter's extensions	weak regularity	hazardous applicability
Lukaszewicz' m-extensions	weak regularity	cautious applicability
Guerreiro et al. C-extensions [dTGCH90]	weak regularity	cautious applicability
Delgrande and Jackson's J-extensions [DJ91] (for semi normal default theories)	strong regularity	cautious applicability
Schaub constrained extensions [Sch91b]	strong regularity	cautious applicability
Brewka's CDL-extensions [Bre91]	strong regularity	cautious applicability
Brewka's priority preserving CDL-extensions [Bre91]	strong regularity	d is cautiously applicable to U, or $W \cup \text{Cons}(U \cup \{d\}) \cup \{\text{Jus}(d)\}$ is consistent and $W \cup \text{Cons}(U \cup \{d\}) \cup \text{Jus}(U)$ is inconsistent.
Rychlik's c-extension [Ryc91]	concise weak regularity	$U \cup \{d\}$ is weakly regular and d is not subsumed by U.

4 Applications

4.1 Cumulativity

Cumulativity seems to be one of the meta-logical properties that an inference relation should satisfy [Mak89]. It is expressed as follows:

Definition 12. Let $\mid\!\sim$ denote some inference relation, $\mid\!\sim$ is cumulative if for any set of formulas W, and any formulas x and y,

$$\text{if } W \mid\!\sim y \text{ then } W \mid\!\sim x \text{ iff } W \cup \{y\} \mid\!\sim x.$$

As Brewka mentions [Bre91], it has to be reformulated for the arbitrary choice notion of derivability (for which $W \mid\!\sim_D f$ if there is at least one extension of (W,D) which contains f). Brewka gives the folowing formulation of choice cumulativity:

> If there is at least one extension of $\Delta = (W,D)$ containing y, then E is an extension of Δ containing y if and only if E is an extension of $(W \cup \{y\},D)$.

Brewka proves in [Bre91] that default logic in its current form is not cumulative. The example that he produces to illustrate it contains normal defaults only. Thus it applies to all variants of default logic that were mentionned in Table 1.

Example 3. [Bre91] Let $\Delta = (W,D)$, with $W = \{\text{sing}, \text{bird} \vee \text{dog} \Rightarrow \text{pet}, \neg\text{bird} \vee \neg\text{dog}\}$, and $D = \{\frac{\text{sing:bird}}{\text{bird}}, \frac{\text{pet:dog}}{\text{dog}}\}$. This theory has a single extension, which contains $\text{pet} \wedge \text{bird}$. Let $W' = W \cup \{\text{pet}\}$, then (W',D) has two extensions, respectively generated by $\{\frac{\text{sing:bird}}{\text{bird}}\}$ and $\{\frac{\text{pet:dog}}{\text{dog}}\}$. While dog is in no extension of (W,D), it is in one extension of $(W \cup \{\text{pet}\},D)$.

As long as pet is not in W, the default $\frac{\text{pet:dog}}{\text{dog}}$ cannot be used before bird is known (which permits to conclude pet, using bird $\vee$ dog $\Rightarrow$ pet); but if bird is known, dog is inconsistent (for $\neg$bird $\vee$ $\neg$dog is in W), thus the default still cannot be used. Things are different when pet is added to W: it is possible to use $\frac{\text{pet:dog}}{\text{dog}}$ without knowing bird, and then $\frac{\text{sing:bird}}{\text{bird}}$ can no longer be used. Choice cumulativity is not satisfied, because it is not recorded that the conclusion pet was first obtained by using the default $\frac{\text{sing:bird}}{\text{bird}}$, which blocks $\frac{\text{pet:dog}}{\text{dog}}$.

Brewka defines a cumulative variant of default logic, in which all formulas are labelled. This allows one to know, when a conclusion q obtained by default is added to W, which defaults have to be blocked, because otherwise they would contradict the defaults that were used to deduce q. As his variant of default logic is also motivated by the problem that appears with the weak regularity in the broken-arm example, it turns out that it is based on the notion of strong regularity.

We generalize below Brewka's method for defining a cumulative variant of default logic. It can be applied to most variants of default logic that we have seen. In particular, it could be used to define the cumulative variant of Reiter's default logic obtained by Makinson [Bre91].

The idea is that when a formula q is obtained by default, we must keep a track of the defaults used to deduce it, so that we can make sure that when the formula is added to W, we will not use any other default which would have blocked the deduction of q. For this, we first define a logic which can deal with those tracks:

Definition 13. Let $\mathcal{L}$ be some logical language, and Th() be the consequence operator of a logic based on this language. Let $\mathcal{D}_{\mathcal{L}}$ be the set of defaults that can be built on $\mathcal{L}$. We define a **labelled logical language** $\mathcal{L}_l$, whose formulas are of the form $\langle p : P \rangle$, where $p \in \mathcal{L}$ and $P \subseteq \mathcal{D}_{\mathcal{L}}$. We also define a **labelled consequence operator** Th_l: if $\mathcal{A} \subseteq \mathcal{L}_l$, $\mathrm{Th}_l(\mathcal{A})$ is the smallest set $\mathcal{A}' \subseteq \mathcal{L}_l$ such that:

- $\mathcal{A} \subseteq \mathcal{A}'$,
- if $\langle p_1 : P_1 \rangle, \ldots, \langle p_n : P_n \rangle \in \mathcal{A}'$, and $p \in \mathrm{Th}(\{p_1, \ldots, p_n\})$, then $\langle p : \bigcup_{1 \leq j \leq n} P_j \rangle \in \mathcal{A}'$.

For any set $\mathcal{A}$ of labelled formulas, we define the support and the set of classical formulas of $\mathcal{A}$: $\mathrm{Supp}(\mathcal{A}) = \bigcup_{\langle p:P \rangle \in \mathcal{A}} P$, $\mathrm{Form}(\mathcal{A}) = \{p, \langle p : P \rangle \in \mathcal{A}\}$.

The labels attached to formulas deduced by using defaults will record which defaults have been used to deduce these formulas. The definition of Th_l ensures that if a formula is a classical consequence of $p_1, \ldots, p_n$, and if $P_1, \ldots, P_n$ are the sets of defaults that have been used to deduce $p_1, \ldots, p_n$, then p can be deduced using the defaults of the union of the P_is.

Definition 14. A **labelled default theory** is a pair $\Delta = (\mathcal{W}, \mathcal{D})$, where $\mathcal{W}$ is a set of labelled formulas, and $\mathcal{D}$ a set of labelled defaults, i.e. defaults of the form $\frac{\langle a:A \rangle : b}{\langle c : A \cup \{\frac{a:b}{c}\} \rangle}$, where a, b, c are classical formulas and A is a set of classical defaults.

Such a default $\frac{\langle a:A \rangle : b}{\langle c : A \cup \{\frac{a:b}{c}\} \rangle}$ means: "if a can be deduced, using the defaults in A, and if b is consistent, then c can be deduced, using the defaults in $A \cup \{\frac{a:b}{c}\}$".

In the following, we denote by $\overline{\delta}$ the "classical" part of a labelled default δ: $\overline{\frac{\langle a{:}A\rangle : b}{\langle c{:}A \cup \{\frac{a:b}{c}\}\rangle}} = \frac{a:b}{c}$. Similarly, given a set $\mathcal{U}$ of labelled defaults, we denote by $\overline{\mathcal{U}}$ the set $\{\overline{\delta}, \delta \in \mathcal{U}\}$.

The cumulative extensions of a labelled default theory are defined as follows:

Definition 15. Let $\Delta = (\mathcal{W}, \mathcal{D})$ be a labelled default theory. A set $\mathcal{S}$ of labelled formulas is a **cumulative extension** of Δ if there exists a subset $\mathcal{U}$ of $\mathcal{D}$, grounded in $\mathcal{W}$, such that $\mathcal{S} = \mathrm{Th}_l(\mathcal{W} \cup \mathrm{Cons}(\mathcal{U}))$ and:

(i) $\forall d \in \mathrm{Supp}(\mathcal{W}), \mathrm{Q}_{\mathrm{cum}}(\mathrm{Th}_l(\mathcal{W} \cup \mathrm{Cons}(\mathcal{U})), d)$ holds, and
(ii) $\forall \delta \in \mathcal{D}, \delta \in \mathcal{U} \Leftrightarrow$
$\mathcal{U} \cup \{\delta\}$ is grounded in $\mathcal{W}$ and $\mathrm{Q}_{\mathrm{cum}}(\mathrm{Th}_l(\mathcal{W} \cup \mathrm{Cons}(\mathcal{U})), \overline{\delta})$ holds,

where $\mathrm{Q}_{\mathrm{cum}}$ is some predicate. A labelled default theory $(\mathcal{W}, \mathcal{D})$ is **well-based** if $\mathrm{Q}_{\mathrm{cum}}(\mathcal{W}, d)$ holds for any $d \in \mathrm{Supp}(\mathcal{W})$.

Notice that without condition (i), we obtain the usual extensions of a default theory based, not on the classical language $\mathcal{L}$, but on the language $\mathcal{L}_l$ of our labelled logic. The condition (i) acts as a filter on those extensions: if a formula p of Form$(\mathcal{S})$, where $\mathcal{S}$ is a cumulative extension of $(\mathcal{W}, \mathcal{D})$, depends on the defaults in P, i.e. $\langle p : P\rangle \in \mathcal{S}$, then no cumulative extension of $(\mathcal{W} \cup \{\langle p : P\rangle\}, \mathcal{D})$ can contain a default which contradicts the defaults in P.

Example 3 continued. In the case of normal defaults, where all variants of default logic coïncide, we define $\mathrm{Q}_{\mathrm{cum}}(\mathrm{Th}_l(\mathcal{W} \cup \mathrm{Cons}(\mathcal{U})), \overline{\delta}) \equiv \neg\mathrm{Jus}(\overline{\delta}) \notin \mathrm{Form}(\mathrm{Th}_l(\mathcal{W} \cup \mathrm{Cons}(\mathcal{U})))$. Let $\Delta' = (\mathcal{W}, \mathcal{D})$, with $\mathcal{W} = \{\langle \mathrm{sing} : \emptyset\rangle, \langle \mathrm{bird} \vee \mathrm{dog} \Rightarrow \mathrm{pet} : \emptyset\rangle, \langle \neg\mathrm{bird} \vee \neg\mathrm{dog} : \emptyset\rangle\}$, and $\mathcal{D} = \{\frac{\langle \mathrm{sing}:A\rangle : \mathrm{bird}}{\langle \mathrm{bird}:A \cup \{\frac{\mathrm{sing:bird}}{\mathrm{bird}}\}\rangle}, \frac{\langle \mathrm{pet}:A\rangle : \mathrm{dog}}{\langle \mathrm{dog}:A \cup \{\frac{\mathrm{pet:dog}}{\mathrm{dog}}\}\rangle}, A \subseteq D\}$. The As are the possible labels for the formulas that are of interest, given the set of D from which they can be deduced. This theory has a single cumulative extension, generated by the set $\{\frac{\langle \mathrm{sing}:\emptyset\rangle : \mathrm{bird}}{\langle \mathrm{bird}:\{\frac{\mathrm{sing:bird}}{\mathrm{bird}}\}\rangle}\}$, which contains $\langle \mathrm{pet} : \{\frac{\mathrm{sing:bird}}{\mathrm{bird}}\}\rangle$, and where dog does not appear. Let $\mathcal{W}' = \mathcal{W} \cup \{\langle \mathrm{pet} : \{\frac{\mathrm{sing:bird}}{\mathrm{bird}}\}\rangle\}$, then $(\mathcal{W}', \mathcal{D})$ has a single cumulative extension, generated by $\{\frac{\langle \mathrm{sing}:\emptyset\rangle : \mathrm{bird}}{\langle \mathrm{bird}:\{\frac{\mathrm{sing:bird}}{\mathrm{bird}}\}\rangle}\}$. The set $\{\frac{\langle \mathrm{pet}:\{\frac{\mathrm{sing:bird}}{\mathrm{bird}}\}\rangle : \mathrm{dog}}{\langle \mathrm{dog}:\{\frac{\mathrm{sing:bird}}{\mathrm{bird}}, \frac{\mathrm{pet:dog}}{\mathrm{dog}}\}\rangle}\}$, grounded in $\mathcal{W}'$, which verifies (ii) and contains $\langle \mathrm{dog} : \{\frac{\mathrm{sing:bird}}{\mathrm{bird}}, \frac{\mathrm{pet:dog}}{\mathrm{dog}}\}\rangle$, does not verify (i), thus does not generate a cumulative extension of $(\mathcal{W}', \mathcal{D})$.

This definition of cumulative extensions yields cumulative default logics, as the next theorem indicates:

Theorem 6. *Let $\Delta = (\mathcal{W}, \mathcal{D})$ be a labelled default theory. If there exists at least one cumulative extension $\mathcal{F}$ of Δ such that $\langle p : P\rangle \in \mathcal{F}$, then a set $\mathcal{S}$ of labelled formulas is a cumulative extension of Δ containing $\langle p : P\rangle$ if and only if $\mathcal{S}$ is a cumulative extension of $(\mathcal{W} \cup \{\langle p : P\rangle\}, \mathcal{D})$.*

Sketch of the proof. $\Rightarrow$ If $\mathcal{S}$ is a cumulative extension of $(\mathcal{W}, \mathcal{D})$ that contains $\langle p : P\rangle$, there exists a subset $\mathcal{U}$ of $\mathcal{D}$, grounded in $\mathcal{W}$, such that:

(i) $\forall d \in \mathrm{Supp}(\mathcal{W}), \mathrm{Q}_{\mathrm{cum}}(\mathrm{Th}_l(\mathcal{W} \cup \mathrm{Cons}(\mathcal{U})), d)$ holds, and

(ii) $\forall \delta \in \mathcal{D}, \delta \in \mathcal{U} \Leftrightarrow$
$\mathcal{U} \cup \{\delta\}$ is grounded in $\mathcal{W}$ and $\mathrm{Q}_{\mathrm{cum}}(\mathrm{Th}_l(\mathcal{W} \cup \mathrm{Cons}(\mathcal{U})), \overline{\delta})$ holds,

and $\mathcal{S} = \mathrm{Th}_l(\mathcal{W} \cup \mathrm{Cons}(\mathcal{U}))$. Then $\mathcal{U}$ is grounded in $\mathcal{W} \cup \{\langle p : P\rangle\}$. Moreover, $\mathrm{Supp}(\mathcal{W} \cup \{\langle p : P\rangle\}) = \mathrm{Supp}(\mathcal{W}) \cup P$, and $P \subseteq \mathrm{Supp}(\mathcal{W}) \cup \overline{\mathcal{U}}$, thus for any $d \in P$, $\mathrm{Q}_{\mathrm{cum}}(\mathrm{Th}_l(\mathcal{W} \cup \mathrm{Cons}(\mathcal{U})), d)$ holds. Lastly, $\mathrm{Th}_l(\mathcal{W} \cup \mathrm{Cons}(\mathcal{U})) = \mathrm{Th}_l(\mathcal{W} \cup \{\langle p : P\rangle\} \cup \mathrm{Cons}(\mathcal{U}))$, thus $\forall d, \mathrm{Q}_{\mathrm{cum}}(\mathrm{Th}_l(\mathcal{W} \cup \{\langle p : P\rangle\} \cup \mathrm{Cons}(\mathcal{U})), d) \equiv \mathrm{Q}_{\mathrm{cum}}(\mathrm{Th}_l(\mathcal{W} \cup \mathrm{Cons}(\mathcal{U})), d)$. Thus $\mathcal{S}$ is a cumulative extension of $(\mathcal{W} \cup \{\langle p : P\rangle\}, \mathcal{D})$.

$\Leftarrow$ If $\mathcal{S}$ is a cumulative extension of $(\mathcal{W} \cup \{\langle p : P\rangle\}, \mathcal{D})$, there exists a subset $\mathcal{U}$ of $\mathcal{D}$, grounded in $\mathcal{W} \cup \{\langle p : P\rangle\}$, such that $\mathcal{S} = \mathrm{Th}_l(\mathcal{W} \cup \{\langle p : P\rangle\} \cup \mathrm{Cons}(\mathcal{U}))$ and:

(i) $\forall d \in \mathrm{Supp}(\mathcal{W}) \cup P$, $\mathrm{Q}_{\mathrm{cum}}(\mathrm{Th}_l(\mathcal{W} \cup \{\langle p : P\rangle\} \cup \mathrm{Cons}(\mathcal{U})), d)$ holds, and
(ii) $\forall \delta \in \mathcal{D}, \delta \in \mathcal{U} \Leftrightarrow$ ($U \cup \{\delta\}$ is grounded in $\mathcal{W} \cup \{\langle p : P\rangle\}$ and $\mathrm{Q}_{\mathrm{cum}}(\mathrm{Th}_l(\mathcal{W} \cup \{\langle p : P\rangle\} \cup \mathrm{Cons}(\mathcal{U})), \overline{\delta})$ holds).

There exists a subset $\mathcal{V}$ of $\mathcal{D}$ grounded in $\mathcal{W}$ such that $\mathcal{F} = \mathrm{Th}_l(\mathcal{W} \cup \mathrm{Cons}(\mathcal{V}))$. Let $\mathcal{V}_p$ be a minimal subset of $\mathcal{V}$ grounded in $\mathcal{W}$ such that $\langle p : P\rangle \in \mathrm{Th}_l(\mathcal{W} \cup \mathrm{Cons}(\mathcal{V}_p))$, then $\overline{\mathcal{V}_p} \subseteq P$. Hence for any $\delta \in \mathcal{V}_p$, $\mathrm{Q}_{\mathrm{cum}}(\mathrm{Th}_l(\mathcal{W} \cup \{\langle p : P\rangle\} \cup \mathrm{Cons}(\mathcal{U})), \overline{\delta})$ holds (cf. (i)). Define $\mathcal{V}_{p,0} = \emptyset$ and for $n \geq 0$, $\mathcal{V}_{p,n+1} = \{\delta \in \mathcal{V}_p, \mathrm{Pre}(\delta) \in \mathrm{Th}_l(\mathcal{W} \cup \mathrm{Cons}(\mathcal{V}_{p,n}))\}$. We prove, using (ii), that for al n, $\mathcal{V}_{p,n} \subseteq \mathcal{U}$. Thus $\langle p : P\rangle \in \mathrm{Th}_l(\mathcal{W} \cup \mathrm{Cons}(\mathcal{U}))$. Consequently, as $\mathcal{U}$ is grounded in $\mathcal{W} \cup \{\langle p : P\rangle\}$, it is also grounded in $\mathcal{W}$.

□

The next theorem indicates the links between non-cumulative default logic and cumulative default logic.

Theorem 7. *Let $\Delta = (W, D)$ be a classical default theory. Let $\mathcal{W} = \{\langle w : \emptyset\rangle, w \in W\}$ and $\mathcal{D} = \{\frac{\langle a:A\rangle : b}{\langle c : A \cup \{\frac{a:b}{c}\}\rangle}, \frac{a:b}{c} \in D, A \subseteq D\}$. Define the predicate $\mathrm{Q}_{\mathrm{cum}}$ by: $\mathrm{Q}_{\mathrm{cum}}(\mathcal{S}, d) \equiv \mathrm{Q}(\mathrm{Form}(\mathcal{S}), d)$. Then a set of formulas $E \subseteq \mathcal{L}$ is an extension of (W, D) if and only if there exists a cumulative extension $\mathcal{S}$ of $(\mathcal{W}, \mathcal{D})$ such that $E = \mathrm{Form}(\mathcal{S})$.*

Sketch of the proof. We have $\mathrm{Form}(\mathcal{W}) = W$ and $\mathrm{Supp}(\mathcal{W}) = \emptyset$, thus any set of labelled defaults verifies (i). Hence, from (1), the cumulative extensions of $(\mathcal{W}, \mathcal{D})$ are the sets of labelled formulas generated (in labelled logic) by the sets $\mathcal{U} \subseteq \mathcal{D}$ of labelled defaults, grounded in $\mathcal{W}$, such that $\mathcal{U} = \bigcup_{n \geq 0} \mathcal{U}_n$, where $\mathcal{U}_0 = \emptyset$ and for $n \geq 0$, $\mathcal{U}_{n+1} = \{\delta \in \mathcal{D}, \mathrm{Pre}(\delta) \in \mathrm{Th}_l(\mathcal{W} \cup \mathrm{Cons}(\mathcal{U}_n)), \mathrm{Q}(\mathrm{Th}(W \cup \mathrm{Cons}(\overline{\mathcal{U}})), \overline{\delta})\}$. Similarly, from (1), the extensions of (W, D) are generated by the sets of defaults $U \subseteq D$, grounded in W, such that $U = \bigcup_{n > 0} U_n$, where $U_0 = \emptyset$ and for $n \geq 0$, $U_{n+1} = \{d \in D, \mathrm{Pre}(d) \in \mathrm{Th}(W \cup \mathrm{Cons}(U_n)), \mathrm{Q}(\mathrm{Th}(W \cup \mathrm{Cons}(U)), d)\}$.

Suppose that $U = \bigcup_{n \geq 0} U_n$. Let $\mathcal{U}$ be the largest subset of $\{\frac{\langle a:A\rangle : b}{\langle c : A \cup \{\frac{a:b}{c}\}\rangle}, \frac{a:b}{c} \in U, A \subseteq D\}$ grounded in $\mathcal{W}$. We prove that for all $n \geq 0$, $\overline{\mathcal{U}}_n = U_n$. Moreover $\bigcup_{n \geq 0} \mathcal{U}_n$ is grounded, thus $\bigcup_{n \geq 0} \mathcal{U}_n \subseteq \mathcal{U}$. Lastly if $\delta \in \mathcal{U}$, then $\mathrm{Q}(\mathrm{Th}(W \cup \mathrm{Cons}(\overline{\mathcal{U}})), \overline{\delta})$ holds. As $\mathcal{U}$ is grounded, there exists $n \geq 0$, such that $\delta \in \mathcal{U}_n$. Thus $\mathcal{U} = \bigcup_{n \geq 0} \mathcal{U}_n$.

Conversely, suppose that $\mathcal{U} = \bigcup_{n \geq 0} \mathcal{U}_n$. Let $U = \overline{\mathcal{U}}$. Again we prove that for all $n \geq 0$, $U_n = \overline{\mathcal{U}}_n$, thus $\bigcup_{n \geq 0} U_n = \overline{\bigcup_{n \geq 0} \mathcal{U}_n} = \overline{\mathcal{U}} = U$.

□

Brewka gives in [Bre91] a definition of the extensions of a labelled default theory. Although it is expressed in terms of his assertion logic, it can be as well expressed as follows:

Definition 16. [Bre91] Let $(\mathcal{W},\mathcal{D})$ be an labelled default theory, a **CDL-extension** of $(\mathcal{W},\mathcal{D})$ is a set of labelled formulas $\mathcal{S}$ which is a fixed point of the operator Γ which associates to $\mathcal{S}$ the smallest set $\mathcal{S}'$ such that:

P1 $\mathcal{W} \subseteq \mathcal{S}'$,
P2 $\mathrm{Th}_l(\mathcal{S}') = \mathcal{S}'$,
B3$_{\mathcal{S}}$ for all $\delta \in \mathcal{D}$, if $\mathrm{Pre}(\delta) \in \mathcal{S}'$ and $\mathbf{Q}_B(\mathcal{S},\overline{\delta})$ holds then $\mathrm{Cons}(\delta) \in \mathcal{S}'$,

where $\mathbf{Q}_B$ is the predicate defined by: $\mathbf{Q}_B(\mathcal{S},\overline{\delta}) \equiv \mathrm{Form}(\mathcal{S}) \cup \mathrm{Cons}(\mathrm{Supp}(\mathcal{S}) \cup \{\overline{\delta}\}) \cup \mathrm{Jus}(\mathrm{Supp}(\mathcal{S}) \cup \{\overline{\delta}\})$ is consistent.

From Theorem 4, we deduce that the CDL-extensions of a labelled default theory $(\mathcal{W},\mathcal{D})$ are generated by the subsets $\mathcal{U}$ of $\mathcal{D}$, grounded in $\mathcal{W}$, such that for all $\delta \in \mathcal{D}$,

$$\delta \in \mathcal{U}$$
$$\text{if and only if}$$
$$\mathcal{U} \cup \{\delta\} \text{is grounded in } \mathcal{W} \text{ and } \mathbf{Q}_B(\mathrm{Th}_l(\mathcal{W} \cup \mathrm{Cons}(\mathcal{U})),\overline{\delta}) \text{ holds.}$$

Notice that, in the case where $(\mathcal{W},\mathcal{D})$ is not well-based, the only CDL-extension of $(\mathcal{W},\mathcal{D})$ is $\mathrm{Th}_l(\mathcal{W})$, whereas, as condition (i) of Definition 15 cannot be satisfied for any subset $\mathcal{U}$ of $\mathcal{D}$, $(\mathcal{W},\mathcal{D})$ has no cumulative extension.

If $(\mathcal{W},\mathcal{D})$ is well based, then for any set of labelled defaults $\mathcal{U} \subseteq \mathcal{D}$, property (i) of Definition 15 holds, thus the CDL-extensions and cumulative extensions of $(\mathcal{W},\mathcal{D})$ coïncide.

Let (W,D) be a default theory, and define $(\mathcal{W},\mathcal{D})$ as in Theorem 7. Then $\mathrm{Form}(\mathcal{W}) = W$, and $\mathrm{Supp}(\mathcal{W}) = \emptyset$, thus $(\mathcal{W},\mathcal{D})$ is well based. Moreover, for any $\mathcal{U} \subseteq \mathcal{D}$ grounded in $\mathcal{W}$, $\mathrm{Form}(\mathrm{Th}_l(\mathcal{W} \cup \mathrm{Cons}(\mathcal{U}))) = \mathrm{Th}(W \cup \mathrm{Cons}(\overline{\mathcal{U}}))$, and $\mathrm{Supp}(\mathrm{Th}_l(\mathcal{W} \cup \mathrm{Cons}(\mathcal{U}))) = \overline{\mathcal{U}}$, thus in this case $\mathbf{Q}_B(\mathrm{Th}_l(\mathcal{W} \cup \mathrm{Cons}(\mathcal{U})),d) \equiv W \cup \mathrm{Cons}(\overline{\mathcal{U}} \cup \{d\}) \cup \mathrm{Jus}(\overline{\mathcal{U}} \cup \{d\})$ is consistent. Hence, from Theorem 7, we conclude that the CDL-extensions of $(\mathcal{W},\mathcal{D})$ correspond to the subsets U of D such that for any $d \in D$, $d \in U$ if and only if (d is U-active and $W \cup \mathrm{Cons}(U \cup \{d\}) \cup \mathrm{Jus}(U \cup \{d\})$ is consistent), i.e. the strongly regular and cautiously saturated grounded sets of (W,D).

4.2 Operational Definition of the Extensions

Our operational approach is in the same spirit as Lukaszewicz' approach for a semantical account of default logic [Luk88] and as Besnard's method for proving some properties of Lukaszewicz' formalism [Bes89].

Let $\Delta = (W,D)$ be a default theory and let U be a grounded set of defaults. We define an **operator** $\mathrm{APP}_\Delta()$ which associates with each grounded set of defaults U the set of defaults which are applicable to U and which are not already in U, as follows:

Definition 17. $\mathrm{APP}_\Delta(U) = \{d \in D, d \notin U \text{ and } d \text{ is applicable to } U\}$.

We also define a **nondeterministic choice operator** TC_Δ which, given a grounded set of defaults U, selects a default among the defaults applicable to U which are not already in U:

Definition 18.

$$TC(U) = \begin{cases} U \text{ if } APP_\Delta(U) = \emptyset, \\ U \cup \{d\} \text{ for some } d \in APP_\Delta(U) \text{ otherwise.} \end{cases}$$

Definition 19. The ordinal[8] powers of TC are inductively defined as follows:

$TC^0 = \emptyset$,
$TC^\alpha = TC(TC^{\alpha-1})$ if α is a successor ordinal,
$TC^\alpha = Sup_\subseteq\{TC^\beta, \beta < \alpha\} = \cup_{\beta<\alpha} TC^\beta$ if α is a limit ordinal.

TC^α is a **limit** if for all $\beta \geq \alpha$, $TC^\beta = TC^\alpha$.

We denote by $(TC^\alpha_\rho)_{\alpha \leq \beta}$ the transfinite sequence of ordinal powers until β, under some *default rule selection strategy* ρ, which gives the default rule to be selected in order to get the successor element in the sequence.

It can be proved that such a sequence TC^α always has a limit (which corresponds to the smallest γ such that $APP(TC^\gamma) = \emptyset$), and that this limit is a saturated grounded set of defaults.

It is worth noticing that the existence of a limit does not depend on the strategy ρ chosen. However, we are not sure that the limit is an extension, as the regularity condition might not be verified. Thus, in the general case there remains to check that limits are regular grounded sets of defaults. This is not always the case, but if the applicability condition is the cautious one, it clearly is.

Theorem 8. *Let $\Delta = (W, D)$ be a default theory. Assume that* TC *is a choice operator that selects defaults under a* <u>*cautious*</u> *applicability condition, and that extensions of Δ are defined using this same notion of applicability (whatever the notion of regularity may be). Then a set E of formulas is an extension of Δ if and only if there exists a grounded set of defaults U, limit of* TC, *such that* $E = Th(W \cup Cons(U))$.

Theorem 8 holds for Lukaszewicz' extensions and for Brewka's extensions, but does not hold for Reiter's extensions. However, we can use the operational approach for Reiter's extensions, insofar as we can construct Lukaszewicz' extensions as limits of TC with respect to weak regularity, and then verify whether the limits obtained are hazardously saturated or not.

From the point of view of logic programming, [BF87], [BF91] have shown how logic programs with negation in the premises can be seen as particular default theories associated with them. The semantics obtained (default models semantics or equivalently stable semantics [GL88]) is defined by Herbrand models of the Reiter's extensions. Our operational approach provides a way of defining these default models by iteration. We can take into account any restriction on the order in which logic rules have to be selected by considering a well-suited strategy. For example, stratified logic programs lead us to select default rules in one stratum after the other.

[8] see e.g. [Llo87] for details about transfinite arithmetic

Rule-based systems have also been formally specified by means of default logic, by considering some particular default theories [KBHC91], [Fro92]. The operational definition is closely related to the way of applying production rules: rules are fired only one at a time.

4.3 Proof Theory

We can extend Lukaszewicz' default proof theory to the extensions defined by regular and cautiously saturated grounded sets of defaults.

Definition 20. Let $\Delta = (W, D)$ be a default theory. Let f be a formula. A finite sequence of defaults $d_1, \ldots, d_n$ of defaults of D is *a default proof* for f with respect to Δ if and only if $\{d_1, \ldots, d_n\}$ is a regular grounded set of defaults such that $W \cup \mathrm{Cons}(\{d_1, \ldots, d_n\}) \vdash f$.

Theorem 9. *Let Δ be a default theory, where extensions are generated by regular and cautiously saturated grounded sets of defaults. A formula f has a default proof with respect to Δ if and only if there is some extension of Δ that contains f.*

4.4 Semantics

Following Lukaszewicz' approach again, we provide a semantical account of a default theory $\Delta = (W, D)$, by considering defaults of a grounded set of defaults U that generates an extension E for Δ as restricting the family of models of W, so that we obtain exactly all models of the extension E.

The following nondeterministic operator restricts the set of interpretations and constructs a set of defaults, by applying some default. We consider that some notion of regularity has been defined.

Definition 21. Let $\Delta = (W, D)$ be a default theory. Let Φ be a family of interpretations and let U be a set of defaults of D, and let $\mathrm{APP}(U)$ denote the set of defaults that are cautiously applicable to U. We define $\mathrm{CC}(\Phi, U) = (\mathrm{CC}_1(\Phi, U), \mathrm{CC}_2(\Phi, U))$, where

$$\begin{cases} \mathrm{CC}_1(\Phi, U) = \{\phi \in \Phi, \phi \models \mathrm{Cons}(U \cup \{d\})\} \\ \mathrm{CC}_2(\Phi, U) = U \cup \{d\}, \end{cases}$$

for some $d \in \mathrm{APP}(U)$, and $\mathrm{CC}_1(\Phi, U) = \Phi$, $\mathrm{CC}_2(\Phi, U) = U$ if $\mathrm{APP}(U) = \emptyset$.

The **ordinal powers of** CC are defined as usual:

$$\mathrm{CC}^0(\Phi, U) = (\Phi, U),$$
$$\mathrm{CC}^\alpha(\Phi, U) = \mathrm{CC}(\mathrm{CC}^{\alpha-1}(\Phi, U)), \text{ if } \alpha \text{ is a successor ordinal,}$$
$$\mathrm{CC}^\alpha(\Phi, U) = (\textstyle\bigcap_{\beta<\alpha} \mathrm{CC}_1^\beta(\Phi, U), \bigcup_{\beta<\alpha} \mathrm{CC}_2^\beta(\Phi, U)) \text{ if } \alpha \text{ is a limit ordinal.}$$

$\mathrm{CC}^\alpha(\Phi, U)$ is a **limit** if $\forall \beta \geq \alpha$, $\mathrm{CC}^\beta(\Phi, U) = \mathrm{CC}^\alpha(\Phi, U)$.

As previously, we can use a strategy for applying defaults. We will denote by $(\mathrm{CC}_\rho^n(\Phi, U))_{n \geq 0}$ the sequence obtained by iteratively mapping CC on (Φ, U), for a sequence of defaults $(d_1, \ldots, d_n, d_{n+1}, \ldots)$ chosen under the strategy ρ. The following completeness result holds:

Theorem 10. *Let $\Delta = (W, D)$ be a default theory. Let Ψ be the family of all models of W. The family of interpretations Φ is the family of all models of some extension for Δ generated by a regular and cautiously saturated grounded set of defaults U if and only if there is some strategy ρ such that $(\Phi, U) = CC_\rho^\omega(\Psi, \emptyset)$ is a limit.*

Sketch of the proof. Let $(\Phi_n, U_n) = CC_\rho^n(\Psi, \emptyset)$. By a straightforward induction on n, we can show that Φ_n is the class of all models of $W \cup \text{Cons}(U_n)$. $(U_n)_{n\geq 0}$ is an increasing sequence while $(\Phi_n)_{n\geq 0}$ is a decreasing one. Therefore, $CC_\rho^\omega(\Psi, \emptyset) = (\bigcap_{n\geq 0} \Phi_n, \bigcup_{n\geq 0} U_n)$ and $\bigcap_{n\geq 0} \Phi_n$ is the class of all models of $W \cup \text{Cons}(\bigcup_{n\geq 0} U_n)$. Moreover, since for each n, U_n is a regular grounded set of defaults, $\bigcup_{n\geq 0} U_n$ is a regular grounded set of defaults.

If U is a regular and cautiously saturated grounded set of defaults, let $(d_1, \ldots, d_n, \ldots)$ be some enumeration of U and let ρ be the strategy that selects d_n at step n, such that $W \cup \text{Cons}(\{d_1, \ldots, d_i\}) \vdash \text{Pre}(d_{i+1})$ for $i \geq 0$. Every default d_n of U is applicable for CC. Clearly, $U = \bigcup_{n\geq 0} U_n$ and $\bigcap_{n\geq 0} \Phi_n$ is the class of all models of $W \cup \text{Cons}(U)$, so that $\Phi = \bigcap_{n\geq 0} \Phi_n$. Moreover, $CC_\rho^\omega(\Psi, \emptyset)$ is a limit for CC, since U is cautiously saturated.

Conversely, suppose that there is some strategy ρ such that $(\Phi, U) = CC_\rho^\omega(\Psi, \emptyset)$ is a limit. Since $\bigcap_{n\geq 0} \Phi_n$ is the class of all models of $W \cup \text{Cons}(\bigcup_{n\geq 0} U_n)$, Φ is the class of all models of $W \cup \text{Cons}(U)$. Moreover, every U_n is regular, so that U is also regular. Furthermore, since $CC_\rho^\omega(\Psi, \emptyset)$ is a limit, U is cautiously saturated. U is clearly grounded, therefore, U generates an extension.

□

From this theorem, we have a way of providing a semantics for extensions generated by strongly regular and cautiously saturated grounded sets of defaults, underlying Brewka's assertional default logic, and which generate Schaub's [Sch91b] constrained extensions.

Notice that if we define an ordering $\preceq$ on the sets of pairs of the form (Φ, U) by $(\Phi, U) \preceq (\Phi', U')$ if there is some default d such that $(\Phi', U') = CC(\Phi, U)$, then the extensions are characterized by the maximal pairs which are greater than $(\Psi, \emptyset)$.

Another semantics for constrained extensions can be found in [Sch91a]. It involves pairs of sets of models of the form (Π, H): Π has a meaning similar to our Φ in (Φ, U), since it is the set of models that one obtains after the application of some defaults to W. Furthermore, H enables to record the justifications of the defaults that have been used, and can thus be compared to our U.

Etherington's [Eth88] semantics for R-extensions involves a relation among sets of models (instead of pairs of sets of models). However a set Γ of models maximal for this relation characterizes hazardously saturated grounded sets of defaults. Some stability condition, involving some set of defaults, has then to be verified to check whether Γ corresponds to a weakly regular grounded set of defaults.

In fact, as there are several 'levels' of truth that appear in default logic (the justifications of the defaults do not have the same status as the the consequents, with either weak or strong regularity), it seems that a simple relation among sets of models of classical logic cannot give a semantics for default logic in general. However, it works for normal default logic, where there is no difference betweeen the justifications and the consequents of the defaults: let (W, D) be some normal default theory, and define

$$CC_n(\Phi) = \{\phi \in \Phi, \phi \models \text{Cons}(\{d\})\}$$

for some $d \in D$ such that $\forall \phi \in \Phi, \phi \models \text{Pre}(d)$ and $\exists \phi \in \Phi, \phi \models \text{Cons}(d)$. Then the limits $CC_n^\omega(\Psi)$, where Ψ is the set of models of W, characterize the extensions of (W, D).

For the same reason, Besnard and Schaub [BS92] introduce an ordering among sets of Kripke structures: let D be some set of defaults, if $\mathcal{M}$ and $\mathcal{M}'$ are sets of Kripke structures, $\mathcal{M} \geq_D \mathcal{M}'$ if there exists $d \in D$ such that $\mathcal{M} = \{m \in \mathcal{M}', m \models \mathrm{Cons}(d) \wedge \Box\mathrm{Cons}(d) \wedge \Diamond\mathrm{Jus}(d)\}$, $\mathcal{M}' \models \mathrm{Pre}(d)$, and $\mathcal{M}' \not\models \Box\neg\mathrm{Jus}(d)$, where $\Box$ (respectively $\Diamond$) denotes the necessity operator (respectively its dual, the possibility operator). They show that the $\geq_D$-maximal sets of Kripke structures, above the set $\mathcal{M}_W$ of Kripke structures that satisfy W, characterize the R-extensions of (W,D). They give a similar relation among sets of Kripke structures to characterize Schaub's constrained extensions.

Conclusion

The results of the last section (cumulativity, operational definition of the extensions, proof theoretical and semantical aspects), show some of the interesting possibilities offered by our approach to default logic. Notice however that these results apply mainly when using the notion of cautious applicability. This approach, because it encompasses several other definitions of extensions, and because it makes clear some of the fundamental concepts of default logic, provides some grounds on which to compare these other definitions. We have also described how it is possible to modify these definitions in order to obtain a cumulative inference relation, generalizing Brewka's work on cumulative default logic. There remains to see where the approach of [GLPT91] stands in our framework. The approach in terms of grounded sets of defaults is especially interesting from the theorem-proving point of view. It has already given rise to two theorem provers: the one, based on CAT-resolution, is presented in [Lev91], the other, based on the tableaux method, can be found in [Sch90].

Although we have based our definitions on the classical propositional calculus, all results still hold if we use another logic to define $\vdash$, as long as the associated consequence operator is monotonic, reflexive and compact. In this case the formulas that appear in the defaults are elements of the corresponding logical language. For example, for the proof that Brewka's CDL-extensions correspond to strongly regular and cautiously saturated grounded sets of defaults, we have used the deduction system of his assertion logic. It could also be applied to the graded logic of Chatalic and Froidevaux [CF91], thus redefining graded default logic [FG90].

As future work, a study of necessary and sufficient conditions on the regularity and saturation conditions in order to have properties like semi-monotonicity, cumulativity, some kind of proof theory could give some further information on the nature of default entailment.

The authors wish to thank DRUMS people for their support. Especially, J. Mengin wishes to thank Mike Clarke for his support, and Mike Hopkins and Nick Wilson for fruitful discussions.

References

[AFS91] J. Allen, R. Fikes, and E. Sandewall, editors. *Proceedings of the 2nd International Conference on Principles of Knowledge Representation and Reasoning (KR'91)*. Morgan Kaufmann, 1991.

[Aie90] L. C. Aiello, editor. *Proceedings of the 9th European Conference on Artificial Intelligence (ECAI 90)*. ECCAI, August 1990.

[Bes89] P. Besnard. *An Introduction to Default Logic*. Springer-Verlag, New York, 1989.

[BF87] N. Bidoit and C. Froidevaux. Minimalism subsums default logic and circumscription in stratified logic programming. In *Proceedings of LICS-87, Ithaca*, pages 89–97, 1987.

[BF91] N. Bidoit and C. Froidevaux. General logic databases and programs: Default logic semantics and stratification. *Information and Computation*, 91:15–54, 1991.

[BL89] E. Bonté and F. Levy. Une procédure complète de calcul des extensions pour les théories de défauts en réseaux. In *Actes du 7ème Congrès AFCET-RFIA*, 1989.

[Bre91] G. Brewka. Cumulative default logic: In defense of nonmonotonic inference rules. *Artificial Intelligence*, 50:183–205, 1991.

[BS92] P. Besnard and T. Schaub. Possible world semantics for default logic. In *Proceedings of the Canadian conference on artificial intelligence*, 1992. to appear.

[CF91] P. Chatalic and C. Froidevaux. Graded logics: a framework for uncertain and defeasible knowledge. In *Proceedings of the International Symposium on Methodologies for Intelligent Systems*, 1991.

[DJ91] J. P. Delgrande and W. K. Jackson. Default logic revisited. In Allen et al. [AFS91], pages 118–127.

[dTGCH90] R. A. de T. Guerreiro, M. A. Casanova, and A. S. Hemerly. Contribution to a proof theory for generic defaults. In Aiello [Aie90], pages 213–218.

[Eth88] D. W. Etherington. *Reasoning with Incomplete Information*. Pitman, London, 1988.

[FG90] C. Froidevaux and C. Grossetête. Graded default theories for uncertainty. In Aiello [Aie90], pages 283–288.

[FM92] C. Froidevaux and J. Mengin. A framework for default logics. Rapport de Recherche 755, L.R.I., Université Paris Sud, 1992.

[Fro92] C. Froidevaux. Default logic for action rule-based systems. In *Proceedings of the 10th European Conference on Artificial Intelligence (ECAI 92)*. ECCAI, 1992. to appear.

[GL88] M. Gelfond and V. Lifschitz. The stable model semantics for logic programming. In *Proceedings of the International Conference on Logic Programming, Seattle*, pages 1070–1080. MIT Press, 1988.

[GLPT91] M. Gelfond, V. Lifschitz, H. Przymusińska, and M. Truszczyński. Disjunctive defaults. In Allen et al. [AFS91], pages 230–237.

[KBHC91] P. J. Krause, P. Byers, S. Hajnal, and J. Cozens. The formal specification of a database extension management system. Technical Report 116, Biomedical Computing Unit, Imperial Cancer Research Fund, 1991.

[KS91] R. Kruse and P. Siegel, editors. *Symbolic and Quantitative Approaches for Uncertainty, European Conference ECSQAU*, volume 548 of *Lecture Notes in Computer Science*. Springer Verlag, 1991.

[Lev91] F. Levy. Computing extensions of default theories. In Kruse and Siegel [KS91], pages 219–226.

[Llo87] J.W. Lloyd. *Foundations of logic programming*. Springer-Verlag, New York, 2nd edition, 1987.

[Luk88] W. Lukaszewicz. Considerations on default logic - an alternative approach. *Computational Intelligence*, 4:1–16, 1988.

[Mak89] D. Makinson. General theory of cumulative inference. In *Proceedings of the 2nd International Workshop on Nonmonotonic Reasoning*, pages 1–18, Grassau, FRG, 1989. Springer-Verlag. Lecture Notes in A.I. 346.

[Poo89] D. L. Poole. What the lottery paradox tells us about default reasoning. In *Proceedings of the 1st International Conference on Principles of Knowledge Representation and Reasoning (KR'89)*, pages 333–340, Toronto, 1989.

[Rei80] R. Reiter. A logic for default reasoning. *Artificial Intelligence*, 13:81–132, 1980.

[Ryc91] P. Rychlik. Some variations on default logic. In *Proceedings of the National Conference on Artificial Intelligence (AAAI 91)*, pages 373–378. AAAI, 1991.

[Sch90] C. Schwind. A tableau-based theorem prover for a decidable subset of default logic. In *Proceedings of the 10th International Conference on Automated deduction*, volume 449 of *Lecture notes in computer science*, pages 541–546. Springer Verlag, 1990.

[Sch91a] T. Schaub. Assertional default theories: a semantical view. In Allen et al. [AFS91], pages 496–506.

[Sch91b] T. Schaub. On commitment and cumulativity in default logic. In Kruse and Siegel [KS91], pages 305–309.

[SR91] C. Schwind and V. Risch. A tableau-based characterization for default logic. In Kruse and Siegel [KS91], pages 310–317.

[Wil90] N. Wilson. Rules, belief functions and default logic. In *Proceedings of the Conference on Uncertainty in Artificial Intelligence*, 1990.

[ZM89] A. Zhang and W. Marek. On the classification and existence of structures in default logic. In J.P. Martin and E.M. Morgado, editors, *Proceedings of the 4th Portuguese Conference on Artificial Intelligence*, volume 390 of *Lecture Notes in Computer Science*. Springer-Verlag, 1989.

This article was processed using the LaTeX macro package with LLNCS style

A conceptualization of preferences in non-monotonic proof theory

Anthony Hunter

Department of Computing, Imperial College,
180 Queen's Gate, London SW7 2BZ, UK

Abstract. Formalizing non-monotonic reasoning is a significant problem within artificial intelligence. A number of approaches have been proposed, but a clear understanding of the problem remains elusive. Given the diversity of proof theoretic approaches, we argue the need for frameworks for elucidating key concepts within non-monotonic reasoning. In this paper we consider the preferences, implicit and explicit, that can be seen in a disparate range of non-monotonic logics. In particular, we argue the case for an analysis based on Labelled Deductive Systems for viewing existing approaches to formalizing non-monotonic reasoning, and for identifying new approaches. For this we introduce the family of prioritized logics - each member being a defeasible logic defined in terms of labelled deduction - that forms the basis of a framework for viewing the nature and mechanization of non-monotonic reasoning.

Keywords: Non-monotonic logics, Labelled Deductive Systems

1 Introduction

There is a wide variety of approaches to formalising non-monotonic reasoning, and even though some inter-relationships between formalisms have been established, there is a lack of appropriate general frameworks to support such analyses. There has been some success such as using general properties of the consequence relation as a framework (Gabbay 1985, Makinson 1989, Kraus 1990), but these constitute a relatively high level view on the formalisms. We require lower level, higher resolution, frameworks for analysing the proof-theoretic approaches. Furthermore, we require the range of concepts that we use as the basis of actual frameworks to be augmented and clarified. In this paper we focus on the concept of preferences in non-monotonic reasoning, and use it as the basis of a framework to compare a range of non-monotonic logics.

In human reasoning, it is usually the case that there is a preference for some inferences over others. We argue that such preference is based partly on preference of some data over other data, and also on some reasoning strategies over others. The use of preference is reflected in non-monotonic logics. In some approaches, such as default logic (Reiter 1980), and negation-as-failure (Reiter 1978) the prioritization is implicit, for other approaches such as inheritance hierarchies (Touretsky 1984), and ordered logic (Laneans 1990), the prioritization is explicit.

In order to analyse the concept of preferences, the framework captures both preferences of data and preferences of proof strategy, and the formalization is influenced by the approach of Labelled Deductive Systems (Gabbay 1991). The framework is based on a family of logics called prioritized logics. In this family of logics each member is such that (1) each formula of the logic has an associated label; (2) the rules of manipulation for the logic are augmented with rules of manipulation for the labels; and (3) the labels correspond to a partially-ordered structure. A prioritized logic can be used for non-monotonic reasoning by defining a second consequence relation that allows the derivation of the formula with a label that is 'most preferred' according to some preference criteria.

2 Implicit and explicit preferences in existing logics

Preferences in proof theory seem to be a feature of a wide variety of approaches. We consider inheritance hierarchies, ordered logic, LDR, probabilistic reasoning, negation-as-failure, and normal defeault logic. For inheritance hierarchies, a network representation is used to represent explicitly general cases, and specializations of general cases. For example in the following figure a general case is animal, and a specialization is a bird. With regard to the property of flying, we see that an animal does not fly in general, but the specialization bird does fly:

animal - animals don't fly
↑
bird - birds do fly

A similar, but more sophisticated system, is ordered logic. A database for ordered logic is partially-ordered set of of sets of formulae. For example, in the following figure the set of formulae for 'bird' are preferred over the set of formulae for 'animal':

animal - {animal(x) → ¬fly(x) }
↑
bird - {bird(x) → fly(x) }

Defeasible logics, such as LDR (Nute 1988), support the represention of defeasible information such as the following, where <<- is the LDR defeasible implication operator:

fly(x) <<- bird(x)

¬fly(x) <<- bird(x) ∧ baby(x)

This approach supports a preference for more specific information. In this case there is a preference for the rule with the more specialized antecedent. In probabilistic reasoning, the Principle of Total Evidence is related to the notion of specificity as used in defeasible logics (Cussens 1991). Using this principle, there is a preference for the more specific information:

p('Match lights' | 'Match is struck') = 0.8

p('Match lights' | 'Match is struck' ∧ 'Match is wet') = 0.1

Again, this is an implicit preference, but it is a preference that can be identified by a straightforward syntactical analysis of the probability statements.

For negation-as-failure there is also an implicit notion of preference. For a database Δ, we denote the set of atomic propositions used in the language of Δ as $A(\Delta)$, and denote the set of complements of $A(\Delta)$ as $C(\Delta)$ where $C(\Delta) = \{\neg\alpha \mid \alpha \in A(\Delta)\}$. We can regard $C(\Delta)$ as the set of default values for atoms. The preference implict for negation-as-failure (NAF) is that formulae in Δ are preferred over all formulae in $C(\Delta)$.

For normal default logic we consider normal defaults, in a default theory (D,W), of the form $\alpha: \beta / \beta$ as object-level rules of the form $\alpha \rightarrow \beta$. Furthermore, we can consider there is an implicit preference for information from the non-defeasible information over the default information in D.

These are just examples of a wide variety of approaches to non-monotonic reasoning where there is an implicit or explicit preference on data.

3 Languages for prioritized logics

For a prioritized logic, a database Δ is a set of labelled formulae, where the label is juxtaposed to the formula. When we denote the labels using i, j, k, i_1, i_2, ... and denote unlabelled formulae using the symbols α, β, γ, δ, ... , then i: α, k: β, i_1: δ, . . . denote labelled formulae. For this paper, the formulae are restricted to the form $\alpha_0 \rightarrow (\ldots (\alpha_n \rightarrow \beta))$, where $\alpha_0, \ldots, \alpha_n$, β are literals, and $n \geq 0$.

Associated with every prioritized logic database Δ is a database frame which provides the ordering on the labels of the formulae in Δ. The database frame, denoted $(\Omega, \leq, \sigma)$, is a poset $(\Omega, \leq)$, together with a function, termed the database frame function σ: $\Lambda \rightarrow \Omega$, where Λ is the set of labels used to label formulae. The labels used in the database Δ, are denoted labels(Δ), and defined as labels(Δ) = $\{ i \mid i: \alpha \in \Delta \}$. We term the ordering relation the atomic ordering relation.

A partitioning of a database Δ is a collection of pairwise disjoint non-empty subsets of Δ. In other words the intersection of any pair of partitions (i.e. subsets of Δ) is the empty set. Furthermore, the union of all partitions of Δ is Δ. A head-partitioning of a database Δ into partitions Hp(Δ, α_1), . . . , Hp(Δ, α_k), is a partitioning such that for all i, where $1 \leq i \leq k$, each partition Hp(Δ, α_i) is defined as follows:

$$\begin{aligned} \mathrm{Hp}(\Delta, \alpha) &= \{ k: \beta_0 \rightarrow (\ldots \rightarrow (\beta_n \rightarrow \alpha)) \mid k: \beta_0 \rightarrow (\ldots \rightarrow (\beta_n \rightarrow \alpha)) \in \Delta \} \\ &\cup \{ k: \beta_0 \rightarrow (\ldots \rightarrow (\beta_m \rightarrow \neg\alpha)) \mid k: \beta_0 \rightarrow (\ldots \rightarrow (\beta_m \rightarrow \neg\alpha)) \in \Delta \} \end{aligned}$$

Essentially each partition contains all the rules with the same head, or its negation. We use head-partitioning to define certain classes of database structure. For many existing non-monotonic logics the preferences over data are such that for formulae α and β, α is preferred over β only if α and β are rules with complementary heads.

To view a non-monotonic formalism K as a particular prioritized logic L, we need to rewrite any database Δ_K in the language K as a database Δ_L in the language of L. We assume that any database, the rewrite will be a one-to-one mapping from the formulae of K to the formulae of L. If the rewrite is appropriate, and the consequence relation of L is appropriate, then the inferencing from Δ_K using K is equivalent to inferencing from Δ_L using L.

In the following we consider examples of database rewrite functions for logics including ordered logic, the skeptical form of inheritance hierarchies (Horty 1987), NAF with general logic programs, normal default logic, and LDR . We suggest that the reader only needs to consider one, or two, examples on the first reading.

3.1 Ordered logic as prioritized logic data

Ordered logic is based on a partially ordered structure of logical theories termed objects. An object is a finite set of rules of the form $C \rightarrow p$ where C is a finite set of literals, and p is a literal. C is the antecedent and p is the consequent. A knowledgebase is a tuple $\Delta_K = (O, \leq_k, R, k)$ where $(O, \leq_k)$ is a poset of objects, R is a finite set of rules, and $k: O \rightarrow \wp(R)$ is a function assigning a set of rules to each object.

For a prioritized formulation of ordered logic knowledgebase $\Delta_K = < O, \leq_k, R, k >$, we associate a label o_i^j, for each object $o_i \in O$. Hence for all ordered logic rules $\{\beta_1, .. \beta_n\} \rightarrow \alpha \in k(o_i)$, the database rewrite gives rules of the form o_i^j: $\beta_1 \rightarrow (.. \rightarrow (\beta_n \rightarrow \alpha)) \in \Delta$, where o_i^j is a unique label, and the frame function σ defined as follows: $\sigma(o_i^j) = o_i$, and the frame poset $(\Omega, \leq)$ defined as $\Omega = O$, and $\leq = \{ (j,i) \mid (i,j) \in \leq_k \}$.

For the database rewrite proposed for ordered logic, the rewrite takes the ordered logic formulae, and represents them as prioritized logic formulae. The labels of prioritized logic formulae correspond to the name of the object to which it belongs in the original form. For example if $K = < O, \leq_k, \{ \beta \rightarrow \alpha, \beta, \neg\beta \}, \{(o_1, \beta \rightarrow \alpha), (o_2, \neg\beta), (o_3, \beta)\} >$, where $\{o_1, o_2, o_3\} \subseteq O$, and $o_1 <_k o_2$, and $o_2 <_k o_3$ hold, then $\Delta = \{o_1^1: \beta \rightarrow \alpha, o_2^1: \neg\beta, o_3^1: \beta \}$, where $o_1 > o_2$, and $o_2 > o_3$.

3.2 Negation-as-failure data as prioritized logic data

For NAF, given a general logic program Γ, with formulae of the form $\alpha_0 \wedge .. \wedge \alpha_n \rightarrow \beta$, we form a prioritized logic database with formulae of the form i: $\alpha_0 \rightarrow (.. \rightarrow (\alpha_n \rightarrow \beta))$. For the prioritized logic formulation, the database is partitioned into two, where the first part is labelled from the set $\{ a_1, a_2, a_3, .. \}$, and the second part from the set $\{ b_1, b_2, b_3, ... \}$. One of the partitions reflects the actual program Γ, the other partition reflects the default negation-as-failure values for atoms. For this we define the default completion of the database Γ, denoted $DC(\Gamma)$. If α is an atomic proposition symbol in the language, then $b_i: \neg\alpha \in DC(\Gamma)$, and if $\alpha \in \Gamma$, then $a_j: \alpha \in DC(\Gamma)$, where b_i and a_j are unique for each formula. We denote $\Omega = \{ a_1, a_2, a_3, . \} \cup \{ b_1, b_2, b_3, . \}$ and we define $\leq$ as follows in terms of each head-partition $Hp(\Gamma, \alpha)$:

$\forall a_i, b_j \in \text{Labels}(\text{Hp}(\Gamma, \alpha))\ (\ \sigma(a_i) > \sigma(b_j)\ )$

$\forall i,j \in \text{Labels}(\text{DC}(\Gamma)),\ \sigma(i) \parallel \sigma(j) \text{ if } i \neq j$

For example, if Γ is a NAF general logic program, where $\Gamma = \{ \beta \to \alpha, \delta \to \beta \}$, then $\text{DC}(\Gamma) = \{ a_1: \beta \to \alpha, a_2: \delta \to \beta, b_1: \neg\alpha, b_2: \neg\beta, b_3: \neg\delta \}$.

3.3 Normal default logic data as prioritized logic data

For rewriting normal default logic data, we restrict our consideration to default theories (D,W) where W is data of the form $\alpha_0 \to (\ldots (\alpha_n \to \beta))$, where $\alpha_0, \ldots, \alpha_n, \beta$ are literals, and defaults are of the form $\alpha_0 \wedge \ldots \wedge \alpha_n{:}\beta/\beta$ where $\alpha_0, \ldots, \alpha_n, \beta$ are literals. We rewrite (D,W) to the prioritized logic database Δ, so that we label all items in W with a label a_i, and all items in D to the form of $b_j: \alpha_0 \to (\ldots(\alpha_n \to \beta))$, where the labels are unique. Hence if we denote $\Omega = \{ a_1, a_2, a_3, \ldots \} \cup \{ b_1, b_2, b_3, \ldots \}$, then we define $\leqslant$ as follows in terms of each head-partition $\text{Hp}(\Delta, \alpha)$:

$\forall a_i, b_j \in \text{Labels}(\text{Hp}(\Delta, \alpha))\ (\ a_i > b_j\ )$

$\forall i,j \in \text{Labels}(\Delta),\ i \parallel j \text{ if } i \neq j$

For example for the normal default theory (D, W), where $D = \{ \alpha{:}\ \beta/\ \beta, {:}\neg\delta/\neg\delta \}$, $W = \{ \gamma \to \beta, \alpha \}$, then the regular NDL rewrite gives the database Δ. The associated ordering is represented by the following Hasse diagram, where σ is the identity map, and for any nodes i, j if i is above j, then $i > j$ holds, and if i is not above j, and j is not above i, then $i \parallel j$ holds

$$\begin{array}{ccc} \sigma(a_1) & \sigma(a_2) & \sigma(b_3) \\ | & & \\ \sigma(b_1) & & \end{array}$$

and $\Delta = \{ a_1: \gamma \to \beta, b_1: \alpha \to \beta, a_2: \alpha, b_3: \neg\delta \}$.

3.4 Inheritance hierarchies as prioritized logic data

For the skeptical version of inheritance hierarchies, a database is a set of rules Γ, where all elements of Γ are of the form $x \to y$ or $x \not\to y$. We assume that Γ constitutes an acyclic graph. A positive path in Γ is defined inductively: A rule $x \to y \in \Gamma$ is a positive path in Γ; and (2) if $x_1 \to \ldots \to x_n$ is a positive path, then $x_1 \to \ldots \to x_n \to x_{n+1}$ is a positive path in Γ.

We form a prioritized logic database Δ from Γ as follows: For each rule $x \rightarrow y \in \Gamma$, we have a clause $i: x \rightarrow y \in \Delta$, and for each rule $x \not\rightarrow y$, we have a clause $i: x \rightarrow \neg y \in \Delta$. Each clause in Γ is labelled uniquely. We order formulae in Γ as follows: For any pair of clauses $i: x \rightarrow y$, $j: v \rightarrow w \in \Delta$, if y and w are complements, or $y = w$, and $x \rightarrow .. \rightarrow v$ is a positive path in Γ, then $i > j$ holds.

For example let a = tweety, p = bird, q = flying-things, and r = ostrich, then the following is the corresponding graph for the 'birds fly' example:

The inheritance hierarchy database for this graph is $\{ a \rightarrow r, r \rightarrow p, p \rightarrow q, r \not\rightarrow q \}$. From this database, we form the following prioritized logic database $\{ i_1: a \rightarrow r, i_2: r \rightarrow p, i_3: p \rightarrow q, i_4: r \rightarrow \neg q \}$. The associated ordering is captured by the following diagram, where σ is the identity map:

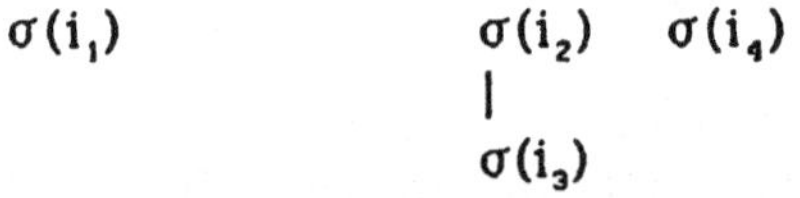

Intuitively, for two rules in an inheritance hierarchy that enter into the same node, we prefer the rules that eminates from the more specialized sub-class.

4 The class of strongforward logics

For the prioritized logic databases, we can define proof rules that manipulate the labelled formulae. For example, the simple naive global relation, denoted $\vdash_S$, is defined as follows where Δ is a database of prioritized logic formulae, and $i: \alpha$, $j: \alpha \rightarrow \beta$, and $ij: \beta$ are prioritized logic formulae such that ij is the concatenation of the labels i and j, and the concatenatied labels are termed complex labels:

If $\Delta \vdash_S i: \alpha$ and $\Delta \vdash_S j: \alpha \rightarrow \beta$ then $\Delta \vdash_S ij: \beta$

If $i: \alpha \in \Delta$ then $\Delta \vdash_S i: \alpha$

Using this simple naive relation on the database $\Delta = \{$ a: α, b: $\alpha \rightarrow \beta$, d: $\beta \rightarrow \delta$ $\}$, we have naive inferences including, the following set $\{$ a: α, d: $\beta \rightarrow \delta$, abd: δ $\}$.

The naive inferences constitute "plausible" or "possible" inferences from a database. If we want to derive "the best possible" inferences, we have to define a second consequence relation, termed a global relation. First, we consider the simple strongforward global relation. This allows an inference α if and only if α is 'proposed' and unchallenged'. The inference α is proposed if and only if there is a naive argument for α, and that all conditions for α are satisfied globally. The inference α is unchallenged if and only if there is no more preferred naive argument for the complement of α, s.t. all the conditions are satisfied globally, or if there is a more preferred naive argument for the complement of α, then there is an even more preferred globally-satisfied argument for α.

The **simple strongforward** global relation, denoted $|\!\sim_S$ is defined as follows, where i is atomic, and $(\Omega, <, \sigma)$ is the database frame, α^ denotes the complement of α, and m, n, i ≥ 0:

$$\Delta \mid\!\sim_S \alpha \text{ iff } \exists i\,[\text{ proposed}(\Delta, i, \alpha) \text{ and unchallenged}(\Delta, i, \alpha)]$$

proposed(Δ, i, α)
iff
$\exists \beta_1, .., \beta_n$ [i: $\beta_0 \rightarrow (\ldots \rightarrow(\beta_n \rightarrow \alpha)) \in \Delta$
and $\Delta \mid\!\sim_S \beta_0$ and, .., and $\Delta \mid\!\sim_S \beta_n$]

unchallenged(Δ, i, α)
iff
$\forall$j [proposed(Δ, j, α^)
implies [(σ(i) > σ(j))
or $\exists$k(proposed(Δ, k, α) and σ(k) > σ(j))]]

For example, if $\Delta = \{ a_1: \beta \rightarrow \alpha, a_2: \neg\alpha, b_1: \beta \}$, where the associated ordering is represented by $\sigma(a_1) > \sigma(a_2)$, $\sigma(a_1) \parallel \sigma(b_1)$ and $\sigma(a_2) \parallel \sigma(b_1)$. For Δ, the naive inference b_1a_1: α would allow a global inference by the simple strongforward global relation. As another example take a structured database as in the following Hasse diagram which is annotated with the respective formulae:

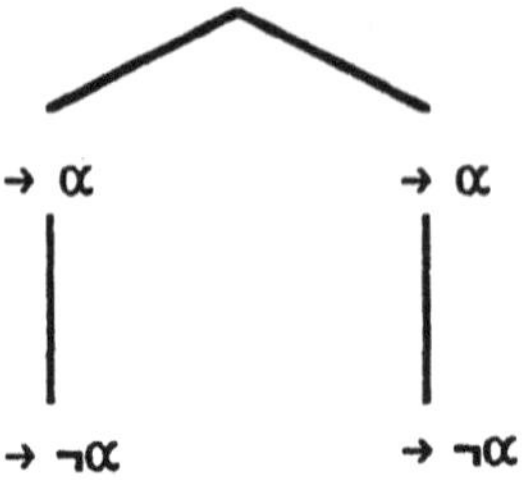

In this situation, we have a naive proof of α that is not more preferred to all naive proofs of $\neg\alpha$. However, for all naive proofs of $\neg\alpha$, we have a more preferred naive proof of α.

In the following we outline defeasible logics that can be viewed as simple strongforward prioritized logics.

4.1 Negation-as-failure as a prioritized logic

For NAF there is the rule of inference such that if α is atomic, and all ways of proving α from Γ fail in finite time, then $\neg\alpha$ follows from Γ. This is captured in strongforward reasoning by the notion of a default completion as given in section 3.2. For example, if the NAF database $\Gamma = \{ \beta \to \alpha, \delta \to \beta \}$ is rewriten to the database Δ, using the regular NAF rewrite, then $DC(\Gamma) = \{ a_1: \beta \to \alpha, a_2: \delta \to \beta, b_1: \neg\alpha, b_2: \neg\beta, b_3: \neg\delta \}$ and $\neg\alpha$, $\neg\beta$, and $\neg\delta$ are global inferences by the strongforward global relation.

4.2 Ordered logic as a prioritized logic

Ordered logic proof rules allow the inference of knowledge at an object on the basis of knowledge in a higher object. The notion of inheritance is supported by allowing knowledge to filter from a higher object to a lower object, and may be blocked by either being overruled by more specific information, that is contradictory, at the lower object; or defeat which occurs when an an object o_i inherits contradictory information from two different objects o_j and o_k s.t. not $o_j \,{}_k\!\geq o_k$ and not $o_k \,{}_k\!\geq o_j$. Given the rewrite in section 3.1, there is an obvious equivalence between ordered logic proof theory and strongforward proof theory.

For example if $K = < O, \leq_k, \{\beta \to \alpha, \beta, \neg\beta\}, \{(o_1, \beta \to \alpha), (o_2, \neg\beta), (o_3, \beta)\} >$, where $\{o_1, o_2, o_3\} \subseteq O$, and $o_1 <_k o_2 <_k o_3$, then if K is rewritten to $\Delta = \{o_1^1: \beta \to \alpha, o_2^1: \neg\beta, o_3^1: \beta\}$, and $o_3 < o_2 < o_1$ holds, then the only global inference is $\neg\beta$ by the strongforward global relation.

4.3 Inheritance hierarchies as a prioritized logic

Inheritance hierarchies - For any inheritance hierarchy database Γ, the reasoning from Γ is such that an allowed path from an individual to a class holds, then the class proposition is a conclusion of Γ.

A database Γ for the skeptical version of inheritance hierarchies is a set of rules where all rules are of the form $\alpha \rightarrow \beta$, or $\alpha \Rightarrow \beta$, and a set of propositions that refer to individuals. The consequence relation, denoted $\vdash_{IHS}$, is defined as follows, where α is a proposition:

$$\Gamma \vdash_{IHS} \alpha \text{ iff } (\alpha \in \Gamma) \text{ or } [(\beta \in \Gamma) \text{ and } (\beta \rightarrow \ldots \rightarrow \alpha \text{ is an allowed inference in } \Gamma)]$$

Given a skeptical inheritance hierarchy rewrite of Γ, if two rules have complementary heads, then the preference for one of the rules holds iff the antecedent of the rule is a sub-class of the antecedent of the other rule. For example for $\Gamma = \{ a \rightarrow r, r \rightarrow p, p \rightarrow q, r \Rightarrow q \}$ then the skeptical IHS rewrite gives $\Delta = \{ i_1: a \rightarrow r, i_2: r \rightarrow p, i_3: p \rightarrow q, i_4: r \rightarrow \neg q \}$, where the associated ordering is represented by the following diagram, and σ is the identity map:

$\sigma(i_1)$ $\quad\quad$ $\sigma(i_2)$ $\quad\quad$ $\sigma(i_4)$

$\quad\quad\quad\quad\quad$ |

$\quad\quad\quad\quad$ $\sigma(i_3)$

Assuming the assertion a with some label j, then from $\Delta \cup \{ j: a\}$, the global inferences a, r, p and $\neg q$ hold by the simple strongforward global relation.

4.4 Using strongforward logic as a classification

A non-monotonic logic K is in the class of simple strongforward logics if there is a database rewrite s.t. given any database Γ in the language of K, the prioritized logic database Δ is the rewrite of Γ, and the following equivalence holds, where $\vdash_K$ is the consequence relation of K, $|\!\sim$ is the simple strongforward global relation, and α is a literal:

$$\Gamma \vdash_K \alpha \quad \text{iff} \quad \Delta \mid\!\sim \alpha$$

[Result] The class of simple strongforward logics includes ordered logic, negation-as-failure with general logic programs and the skeptical version of inheritance hierarchies. For a proof see Hunter (1991).

5 Capturing a restricted version of default logic

We now consider a global relation that captures a restricted subset of normal default logic. We assume the rewrite given in section 3.3. We also add to the rewritten database the following data: $\{ a_j: \alpha \mid \alpha$ is a theorem of classical logic $\}$. We partition the rewritten database Δ as follows: $\Delta_a = \{ a_i: \alpha \mid a_i: \alpha \in \Delta \}$ and $\Delta_b = \{ b_j: \alpha \mid b_j: \alpha \in \Delta \}$. For naive inferencing, default logic also allows non-defeasible inferences to be used together with the simple naive relation, but only allows atomic defeasible inferences to be used with Δ_a to derive further inferences. We define the default naive relation, denoted $\vdash_d$, as follows:

$\Delta \vdash_s i: \alpha$ if $i: \alpha \in \Delta_b$
$\Delta \vdash_s ij: \alpha$ if $\Delta \vdash_s i: \beta \rightarrow \alpha$ and $\Delta \vdash_s j: \beta$

$\Delta \vdash_d i: \alpha$ if $\Delta \vdash_s i: \alpha$ and α is a literal
$\Delta \vdash_d i: \alpha$ if $i: \alpha \in \Delta_a$
$\Delta \vdash_d i: \alpha$ if $\Delta \vdash_d i: \beta \rightarrow \alpha$ and $\Delta \vdash_d j: \beta$

Furthermore, we require the function Form(Δ_b, i) which is the set of labelled formulae that correspond to the b labels in the complex label i: Form(Δ_b, i) = $\{ b_j: \alpha \mid b_j \in$ Labels(i) and $b_j\ \alpha \in \Delta \}$ where Labels(ij) $=_{def}$ {i} ∪ Labels(j) if i is atomic, and Labels(i) $=_{def}$ {i} if i is atomic. We define the DL global relation, denoted $\mid\sim_{DL}$, for the prioritized logic version of NDL as follows:

$\Delta \mid\sim_{DL} \alpha$ if $\Delta \vdash_d i: \alpha$ and α is atomic and not inconsistent(i, α)

inconsistent(i, α)
if $\exists\, j, k, \beta$ ([$\Delta_a \cup$ Form(Δ_b, i) $\vdash_d j: \beta$]
and [$\Delta_a \cup$ Form(Δ_b, i) $\vdash_d k: \beta$^])

$\Delta \mid\sim_{DL} \alpha$ if $\Delta \vdash_c i: \alpha$ and α is atomic

For example, for the following database $\{ a_1: \neg\alpha, a_2: \alpha \rightarrow \neg\beta, b_1: \alpha, b_2: \neg\alpha \rightarrow \beta \}$ where the following is the associated ordering:

a_1 a_2
| |
b_1 b_2

Then the following naive inferences follow, $\Delta \vdash_d \; a_1b_2{:}\ \beta$ and $\Delta \vdash_d \; b_1a_2{:}\ \neg\beta$. The global inference relation infers β in preference to $\neg\beta$. This is the same inference as with a strongforward relation. However, the following example shows that this global relation is not a strongforward global relation: { a_1: α, a_2: $\delta \rightarrow \neg\beta$, b_1: δ, b_2: $\alpha \rightarrow \beta$ } where the ordering is defined by the following Hasse diagram:

$$\begin{array}{ccc} a_1 & a_2 & b_1 \\ & | & \\ & b_2 & \end{array}$$

Then a_1b_2: β and b_1a_2: $\neg\beta$ naive inferences by the default naive relation and the DL global relation allows either β or $\neg\beta$ as inferences. This is in constrast to the strongforward ordering relation which gives only $\neg\beta$ as the inference. We show in the next section that the failure of the normal default logic consequence relation to adhere to strongforward reasoning is due to the fact that the preference criteria for prioritized logic version of normal default logic is weaker than that for strongforward. We also define this this weaker preference which we term cover-preference.

6 The class of cover-preference logics

The notion we want to define is called cover-preference. A cover is a subset of a prioritized database from which the proof theory allows inferences to be made. For example consider the following normal default theory where D = { $:\neg\alpha / \neg\alpha$ }, and W = { α }. The proof theory of normal default logic is such that inferences can only be made from the following subsets, denoted D", W" of the default theory, where D" = { } and W" = { α }. This kind of restriction to a subset of a database is to avoid inconsistencies resulting from reasoning with all information from the database on equal terms. The proof theory only allows inferencing from a preferred subset.

A cover is a particular sort of subset that results from a preference for data higher in the ordering over data lower in the ordering. Hence covers are closed upwards. The intuition behind this condition is that if we wish to draw some inferences from a prioritized database we attempt to draw them from the most preferred data. If we fail to draw our inferences on that basis, we widen the set of data we use by using the next most preferred data. We continue this process until either we satisfy all our queries, or until any further expansion would cause inconsistency. This process dictates a certain preference for some subsets of the database over others. We term a logic for which such a preference holds a cover-preference logic:

[Definition] A logic L, with a global relation $|\!\sim$, and naive relation $\vdash_L$, is a **cover-preference** logic iff for all databases Δ, and queries α, in the language of L, $\Delta \mid\!\sim_L \alpha$ iff there is a subset Δ^* of Δ such that the following conditions hold, where $(\Omega, \leqslant, \sigma)$ is the database frame. We term Δ^* a **cover** for α from Δ:

(1) $\Delta^* \mid\!\sim \alpha$

(2) $\forall\, i{:}\,\delta \in \Delta^* \;\; \forall\, j{:}\,\beta \in \Delta$ [If $\sigma(j) \geqslant \sigma(i)$ then $j{:}\,\beta \in \Delta^*$]

(3) not $\exists i,j,\beta$ [$\Delta^* \vdash_L i{:}\,\beta$ and $\Delta^* \vdash_L j{:}\,\neg\beta$]

(4) not $\exists\Delta''$ [$\Delta^* \subseteq \Delta'' \subseteq \Delta$ and not $\Delta^* = \Delta''$
and not $\exists i,j,\beta$ [$\Delta'' \vdash_L i{:}\,\beta$ and $\Delta'' \vdash_L j{:}\,\neg\beta$] □

Condition (1) checks that our subset contains enough data to answer our queries, and condition (2) checks that for all items in our subset, all more preferred items in the database are also in the subset, and conditions (3) and (4) check that the subset is maximally consistent.

If we return to the following database: { $a_1{:}\,\alpha$, $a_2{:}\,\delta \rightarrow \neg\beta$, $b_1{:}\,\delta$, $b_2{:}\,\alpha \rightarrow \beta$ } where for all i, j $b_i < a_j$ and $b_i \parallel b_j$, then the following are covers, assuming the simple strongforward global relation: cover_1 = { $a_1{:}\,\alpha$, $a_2{:}\,\delta \rightarrow \neg\beta$, $b_1{:}\,\delta$ } and cover_2 = { $a_1{:}\,\alpha$, $a_2{:}\,\delta \rightarrow \neg\beta$, $b_2{:}\,\alpha \rightarrow \beta$ }

Note, however, that the definition of a cover logic does not imply that if Δ^* is a cover then all queries α that follows from Δ must also follow from Δ^*. In other words, there may be more than one cover for a database, and that to obtain all queries from Δ requires consideration of all covers for the database. Furthermore, the set of covers for a database can change if the databases changes. For example, take the database Γ = {$b_1{:}\,\alpha \rightarrow \beta$, $b_2{:}\,\delta \rightarrow \neg\beta$, $b_3{:}\,\gamma \rightarrow \beta$, $b_4{:}\,\neg\beta$ } with the ordering $\sigma(a_1) = \sigma(a_2) = \sigma(a_3)$, and $\sigma(a_1) > \sigma(b_1)$, $\sigma(b_1) > \sigma(b_2)$, $\sigma(b_2) > \sigma(b_3)$, $\sigma(b_3) > \sigma(b_4)$. We can form a table of the covers that hold for any set of non-defeasible assertions that can be added to Γ. We provide a few lines of such a table below assuming the simple strongforward global relation:

Non-defeasible assertions	Covers
{ }	{ } ∪ {b_1, b_2, b_3, b_4}
{$a_1{:}\,\alpha$}	{a_1} ∪ {b_1, b_2, b_3}
{$a_1{:}\,\alpha$, $a_2{:}\,\delta$}	{a_1, a_2} ∪ {b_1}
{$a_3{:}\,\gamma$}	{a_3} ∪ {b_1, b_2, b_3}

[Result] The class of cover-preference logics includes: (1) Simple strongforward; and (2) The prioritized logic version of normal default logic, if the default theories(D, W) for the logic are restricted such that W is classically consistent. For a proof see Hunter (1991). □

Discussion

In this paper we have seen how a disparate range of non-monotonic logics incorporate a notion of either implicit or explicit preference in the proof theory. Given the apparent ubiquity of the phenomenon, there is a need to clarify the concept within an appropriate framework. To address this we have presented a framework based on the family of prioritized logics. This family constitutes a new sort of logic that has labelled formulae where the labels are explcitly ordered. Since the family uses labelled formulae, it can also be considered to be part of the approach of Labelled Deductive Systems (Gabbay 1991a, 1991b).

Using prioritized logics, we have identified two interesting classes: (1) The class of strongforward logics - which can be considered to be those logics that use the best possible information first when going from data to query; and (2) The class of cover-preference logics - which is based on answering queries form maximally consistent subset of the database.

The benefit of this work is the characterization of priorities in non-monotonic proof theory. For example, by representing preferences on labels, and not on formulae, we are now in a position to analyse the algebra on the labels in abstraction from the logical operations on the formulae. We pursue this direction in subsequent work (Hunter 1992). Subsidiary benefits include the notions defined in the framework suggest alternative non-monotonic logics. Using the structure that prioritized logics provide, together with the notions of naive relations and global relations, we can see a variety of new intuitive non-monotonic logics. We pursue these in further work (Hunter 1992).

The selection of logics we have discussed here is only illustrative of the range that can be viewed in this framework. We show, in further work, a far wider spectrum of logics that can fit in the framework including graded default logics (Froidevaux 1990), prioritized default logic (Brewka 1989), Theorist (Poole 1988), and defeasible argument systems (Loui 1987).

Acknowledgements

This work is currently being funded by UK SERC under grants GR/G 29861 and GR/G 46671. Special thanks are due to Dov Gabbay. Also thanks are due to James Cussens, Marcelo Finger, Ian Hodkinson, Jeremy Pitt, and Mark Reynolds.

References

Brewka G (1989) Preferred subtheories: An extended logical framework for default reasoning, IJCAI'89, Morgan Kaufmann,

Cussens J & Hunter A (1991) Using defeasible logic for a window on a probabilistic database: Some preliminary notes, Proceedings of European Conference on Symbolic and Quantitaive Approaches to Uncertainty, LNCS 548, Springer

Froidevaux C & Grossetete C (1990) Graded default theories for uncertainty, Proceedings ECAI'90, Pitman

Gabbay D (1985) Theoretical foundations for non-monotonic reasoning in expert systems, in Apt K, Logics and Models of Concurrent Systems, LNCS, Springer

Gabbay D (1991a) Labelled Deductive Systems: A position paper, in proceedings of Logic Colloquium 90, Lecture Notes in Logic 1, Springer

Gabbay D (1991b) Labelled deductive systems, Technical report, Centrum fur Informations und Sprachverarbeitung, Universitat Munchen

Hunter A (1991) Use of priorities in non-monotonic proof theory, Technical report, Department of Computing, Imperial College, London

Hunter A (1992) Labels theory for prioritized logics, Technical report, Department of Computing, Imperial College, London

Kraus S, Lehmann D & Magidor M (1990) Non-monotonic reasoning, preferential models, and cumulative logics, Artificial Intelligence, 44, 167 - 207

Laenens E & Vermier D (1990) A fixpoint semantics for ordered logic, J Logic & Computation, 1, 159 - 185

Loui R (1986) Defeat amongst arguments: a system of defeasible inference, Technical report No 130, Department of Computer Science, University of Rochester, New York

Makinson D (1989) A general theory of cumulative inference, in Reinfrank M, Proceedings of Second International Workshop on Non-monotonic Logic, LNCS 346, Springer

Nute D (1988) Defeasible reasoning and decision support systems, Decision Support Systems, 4, 97-110

Poole D (1988) Logical framework for default reasoning, Artificial Intelligence, 36, 27 - 48

Reiter R (1978) On the closed-world databases, in Gallaire H & Minker J, Logic and Databases, Plenum Press

Reiter R (1980) A logic for default reasoning, Artificial Intelligence, 13, 81 - 13

Reasoning with Defeasible Arguments: Examples and Applications

Gerard Vreeswijk

Vrije Universiteit Amsterdam,
Faculty of Mathematics and Informatics,
De Boelelaan 1081a, NL-1081 HV Amsterdam.

This paper attempts to demonstrate the wide variety of characteristic properties of defeasible argumentation, of which nonmonotonicity is one. To do so, we introduce a simple formalism, called abstract argumentation system, with which we discuss different methods for raising arguments: forward reasoning, backward reasoning and, in particular, *combinations* thereof. Resource bounded defeasible reasoning is also briefly dealt with (the prefix 'defeasible' is essential here). The main goal of this paper is not to tell how this should be done, but to show how these simple procedures suddenly may change in the context of defeasible argumentation.

1. Introduction

A defeasible argument is a piece of reasoning that might be defeated, i.e. turned down, by other defeasible arguments. In form, a defeasible argument looks like a proof tree in classical logic. Reasoning with defeasible arguments, however, radically differs from classical theorem proving. Unlike a proof, an argument does not establish warrant for a conclusion once and for all, but may be defeated by counterarguments which, on their turn, may be defeated by counter-counterarguments, and so on. So an argument may be raised, defeated, and reinstated several times in the course of reasoning. Therefore, reasoning with defeasible arguments, or *defeasible argumentation*, is not as easy as generating proofs in classical proof theory.[1]

Upon further consideration, the theory of defeasible argumentation has little in common with classical proof theory. This is because the mere possibility of raising a defeasible proof has a strong impact on established logical formalism, and enforces a redefinition of basic logical concepts. Inevitably, it follows that almost every constructive treatment of this issue is bound to start from scratch.[2] This will be done here too, but without the aspiration of presenting the ultimate defeasible logic. Rather, we touch upon

[1] For further discussion of this issue, see the paper 'Critical Issues in Nonmonotonic Reasoning,' (Etherington *et al.* 1989).

[2] If not *logically*, then certainly *methodologically*. Poole's suggestion of how default reasoning can be done entirely within the framework classical logic is a good example of proposing an alternative methodological approach. Poole: "instead of changing the logic, the way in which the logic is used is changed" (1989).

characteristic features of defeasible argumentation (of which nonmonotonicity is one), without pushing formalisation too far, without boasting theorems, and without considering every aspect of the matter into full detail. We purposively try to be non-committal.

Yet, to come to grips with the basic notions and terminology of defeasible argumentation, the overall discussion will be set up around a simple formalism. This formalism—called abstract argumentation system—might be considered as the defeasible counterpart of a classical proof system. The main difference is that proofs may now differ in conclusive force. For this reason, these defeasible proofs are called *arguments*. The prefix *abstract* denotes that the notion of conclusive force is already assumed to be determined, and need not be defined within the system.

For properly assessing the use and function of an abstract argumentation system, it is first of all important to know that it is a static framework on which different ideas on argumentation can be developed. In particular, it does not in and of itself prescribe how argumentation should be performed, or how defeasible information should be managed with. Instead, it will serve as the starting point for methodological considerations. Another important feature is that, unlike most of its 'rivals,' an argumentation system allows for different notions of conclusive force, thus being able to bring about different phenomena of defeat. In this manner, we are not committed to a specific machinery of defeat, so that we maintain flexibility on this point as well.

Over the years, a variety of systems has been proposed and discussed. Among the contributions that are relevant to this paper, we find work of (Pollock, 1987), (Loui, 1987), (Nute, 1988), (Lin and Shoham, 1989) and, again, Pollock (1991). The question might arise, then, as to why we do not use any of these systems in our present treatment. Pollock's system, for instance, is a much more detailed and well-thought-out exposition of defeasible reasoning than what is offered here. In this respect, also Loui's system needs to be mentioned, as it provides clear and well-defined criteria of conclusive force, and an accurate prescription of defeat. We have our reasons, however, for introducing yet another system of defeasible reasoning. First—and we do not hesitate to mention this—we find this a suitable opportunity for presenting several examples and applications of our notion of abstract argumentation system. We have not done this in other papers and, again, this paper is a good opportunity to compensate for. A reason that is much more important, is the following. If we use Pollock's or Loui's system, we are fully committed to a specific implementation, where all choices has been made, and in which there is no room for methodological considerations. Therefore, in order to have some free space in which we can 'play around' with the basic problems of defeasible argumentation, we have found it necessary to come up with our own formalism.

The paper proceeds as follows. After the presentation of the basic concepts, we first give a canonical example of an abstract argumentation system. Thereafter, we will consider different modes of defeasible argumentation in turn: forwards reasoning, backwards reasoning, and combinations thereof. Also defeasible reasoning with limited resources is briefly dealt with. The paper concludes with a short reference to the work of Lin and Shoham.

2. Basic concepts

In order to have a handle on basic notions of argumentation, the entire discussion will be set up around a formalism that will be called an abstract argumentation system. It will be defined in a top-down manner.

DEFINITION 2.1. An *abstract argumentation system* $\mathcal{A}$ is a triple $(\mathcal{L}, R, \preceq)$ where $\mathcal{L}$ is a language, R is a set of rules of inference, and $\preceq$ is a reflexive and transitive order on arguments.

The prefix *abstract* will often be omitted for the sake of brevity. Sometimes we wish to read the letter $\mathcal{A}$ *in extension*, meaning that we do not want to denote the argumentation system itself, but rather the arguments enabled by it, positioned in accordance with $\preceq$.

DEFINITION 2.2. A *language* is a set $\mathcal{L}$, containing a distinguished element $\perp$.

The symbol $\perp$ is used to denote contradictions. The language $\mathcal{L}$ is not subject to particular constraints. Any set will do, provided it contains a distinguished element representing the contradiction.

The rules R are given in terms of $\mathcal{L}$ and determine what inferences are possible.

DEFINITION 2.3. Let $\mathcal{L}$ be a language.

1. A *strict rule* is a formula of the form $\phi_1,\ldots,\phi_n \rightarrow \phi$ with $\phi_1,\ldots,\phi_n$ a finite sequence in $\mathcal{L}$ and ϕ a member of $\mathcal{L}$.
2. A *defeasible rule* is a formula of the form $\phi_1,\ldots,\phi_n \Rightarrow \phi$ with $\phi_1,\ldots,\phi_n$ a finite sequence in $\mathcal{L}$ and ϕ a member of $\mathcal{L}$.

A *rule* is a strict rule or a defeasible rule.

It is essential to understand that rules, being *rules of inference* are not part of the language itself.

EXAMPLE 2.4. Let $\mathcal{L} = \{p, q, r\} \cup \{\perp\}$. Then $p, q \rightarrow r$, $p \rightarrow \perp$, $p, q \Rightarrow r$, $\Rightarrow q$, $\perp, p, \perp \rightarrow q$, $p, q, r \rightarrow \perp$, $p \Rightarrow q$, and $\rightarrow \perp$ all are rules in $\mathcal{L}$.

Chaining rules together into trees, we get arguments.

DEFINITION 2.5. Let R be a set of rules. An argument has *premises*, a *conclusion*, *sentences*, *subarguments*, *top arguments*, a *length*, and a *size*. These will be abbreviated by corresponding key strings. An *argument* σ is

1. A member of $\mathcal{L}$. In that case, $prem(\sigma) = \{\sigma\}$, $conc(\sigma) = \sigma$, $sent(\sigma) = \{\sigma\}$, $sub(\sigma) = \{\sigma\}$, $top(\sigma) = \{\sigma\}$, $length(\sigma) = 1$, and $size(\sigma) = 1$; or
2. A formula of the form $\sigma_1,\ldots,\sigma_n \rightarrow \phi$ with $\sigma_1,\ldots,\sigma_n$ a finite sequence of arguments, ϕ not in $sent(\sigma_1) \cup \ldots \cup sent(\sigma_n)$, and $conc(\sigma_1),\ldots,conc(\sigma_n) \rightarrow \phi$ a rule in R.
   In that case, $prem(\sigma) = prem(\sigma_1) \cup \ldots \cup prem(\sigma_n)$, $conc(\sigma) = \phi$, $sent(\sigma) = sent(\sigma_1) \cup \ldots \cup sent(\sigma_n) \cup \{\phi\}$, $sub(\sigma) = sub(\sigma_1) \cup \ldots \cup sub(\sigma_n) \cup \{\sigma\}$, $top(\sigma) = \{\tau_1,\ldots,\tau_n \rightarrow \phi \mid \tau_1 \in top(\sigma_1),\ldots,\tau_n \in top(\sigma_n)\} \cup \{\phi\}$. $length(\sigma) = max\{length(\sigma_1),\ldots,length(\sigma_n)\} + 1$, and $size(\sigma) = size(\sigma_1) + \ldots + size(\sigma_n) + 1$; or
3. A formula of the form $\sigma_1,\ldots,\sigma_n \Rightarrow \phi$ with $\sigma_1,\ldots,\sigma_n$ a finite sequence of arguments, ϕ not in $sent(\sigma_1) \cup \ldots \cup sent(\sigma_n)$, and $conc(\sigma_1),\ldots,conc(\sigma_n) \Rightarrow \phi$ a rule in R. In that case, premises, conclusions, etc. are defined as in (2).

Very often, we will abuse notation and write $\sigma_1,\ldots,\sigma_n \rightarrow \sigma$ (or $\sigma_1,\ldots,\sigma_n \Rightarrow \sigma$) to denote that σ is an argument, constructed from $\sigma_1,...,\sigma_n$ using the rule $conc(\sigma_1),...,conc(\sigma_n) \rightarrow conc(\sigma)$ (or the rule $conc(\sigma_1),...,conc(\sigma_n) \Rightarrow conc(\sigma)$, respectively).

An argument has, besides subarguments, also *super arguments*, and besides top arguments, also *bottom arguments*.

DEFINITION 2.6. Let σ be an argument. Then $sup(\sigma) = \{\tau \mid \sigma \in sub(\tau)\}$. Instead of $\sigma \in sub(\tau)$ or $\tau \in sup(\sigma)$ we will sometimes write $\sigma \sqsubseteq \tau$. Furthermore, $bot(\sigma) = \{\tau \mid \sigma \in top(\tau)\}$.

DEFINITION 2.7. Let Σ be a set of arguments. Then $min(\Sigma) = \{\sigma \in \Sigma \mid \tau \in \Sigma, \tau \preceq \sigma \Rightarrow \sigma = \tau\}$, and $max(\Sigma) = \{\sigma \in \Sigma \mid \tau \in \Sigma, \sigma \preceq \tau \Rightarrow \sigma = \tau\}$.

DEFINITION 2.8. An argument σ is *strict* if $\sigma \in \mathcal{L}$ or $\sigma_1,...,\sigma_n \rightarrow \sigma$ where $\sigma_1,...,\sigma_n$ are strict arguments.[3] An argument is *defeasible* if it is not strict.

With this terminology, we have that every argument is either strict or defeasible, but never both.

DEFINITION 2.9. Let $\mathcal{L}$ be a language and let P be a subset of $\mathcal{L}$. An argument σ is *based* on P if $prem(\sigma)$ is included in P. The set of all arguments that are based on P is denoted by $arguments(P)$. A member of $\mathcal{L}$ is *strictly based* on P if it is the conclusion of some strict argument based on P. The set of all arguments that are strictly based on P is denoted by $strict(P)$. A member of $\mathcal{L}$ is *defeasibly based* on P if it is the conclusion of some defeasible argument based on P. The set of all arguments that are defeasibly based on P is denoted by $defeasible(P)$.

The third and last slot of an argumentation system is the order on arguments. All defeasible logics operate by means of a mechanism of defeat that computes relationships of conclusive force among arguments. But, no matter how complicated such a mechanism of defeat might be, in the end all comes down by telling which argument should overrule the other.

DEFINITION 2.10. Let σ and τ be arguments. If $\sigma \preceq \tau$, then τ is *as good as* σ, and if $\sigma \prec \tau$, then τ is *better* than σ. The order of conclusive force should satisfy, besides reflexivity and transitivity, two additional specific constraints.

1. Upwards-well-foundedness (informally: principle of finite outbidding). There are no infinite chains $\sigma_1 \prec \sigma_2 \prec ... \prec \sigma_n \prec ...$;
2. Monotonic non-increasement in conclusive force. If $\sigma \sqsubseteq \tau$, then $\tau \preceq \sigma$, for all σ and τ;

These two constraints together ensure that, roughly, defeat will be a finite process.

Once more, it is emphasised that we purposively abstain from telling how and why any particular argument should overrule the other. That is why we speak of *abstract* argumentation systems. Nevertheless, the following example presents some specific instantiations.[4]

EXAMPLE 2.11. Here are some examples of how the parameter $\preceq$ actually might be filled in.

[3]Cf. definition 2.5.2, last sentence.

[4]For an extensive treatment on this complicated issue, we refer to (Prakken, 1991).

1. *Simple order*. For all σ and τ, we set $\sigma \preceq \tau$ if and only if σ is defeasible or τ is strict. It can easily be understood, at least intuitively, that a simple order indeed is the most simple order that is still in line with the ideas of defeasible argumentation. (Note that it trivially obeys the constraints of definition 2.10.)
2. *Number of defeasible steps*. Not very realistic, but this order proves to be handy in examples and counterexamples: $\sigma \preceq \tau$ if and only if $|\tau|_{\Rightarrow} \leq |\sigma|_{\Rightarrow}$, where $|\cdot|_{\Rightarrow}$ stands for the number of defeasible arrows in an argument.
3. *Weakest link*. With this order, an argument is as strong as its weakest link. Formally, this amounts to an extension of an order $\preceq$ on rules to an order $\preceq$ on arguments, in a conservative manner. To ensure that arguments are as strong as their weakest link, the following additional constraints should be satisfied. If $\sigma_1,...,\sigma_n \Rightarrow \sigma$ with $\phi_1,...,\phi_n \Rightarrow \phi$ as a last rule of inference, then $\sigma \preceq (\phi_1,...,\phi_n \Rightarrow \phi)$, and $(\phi_1,...,\phi_n \Rightarrow \phi) \preceq \sigma$ whenever there is no $1 \leq i \leq n$ such that $\sigma_i \preceq \sigma$. For the rest, definition 2.10 further ensures that the initial order is extended properly. Note that, with this order, atomic arguments (i.e. arguments in $\mathcal{L}$) are incomparable.
4. *Preferring the most specific argument*. For all σ and τ we have $\sigma \prec \tau$ if σ is defeasible and τ is strict, or σ and τ are both defeasible, but the premises of σ are based on the conclusions of subarguments of τ, i.e. *prem* (σ) is based on *conc* $\circ$ *sub* (τ). As it stands, this order does not fit in the shackles of definition 2.10, but if arguments are bound to be finite, then all constraints will be met.

This example completes the definition of abstract argumentation systems.

We now turn to the notion of compatibility. Compatibility is the shadow concept of consistency. However, we are not working in a traditional context, so the term consistency would be abused if we decided to use it.

DEFINITION 2.12. An argument σ is *in contradiction* if *conc* $(\sigma) = \perp$.

DEFINITION 2.13. A subset P of $\mathcal{L}$ is *incompatible* if there exists a strict argument in contradiction, that is based on P. A subset of $\mathcal{L}$ is *compatible* if it is not incompatible.

EXAMPLE 2.14. Let $\mathcal{L} = \{p, q, r, s\} \cup \{\perp\}$ and let $R = \{p \Rightarrow r, p,q \rightarrow s, r,s \rightarrow \perp\}$. Then all subsets of $\{p, q, s\}$, $\{p, r\}$, and $\{q, r\}$ are compatible, while all supersets of $\{p, q, r\}$, and $\{r, s\}$ are incompatible.

Compatibility naturally extends to sets of arguments. Thus, a set of arguments Σ is compatible if *conc* (Σ) is compatible.

We now turn to the notion of base set.

DEFINITION 2.15. A *base set* is a finite compatible subset of $\mathcal{L}$.

A base set contains 'irreducible information'. It is the point of departure in forward argumentation, and the final stopping place in backward justification. Whether a base set should contain irrefutable truths (e.g. axioms) or fleeting observations, does not matter here. Both viewpoints are admissible. Examples of base sets are data bases, scientific libraries, subjective observations, first principles in a case study or, speaking in terms of dialectics, initial concessions in a debate.[5]

The theory will develop along the framework of argument structures.

[5] Cf. Rescher's short monograph 'Dialectics' (1978).

DEFINITION 2.16. A set of arguments Σ is an *argument structure* if Σ contains all subarguments of its members and is compatible.

An argument structure is a structured set of arguments. It may be the result of deliberating on the information in a specific base set, it may be a specific state of reasoning, or a (partial and unfinished) description of a state of the world. In the last reading, terms like possible world, scenario, state of affairs, etc. are also appropriate. Note that base sets are precisely those argument structures which contain language elements only.

EXAMPLE 2.17. Consider the argumentation system $\mathscr{A}$ with language $\mathscr{L} = \{p, q, r\} \cup \{\bot\}$, with rules

$$R = \{p \Rightarrow q, p \Rightarrow r\} \cup \{q, r \rightarrow \bot\} \tag{1}$$

and with a simple order on arguments $\preceq$. Let $P = \{p\}$. Then $\Sigma_1 = P$ is an argument structure, as are $\Sigma_2 = \{p, p \Rightarrow q\}$ and $\Sigma_3 = \{p, p \Rightarrow r\}$. However, $\Sigma_4 = \{p, p \Rightarrow q, p \Rightarrow r\}$ is not an argument structure because it is incompatible, and $\Sigma_5 = \{p \Rightarrow q\}$ is not an argument structure because it does not contain all subarguments of its members.

3. Canonical example

The following argumentation system attempts to be a concrete and recognizable instantiation of an abstract argumentation system. We therefore start with a natural deduction type system in finite propositional logic, and extend it with a defeasible elimination rule for a defeasible implication. It is not allowed to discharge premises here. (Simply because that is not possible in the present setup. The general reason for not having this option is that it would complicate matters considerably.)

EXAMPLE 3.1. Consider the argumentation system $\mathscr{A}$ with language $\mathscr{L}$ as the smallest set, closed under the following conditions.

1. The set $P_n = \{p_1, \ldots, p_n\} \cup \{\bot\}$ is contained in $\mathscr{L}$;
2. The formulas $\phi \wedge \psi$, $\phi \vee \psi$, $\phi \supset \psi$, and $\neg\phi$ are in $\mathscr{L}$ whenever the formulas ϕ and ψ are members of $\mathscr{L}$;
3. The formula $\phi > \psi$ is in $\mathscr{L}$ whenever the formulas ϕ and ψ are $>$-free members of $\mathscr{L}$.

Thus, $\mathscr{L}$ is the language of finite propositional logic, augmented with a binary connective $>$. The argumentation system $\mathscr{A}$ has the following rules.

1. Introduction and elimination rules for the $\wedge$-connective: $\phi, \psi \rightarrow \phi \wedge \psi$; $\phi \wedge \psi \rightarrow \phi$; $\phi \wedge \psi \rightarrow \psi$. Introduction and elimination rules for the $\vee$-connective: $\phi \rightarrow \phi \vee \psi$; $\psi \rightarrow \phi \vee \psi$; $\phi \vee \psi, \phi \supset \chi, \psi \supset \chi \rightarrow \chi$. Elimination rule for the $\supset$-connective: $\phi, \phi \supset \psi \rightarrow \psi$. Elimination rule for the $\neg$-connective: $\phi, \neg\phi \rightarrow \bot$.
2. Combination rules for the $>$-connective: $\phi > \psi, \phi > \chi \rightarrow (\phi \wedge \psi) > \chi$; $\phi > \psi, (\phi \wedge \psi) > \chi \rightarrow \phi > \chi$; $\phi > \chi, \psi > \chi \rightarrow (\phi \vee \psi) > \chi$.
3. Elimination rule for the $>$-connective: $\phi, \phi > \psi \Rightarrow \psi$.

Thus, the set R consists of the analytical part of treewise natural deduction, augmented with rules of combination and a defeasible rule for the $>$-connective. It must be said that these rules are selected largely on pragmatic considerations. The first group of rules are taken from classical logic; the second group of rules can be seen as minimal conditions—rationality postulates, for that matter—a defeasible conditional should

comply with; the third rule simply is the defeasible counterpart of classical *modus ponens*. Experience has shown that the set thus obtained serves well in examples.
The argumentation system $\mathcal{A}$ orders its arguments in the following way. Let σ and τ be arguments in $\mathcal{A}$. Then $\sigma \prec \tau$ if and only if

1. The argument σ is a defeasible argument, while the argument τ is a strict argument; or
2. The argument τ is a bottom-defeasible argument, while the argument σ is defeasible, but not bottom-defeasible.

An argument σ is a *bottom-defeasible* argument if for every subargument $\tau_1,\ldots,\tau_n \Rightarrow \tau$ of σ, we have $\tau_1 \in \mathcal{L},\ldots,\tau_n \in \mathcal{L}$. Thus, $\mathcal{A}$ distinguishes three types of arguments: (i) strict arguments, (ii) bottom-defeasible arguments, and (iii) defeasible arguments satisfying neither (i) nor (ii).

To see why bottom-defeasible rules should defeat non-bottom-defeasible rules, consider the following example.

EXAMPLE 3.2. (Wearing trunks.) Let $P = \{d > b, b \supset t, d \supset \neg w, \neg w > \neg t\} \cup \{d\}$. Informally, d stands for 'day-off,' b for 'at-the-beach,' t for 'wearing trunks,' and w for 'working-at-the-office'. Then P enables the arguments $\sigma_1 = d, d > b \Rightarrow b$, $\sigma_2 = \sigma_1, b \supset t \rightarrow t$, $\tau_1 = d, d \supset \neg w \rightarrow \neg w$, and $\tau_2 = \tau_1, \neg w > \neg t \Rightarrow \neg t$. Now, because σ_2 is a bottom-defeasible argument and τ_2 is not, we may conclude that σ_2 defeats τ_2. Hence, the arguments σ_1, σ_2, and τ_1 should be in force, which is in accordance with our intuitions.

What this example attempts to show is that bottom-defeasible arguments are equivalent in conclusive force to defeasible arguments of length one.

4. Modes of reasoning—declarative, progressive, and regressive

Reasoning can be done in various ways, but there are two 'modes of reasoning' which pervade almost every branch of logic. There are the most fanciful descriptions for it but, essentially, all comes down on the commitment to a specific direction in which arguments are going to be developed. In practice, there are two such choices: forward and backward reasoning.[6]

Insofar the theory of abstract argumentation systems is concerned, these two types of reasoning are named progressive argumentation and regressive argumentation, respectively.

1. *Progressive argumentation.* This is forward reasoning in which arguments are constructed bottom up. The intention is to construct new arguments out of existing ones by repeatedly applying the appropriate rules of inference.
2. *Regressive argumentation.* This is backward reasoning in which arguments are retraced top down. Here, the intention is to find a way back to the premises for one

[6]Here are some synonyms. Progressive reasoning vs. regressive reasoning, prediction vs. explanation (common terminology in the theory of science), deduction vs. retroduction, planning vs. diagnosis (common terminology in AI), proof vs. resolution (common terminology in logic programming), bottom up vs. top down (used in, e.g., the theory of computer programming), and. finally, but by no means unimportant, synthesis vs. analysis (classical terminology).

specific conclusion (mostly) or a set of specific conclusions (rarely).

In addition, there is a third manner in which valid arguments can be obtained, but which involves no reasoning at all. This is the *declarative* mode of 'reasoning,' in which, by means of an effective procedure—that is, an effective procedure on the metalevel—some arguments are simply declared to be in force.

4.1. Declarative arrangement of arguments

A declarative definition simply *declares* which arguments are in force, and which conclusions are warranted. An immediate consequence is that there is no reasoning and, hence, no appeal to argumentation policies or formal dialectics. The entire theory is set at once.[7]

DEFINITION 4.1.1. Let P be a base set. An argument σ is *in force on basis of P*, written $P \vdash \sigma$, if

1. The set P contains σ; or
2. For some arguments $\sigma_1,...,\sigma_n$ we have $P \vdash \sigma_1,...,\sigma_n$ and $\sigma_1,...,\sigma_n \rightarrow \sigma$; or
3. For some arguments $\sigma_1,...,\sigma_n$ we have $P \vdash \sigma_1,...,\sigma_n$ and $\sigma_1,...,\sigma_n \Rightarrow \sigma$ and every set of arguments Σ that is in force on basis of P and such that σ is not better than any member of Σ, is compatible with σ.[8]

DEFINITION 4.1.2. Let P be a base set. Then *info*(P) denotes the set of arguments $\{\sigma \mid P \vdash \sigma\}$.

Let us concentrate on definition 4.1.1 as it stands. We see that it consists of three clauses. The first clause simply ensures that P is in force. The second clause, though simple as well, is imposing an important condition on *info*(P): it requires that *info*(P) is closed with respect to strict argumentation. In this manner, the property of being in force propagates through strict arguments. The third clause forms the heart of definition 4.1.1. Note that it is similar to the second clause, upto the additional condition on P and σ. The rationale behind this condition is that σ cannot be in force if it is incompatible with a set Σ that is in force, and of which there is no element that is properly weaker than σ.

For certain sets of rules, definition 4.1.1 is circular, so that *info*(P) is not uniquely determined. What is more, the set *info*(P) might not exist at all. This is not a flaw in the theory, but simply a manifestation of a restrained mechanism of defeat. In fact, theory formation on argumentation systems would be a lot easier if we could depart from the assumption that there is always rigorous defeat. Unfortunately, that is not the case.[9]

An element of $\mathcal{L}$ is said to be warranted if it is the conclusion of an argument that is in force.

---

[7] Two suitable examples here are (Loui, 1987) and (Horty and Thomason, 1988).

[8] Clearly, with $P \vdash \sigma_1,...,\sigma_n$ we mean $P \vdash \{\sigma_1,...,\sigma_n\}$ which, on its turn, means $P \vdash \sigma_1, \ldots, P \vdash \sigma_n$.

[9] Two excellent examples of theories that rest on the assumption of rigorous defeat are (i) the logic for defeasible reasoning LDR (Nute, 1988), and (ii) the semi-formal but ingenious formalism of defeat proposed in (Loui, 1987). Because these theories presuppose unique defeat, there are no difficulties in telling which argument is in force and which is not. Both systems may be classified as *bellicose* (i.e. eager to defeat). A good example of *pacifistic* argumentation is (Naess, 1966). According to Naess, arguments *pro* and *contra* should exist side by side, not to sow discord, but simply to tabulate as many arguments for and against as possible.

DEFINITION 4.1.3. Let P be a base set. Then *warrant* (P) denotes the set $conc \circ info\,(P)$.

In a way, this definition is paramount in a theory of defeasible argumentation because, in the end, we are merely interested in the conclusions of the arguments which are in force.

EXAMPLE 4.2. (Primitive case.) Consider the argumentation system with language $\mathscr{L} = \{p, q, r, s\} \cup \{\bot\}$, with rules

$$R = \{p \Rightarrow q, p \Rightarrow r, r \Rightarrow s\} \cup \{q, r \rightarrow \bot\} \tag{2}$$

and with an order on arguments $\preceq$ such that the argument $p \Rightarrow q$ is better than the argument $p \Rightarrow r$, i.e. $p \Rightarrow r \prec p \Rightarrow q$. Let $P = \{p\}$. Our goal is to determine which elements are warranted on basis of P. According to definition 4.1.3, this amounts to finding the arguments that are in force on basis of P. To begin with, definition 4.1.1 (i) together with $p \in P$ immediately yields $P \vdash p$. So p, considered as an argument, is in force. Next, let us examine whether the argument $p \Rightarrow q$ is in force. According to definition 4.1.1 we have $P \vdash p \Rightarrow q$ if (i) $P \vdash p$, (ii) the rule $p \Rightarrow q$ is in R, and (iii) every set of arguments Σ that is in force (on basis of P) and such that $p \Rightarrow q$ is not better than any member of Σ, is compatible with $p \Rightarrow q$. Since conditions (i) and (ii) are clearly fulfilled, condition (iii) remains to be examined. We begin by mentioning that $p \Rightarrow r$ and, hence, $p \Rightarrow r \Rightarrow s$ are strictly weaker than $p \Rightarrow q$. As a result, the only set of arguments Σ that is in force and such that the argument $p \Rightarrow q$ is not better than any member of Σ, is $\Sigma = \{p\}$. And this particular Σ clearly is compatible with $p \Rightarrow q$. Hence, condition (iii) is fulfilled as well. We may conclude that $p \Rightarrow q$ is in force on basis of P, i.e. $P \vdash p \Rightarrow q$. From this, it follows that the argument $p \Rightarrow r$ cannot be in force, since $\Sigma = \{p \Rightarrow q\}$ is a set of arguments that is in force, and such that $p \Rightarrow r$ is not better than any member of Σ. At the same time, the set Σ is incompatible with $p \Rightarrow r$. Hence, the argument $p \Rightarrow r \Rightarrow s$ cannot be in force, as it fails to fulfill condition (iii). We end up with *info* $(P) = \{p, p \Rightarrow q\}$.

EXAMPLE 4.3. (Tie.) Consider the argumentation system $\mathscr{A}$ with language $\mathscr{L} = \{p, q, r\} \cup \{\bot\}$, with rules

$$R = \{p \Rightarrow q, p \Rightarrow r, q, r \rightarrow \bot\} \tag{3}$$

and with a simple order on arguments (cf. example 2.11). Let $P = \{p\}$. Our goal is to determine which elements are warranted on basis of P. According to definition 4.1.3 it suffices to determine which arguments are in force on basis of P. To begin with, definition 4.1.1 (i) together with $p \in P$ immediately yields $P \vdash p$. So p, considered as an argument, is in force. Next, let us examine whether the argument $p \Rightarrow q$ is in force. According to definition 4.1.1 we have $P \vdash p \Rightarrow q$ if (i) $P \vdash p$, (ii) the rule $p \Rightarrow q$ is in R, and (iii) every set of arguments Σ that is in force (on basis of P) and such that $p \Rightarrow q$ is not better than any member of Σ, is compatible with $p \Rightarrow q$. Since conditions (i) and (ii) are clearly fulfilled, condition (iii) remains to be examined. To begin with, (iii) can be simplified by observing that (iii) holds if and only if every set Σ of defeasible arguments based on P that is in force, is compatible with $p \Rightarrow q$. Since $p \Rightarrow r$ is the only argument which may cause (iii) not to hold if it is in Σ, condition (iii) can be simplied further by stating that (iii) holds if and only if not $P \vdash p \Rightarrow r$. Consequently, $P \vdash p \Rightarrow q$ if not $P \vdash p \Rightarrow r$. Because the argumentation system $\mathscr{A}$ considered here is symmetric in the variables q and r, the same line of reasoning goes through for $p \Rightarrow r$.[10] Hence, $P \vdash p \Rightarrow r$

[10]For obvious reasons I sometimes deliberately circumvent the word *argument* in referring to arguments *about* argumentation systems.

if not $P \vdash p \Rightarrow q$. Combining this with the previous result yields $P \vdash p \Rightarrow q$ if and only if not $P \vdash p \Rightarrow r$. Put differently, we have either $info(P) = \{p, p \Rightarrow q\}$ or $info(P) = \{p, p \Rightarrow r\}$. As a result we have that p is warranted and, moreover, q is warranted if and only if r is not warranted.

EXAMPLE 4.4. (Lottery paradox.) Consider the argumentation system $\mathcal{A}$ with $\mathcal{L} = \{p_1,...,p_n\} \cup \{\perp\}$, with

$$R = \{\Rightarrow p_1, ..., \Rightarrow p_n\} \cup \{p_1,...,p_n \rightarrow \perp\} \quad (4)$$

and with a simple order on arguments $\preceq$. Let $P = \emptyset$. If, for every $1 \leq i \leq n$, the argument σ_i is defined as $\sigma_i = \Rightarrow p_i$, then all arguments in $\mathcal{A}$ are $\sigma_1,...,\sigma_n$, and $\sigma_1,...,\sigma_n \rightarrow \perp$. As can easily be verified, $info(P) = \{\sigma_1,...,\sigma_{i-1}, \sigma_{i+1},...,\sigma_n\}$, for some $1 \leq i \leq n$.

To obtain the lottery paradox, the element p_i should be read 'ticket i will not win'.

Digression: declarative semantics. When definition 4.1.1 is formulated within another vocabulary, we arrive at what may be called a *declarative semantics*, as follows.

DEFINITION 4.4.1. Let $\mathcal{A}$ be an abstract argumentation system. An *interpretation* of $\mathcal{A}$ is a function $w: \mathcal{A} \rightarrow \{\text{I}, \text{O}\}$.

Informally, the equations $w(\sigma) = \text{I}$ and $w(\sigma) = \text{O}$ denote that σ is 'in' and 'out,' respectively.

DEFINITION 4.4.2. Let $\mathcal{A}$ be an abstract argumentation system, and let P be a base set. Let w be an interpretation of $\mathcal{A}$. Then w is a *model* of P, written $w \models P$, if $info(P) = \{\sigma \mid w(\sigma) = \text{I}\}$.

Note that a single base set can have more than one model. In example 4.3, for instance, we have $w_1 \models P$ as well as $w_2 \models P$ if $w_1^{-1}(\text{I}) = \{p, p \Rightarrow q\}$ and $w_2^{-1}(\text{I}) = \{p, p \Rightarrow r\}$.

DEFINITION 4.4.3. Let $\mathcal{A}$ be an abstract argumentation system and let P be a base set. Then $extension(P) = \{w \mid w \models P\}$ and $degree(P) = |extension(P)|$.

In example 4.3 we have $degree(P) = 2$. Generally, a refined order in conclusive force results in having base sets with a low degree. This is because, with a refined order, most arguments differ in conclusive force, so that in case of conflicting arguments, it is often clear which argument has to be withdrawn.

PROPOSITION 4.4.4.[11] *Let* $\mathcal{A}_1 = (\mathcal{L}, R, \preceq_1)$ *and* $\mathcal{A}_2 = (\mathcal{L}, R, \preceq_2)$ *be abstract argumentation systems, such that* $\preceq_1$ *is a refinement of* $\preceq_2$*—i.e.* $\sigma \preceq_1 \tau$ *whenever* $\sigma \preceq_2 \tau$. *Then, for every base set P, we have that* $degree_1(P) \preceq degree_2(P)$.

In the borderland of nonmonotonic reasoning and traditional logic, the term 'extension' has two denotations (at least). On the one hand, it stands for a largest set of compatible arguments that contains P and, on the other hand, from a semantical point of view, it stands for the collection of all models in which P is true.[12] Remarkably enough, these two readings coincide here. As a result, there is no danger of misinterpretation. This, in turn, justifies reading the degree of a base set as the number of extensions that are enabled by it. There is no deeper meaning behind these matching notions and in fact, in the final analysis, the coincidence proves to be accidental.

[11]The full proofs of this and following propositions can be found in (Vreeswijk, 1991).

[12]In (Vreeswijk, 1991) it is proved that *info*(P) is a maximal complete argument structure that is based on P.

We conclude this digression with a brief remark on the use of the term 'semantics,' and more in particular whether this is the right word to use here. We immediately admit that, unlike classical logic, there is no clear distinction between interpretation (semantics) and computation (proofs). Indeed, the computational model-check of an interpretation is not straightforward.[13] On the other hand, this check is, irrespective of its computational complexity, a combinatorial procedure that has nothing to do with argumentation itself. Actually, the declarative mode of reasoning has more to do with declaration, than with reasoning.

4.5. Progressive argumentation

Progressive argumentation is forward reasoning. This section is concerned with one specific form of progressive argumentation, namely *procedural argumentation*.

Generally, the term procedural argumentation stands for any form of *methodological* forward reasoning—a systematic *modus operandi* for raising new arguments. Here the methodology is, by working from one state of reasoning to the next, a process of gradually constructing a stable and coherent collection of undefeated arguments.

The idea is simple, but to obtain tangible results, there are a number of notions that need formal explication. In particular, the following questions are relevant. What is the point of departure in procedural argumentation? What is a state of reasoning? What is considered to be a legal move from one state of reasoning to the next? How is the procedure to be implemented? And, finally, where does it stop?

We make the following choices. Procedural argumentation is setting up an *argumentation sequence* that begins with a base set. A state of reasoning is represented by a *finite argument structure*. (Note that a base set is a finite argument structure as well.) At this level of abstraction it is impossible to spell out in detail how the actual process of argumentation itself should be performed. Instead, we confine the space of legal moves between states of reasoning by putting an upper bound on the size of an elementary argumentation step.

DEFINITION 4.5.1. An *elementary argumentation step* from Σ_1 to Σ_2 is a pair of argument structures Σ_1, Σ_2 such that all proper subarguments of all members of Σ_2 are contained in Σ_1, and $\Sigma_2 - \Sigma_1$ consists of precisely one argument.

Often, in an elementary argumentation step from Σ_1 to Σ_2, the argument in $\Sigma_2 - \Sigma_1$ will not be mentioned, because the fact that an elementary argumentation step is performed is of greater importance than what is constructed precisely at that step.

DEFINITION 4.5.2. An *argumentation sequence* $(\Sigma_n)_{n=1}^{\infty}$ is a sequence of finite argument structures $\{\Sigma_n\}_{n=1}^{\infty}$ such that, for every $n \geq 1$, the pair Σ_n, Σ_{n+1} is an elementary argumentation step. The sequence $(\Sigma_n)_{n=1}^{\infty}$ *begins with* P if $P = \Sigma_1$.

In the course of argumentation, some arguments ultimately prevail in the long run.[14]

[13] As a matter of fact, a single model check can be reduced to the problem of finding a valid labeling in a so-called *truth maintenance system* (TMS) which, for the general case, is known to be NP-complete. (However, we don't know whether this reduction yields a specific class of TMS-networks. If it does, then the check may be of lower complexity. (Cf. Provan, 1988)).

[14] In fact, this is the sole criterion of Truth for the pragmaticist and, indeed, if one is willing to approach this matter in the 'philosophical mode,' there is a lot more to be said about the connections with *pragmatics* and *methodological pragmatism*. This, however, will not be done here. (Cf., e.g., Rescher, 1978).

DEFINITION 4.5.3. Let $(\Sigma_n)_{n=1}^{\infty}$ be an argumentation sequence. This sequence has a *limit*, denoted by $lim_n\Sigma_n$. An argument σ is in $lim_n\Sigma_n$ if σ is in all but a finite number of terms of $(\Sigma_n)_{n=1}^{\infty}$.

The following result ensures that, even if we take limits, our theory remains within the framework of argument structures.

PROPOSITION 4.5.4. *Let* $(\Sigma_n)_{n=1}^{\infty}$ *be an argumentation sequence, beginning with a base set P. Then* $lim_n\Sigma_n$ *is an argument structure containing P.* □

Ideally, an argumentation sequence beginning with a base set P should end up with *info* (P). However, since an elementary argumentation step is only bounded from above by the notion of elementary argumentation step, it generally does not follow that efficient argumentation is secured. Therefore, the relevant question at this point is whether there exists an argumentation policy that makes true progress, and if so, whether it approximates the collection of arguments which are in force. One possible answer is the following.

PROPOSITION 4.5.5. *Let* $\mathcal{A}$ *be a finite argumentation system, and let* $(\Sigma_n)_{n=1}^{\infty}$ *be an argumentation sequence, starting with a base set P such that, for every* $n \geq 1$*, the set* Σ_{n+1} *is a successor of* Σ_n*. Then* $lim_n\Sigma_n = info(P)$ *is a valid equation.* □

An argumentation system is said to be *finite* if it enables at most a finite number of arguments. A *successor* is a specific implementation of an elementary argumentation step, satisfying the constraints of definition 4.5.1. It is a *one step argumentation procedure*, meaning that, for each $n \geq 1$, the set Σ_{n+1} is determined entirely by the set Σ_n. For reasons of space we do not spell out exactly what is a successor—it is not so important here, anyway.[15]

4.6. Regressive argumentation

As was noticed at the beginning of section 4, regressive argumentation manifests itself in various ways, ranging from problem analysis, medical diagnosis and scientific explanation, to resolution and backward chaining in logic programming.

Here we focus on the problem of *abduction*, the name of which is originally due to the American philosopher C.S. Peirce (1839-1914).

The problem of abduction originally reads as follows.

1. A surprising fact ϕ is observed;
2. If H is true, then ϕ would be a matter of course;
3. This is a reason to suppose that H is the case.

In a scientific context, abduction is considered to be a method for creating new *hypothesis* to *explain* unexpected observations.

The most prominent and well-known account of abduction is the *hypothetico-deductive* scheme of explanation, originally proposed by Hempel and Oppenheim in 1948 as an idealised model of scientific enquiry. Here the abduction problem is formulated in classical logic, as follows. First, it is assumed that there is some *background theory T* that contains lawlike statements which are relevant for the problem at issue. Further, there is

---

[15] We refer to the paper on abstract argumentation systems (Vreeswijk, 1991).

a collection of possible hypothesis $\mathcal{H}$. Finally, if $H \subseteq \mathcal{H}$ and $T, H \vdash \phi$, then H is said to *explain* ϕ.

1. Now the *real* problem of hypothetico-deductive abduction is in finding an explanation that is both *parsimonious* and *consistent* at the same time, i.e. H should be $\subseteq$-minimal to prevent needlessly watering down an explanation, and H should be consistent with T to prevent the explanation from explaining everything. Herewith, the procedure encounters two intractable problems at once.[16]
2. Another disadvantage of the hypothetico-deductive approach is that explanations simply are additional 'gap-filling' premises, rather than rational reconstructions of the observation at hand. Indeed, in a classical deductive context, a rational (here: deductive) reconstruction does not provide new information, because it is already known how the fact to be explained logically follows from T and H. A rational reconstruction is simply redundant here. Generally, however, a rational reconstruction seems closer to what is considered to be an explanation, than a set of additional premises.
3. Finally, there is yet another complication. From a purely formal point of view it is perfectly possible to explain every observation ϕ by letting the implication $T \supset \phi$ be a member of $\mathcal{H}$. Of course we could stipulate that $T \supset \phi$, as a spurious explanation, is not in $\mathcal{H}$, but that does not guarantee that $\mathcal{H}$ is now free from spurious explanations. The question remains: how are spurious explanations to be distinguished from the proper ones?

Here follows an alternative account of abduction that naturally fits in the overall framework of abstract argumentation systems. (To be sure, our aim is not to present a 'better' formalisation of abduction, but to provide an alternative point of view, hoping to put matters in perspective.)

Before turning to the formal definitions, we briefly discuss the main differences with the hypothetico-deductive account of abduction. Let us mention three. First, the fact that is to be explained already follows from the base set, so that there is no need to seek for additional premises. Also, checking whether an explanation is both parsimonious and consistent is not necessary. Secondly, in the theory of argumentation systems, an explanation is an argument, and therefore a rational reconstruction of the observation at issue. Finally, with the notion of conclusive force (the third slot of an argumentation system) it is straightforward to select a 'best' explanation, if needed.

Formally, we proceed as follows. Let P be a base set. Only a few members of *arguments* (P) cannot serve as explanations, as they already interfere with strict consequences of P. These are dropped. The rest of the arguments are all candidate explanations.

DEFINITION 4.6.1. Let P be a base set. Then an argument is in the set *explanations* (P) if it is based on P, and is compatible with *strict* (P).

[16]Cf. Bylander *et al.* (1989). Further, this implies that Poole's "architecture for abductive reasoning," for the rest being a cogent demonstration of hypothetico-deductive abduction, suffers from intractability also. Poole is aware of this, and replies: "In our, albeit limited, experience, this has not been a problem. This is because the complexity more closely resembles the complexity of the problem being solved rather than the abstract complexity of the representation system. This is the topic of another paper, however." (Poole, 1989).

Note that non-singleton subsets of *explanations* (P) may very well be incompatible with *strict* (P)—which is right, because alternative explanations may or may not be compatible.

In the hierarchy of arguments, the explanations can be localised as follows.

$$P \rightarrow strict(P) \rightarrow info(P) \rightarrow explanations(P) \rightarrow arguments(P) \tag{5}$$

Thus, there are things that can be argued, but not explained, and there are things that can be explained, although they were not predicted.

DEFINITION 4.6.2. Let P be a base set, and let ϕ be an element of $\mathcal{L}$. Then

$$explain(P,\phi) = \{\sigma \mid prem(\sigma) \subseteq P, conc(\sigma) = \phi\} \cap explanations(P) \tag{6}$$

In a scientific vocabulary, an argument σ is said to be the *explanans*, explaining the *explanandum* ϕ. If one seeks for the 'best' explanation, i.e. an explanation having superior conclusive force, then the set $max \circ explain(P,\phi)$ should be considered. And for the location of conflicts, the set $explain(P,\perp)$ contains the relevant information.

It is fundamental to realise that the process of finding an explanation is, by the very nature of hindsight, much easier than performing forward argumentation. Because $explain(P,\phi)$ presents a collection of *alternative* rational reconstructions, rather than a set of *competitive* predictions, there is no need for doing laborious defeat. And indeed, also in daily practice, it is easy to be wise after the event.

4.7. Interaction of progressive and regressive argumentation

Various forms of forward reasoning as well as backward reasoning have been studied by many researchers for a long time. The issue of how these two modes of reasoning *interact*, however, is considered less often.

The basic idea of interaction that is propagated here is that alternating forward and backward reasoning yields new information. However. only specific *combinations* of back-and-forth reasoning prove to be sound. Consequently, the problem here is to identify sensible combinations.

As a first appraisal of what might be possible, we will consider a non-trivial example that elaborates on this problem in more detail.

EXAMPLE 4.7.1. Consider the argumentation system $\mathcal{A}$ of example 3.1, and let $P = \{f > (s \wedge l), f > t, t > s\} \cup \{f\}$. Informally, f stands for 'it's foggy,' s for 'cars drive at slow pace,' l for 'cars lighten their head-lamps' and t for 'a tailback of cars comes about'. Let $\sigma_1 = f, f > t \Rightarrow t$, $\sigma_2 = \sigma_1, t > s \Rightarrow s$, $\tau_1 = f, f > (s \wedge l) \Rightarrow s \wedge l$, $\tau_2 = \tau_1 \rightarrow s$, and $\tau_3 = \tau_1 \rightarrow l$. Note that P does not generate conflicting arguments, and that it contains redundant information. Suppose we observe l. The only argument that explains l is τ_3. Because $\tau_1 \sqsubseteq \tau_3$ and $\tau_1 \rightarrow s$, we predict s. Thus, a particular explanation may generate new predictions. Now, at this stage, it is very important to note that s, being predicted by τ_2, can be explained by σ_2 too. In that case, we could also predict t, since $\sigma_1 \sqsubseteq \sigma_2$ and $conc(\sigma_1) = t$. This, however, would not be right, because one explanation explaining l is enough. (Using *Ockham's razor*, hypothesis are not to be multiplied beyond necessity.) As a result, the argument σ_2 is 'explained away' by τ_2. In general, this means that a prediction is not allowed to generate new explanations, other than itself.

Actually, we have constructed a non-problematic example in which the alternative explanations neither support nor weaken each other, for it depends on this relationship whether it makes sense to include alternative explanations. Supportive explanations for a car-accident, for example, are driving too fast and being an unexperienced driver. Mutually weakening explanations are driving too fast and driving at slug speed. In the first case there is no problem in explaining the conclusion of a prediction with the alternative explanation. In the second case—the case 'where it goes wrong'—the same maneuver would lead to the rather odd conclusion that driving at slug speed may result in driving too fast.

The general conclusion is that generating predictions with the help of an explanation makes sense, while the other way around it does not.

Formally, this can be worked out as follows. If, given a base set P, there is nothing to be explained, then all predictions are set, as usual, to the set $info(P)$. If there are fixed explanations, then the set of arguments that is in force should be defined in such a way that it contains all these explanations.

DEFINITION 4.7.2. Let P be a base set, and let Σ be an argument structure that is based on P. An argument is in force on basis of $P(\Sigma)$, written $P(\Sigma)\vdash\sigma$, if $P\vdash\sigma$ or $\sigma\in\Sigma$. Accordingly, $info_{\Sigma}(P)$ is set equal to $\{\sigma \mid P(\Sigma)\vdash\sigma\}$. The set Σ is said to *span* the set $info_{\Sigma}(P)$.

Intuitively, Σ is a set of explanations. Note that Σ may resolve possible ambiguities in the definition of $info(P)$. In example 4.3 for instance, explaining q unwarrants r, and vice versa. Generally, we will have the following proposition.

PROPOSITION 4.7.3. *Let P be a base set. If $degree(P)=n$, then it takes at most $n-1$ explanations to resolve a possible ambiguity in the definition of $info(P)$.*

PROOF. Let $extension(P)=\{w_1,\ldots,w_n\}$, and choose for each $1\leq i\leq n$ and for each $1\leq j\leq n$, an argument $\sigma_{i,j}$ that is based on P, such that $w_i\models\sigma_{i,j}$ but $w_j\not\models\sigma_{i,j}$. Then $\Sigma_i=\{\sigma_{i,j}\mid 1\leq i\leq n,\ i\neq j\}$ is a set of arguments that is satisfied by w_i alone, so that, for each $1\leq i\leq n$, the set $info_{\Sigma(i)}(P)$ is defined unambiguously. Clearly, $|\Sigma_i|\leq n-1$. □

The lottery paradox (cf. example 4.4) shows us that, in some cases, it takes at least $n-1$ explanations to resolve an ambiguity. In other cases, one single explanation may be sufficient. Let us conclude with an abstract example of the latter, and show a single explanation may rearrange the set of arguments which are in force.

EXAMPLE 4.7.4. (Cascaded defeat.) Consider the abstract argumentation system $\mathcal{A}$ with $\mathcal{L}=\{p_1^n,\ldots,p_{n+1}^n\}_{n=1}^{\infty}\cup\{\bot\}$, with rules

$$R=\{p_1^n\Rightarrow p_2^n,\ldots,p_{n-1}^n\Rightarrow p_n^n\}_{n=2}^{\infty}\cup\{p_n^n\rightarrow p_{n+1}^n\}_{n=1}^{\infty}\cup\{p_{n+1}^n,p_{n+1}^{n+1}\rightarrow\bot\}_{n=1}^{\infty} \tag{7}$$

and with the $\Rightarrow$-count order on arguments $\preceq$ (cf. 2.11, 2nd example). In this manner, the letter p_j^i is the jth element of the ith argument. Let $P_n=\{p_1^1,\ldots,p_1^n\}$. Let $\sigma_n=p_1^n\Rightarrow\ldots\Rightarrow p_n^n$, and let $\tau_n=\sigma_n\rightarrow p_{n+1}^n$. Thus, for every $n\geq 1$, the argument τ_n is a counterargument of σ_{n+1} and, since $|\tau_n|_{\Rightarrow}<|\sigma_{n+1}|_{\Rightarrow}$, we have that τ_n is better than σ_{n+1}. From definition 4.1.1 it follows that, for every $n\geq 1$, the set $info(P_n)=sub(\tau_{2k-1}\mid 1\leq 2k-1\leq n)$. Now suppose that, for some $k\geq 1$, and for some $n\geq 2k$, we want to explain p_{2k+1}^{2k} on basis of P_n. In that case, $explain(P_n,p_{2k+1}^{2k})=\{\tau_{2k}\}$ because τ_{2k} is the only explanation of p_{2k+1}^{2k}. Since τ_{2k} is not in $info(P)$, we have that $info_{\tau(2k)}(P_n)=sub(\tau_{2k}\mid 1\leq 2k\leq n)$, so that both sets do not even intersect.

Example 4.7.4 is of course an example in the extreme of how the introduction of one single argument may completely rearrange the set *info*(P). Generally, an explanation compels to minor shifts only.

5. Resource bounded argumentation

The study of resource bounded reasoning seeks to provide a realistic account of the type of reasoning that is bound to use only a limited amount of resources. Within this theory, there are several approaches, of which two prominent ones will now be mentioned.

1. The first approach to resource bounded reasoning is based on the theory of *intractability*, in which the reasoning procedure (the algorithm that computes the answer) should be in a specific complexity class (cf., among others, Moses, 1988).
2. The second approach is based on what may be called the principle of *anytime reasoning*, which means that partial or approximate answers can be provided in the presence of computational resource constraints (cf., among others, Dean and Boddy, 1988).

The discussion here belongs to the second category.

It is essential to realise that resource bounded *defeasible* reasoning causes a particular form of nonmonotonicity that is the result of drawing conclusions on the basis of partially developed arguments. (To be sure, this is only one out of many reasons for withdrawing tentative conclusions.) Ideally, resource bounded defeasible reasoners are governed by a process that constructs arguments over time and determines what is justified relative to what arguments have been constructed. In this manner, conclusions are nonmonotonic, not only in evidence (which is the familiar desideratum), but also in computation.

Actually, we already have all the material to perform resource bounded reasoning at our disposal, if we assume that resource bounded reasoning is simply the possibility to *interrupt* an argumentation sequence at any time we would like to do so.

However, this implies that, except from making statements about argumentation systems ourselves, there should be a language in which is is possible to this as well. Such a language will be introduced in the next section. Thereafter, we will present simple formalisations of explicit and implicit conclusions, with which we have a simple theory of resource bounded defeasible reasoning.

5.1. Attachment semantics

The idea behind attachment here is that we formalise the language in which we speak about argumentation systems. This meta language will be, roughly, propositional logic with two modal connectives E and I, representing implicit and explicit adherence to language elements, respectively.

First, we fix the context by giving the rules and mechanism of defeat of possible inferential structures.

DEFINITION 5.1.1. A *frame* $\mathcal{F}$ is an abstract argumentation system $\mathcal{A}$.

Thus, from a purely syntactical point of view, frames and argument structures are just the same.[17]

[17] Why, then, are we renaming existing notions? There are two reasons for doing so. The first reason is that we simply want to keep in line with the terminology of Konolige, who used the term 'frame' to fix "the possi-

Let us now define the language that enables reasoning about a particular frame.

DEFINITION 5.1.2. The *propositional metalanguage* $\mathcal{L}_p$ contains propositional atoms $\mathcal{L}$, modal atoms $\{E\phi \mid \phi \in \mathcal{L}\}$, $\{I\phi \mid \phi \in \mathcal{L}\}$, and Boolean combinations thereof.

DEFINITION 5.1.3. A *derivational structure* D is a pair (P, Σ), where P is a base set and Σ is an argument structure based on P.

Informally, a derivational structure represents an 'inferential snapshot,' that is, a state of reasoning in which the base set and the arguments constructed thusfar form a pair.

DEFINITION 5.1.4. An *interpretation* is a pair $m = (D, v)$, where D is a derivational structure and v is a truth value assignment on elements of the language $\mathcal{L}$.

Semantic notions like *model*, *satisfaction*, and *entailment* are defined relative to a frame $\mathcal{F}$. The essential clauses read as follows.

- $(D,v) \models_{\mathcal{F}} \phi$ if and only if $v(\phi) = \text{true}$, where ϕ is a propositional atom.
- $(D,v) \models_{\mathcal{F}} E\phi$ if and only if $\phi \in conc(\Sigma)$, where $D = (P, \Sigma)$ is a derivational structure.
- $(D,v) \models_{\mathcal{F}} I\phi$ if and only if $\phi \in warrant(P)$, where $D = (P, \Sigma)$ is a derivational structure.

In this manner, we may write $\Gamma \models_{\mathcal{F}} \phi$ meaning that all interpretations satisfying all members of Γ are satisfying ϕ.

Digression: a multi-agent epistemic logic. Till now, the discussion takes place in an objective context. That is, a derivational structure is supposed to represent a specific information state that is 'public property,' for instance a scientific library.

A derivational structure, however, also directly fits in the subjective reading of information states, where each derivational structure represents a specific state of knowledge (or belief) of a particular agent. And then, of course, the entire formalism naturally extends to a multi-agent epistemic logic.

DEFINITION 5.1.5. A *multi-agent epistemic environment* is a collection $\mathcal{D} = \{D_i \mid i \in I\}$ of derivational structures in the context of a fixed frame $\mathcal{F}$.

In the multiple-agent formalism, there are as many modal connectives as there are agents, viz. $\{E_i \mid i \in I\}$, and $\{I_i \mid i \in I\}$. Moreover, an interpretation m now is a pair $(\mathcal{D}, v)$. The rest of the semantical terminology naturally extends to the multiple-agent formalism as well.

Perhaps somewhat surprisingly, it follows from definition 5.1.5 that, in contrast with, e.g., Konolige (1986), all agents are committed to the same set of rules. To see why this is reasonable after all, it may be clarifying to consider the following example.

EXAMPLE 5.1.6. Let $\mathcal{F} = (\mathcal{L}, R, \preceq)$ be the frame of classical propositional logic. Thus, $\mathcal{L}$ is the language of propositional logic, R is the set that contains all instantiations of modus ponens, and $\preceq$ is the empty set. Let P be a suitable axiomatisation of propositional logic, and let each agent have a base set $P \cup P_i$, where P_i is 'private knowledge'.

---

ble beliefs and derivations among beliefs an agent can have" (Konolige, 1990). We feel that there is sufficient resemblance to justify coinciding terminology. The second reason is that a frame plays a different role in our theory than an argumentation system does. The latter is supposed to be the object of our study, while a frame merely is the background framework of the attachment semantics.

(Note that, by definition, a base set is compatible, so that $P \cup P_i$ is consistent.)

In this example, each agent reasons by means of R, and it is clear that it cannot be otherwise. An agent cannot have a set of private rules of inference. On the other hand, the theory allows an agent to have its own strict and defeasible conditionals on the language level.

This example concludes our digression.

Before it is demonstrated how these notions are directly leading, via explicit and implicit knowledge, to resource bounded defeasible reasoning, let us present two illustrative examples of the attachment semantics first.

EXAMPLE 5.1.7. Consider the frame $\mathcal{F}$ with language $\mathcal{L} = \{p, q, r\} \cup \{\bot\}$, with rules

$$R = \{p \Rightarrow q, p \Rightarrow r, q,r \rightarrow \bot\} \tag{8}$$

and with a *simple order* on arguments: $\sigma \precsim \tau$ if and only if σ is defeasible or τ is strict. So $\mathcal{F}$ equals the argumentation system $\mathcal{A}$ of example 4.3—that is, from a purely syntactical point of view. Let $D = (P, \Sigma)$ with $P = \{p\}$ and $\Sigma = \{p\}$. Recall that the conclusion of example 4.3 was that either $\sigma = p \Rightarrow q$ or $\tau = p \Rightarrow r$ is in force, but not both. Thus, either q or r is warranted, but not both. The means are now available to express this disjunction in the propositional language that can make statements about argumentation systems, as follows.

$$(D,v) \models_{\mathcal{F}} Iq \vee Ir \tag{9}$$

Note that we cannot state either $(D,v) \models_{\mathcal{F}} Iq$ or $(D,v) \models_{\mathcal{F}} Ir$, although one of them actually must hold. (The choice of the truth assignment v does not really matter here.)

EXAMPLE 5.1.8. *Explanatory belief ascription.* (Konolige, 1990). In principle, belief ascription is doing abduction in the logic that reasons *about* beliefs, rather than doing abduction in the logic that reasons *with* beliefs. The latter was already done in section 4.6, where abduction is performed at the argument level. In using belief ascription, one may think of the problem of ascribing beliefs to an agent, giving partial knowledge of his beliefs.
Consider for example the frame $\mathcal{F}$ with language $\mathcal{L} = \{p, q, r\}$, with rules $R = \{p \rightarrow q, q \rightarrow r\}$, and with a simple order $\precsim$. Suppose furthermore that, already knowing that Ep, we come to know that Eq holds as well. Because $Ep \models_{\mathcal{F}} Eq$ does not hold, we are looking for an *explanation* X, such that $Ep, X \models_{\mathcal{F}} Eq$. Out of many possible explanations, perhaps chosen from a fixed set of hypothesis $\mathcal{H}$, one would be $X = Er$, ascribing explicit belief in r to explain explicit belief in q.

5.2. Explicit and implicit belief

Ever since Levesque introduced the notions of explicit and implicit belief to circumvent the problem of logical omniscience, these notions have become pervasive in contemporary research (Levesque, 1984). Instead of giving way to a laborious reformulation, let me quote Lakemeyer, who has spent considerable effort on these matters.

> "Logics of explicit and implicit belief have been proposed to account for the fact that agents generally believe only some but not all of the (...) consequences of their initial beliefs. While the explicit beliefs refer to the beliefs an agent holds given his limited reasoning capabilities, the implicit belief

can be thought of as those the agent would hold if he were an ideally rational agent with unlimited resources." (Lakemeyer, 1991).

Given this informal description of explicit and implicit belief, a number of questions concerning the relation among those two notions immediately come to the fore.

1. Should explicit belief be subsumed in implicit belief, or is it possible that something is explicitly, but not implicitly believed?
2. Should explicit belief be an approximation of implicit belief, or should it simply represent the 'hardware' of some state of belief?
3. Should explicit belief completely determine implicit belief, or not?

And so on.

Because this paper does not aim to provide an exhaustive study of explicit and implicit belief, we conclude the discussion of these notions by giving short answers to each of these questions in turn.

1. Perhaps somewhat surprisingly, the entailment $E\phi \models_{\mathcal{F}} I\phi$ generally does not hold in our theory. This might be surprising, because it is often taken for granted that implicit belief, as an elaboration of explicit belief, is simply an extension of it. This stance towards explicit and implicit belief can easily be refuted. Indeed, explicit belief that is not implicit belief is already encountered if a defeasible reasoner elaborates on—and commits itself to—arguments that are defeated in subsequent stages.

   The point is that resource bounded reasoning introduces yet another form of nonmonotonicity. A belief may come and go as the search for arguments becomes more complete. This principle of 'reasoning on second thoughts' pervades practical reasoning, where mere reflection may lead us to adopt and reject the same belief several times.

2. Another matter is to what extend explicit belief should approximate implicit belief. Let us define a relation of accessibility on interpretations, as follows.

   Let $m_1 = (D_1, v_1)$ and $m_2 = (D_2, v_2)$, where $D_1 = (P_1, \Sigma_1)$, and $D_2 = (P_2, \Sigma_2)$. Then $m_1 \sqsubseteq m_2$ if and only if $P_1 = P_2$ and Σ_2 is obtained from Σ_1 by means of an elementary argumentation step.

   Let $\sqsubseteq^*$ be the reflexive transitive closure of $\sqsubseteq$, let $\Box$ and $\Diamond$ be the modalities corresponding to $\sqsubseteq$, and let $\Box^*$ and $\Diamond^*$ be the modalities corresponding to $\sqsubseteq^*$. Our answer can now be given in an extremely concisive manner, by stating that, at least in the finite case, for some argumentation policy, explicit belief approximates implicit belief in the long run (cf. proposition 4.5.5).

$$\models_{\mathcal{F}} I\phi \equiv \Diamond^* E\phi \tag{10}$$

   This equivalence generally does not hold if frames are allowed to be infinite.

3. The last question seems trivial, and for monotonic reasoning it is. For nonmonotonic reasoning, however, these matters suddenly become less evident (and, therefore, worth studying). Insofar the present definition of explicit and implicit belief is concerned, the answer is, perhaps somewhat surprisingly again, that explicit belief need not completely determine implicit belief.[18]

[18] It might be surprising because one of the main objectives of Lakemeyer is in proving that implicit belief *is* fully determined by explicit belief (Lakemeyer, 1991).

Consider, for example, the frame $\mathcal{F}$ that was introduced in example 5.1.7. Let $P = \{p\}$, and let $D_1 = (P, \Sigma_1)$ and $D_2 = (P, \Sigma_2)$, with $\Sigma_1 = \{p\}$ and $\Sigma_2 = \{p \Rightarrow q\}$. Now, for D_1 it is perfectly clear that neither $(D_1, v) \models Iq$ nor $(D_1, v) \models Eq$ holds. For D_2, however, we have that $(D_2, v) \models Iq$ does not hold as well, in spite of the fact that $(D_2, v) \models Eq$.

To get a full grip on the significance of the last findings, it is essential to understand that, here, unlike the first case, there is no argumentation left to be done. Unlike explicit belief, implicit belief is not committed to the arguments that are constructed thusfar. In particular, even if Eq holds, we only have $Iq \vee Ir$, but not Iq.

To sum up, explicit belief is defined here as a set of arguments that is partially developed due to limited computational resources. In spite of this straightforward definition, we have touched upon three interesting deviations from the Levesque-Lakemeyer type account of explicit knowledge.

6. Related work—argument systems

In this section, we do not so much discuss related work in general, as well as one particular formalism that resembles the notion of abstract argumentation system presented here.

In 1988, Lin and Shoham published a report in which the notion of *argument system* was introduced. Here are two statements, both taken from a later publication that was based on that report (Lin and Shoham, 1989).

> "The key notions of argument systems are *inference rules*, *arguments*, *argument structures*, and *completeness conditions*."
>
> "We show that default logic, autoepistemic logic, negation as failure principles, and circumscription are all special cases of the proposed framework."

These statements are cited for two reasons. First, Lin and Shoham's basic terminology coincides with ours and, indeed, on essential points their theory departs, by and large, from the same principles as the theory that is presented here.[19]

From the second quotation, however, it becomes clear that Lin and Shoham are aiming at quite different goals than we do. While we are particularly concerned with pure theory formation in the isolated framework of abstract argumentation systems, the main virtue of the work of Lin and Shoham is that they have proved many existing formalisms to be special cases of their argument system.

The combination of these two observations yields sufficient material to conclude that the notion of argumentation system, without the third parameter of conclusive force, are directly subsumed in the network of cross-connections as laid out by Lin and Shoham. We might say that, in this way, we take immediate advantage of their work.

---

[19]More specifically, there is agreement on inference rules and arguments. On the other hand, our notion of argument structure is, unlike theirs, not closed under strict argumentation. Our reason to do so is that we feel that arguments should be raised by active argument construction, instead of being obtained automatically by means of a logical closure operator. Finally, a complete argument structure is, either (i) an argument structure having, for each element of some predefined subset of $\mathcal{L}$, an argument for or against that element, or (ii) in our theory, an argument structure that is 'saturated' with respect to argumentation (Lin and Shoham, 1989).

7. Conclusion

The intention of this paper has been to present various examples and applications of defeasible argumentation, without striving for definite theorems, and without committing ourselves to a specific implementation. This approach has been chosen because the mere possibility of raising a defeasible argument turns out to unsettle many of the logical notions that were familiar beforehand, and warrants an exploratory study of the potentials of the corresponding theory.

With the basic theory of abstract argumentation systems on the background, we have done two things.

1. We have explained how different modes of reasoning manifest themselves if argumentation is defeasible. In particular, we have shown how arguments may declared to be in force, and how this way of dealing with arguments may result in a declarative semantics.[20] Further, we have suggested how progressive argumentation can be done, and demonstrated some characteristic features of regressive argumentation (by giving an alternative account of the notion of scientific explanation). We have shown how these modes of reasoning may interact (by showing that how different explanations may disambiguate future predictions), and pointed at specific possibilities that may be pursued here (cf. definition 4.7.2). Finally, we gave a simplified account of resource bounded defeasible reasoning, which, as was demonstrated, differs on essential points with resource bounded non-defeasible reasoning.
2. Secondly, we have briefly described how our theory of abstract argumentation systems is intimately related with that of Lin and Shoham, and therefore directly profits from their efforts in proving that important nonmonotonic logics are special cases of their formalism.

Some modes of reasoning—the declarative mode, and the progressive mode—are already studied in depth in former papers. Some others need further elaboration. The author presently elaborates on the potentialities of regressive argumentation.

Acknowledgements. My sincere thanks to Wiebe van der Hoek, John-Jules Meyer, John Pollock, and Ron Loui for their readiness to discuss these issues with me.

References

BYLANDER, T., D. ALLEMANG, M.C. TANNER, AND J.R. JOSEPHSON, "Some Results Concerning the Computational Complexity of Abduction," *Proceeding of the First International Conference on Principles of Knowledge Representation and Reasoning (KR'89)*, pp. 44-45, Morgan Kaufmann (1989).

DEAN, T. AND M. BODDY, "An Analysis of Time-Dependent Planning," *Proceedings of the Seventh National Conference on Artificial Intelligence*, pp. 49-54 (1988).

ETHERINGTON, D.W., K.D. FORBUS, M.L. GINSBERG, D. ISRAEL, AND V. LIFSCHITZ, "Critical Issues in Nonmonotonic Reasoning," *Proceedings of the 1st International Conference on Knowledge Representation and Reasoning*, pp. 500-504, Morgan Kaufmann Publishers (1989).

HORTY, J.F. AND R.H. THOMASON, "Mixing Strict and Defeasible Inheritance,"

[20]This project has been fully elaborated on by now in the 1991 paper on nonmonotonicity and partiality in defeasible argumentation.

Proceedings of the AAAI, pp. 427-432 (1988).
KONOLIGE, K., *A Deduction Model of Belief*, Morgan Kaufmann Publishers, Los Altos, California (1986).
KONOLIGE, K., "Explanatory Belief Ascription, notes and premature formalization," *Proceedings of the Third Conference on Theoretical Aspects of Reasoning About Knowledge.*, Pacific Grove, CA, pp. 85-96, Morgan Kaufmann Publishers (March 1990).
LAKEMEYER, G., "On the Relation between Explicit and Implicit Belief," *Proceedings of the Second International Conference on Principles of Knowledge Representation and Reasoning (KR'91)*, pp. 368-375, Morgan Kaufmann (1991).
LEVESQUE, H.J., "A Logic of Implicit and Explicit Belief," *Proc of the AAAI*, pp. 198-202 (1984).
LIN, F. AND Y. SHOHAM, "Argument Systems: A Uniform Basis for Nonmonotonic Reasoning," *Proceedings of the 1st International Conference on Knowledge Representation and Reasoning*, pp. 245-255, Morgan Kaufmann Publishers (1989).
LOUI, R.P., "Defeat Among Arguments: A System of Defeasible Inference," *Computational Intelligence* **3**(2), pp. 100-106 (April 1987).
MOSES, Y., "Resource-bounded Knowledge," pp. 261-275 in *Proceedings of the Second Conf. on Theoretical Aspects of Reasoning about Knowledge*, ed. M.Y. Vardi, Morgan Kaufmann, Pacific Grove, California (March 7-9 1988).
NAESS, A., *Communication and Argument: elements of applied semantics*, George Allen and Unwin, London (1966). Translation of: En del elementaere logiske emner.
NUTE, D., "Defeasible Reasoning and Decision Support Systems," *Decision Support Systems* **4**(1), pp. 97-110, North-Holland (1988).
NUTE, D., "Defeasible Reasoning: a philosophical analysis in PROLOG," *Aspects of Artificial Intelligence*, pp. 251-288, Kluwer Academic Publishers (1988).
POLLOCK, J.L., "Defeasible Reasoning," *Cognitive Science* **11**, pp. 481-518 (1987).
POLLOCK, J.L., "A Theory of Defeasible Reasoning," *International Journal of Intelligent Systems* **6**, pp. 33-54, John Wiley & Sons, Inc. (1991).
POOLE, D., "Explanation and Prediction: an Architecture for Default and Abductive Reasoning," *Computational Intelligence* **5**, pp. 97-110 (1989).
PRAKKEN, H., "A Tool in Modelling Disagreement in Law: Preferring the Most Specific Argument," *Proceedings of the Third International Conference on Artifiical Intelligence and Law*, St. Catherine's College, Oxford, England, ACM Press (1991).
PROVAN, G.M., "The Computational Complexity of Assumption Based Truth Maintenance Systems," technical report 88-11, University of British Columbia, Department of Computer Science, Vancouver B.C., Canada (July 1988).
RESCHER, N., *Dialectics: A Controversy-Oriented Approach to the Theory of Knowledge*, State University of New York Press, Albany (1977).
RESCHER, N., *Methodological Pragmatism: A Systems-Theoretic Approach to the Theory of Knowledge*, Basil Blackwell, Oxford (1978).
VREESWIJK, G.A.W., "Abstract Argumentation Systems: Preliminary Report," *Proceedings of the First World Conference on the Fundamentals of Artificial Intelligence*, Paris, pp. 501-510, Angkor (1991). A full version is to appear in Studia Logica.
VREESWIJK, G.A.W., "Interpolation of Benchmark Problems in Defeasible Reasoning," report IR-255, Department of Mathematics and Computer Science, Vrije Universiteit Amsterdam (September 1991). Presented at the Second Symposium on AI and

Mathematics, Fort Lauderdale, Florida.
VREESWIJK, G.A.W., "The Feasibility of Defeat in Defeasible Reasoning," *Proceedings of the 2nd International Conference on Knowledge Representation and Reasoning*, pp. 526-534, Morgan Kaufmann Inc. (1991).
VREESWIJK, G.A.W., "Defeasible Dialectics: A Controversy-Oriented Approach towards Defeasible Argumentation," report IR-265, Department of Mathematics and Computer Science, Vrije Universiteit Amsterdam (December 1991). To appear in The Journal of Logic and Computation
VREESWIJK, G.A.W., "Nonmonotonicity and Partiality in Defeasible Argumentation," pp. 157-180 in *Nonmonotonic Reasoning and Partial Semantics*, ed. W. van der Hoek, J.-J.Ch. Meyer, Y.H. Tan, C. Witteveen, Prentice Hall (1992).

About Deductive Generalization

Ph. Besnard & E. Grégoire

IRISA
Campus de Beaulieu
F-35042 Rennes Cédex
FRANCE

Abstract. In this paper, we compare several forms of deductive generalization from positive and negative examples. We show that there exist cases where the logical conditions under which such paradigms can take place just appear too restrictive and somewhat counter-intuitive. Moreover, some of these conditions, which can be checked in an efficient way, also prove useful when satisfied since they sometimes allow one to prune the space of potential generalizations. Departing from a standard logical framework, we then show how an intuitionistic account of deductive generalization relaxes those conditions to make them more natural.

1 Introduction

Generalization from examples is a central issue in machine learning that can refer to very different paradigms. In this paper, we focus on forms of *deductive generalization*, which allow one to obtain the description of a more general concept by extracting the assertions that can be deduced from all the displayed examples. From a logical point of view, deductive generalization can be interpreted as being the converse of induction. Indeed, induction is usually defined as the problem of finding a formula F such that all the examples can be deduced from F, whereas deductive generalization consists in finding a formula F that can be deduced from all the examples. Although deductive generalization is usually followed by an inductive step, goals and application domains of these two paradigms are different (see e.g. [8]). Let us simply emphasize here that a formula F that is obtained through deductive generalization encodes information that is logically conveyed by the examples. Moreover, interpreting a formula F as a *generalization* of G when F can be deduced from G correctly translates the idea that F is *more general* than G, since the models of G form a subset of the models of F.

Although deductive generalization from positive and negative examples (in short, GPN) is thus of a *logical* nature, A.I. researchers have focused on its algorithmic perspectives, mainly (see e.g. [5, 6], with [7, 8] as a notable exception). Consequently, lacking serious logical foundations, the actual properties and power of GPN remain unclear. In this paper, we contribute to the logical foundations of GPN according to the following directions. We compare several forms of GPN and show that the logical conditions under which they can take place should not be taken for granted because they are very restrictive and somewhat counter-intuitive. Moreover, some of these conditions, which can be checked in an efficient way, also prove useful when satisfied since they sometimes allow one to prune the space of potential generalizations. Departing from a standard logical framework, we then show how an intuitionistic account of GPN relaxes those conditions to make them more natural.

For the clarity of the presentation, proofs of the theorems are given in an appendix.

2 Various forms of deductive generalization

2.1 Logical preliminaries

Let $\mathcal{L}$ be a set of well-formed formulas (thereafter, formulas) closed under classical connectives. We use $\neg$, $\wedge$, $\vee$, $\rightarrow$ and $\leftrightarrow$ to denote the negation, conjunction, disjunction, material implication and equivalence connectives, respectively. The symbol $\vdash$ represents the standard propositional inference relation. A set of formulas will be identified with the conjunctive formula whose conjuncts are the elements of the set. For example, given a formula F and a set $\mathcal{A}$ of formulas, we can speak of the formula $\mathcal{A} \wedge F$. Let $C = \{C_i : i \in [1..m]\}$ be a non-empty set of formulas of $\mathcal{L}$, we use $\vee C_i$ and $\wedge C_i$ to denote $C_1 \vee ... \vee C_m$ and $C_1 \wedge ... \wedge C_m$, respectively.

2.2 Various forms of deductive generalization

Let $\mathcal{A}$ be a consistent subset of $\mathcal{L}$. Let $\mathcal{P} = \{P_i : i \in [1..n]\}$ and $\mathcal{N} = \{N_j : j \in [1..m]\}$ be two non-empty finite sets of consistent formulas of $\mathcal{L}$. $\mathcal{A}$ represents the *domain* where GPN will take place, whereas $\mathcal{P}$ and $\mathcal{N}$ represent sets of *positive* and *negative examples* of the general concept to be discovered. A same language involving no a priori restriction on the use of classical connectives is thus selected to express both the available domain-knowledge in $\mathcal{A}$ and the examples in $\mathcal{P}$ and $\mathcal{N}$. This language will be used to express the generalizations as well.

Intuitively, a (deductive) generalization should be a formula G that can be deduced from the positive examples and that cannot be deduced from the negative ones, or a formula G such that its negation can be deduced from the negative examples and cannot be deduced from the positive ones. Actually, we shall study three different interpretations of the above phrase *"can be deduced from the examples"*, namely *"deduced from each example"*, *"deduced from the disjunction of all the examples"* and *"deduced from the conjunction of all the examples"*.

The following definition of generalizations translates the first interpretation above. It is due to [7, 8] and can be considered as the standard one.

Definition 1

The set $\mathcal{G}_1$ of *generalizations* from $\mathcal{P}$ and $\mathcal{N}$ w.r.t. $\mathcal{A}$ is the set of formulas G of $\mathcal{L}$ such that:

- $\forall P_i \in \mathcal{P} : \mathcal{A} \wedge P_i \vdash G$ or $\forall N_j \in \mathcal{N} : \mathcal{A} \wedge N_j \vdash \neg G$
- $\forall P_i \in \mathcal{P} : \mathcal{A} \wedge P_i \nvdash \neg G$
- $\forall N_j \in \mathcal{N} : \mathcal{A} \wedge N_j \nvdash G$

□

Since there cannot be any generalization whenever $\exists P_h \in \mathcal{P}: \mathcal{A} \wedge P_h$ is inconsistent or $\exists N_k \in \mathcal{N}$: $\mathcal{A} \wedge N_k$ is inconsistent, in the following we shall always assume that $\mathcal{A} \wedge P_i$ and $\mathcal{A} \wedge N_j$ are consistent for every P_i and every N_j.

The generalizations that satisfy the left (resp. right) part of the first condition are called *positive* (resp. *negative*). The set $\mathcal{G}1$ can be interpreted as the union of the set of positive generalizations, noted $\mathcal{G}1+$, with the set of negative generalizations, noted $\mathcal{G}1-$. Let us stress that Definition 1 allows positive and negative examples to play perfectly symmetrical roles. A definition that requires both parts of the first condition to be satisfied at the same time (i.e. that $\mathcal{G}1 = \mathcal{G}1+ \cap \mathcal{G}1-$) would be much too restrictive in the sense that it would not allow one to derive any generalization at all in some simple cases where natural and intuitive generalizations are expected (see e.g. [4]).

We shall also consider and compare variants of this definition, each of them giving rise to an alternative paradigm with a specific scope. A first variant only requires that G be deducible from the disjunction of all the positive examples or that $\neg G$ be deducible from the disjunction of all the negative examples, and that G and $\neg G$ be not deducible from the disjunction of the negative and the positive examples, respectively.

Definition 2
The set $\mathcal{G}2$ of *generalizations* from $\mathcal{P}$ and $\mathcal{N}$ w.r.t. $\mathcal{A}$ is the set of formulas G of $\mathcal{L}$ such that:

- $\mathcal{A} \wedge \vee P_i \vdash G$ or $\mathcal{A} \wedge \vee N_j \vdash \neg G$
- $\mathcal{A} \wedge \vee P_i \nvdash \neg G$
- $\mathcal{A} \wedge \vee N_j \nvdash G$

□

Since we shall compare $\mathcal{G}2$ with $\mathcal{G}1$, we also require $\mathcal{A} \wedge P_i$ and $\mathcal{A} \wedge N_j$ to be consistent for every P_i and every N_j. Note that this is a sufficient condition to ensure that $\mathcal{A} \wedge \vee P_i$ and $\mathcal{A} \wedge \vee N_j$ are consistent. By convention, the formulas of $\mathcal{G}2$ that satisfy the left (resp. right) part of the first condition are called *positive* (resp. *negative*) generalizations. The sets of positive and negative generalizations are noted $\mathcal{G}2+$ and $\mathcal{G}2-$, respectively.

Despite the fact that Definitions 1 and 2 are very close and share similar motivations, we shall see that they behave differently. Conclusions that can be derived from the disjunction of all the examples do not always coincide with the conclusions that can be deduced from each example.

A second variant allows one to extract the logical consequences of the *conjunction* of all the examples.

Definition 3
The set $\mathcal{G}3$ of *generalizations* from $\mathcal{P}$ and $\mathcal{N}$ w.r.t. $\mathcal{A}$ is the set of formulas G of $\mathcal{L}$ such that:

- $\mathcal{A} \wedge \wedge P_i \vdash G$ or $\mathcal{A} \wedge \wedge N_j \vdash \neg G$
- $\mathcal{A} \wedge \wedge P_i \nvdash \neg G$
- $\mathcal{A} \wedge \wedge N_j \nvdash G$

□

In the following, we shall require that $\mathcal{A} \wedge \wedge P_i$ and $\mathcal{A} \wedge \wedge N_j$ be consistent to avoid a too direct cause of emptiness of $\mathcal{G}_3$. The *positive* and *negative* generalizations, together with their corresponding sets $\mathcal{G}_{3+}$ and $\mathcal{G}_{3-}$, are defined as above.

Although Definition 3 is presented and discussed here as a mere formal variant of Definition 1, it is clear that its scope is different. The formulas of $\mathcal{P}$ and $\mathcal{N}$ should no longer represent positive or negative *instances* of the concept to be described. Instead, the formulas of $\mathcal{P}$ should represent *necessary parts* or *facets* of this latter concept. For example, suppose that the P_i's are representing the information encoded in fragments of a map of Paris whereas the N_j's are representing the information encoded in fragments of maps of some other towns. In this context, the goal of Definition 3 is to characterize the map of Paris, possibly enabling one to check in a further step whether a new fragment of a map concerns Paris or not.

From a formal point of view, Definitions 2 and 3 can be interpreted as two special cases of Definition 1 that involve one positive and one negative example, only (namely, $\vee P_i$ and $\vee N_j$ in the case of Definition 2 and $\wedge P_i$ and $\wedge N_j$ in the case of Definition 3). However, it is worth keeping this distinction into account because the three definitions correspond to different reasoning paradigms and exhibit different formal properties. Let us also emphasize that we consider variants of Definition 1 that preserve the symmetrical roles of positive and negative examples. Obviously, other variants can be investigated. For instance, we could also justify a variant of Definition 3 where the right part of the first condition requires *each* negative example to contradict G.

3 Existence of generalizations in the standard logical framework

3.1 Introduction

In this section, we stick to the framework of standard logic to study the logical conditions under which the above forms of deductive generalization can take place. More precisely, we establish the necessary and sufficient conditions for the sets $\mathcal{G}_1$, $\mathcal{G}_2$ and $\mathcal{G}_3$ to be empty. We then examine whether these conditions cover the cases where $\mathcal{P}$ and $\mathcal{N}$ share examples and show how these conditions change when additional examples are taken into account. Such a study allows us to extract very decisive elements of comparison between the three sets. Proceeding further in this comparison, we show how $\mathcal{G}_1$, $\mathcal{G}_2$ and $\mathcal{G}_3$ are actually related in the general case. We then successively examine how the presence of tautologies in the sets of examples and how the existence of positive (resp. negative) examples subsuming negative (resp. positive) ones influence the emptiness of these sets. Finally, we discuss computational techniques that allow one to perform a priori tests of the existence of generalizations. In some cases, these techniques allow one to achieve dramatic simplifications of the actual search for the generalizations.

3.2 Basic existence theorems

The following basic theorem states the a priori conditions for GPN to take place when Definition 1 and standard propositional logic are under consideration.

Theorem 1 $\mathcal{G}_1 = \emptyset$ iff $(\exists N_k \in \mathcal{N} : \mathcal{A} \wedge N_k \vdash \vee P_i)$ and $(\exists P_h \in \mathcal{P} : \mathcal{A} \wedge P_h \vdash \vee N_j)$

Note that the first (resp. second) condition in the right part of the theorem is the necessary and sufficient condition for $\mathcal{G}_{1+}$ (resp. $\mathcal{G}_{1-}$) to be empty. Actually, as $\mathcal{G}_i = \mathcal{G}_{i+} \cup \mathcal{G}_{i-}$ $(i \in \{1,2,3\})$, most of the results that we shall obtain about the emptiness of $\mathcal{G}_i$ will consist of a conjunction of two conditions related to the emptiness of $\mathcal{G}_{i+}$ and $\mathcal{G}_{i-}$, respectively.

The two following corollaries show us that dropping the domain $\mathcal{A}$ in the right part of the above theorem entails a sufficient criterion for $\mathcal{G}_1$ to be empty, which becomes a necessary and sufficient condition only when $\mathcal{A}$ is itself empty.

Corollary 1.1 If $(\exists N_k \in \mathcal{N} : N_k \vdash \vee P_i)$ and $(\exists P_h \in \mathcal{P} : P_h \vdash \vee N_j)$ then $\mathcal{G}_1 = \emptyset$

Corollary 1.2 Let $\mathcal{A} = \emptyset$. $\mathcal{G}_1 = \emptyset$ iff $(\exists N_k \in \mathcal{N} : N_k \vdash \vee P_i)$ and $(\exists P_h \in \mathcal{P} : P_h \vdash \vee N_j)$

As an important consequence of Theorem 1, $\mathcal{G}_1$ is empty when $\mathcal{P}$ and $\mathcal{N}$ share examples, and more generally, when $\mathcal{P}$ and $\mathcal{N}$ share semantically equivalent examples.

Corollary 1.3 If $\mathcal{P} \cap \mathcal{N} \neq \emptyset$ then $\mathcal{G}_1 = \emptyset$

The following theorem shows us that any collapse of $\mathcal{G}_1$ is *definitive*: whenever we have $\mathcal{G}_1 = \emptyset$, enlarging the domain $\mathcal{A}$ or introducing additional examples in $\mathcal{P}$ or $\mathcal{N}$ does not change this result. This is very useful in the sense that if we manage to prove $\mathcal{G}_1 = \emptyset$ for a subset of $\mathcal{P}$, a subset of $\mathcal{N}$ and a subset of the domain $\mathcal{A}$, then we keep having $\mathcal{G}_1 = \emptyset$ whatever the actual domain $\mathcal{A}$, sets $\mathcal{P}$ and $\mathcal{N}$ are, and however these sets *could* be enlarged.

Theorem 2 Let $\mathcal{A} \subseteq \mathcal{A}_b$, $\mathcal{P} \subseteq \mathcal{P}_b$ and $\mathcal{N} \subseteq \mathcal{N}_b$.
Let $\mathcal{G}_1$ be the set of generalizations associated with a domain $\mathcal{A}$ and with sets of examples $\mathcal{P}$ and $\mathcal{N}$.
Let $\mathcal{G}_{1b}$ be the set of generalizations associated with a domain $\mathcal{A}_b$ and with sets of examples $\mathcal{P}_b$ and $\mathcal{N}_b$. If $\mathcal{G}_1 = \emptyset$ then $\mathcal{G}_{1b} = \emptyset$

Let us replay the previous analysis for the alternative definitions of GPN. The a priori logical conditions for GPN to take place when Definition 2 is under consideration are expressed by means of the following theorem.

Theorem 3 $\mathcal{G}_2 = \emptyset$ iff $(\mathcal{A} \wedge \vee N_j \vdash \vee P_i)$ and $(\mathcal{A} \wedge \vee P_i \vdash \vee N_j)$

Corollary 3.1 If $\vdash \vee N_j \leftrightarrow \vee P_i$ then $\mathcal{G}_2 = \emptyset$

Corollary 3.2 Let $\mathcal{A} = \emptyset$. $\mathcal{G}2 = \emptyset$ iff $\vdash \vee N_j \leftrightarrow \vee P_i$

Contrary to $\mathcal{G}1$, the presence of examples in common in $\mathcal{P}$ and $\mathcal{N}$ is not a sufficient condition to obtain an empty set of generalizations when $\mathcal{G}2$ is under consideration. Moreover, unlike any collapse of $\mathcal{G}1$, a collapse of $\mathcal{G}2$ does not need to be definitive (this is easily understood as e.g. the equivalence $\vee N_j \leftrightarrow \vee P_i$ can be altered by introducing an additional example N_s or P_t). This shows us that $\mathcal{G}1$ and $\mathcal{G}2$ should be carefully distinguished although their motivations can appear very similar.

A similar analysis can be undertaken for $\mathcal{G}3$. A collapse of $\mathcal{G}3$ is neither definitive, nor ensured by examples in common in $\mathcal{P}$ and $\mathcal{N}$.

Theorem 4 $\mathcal{G}3 = \emptyset$ iff $(\mathcal{A} \wedge \wedge N_j \vdash \wedge P_i)$ and $(\mathcal{A} \wedge \wedge P_i \vdash \wedge N_j)$

Corollary 4.1 If $\vdash \wedge N_j \leftrightarrow \wedge P_i$ then $\mathcal{G}3 = \emptyset$

Corollary 4.2 Let $\mathcal{A} = \emptyset$. $\mathcal{G}3 = \emptyset$ iff $\vdash \wedge N_j \leftrightarrow \wedge P_i$

3.3 Comparison of $\mathcal{G}1$, $\mathcal{G}2$ and $\mathcal{G}3$

Let us proceed further in the comparison of $\mathcal{G}1$, $\mathcal{G}2$ and $\mathcal{G}3$. The following lemma shows that Definition 2 *subsumes* Definition 1 in the sense that every generalization obtained according to Definition 1 can be derived according to Definition 2. In other words, the generalizations extracted from each example are generalizations extracted from the disjunction of the examples.

Lemma 5.1 $\mathcal{G}1 \subseteq \mathcal{G}2$

The following theorem completes the comparison of the existence (in the general case) of generalizations according to Definition 1, 2 and 3. It shows that, with the exception of the required relation $\mathcal{G}1 \subseteq \mathcal{G}2$, the three sets are unrelated in the general case. Accordingly, the converse of Lemma 5.1 does not hold: generalizations from the disjunction of all the examples are not always generalizations from each example. Such a result is even more important since it departs from the obvious equivalence $\mathcal{G}1 = \mathcal{G}2$ obtained when deductive generalization is performed when $\mathcal{P} = \emptyset$ or $\mathcal{N} = \emptyset$.

Theorem 5

The following table considers every combination of empty sets taken from $\mathcal{G}1$, $\mathcal{G}2$ and $\mathcal{G}3$ and states whether this combination is possible or not (in the general case).
(NE stands for 'non-empty'; P for 'possible' and I for 'impossible combination')

$\mathcal{G}1$	$\mathcal{G}2$	$\mathcal{G}3$		
$\emptyset$	$\emptyset$	$\emptyset$	P	(see Example 1)
$\emptyset$	$\emptyset$	NE	P	(see Example 2)

Ø	NE	Ø	P	(see Example 3)
Ø	NE	NE	P	(see Example 4)
NE	Ø	Ø	I	(Lemma 5.1)
NE	Ø	NE	I	(Lemma 5.1)
NE	NE	Ø	P	(see Example 5)
NE	NE	NE	P	(see Example 6)

The following examples illustrate all the different classes of possible situations described by Theorem 5. They also allow us to point out some particular behaviours of $\mathcal{G}1$, $\mathcal{G}2$ and $\mathcal{G}3$ that will be discussed in the next paragraphs.

Example 1 Let $\mathcal{P} = \{(P \to Q) \vee (Q \to R)\}$, $\mathcal{N} = \{(S \to T) \vee (T \to U)\}$ and $\mathcal{A} = \emptyset$.
We have $\mathcal{G}1 = \mathcal{G}2 = \mathcal{G}3 = \emptyset$.
At first glance, $\mathcal{P}$ and $\mathcal{N}$ could be interpreted as reasonable sets of positive and negative examples since they are apparently completely unrelated. However, both $(P \to Q) \vee (Q \to R)$ and $(S \to T) \vee (T \to U)$ are tautologies and are thus semantically equivalent. Let us stress that the presence of a tautology in either $\mathcal{P}$ or $\mathcal{N}$ does not necessary entail $\mathcal{G}1 = \mathcal{G}2 = \mathcal{G}3 = \emptyset$, as the following example illustrates it.

Example 1' Let $\mathcal{P} = \{P \vee Q,\ Q \vee \neg Q\}$, $\mathcal{N} = \{P\}$ and $\mathcal{A} = \emptyset$.
We have $\mathcal{G}1 \neq \emptyset$, $\mathcal{G}2 \neq \emptyset$ and $\mathcal{G}3 \neq \emptyset$.
The influence of the presence of tautologies in $\mathcal{P}$ and $\mathcal{N}$ on the sets of generalizations will be studied in further theorems.

Example 2 Let $\mathcal{P} = \{P, Q\}$, $\mathcal{N} = \{P \vee Q\}$ and $\mathcal{A} = \emptyset$.
We have $\mathcal{G}1 = \mathcal{G}2 = \emptyset$, $\mathcal{G}3 \neq \emptyset$.

Example 3 Let $\mathcal{P} = \{R, T \vee \neg T\}$, $\mathcal{N} = \{R\}$ and $\mathcal{A} = \emptyset$.
We have $\mathcal{G}1 = \mathcal{G}3 = \emptyset$, $\mathcal{G}2 \neq \emptyset$.
This example shows that the converse of Lemma 5.1 does not hold. In other words, Definition 2 gives rise to larger sets of generalizations than Definition 1 does. This does not mean that Definition 2 should be preferred over Definition 1; in the example, we could expect an empty set of generalizations since $\mathcal{P}$ and $\mathcal{N}$ have an element in common.

Example 4 Let $\mathcal{P} = \{P \vee Q, R \vee P\}$, $\mathcal{N} = \{Q \vee R,\ S \vee P, S \vee R\}$ and $\mathcal{A} = \emptyset$.
We have $\mathcal{G}1 = \emptyset$, $\mathcal{G}2 \neq \emptyset$, $\mathcal{G}3 \neq \emptyset$.

Example 5 Let $\mathcal{P} = \{P \wedge Q, R \wedge S\}$, $\mathcal{N} = \{P \wedge S,\ Q \wedge R\}$ and $\mathcal{A} = \emptyset$.
We have $\mathcal{G}1 \neq \emptyset$, $\mathcal{G}2 \neq \emptyset$, $\mathcal{G}3 = \emptyset$.

Example 6 Let $\mathcal{P} = \{P\}$, $\mathcal{N} = \{Q\}$ and $\mathcal{A} = \emptyset$.
We have $\mathcal{G}1 \neq \emptyset$, $\mathcal{G}2 \neq \emptyset$, $\mathcal{G}3 \neq \emptyset$.

3.4 Role of the tautologies

Let us now examine how the presence of tautologies in either $\mathcal{P}$ or $\mathcal{N}$ influence the existence of generalizations. As we have seen in Examples 1, 1' and 3, tautologies in $\mathcal{P}$ and $\mathcal{N}$ play a very particular role. The following results allow one to (sometimes) simplify the generalization process by restricting the search procedure for generalizations to either positive or negative ones. Note that the following results do not refer to the domain $\mathcal{A}$.

Theorem 6 If $(\exists P_h \in \mathcal{P}: \vdash P_h)$ then $\mathcal{G}_{1+} = \emptyset$
If $(\exists N_k \in \mathcal{N}: \vdash N_k)$ then $\mathcal{G}_{1-} = \emptyset$

Corollary 6.1 If $(\exists P_h \in \mathcal{P}\ \exists N_k \in \mathcal{N} : \vdash P_h$ and $\vdash N_k)$ then $\mathcal{G}_1 = \emptyset$

The presence of tautologies in the examples is not a sufficient condition to ensure that $\mathcal{G}_{2+}$ or $\mathcal{G}_{2-}$ is empty. However, these sets are empty when the disjunction of the examples is a tautology. Since $\mathcal{G}_1 \subseteq \mathcal{G}_2$, conditions ensuring the emptiness of $\mathcal{G}_2$ also apply for $\mathcal{G}_1$.

Theorem 7 If $\vdash \vee P_i$ then $\mathcal{G}_{1+} = \mathcal{G}_{2+} = \emptyset$
If $\vdash \vee N_j$ then $\mathcal{G}_{1-} = \mathcal{G}_{2-} = \emptyset$

Corollary 7.1 If $\vdash \vee P_i$ and $\vdash \vee N_j$ then $\mathcal{G}_1 = \mathcal{G}_2 = \emptyset$

When each example is a tautology, results expressing the emptiness of $\mathcal{G}_3$ can be derived.

Theorem 8 If $\vdash \wedge P_i$ then $\mathcal{G}_{3+} = \emptyset$
If $\vdash \wedge N_j$ then $\mathcal{G}_{3-} = \emptyset$

Corollary 8.1 If $\vdash \wedge P_i$ and $\vdash \wedge N_j$ then $\mathcal{G}_3 = \emptyset$

3.5 Role of subsuming examples

We have already examined under which conditions the presence of positive examples logically equivalent to negative ones entails the absence of generalizations. Let us now examine the weaker situations where members of one set of examples are subsumed (i.e. entailed) by a member of the other set. Note that the following results do not refer to the domain $\mathcal{A}$.

A first result can be considered as a strenghtening of the general conditions expressed by Corollary 1.1.

Theorem 9 If $(\exists P_h \in \mathcal{P}\ \exists N_k \in \mathcal{N} : P_h \vdash N_k)$ then $\mathcal{G}_{1-} = \emptyset$
If $(\exists P_h \in \mathcal{P}\ \exists N_k \in \mathcal{N} : N_k \vdash P_h)$ then $\mathcal{G}_{1+} = \emptyset$

Corollary 9.1
If $(\exists P_h \in \mathcal{P}\ \exists N_k \in \mathcal{N} : P_h \vdash N_k)$ and $(\exists P_s \in \mathcal{P}\ \exists N_t \in \mathcal{N} : N_t \vdash P_s)$ then $\mathcal{G}_1 = \emptyset$

The following sufficient condition for the emptiness of $\mathcal{G}_1$ is dual to the condition expressed by Corollary 1.1.

Theorem 10 If $(\exists N_k \in \mathcal{N} : \vee P_i \vdash N_k)$ then $\mathcal{G}_{1-} = \mathcal{G}_{2-} = \emptyset$
If $(\exists P_h \in \mathcal{P} : \vee N_j \vdash P_h)$ then $\mathcal{G}_{1+} = \mathcal{G}_{2+} = \emptyset$

Corollary 10.1
If $(\exists N_k \in \mathcal{N} : \vee P_i \vdash N_k)$ and $(\exists P_h \in \mathcal{P} : \vee N_j \vdash P_h)$ then $\mathcal{G}_1 = \mathcal{G}_2 = \emptyset$

As far as $\mathcal{G}_3$ are concerned, the absence of generalization can be obtained when an example in one set subsumes the conjunction of all the examples in the other set.

Theorem 11 If $(\exists N_k \in \mathcal{N} : N_k \vdash \wedge P_i)$ then $\mathcal{G}_{3+} = \emptyset$
If $(\exists P_h \in \mathcal{P} : P_h \vdash \wedge N_j)$ then $\mathcal{G}_{3-} = \emptyset$

Corollary 11.1 If $(\exists N_k \in \mathcal{N} : N_k \vdash \wedge P_i)$ and $(\exists P_h \in \mathcal{P} : P_h \vdash \wedge N_j)$ then $\mathcal{G}_3 = \emptyset$

3.6 Computational issues

The ultimate goal of GPN is to find out the generalizations or pick up the most interesting ones. In this respect, the above results are useful since

-they can avoid us searching for non-existent generalizations

-they can entail dramatic simplifications of the actual search for generalizations.

First, let us consider $\mathcal{G}_1$. As we have shown it, $\mathcal{G}_1$ exhibits several interesting properties with respect to its potential emptiness:

-examples in common in $\mathcal{P}$ and $\mathcal{N}$ entails $\mathcal{G}_1 = \emptyset$.

-a test of the emptiness of $\mathcal{G}_1$ consists of a test of the emptiness of both $\mathcal{G}_{1+}$ and $\mathcal{G}_{1-}$. Even if the final result of the global test is negative, it can indicate that $\mathcal{G}_{1+}$ or $\mathcal{G}_{1-}$ is empty and thus simplify the actual generalization process.

-the discovery of tautologies in either $\mathcal{P}$ or $\mathcal{N}$ and results obtained about subsuming examples allow us also to reduce the search process to either $\mathcal{G}_{1+}$ or $\mathcal{G}_{1-}$.

-moreover, whenever a test of the emptiness of $\mathcal{G}_1$ (or $\mathcal{G}_{1+}$ or $\mathcal{G}_{1-}$) succeeds with subsets of $\mathcal{A}$, $\mathcal{P}$ and $\mathcal{N}$, this result applies also for the actual domain and sets of examples. Accordingly, we can pick up *suspicious* examples and apply the test on them, only. A positive result will apply for the actual problem. There are of course many ways to define what suspicious examples could be. For instance, suspicious examples can be all the examples sharing some common predicates.

A very efficient implementation of the above findings can be achieved when all P_i's and N_j's are consistent propositional clauses. In such a case, the following theorem entails a general procedure testing sufficient conditions for $\mathcal{G}_1$, $\mathcal{G}_{1+}$ and $\mathcal{G}_-$ to be empty, that runs in $O(nm \log(nm))$, where $2n$ is the number of considered examples and m is the average number of literals occurring in an example.

(This test could be improved in several ways: e.g. by inserting a procedure detecting common examples in $\mathcal{P}$ and $\mathcal{N}$ in a purely syntactical way.)

Theorem 12 Let every P_i and N_j be a consistent propositional clause.

If ($\vee P_i$ contains opposite literals) or ($\exists N_k \in \mathcal{N}$ such that every literal in N_k occurs also in $\vee P_i$)
  then $\mathcal{G}_{1+} = \emptyset$

If ($\vee N_j$ contains opposite literals) or ($\exists P_h \in \mathcal{P}$ such that every literal in P_h occurs also in $\vee N_j$)
  then $\mathcal{G}_{1-} = \emptyset$

Note that in the case where one of the left parts of these conditions is not satisfied, we can conclude that, in the general case, the corresponding conclusion is false only when $\mathcal{A} = \emptyset$ and all the examples have been taken into account.

Whenever $\mathcal{G}_2$ or $\mathcal{G}_3$ is under consideration, all the P_i's and the N_j's must be taken into account in the global emptiness tests. Moreover, the presence of common examples in $\mathcal{P}$ or $\mathcal{N}$ does not entail the emptiness of the set of generalizations or even simplifications in the further search for generalizations.

Theorem 13 Let every P_i and N_j be a consistent propositional clause.
If (every literal in $\vee N_j$ occurs in $\vee P_i$) then $\mathcal{G}_{2+} = \emptyset$
If (every literal in $\vee P_i$ occurs in $\vee N_j$) then $\mathcal{G}_{2-} = \emptyset$

Theorem 14 Let every P_i and N_j be a consistent conjunction of literals.
If (every literal in $\wedge P_i$ occurs in $\wedge N_j$) then $\mathcal{G}_{3+} = \emptyset$
If (every literal in $\wedge N_j$ occurs in $\wedge P_i$) then $\mathcal{G}_{3-} = \emptyset$

Note that in these two last theorems, when one of the sufficient conditions is not satisfied, we can conclude that the corresponding conclusion does not hold in the general case only when $\mathcal{A} = \emptyset$.

4 Existence theorems in the intuitionistic framework

The above study has been conducted within a standard logical framework. Let us now examine how the existence conditions for GPN change when the logical system under consideration is intuitionistic. Let us simply recall here that proofs are of constructive nature in intuitionistic logic, $\alpha \vee \neg\alpha$ is not an axiom schema, the theorems are a subset of the standard theorems and the logical connectives are not definable in terms of each other. Actually, the intuitionistic interpretation of the logical connectives is often claimed to be closer to their naturally intuitive reading, than the standard one is. For an introduction to intuitionistic logic, see [3].

Let $\vdash_i$ denote the standard propositional intuitionistic inference relation and let $\mathcal{G}_1'$, $\mathcal{G}_2'$ and $\mathcal{G}_3'$ denote the sets of generalizations obtained through Definitions 1, 2 and 3, when these definitions resort to $\vdash_i$ instead of $\vdash$. Let us write $\mathcal{G}_j \subseteq_i \mathcal{G}_j'$ ($j \in \{1,2,3\}$) when $\forall G \in \mathcal{G}_j : \neg\neg G \in \mathcal{G}_j'$.

Theorem 15 $\qquad \mathcal{G}_1 \subseteq_i \mathcal{G}_1' \qquad \mathcal{G}_2 \subseteq_i \mathcal{G}_2' \qquad \mathcal{G}_3 \subseteq_i \mathcal{G}_3'$

The converse of Theorem 15 does not hold, as illustrated by the following example.

Example 7
Let $\mathcal{P} = \{P \to Q, R \to S\}$, $\mathcal{N} = \{P \to S, R \to Q\}$ and $\mathcal{A} = \emptyset$. We have $\mathcal{G}_1 = \emptyset$.
But $(P \to Q) \vee (R \to S) \in \mathcal{G}_1'$. Indeed, $P \to Q \vdash_i (P \to Q) \vee (R \to S)$, $R \to S \vdash_i (P \to Q) \vee (R \to S)$, $P \to S \nvdash_i (P \to Q) \vee (R \to S)$ and $R \to Q \nvdash_i (P \to Q) \vee (R \to S)$.
Observe that, similarly, $\mathcal{G}_2 = \emptyset$ and $(P \to Q) \vee (R \to S) \in \mathcal{G}_2'$. Also, $(P \wedge \neg\neg R) \to Q \notin \mathcal{G}_3$ but $(P \wedge \neg\neg R) \to Q \in \mathcal{G}_3'$ (however, $\neg\neg((P \wedge \neg\neg R) \to Q) \notin \mathcal{G}_3'$).

These results give grounds for considering intuitionistic logic as being more fruitful than standard logic when the definition of sets of naturally expected generalizations is under consideration. Remarkably enough, sets of intuitionistic generalizations include corresponding sets of standard logic ones. Moreover, as illustrated by Example 7, intuitionistic definitions of GPN can allow sets of examples to exhibit generalizations when standard definitions of GPN gave rise to no generalization at all. To illustrate the actual significance of this latter result, let us revisit Example 1. No (standard logic) generalizations at all were to be expected from $\mathcal{P} = \{(P \to Q) \vee (Q \to R)\}$, $\mathcal{N} = \{(S \to T) \vee (T \to U)\}$ and $\mathcal{A} = \emptyset$, since both these positive and negative example are logically equivalent. Under an intuitionistic interpretation, $(P \to Q) \vee (Q \to R)$ and $(S \to T) \vee (T \to U)$ are no longer theorems and are not even longer logically equivalent. Indeed, the (intuitive) intuitionistic interpretation of e.g. the positive example is *having a proof of P yields a proof of Q or having a proof of Q yields R does hold*, which involves a natural reading of the $\to$ connective. Not surprisingly, $(P \to Q) \vee (Q \to R)$ is an intuitionistic generalization of the above sets of examples.

The following results express the general conditions for intuitionistic generalization to take place.

Theorem 16

$$\mathcal{G}_1' = \emptyset$$
$$\text{iff}$$
$$(\exists N_k \in \mathcal{N} : \mathcal{A} \wedge N_k \vdash_i \vee P_i) \text{ and } (\exists P_h \in \mathcal{P} : \mathcal{A} \wedge P_h \vdash_i \neg\neg \vee N_j)$$

Corollary 16.1 If $(\exists N_k \in \mathcal{N} : N_k \vdash_i \vee P_i)$ and $(\exists P_h \in \mathcal{P} : P_h \vdash_i \neg\neg \vee N_j)$ then $\mathcal{G}_1' = \emptyset$

Corollary 16.2 Let $\mathcal{A} = \emptyset$.

$$\mathcal{G}_1' = \emptyset$$
$$\text{iff}$$
$$(\exists N_k \in \mathcal{N} : N_k \vdash_i \vee P_i) \text{ and } (\exists P_h \in \mathcal{P} : P_h \vdash_i \neg\neg \vee N_j)$$

As an important consequence of Theorem 16, $\mathcal{G}_1'$ is empty when $\mathcal{P}$ and $\mathcal{N}$ share common examples, and more generally, when $\mathcal{P}$ and $\mathcal{N}$ share logically equivalent examples.

Corollary 16.3 If $\mathcal{P} \cap \mathcal{N} \neq \emptyset$ then $\mathcal{G}_1' = \emptyset$

The following theorem shows us that, like any collapse of $\mathcal{G}_1$, any collapse of $\mathcal{G}_1'$ is *definitive*. Accordingly, if we manage to prove $\mathcal{G}_1' = \emptyset$ for a subset of $\mathcal{P}$, a subset of $\mathcal{N}$ and a subset of the domain $\mathcal{A}$, then we keep having $\mathcal{G}1' = \emptyset$ whatever the actual domain $\mathcal{A}$, sets $\mathcal{P}$ and $\mathcal{N}$ are, and however these sets *could* be enlarged.

Theorem 17 Let $\mathcal{A} \subseteq \mathcal{A}_b$, $\mathcal{P} \subseteq \mathcal{P}_b$ and $\mathcal{N} \subseteq \mathcal{N}_b$.
Let $\mathcal{G}1'$ be the set of (intuitionistic) generalizations associated with a domain $\mathcal{A}$ and sets of examples $\mathcal{P}$ and $\mathcal{N}$.
Let $\mathcal{G}1_b'$ be the set of (intuitionistic) generalizations associated with a domain $\mathcal{A}_b$ and sets of examples $\mathcal{P}_b$ and $\mathcal{N}_b$. If $\mathcal{G}1' = \emptyset$ then $\mathcal{G}1_b' = \emptyset$

The results for $\mathcal{G}2'$ and $\mathcal{G}3'$ are as follows. The presence of examples in common in $\mathcal{P}$ and $\mathcal{N}$ does not entail $\mathcal{G}2' = \emptyset$ or $\mathcal{G}3' = \emptyset$, nor is any collapse of $\mathcal{G}2'$ and $\mathcal{G}3'$ definitive.

Theorem 18

$$\mathcal{G}2' = \emptyset \quad \text{iff} \quad (\mathcal{A} \wedge \vee N_j \vdash_i \vee P_i) \text{ and } (\mathcal{A} \wedge \vee P_i \vdash_i \neg\neg \vee N_j)$$

Corollary 18.1 If $\vdash_i \vee N_j \leftrightarrow \vee P_i$ then $\mathcal{G}2' = \emptyset$

Corollary 18.2 Let $\mathcal{A} = \emptyset$. $\mathcal{G}2' = \emptyset$ iff $\vdash_i \vee N_j \leftrightarrow (\vee P_i \wedge (\neg\neg \vee N_j \rightarrow \vee N_j))$

Theorem 19 $\mathcal{G}3' = \emptyset$ iff $(\mathcal{A} \wedge \wedge N_j \vdash_i \wedge P_i$ and $\mathcal{A} \wedge \wedge P_i \vdash_i \neg\neg \wedge N_j)$

Corollary 19.1 If $\vdash_i \wedge N_j \leftrightarrow \wedge P_i$ then $\mathcal{G}3' = \emptyset$

Corollary 19.2 Let $\mathcal{A} = \emptyset$. $\mathcal{G}3' = \emptyset$ iff $\vdash_i \wedge N_j \leftrightarrow (\wedge P_i \wedge (\neg\neg \wedge N_j \rightarrow \wedge N_j))$

Actually, we can show that our previous comparison of the three forms of generalizations within a standard logic framework can be replayed in the intuitionistic setting to yield similar results.

Lemma 20.1 $\mathcal{G}1' \subseteq \mathcal{G}2'$

The following theorem completes the comparison of the existence (in the general case) of generalizations through the three intuitionistic definitions of GPN.

Theorem 20
The following table considers every combination of empty sets taken from $\mathcal{G}1'$, $\mathcal{G}2'$ and $\mathcal{G}3'$ and states whether this combination is possible or not (in the general case).
(NE stands for 'non-empty'; P for 'possible' and I for 'impossible combination')

$\mathcal{G}1'$	$\mathcal{G}2'$	$\mathcal{G}3'$	
Ø	Ø	Ø	P
Ø	Ø	NE	P
Ø	NE	Ø	P
Ø	NE	NE	P

NE	Ø	Ø	I
NE	Ø	NE	I
NE	NE	Ø	P
NE	NE	NE	P

Actually, all results we obtained in the classical framework have corresponding results in the intuitionistic one. Results about tautologies are as follows.

Theorem 21 If $(\exists P_h \in \mathcal{P}: \vdash_i P_h)$ then $\mathcal{G}_1'{}_+ = \emptyset$
If $(\exists N_k \in \mathcal{N}: \vdash_i \neg\neg N_k)$ then $\mathcal{G}_1'{}_- = \emptyset$

Corollary 21.1 If $(\exists P_h \in \mathcal{P}\ \exists N_k \in \mathcal{N} : \vdash_i P_h$ and $\vdash_i \neg\neg N_k)$ then $\mathcal{G}_1' = \emptyset$

Theorem 22 If $\vdash_i \vee P_i$ then $\mathcal{G}_1'{}_+ = \mathcal{G}_2'{}_+ = \emptyset$
If $\vdash_i \neg\neg\vee N_j$ then $\mathcal{G}_1'{}_- = \mathcal{G}_2'{}_- = \emptyset$

Corollary 22.1 If $\vdash_i \vee P_i$ and $\vdash_i \neg\neg\vee N_j$ then $\mathcal{G}_1' = \mathcal{G}_2' = \emptyset$

Theorem 23 If $\vdash_i \wedge P_i$ then $\mathcal{G}_3'{}_+ = \emptyset$
If $\vdash_i \neg\neg\wedge N_j$ then $\mathcal{G}_3'{}_- = \emptyset$

Corollary 23.1 If $\vdash_i \wedge P_i$ and $\vdash_i \neg\neg\wedge N_j$ then $\mathcal{G}_3' = \emptyset$

Results about subsuming examples are as follows.

Theorem 24 If $(\exists P_h \in \mathcal{P}\ \exists N_k \in \mathcal{N} : P_h \vdash_i \neg\neg N_k)$ then $\mathcal{G}_1'{}_- = \emptyset$
If $(\exists P_h \in \mathcal{P}\ \exists N_k \in \mathcal{N} : N_k \vdash_i P_h)$ then $\mathcal{G}_1'{}_+ = \emptyset$

Corollary 24.1
If $(\exists P_h \in \mathcal{P}\ \exists N_k \in \mathcal{N} : P_h \vdash_i \neg\neg N_k)$ and $(\exists P_s \in \mathcal{P}\ \exists N_t \in \mathcal{N} : N_t \vdash_i P_s)$ then $\mathcal{G}_1' = \emptyset$

Theorem 25 If $(\exists N_k \in \mathcal{N} : \vee P_i \vdash_i \neg\neg N_k)$ then $\mathcal{G}_1'{}_- = \mathcal{G}_2'{}_- = \emptyset$
If $(\exists P_h \in \mathcal{P} : \vee N_j \vdash_i P_h)$ then $\mathcal{G}_1'{}_+ = \mathcal{G}_2'{}_+ = \emptyset$

Corollary 25.1
If $(\exists N_k \in \mathcal{N} : \vee P_i \vdash_i \neg\neg N_k)$ and $(\exists P_h \in \mathcal{P} : \vee N_j \vdash_i P_h)$ then $\mathcal{G}_1' = \mathcal{G}_2' = \emptyset$

Theorem 26 If $(\exists N_k \in \mathcal{N} : N_k \vdash_i \wedge P_i)$ then $\mathcal{G}_3'{}_+ = \emptyset$
If $(\exists P_h \in \mathcal{P} : P_h \vdash_i \neg\neg\wedge N_j)$ then $\mathcal{G}_3'{}_- = \emptyset$

Corollary 26.1 If $(\exists N_k \in \mathcal{N} : N_k \vdash_i \wedge P_i)$ and $(\exists P_h \in \mathcal{P} : P_h \vdash_i \neg\neg\wedge N_j)$ then $\mathcal{G}_3' = \emptyset$

The computational results obtained within the standard logic framework also apply for intuitionistic deductive generalizations, through the following theorems.

Theorem 27 Let every P_i and N_j be a consistent propositional clause.
If ($\exists N_k \in \mathcal{N}$ such that every literal in N_k occurs also in $\vee P_i$) then $\mathcal{G}_1'_+ = \emptyset$
If ($\exists P_h \in \mathcal{P}$ such that for every literal L in P_h, either L or $\neg\neg L$ occurs also in $\vee N_j$) then $\mathcal{G}_1'_- = \emptyset$

Theorem 28 Let every P_i and N_j be a consistent propositional clause.
If (every literal in $\vee N_j$ occurs in $\vee P_i$) then $\mathcal{G}_2'_+ = \emptyset$
If (for every literal L in $\vee P_i$, either L or $\neg\neg L$ occurs in $\vee N_j$) then $\mathcal{G}_2'_- = \emptyset$

Theorem 29 Let every P_i and N_j be a consistent conjunction of literals.
If (every literal in $\wedge P_i$ occurs in $\wedge N_j$) then $\mathcal{G}_3'_+ = \emptyset$
If (for every literal L in $\wedge N_j$, either L or $\neg\neg L$ occurs in $\wedge P_i$) then $\mathcal{G}_3'_- = \emptyset$

5 Summary of the technical results

Let us summarize the technical results described in this paper. Three forms of deductive generalization from positive and negative examples have been introduced and compared. The logical conditions under which these paradigms can take place have been investigated thoroughly. This analysis has been undertaken from both a standard and an intuitionistic logic points of view.

The three forms of deductive generalization give rise to three sets of generalizations, namely $\mathcal{G}_1$, $\mathcal{G}_2$ and $\mathcal{G}_3$. The motivations behind the definitions of $\mathcal{G}_1$ and $\mathcal{G}_2$ are similar. Accordingly, $\mathcal{G}_1$ and $\mathcal{G}_2$ coincide when one set of examples (either the set of positive or the set of negative ones) is under consideration. However, $\mathcal{G}_1$ and $\mathcal{G}_2$ can differ from one another when both kinds of examples are taken into account. Mainly, $\mathcal{G}_2$ subsumes $\mathcal{G}_1$, in both the standard logic and intuitionistic settings. In [2], the two sets have been shown unrelated (in the general case) in a default logic framework [9]. When the sets of positive and negative examples intersect one another, $\mathcal{G}_1$ collapses. A collapse of $\mathcal{G}_1$ is always definitive no matter how the sets of examples and the domain-knowledge are enlarged. These properties do not hold neither for $\mathcal{G}_2$ nor for $\mathcal{G}_3$. More generally, the conditions for $\mathcal{G}_1$, $\mathcal{G}_2$ and $\mathcal{G}_3$ to take place have been established and the role of tautologies and subsuming examples have been studied in this context. Related computational results have also been obtained, allowing one to check whether $\mathcal{G}_1$, $\mathcal{G}_2$ and $\mathcal{G}_3$ can take place and whether the actual search for generalizations can be simplified.

Next, we have shown that intuitionistic definitions of $\mathcal{G}_1$, $\mathcal{G}_2$ and $\mathcal{G}_3$ subsume the corresponding standard ones. Actually, all the results described for the classical setting can be adapted to the intuitionistic one. Interestingly enough, switching to an intuitionistic framework can allow us to find out intuitively natural generalizations in situations where no standard logic generalization could be extracted.

6 Significance of the work and perspectives

By definition, *deductive* generalization is a paradigm that relies on logic. In this respect, this paper is a contribution to the hardly studied logical foundations of this important paradigm of machine learning. Focusing on the existence conditions of several forms of deductive generalization, we have been able to compare these paradigms in a successful manner. We have shown that existence conditions should not be taken for granted since very casual sets of examples exhibit no generalizations at all. Moreover, testing these conditions may allow for simplifications of the actual search process.

Obviously enough, this study is only a first step towards a comprehensive logical characterization of deductive generalization processes. Further investigations should be undertaken about the logical properties of the generalizations themselves, the sets of generalizations, the criteria for selecting the *preferred* generalizations in these sets and about the actual search of generalizations (see e.g. [7, 8]).

As examined in this paper, intuitionistic logic can play a useful role in the foundations of deductive generalization. Its interpretation of the logical connectives being very close to a direct common-sense reading of these latter ones, it allows one to explain naturally expected generalizations. In this respect, future studies could examine whether relevance logic [1] further restricts the range of situations where no generalizations can be extracted. Finally, we believe that forms of machine learning that have to be performed in the actual world must take into account the time-sensitive aspects and the incompleteness of the available knowledge. In this respect, a natural but complex transition involves switching from standard deduction to some forms of defeasible reasoning. A formal study of such a relaxed form of deductive generalization can be found in [2].

Acknowledgements

Ph. Besnard and E. Grégoire were supported by the CNRS (PRC-IA) and by an EEC fellowship (Science Program), respectively.

References

1. N.D. Belnap: How A Computer Should Think. In: G. Ryle (ed.): *Contempory Aspects of Philosophy*. Oriel Press (1977)
2. Ph. Besnard, E. Grégoire: Deductive Generalization in a Default Logic Setting. In: *Proc. of the Second Int. Workshop on Nonmonotonic and Inductive Logics (NIL91)*. Lecture Notes in Computer Science. Berlin: Springer (1992)
3. M. Dummett: *Elements of Intuitionism*. Oxford University Press (1977)
4. Y. Kodratoff: *Introduction to Machine Learning*. London: Pitman (1988)
5. R. Michalski: A Theory and Methodology of Inductive Learning. In: *Machine Learning: an A.I. Approach*. Tioga (1983)
6. T. Mitchell: *Version Spaces: an Approach to Concept Learning*. Stanford University: Ph.D. Thesis (1978)
7. J. Nicolas: *ALLY: un système logique pour la généralisation en apprentissage automatique*. Université de Rennes I: Thèse d'Université (1987)
8. J. Nicolas: Consistency and Preference Criteria for Generalization Languages Handling Negation

and Disjunction. In: *Proc. ECAI-88*. London: Pitman pp. 402-407 (1988)
9. R. Reiter: A Logic for Default Reasoning. *Artificial Intelligence*, 13, pp. 81-132 (1980)

Appendix. Proofs of the theorems

Theorem 1

$\Rightarrow$

Consider $\vee P_i$. Since $\mathcal{G}_1 = \emptyset$, at least one of the following conditions does not hold:
1. $\forall P_i \in \mathcal{P}: \mathcal{A} \wedge P_i \vdash \vee P_i$ or $\forall N_j \in \mathcal{N}: \mathcal{A} \wedge N_j \vdash \neg\vee P_i$
2. $\forall P_i \in \mathcal{P}: \mathcal{A} \wedge P_i \nvdash \neg\vee P_i$ 3. $\forall N_j \in \mathcal{N}: \mathcal{A} \wedge N_j \nvdash \vee P_i$

Since 1 and 2 are satisfied, 3 does not hold. Thus, $\exists N_k \in \mathcal{N}: \mathcal{A} \wedge N_k \vdash \vee P_i$.
Considering $\neg\vee N_j$, by a similar argument, we derive that $\forall P_i \in \mathcal{P}: \mathcal{A} \wedge P_i \nvdash \neg\neg\vee N_j$ does not hold, and thus that $\exists P_h \in \mathcal{P}: \mathcal{A} \wedge P_h \vdash \vee N_j$.

$\Leftarrow$

Suppose $\exists G \in \mathcal{G}_1$. Then, $\forall P_i \in \mathcal{P}: \mathcal{A} \wedge P_i \vdash G$ or $\forall N_j \in \mathcal{N}: \mathcal{A} \wedge N_j \vdash \neg G$.

Case 1.

$\forall P_i \in \mathcal{P}: \mathcal{A} \wedge P_i \vdash G$ holds. Thus, $\mathcal{A} \wedge \vee P_i \vdash G$. Since $\exists N_k \in \mathcal{N}: \mathcal{A} \wedge N_k \vdash \vee P_i$, we also have that $\mathcal{A} \wedge N_k \vdash \mathcal{A} \wedge \vee P_i$. By transitivity, we obtain $\exists N_k \in \mathcal{N}: \mathcal{A} \wedge N_k \vdash G$, which contradicts the definition of G. Thus, $\mathcal{G}_{1+} = \emptyset$.

Case 2.

$\forall N_j \in \mathcal{N}: \mathcal{A} \wedge N_j \vdash \neg G$ holds. Thus, $\mathcal{A} \wedge \vee N_j \vdash \neg G$. Since $\exists P_h \in \mathcal{P}: \mathcal{A} \wedge P_h \vdash \vee N_j$, we also have that $\mathcal{A} \wedge P_h \vdash \mathcal{A} \wedge \vee N_j$. By transitivity, we obtain $\exists P_h \in \mathcal{P}: \mathcal{A} \wedge P_h \vdash \neg G$, which contradicts the definition of G. Thus, $\mathcal{G}_{1-} = \emptyset$.

Corollary 1.1

Since $\vdash$ is monotonic, the condition ($\exists N_k \in \mathcal{N}: N_k \vdash \vee P_i$ and $\exists P_h \in \mathcal{P}: P_h \vdash \vee N_j$) entails ($\exists N_k \in \mathcal{N}: \mathcal{A} \wedge N_k \vdash \vee P_i$ and $\exists P_h \in \mathcal{P}: \mathcal{A} \wedge P_h \vdash \vee N_j$), which is according to Theorem 1 the necessary and sufficient condition for $\mathcal{G}_1 = \emptyset$.

Corollary 1.2 Direct consequence of Theorem 1.

Corollary 1.3 Direct consequence of Theorem 1.

Theorem 2

Assume $\mathcal{G}_1 = \emptyset$. According to Theorem 1, this is equivalent to (($\exists N_k \in \mathcal{N}: \mathcal{A} \wedge N_k \vdash \vee P_i$) and ($\exists P_h \in \mathcal{P}: \mathcal{A} \wedge P_h \vdash \vee N_j$)). Since $\vdash$ is monotonic and since for any formula R, T and Q of $\mathcal{L}$, when $R \vdash T$ then $R \vdash T \vee Q$, this entails (($\exists N_k \in \mathcal{N}_b : \mathcal{A}_b \wedge N_k \vdash \vee P_i$) and ($\exists P_h \in \mathcal{P}_b : \mathcal{A}_b \wedge P_h \vdash \vee N_j$)) where $\vee P_i$ and $\vee N_j$ are the disjunctions of all the examples in $\mathcal{P}_b$ and $\mathcal{N}_b$, respectively. According to Theorem 1, this is equivalent to $\mathcal{G}_{1b} = \emptyset$.

Theorem 3 Apply Theorem 1 with $\mathcal{P} = \{\vee P_i\}$ and $\mathcal{N} = \{\vee N_j\}$.

Corollary 3.1

Since $\vdash$ is monotonic, $\vee N_j \vdash \vee P_i$ and $\vee P_i \vdash \vee N_j$ entail $\mathcal{A} \wedge \vee N_j \vdash \vee P_i$ and $\mathcal{A} \wedge \vee P_i \vdash \vee N_j$, and thus, according to Theorem 3, $\mathcal{G}_2 = \emptyset$.

Corollary 3.2 Direct consequence of Theorem 3.

Theorem 4 Apply Theorem 1 with $\mathcal{P} = \{\wedge P_i\}$ and $\mathcal{N} = \{\wedge N_j\}$.

Corollary 4.1

Since $\vdash$ is monotonic, $\wedge N_j \vdash \wedge P_i$ and $\wedge P_i \vdash \wedge N_j$ entail $\mathcal{A} \wedge \wedge N_j \vdash \wedge P_i$ and $\mathcal{A} \wedge \wedge P_i \vdash \wedge N_j$, and thus, according to Theorem 4, $\mathcal{G}_3 = \emptyset$.

Corollary 4.2 Direct consequence of Theorem 4.

Lemma 5.1

Let $G \in \mathcal{G}_1$.

$(\forall P_i \in \mathcal{P}: \mathcal{A} \wedge P_i \vdash G$ or $\forall N_j \in \mathcal{N}: \mathcal{A} \wedge N_j \vdash \neg G)$ entails $(\mathcal{A} \wedge \vee P_i \vdash G$ or $\mathcal{A} \wedge \vee N_j \vdash \neg G)$.

Since $G \in \mathcal{G}_1$, we have $\forall P_i \in \mathcal{P}: \mathcal{A} \wedge P_i \nvdash \neg G$. This entails $\mathcal{A} \wedge \vee P_i \nvdash \neg G$. Since we also have $\forall N_j \in \mathcal{N}: \mathcal{A} \wedge N_j \nvdash G$, we obtain $\mathcal{A} \wedge \vee N_j \nvdash G$. Thus, $G \in \mathcal{G}_2$.

Theorem 5 A direct consequence of Lemma 5.1 and of the Examples 1 to 6.

Theorem 6

Assume $(\exists P_h \in \mathcal{P}: \vdash P_h)$ and let $G \in \mathcal{G}_{1+}$. This entails $\mathcal{A} \vdash G$ and, since $\vdash$ is monotonic, that $\forall N_j \in \mathcal{N}: \mathcal{A} \wedge N_j \vdash G$, which contradicts the definition of G. The proof of the second part of this theorem is similar.

Theorem 7

Let $G \in \mathcal{G}_1$.

Case 1.

$\vdash \vee P_i$. Assume G is a positive generalization. Thus, $\forall P_i \in \mathcal{P}: \mathcal{A} \wedge P_i \vdash G$, and thus $\mathcal{A} \wedge \vee P_i \vdash G$. Since we have $\vdash \vee P_i$, this entails $\mathcal{A} \vdash G$. Since standard logic is monotonic, we have $\forall N_j \in \mathcal{N}: \mathcal{A} \wedge N_j \vdash G$, which contradicts the definition of G.

Case 2.

$\vdash \vee N_j$. Similar.

A similar argument applies for $\mathcal{G}_2$.

Theorem 8 Similar.

Theorem 9

Direct strenghtening of the right part of Theorem 1, using the fact that $\vdash$ is monotonic and that for every formula R, T and Q of $\mathcal{L}$, when $R \vdash T$ then $R \vdash T \vee Q$.

Theorem 10

Direct consequence of Corollary 3.1, of the fact that for every formula R, T and Q of $\mathcal{L}$, when $R \vdash T$ then $R \vdash T \vee Q$, and of Lemma 5.1.

Theorem 11 Direct consequence of Corollary 4.1 and of the fact that $\vdash$ is monotonic.

Theorem 12

Consider Corollary 1.1. Whenever every $P_m \in \mathcal{P}$ and every $N_n \in \mathcal{N}$ are consistent propositional clauses, $\vee P_i$ and $\vee N_j$ are two consistent propositional clauses. Accordingly, $N_k \vdash \vee P_i$ iff $\vee P_i$ contains two opposite literals or every literal in N_k occurs also in $\vee P_i$, and $P_h \vdash \vee N_j$ iff $\vee N_j$ contains two opposite literals or every literal in P_h also occurs in $\vee N_j$. According to Theorem 2, we know that when $\mathcal{G}_1 = \emptyset$, such a result applies for all supersets of $\mathcal{A}$, $\mathcal{P}$ and $\mathcal{N}$.

Theorem 13

Direct consequence of Corollary 3.1 and of the fact that whenever every $P_m \in \mathcal{P}$ and every $N_n \in \mathcal{N}$ are consistent propositional clauses, $\vee P_i$ and $\vee N_j$ are two consistent propositional clauses.

Theorem 14 Direct consequence of Theorem 4.

Theorem 15

First, let us prove $\mathcal{G}_1 \subseteq_i \mathcal{G}_1'$. Consider any $G \in \mathcal{G}_1$.

Case 1. The three following conditions are satisfied:

1. $\forall P_i \in \mathcal{P}: \mathcal{A} \wedge P_i \vdash G$ 2. $\forall P_i \in \mathcal{P}: \mathcal{A} \wedge P_i \nvdash \neg G$ 3. $\forall N_j \in \mathcal{N}: \mathcal{A} \wedge N_j \nvdash G$.

Consider any $P_i \in \mathcal{P}$ and any $N_j \in \mathcal{N}$. From 1, we obtain $\vdash (\mathcal{A} \wedge P_i) \rightarrow G$. Since for any propositional formula α, we have $\vdash \alpha$ iff $\vdash_i \neg\neg\alpha$, we derive $\vdash_i \neg\neg((\mathcal{A} \wedge P_i) \rightarrow G)$ and thus $\vdash_i (\mathcal{A} \wedge P_i) \rightarrow \neg\neg G$ and $(\mathcal{A} \wedge P_i) \vdash_i \neg\neg G$ (deduction theorem for intuitionistic logic). From 2, we obtain $\mathcal{A} \wedge P_i \nvdash \neg\neg\neg G$. Since intuitionistic logic is contained in classical logic, we obtain $\mathcal{A} \wedge P_i \nvdash_i \neg\neg\neg G$. By a similar argument, we obtain $\mathcal{A} \wedge N_j \nvdash_i \neg\neg G$ from 3. Thus, $\neg\neg G \in \mathcal{G}_1'$.

Case 2. The three following conditions are satisfied:

1. $\forall N_j \in \mathcal{N}: \mathcal{A} \wedge N_j \vdash \neg G$ 2. $\forall P_i \in \mathcal{P}: \mathcal{A} \wedge P_i \nvdash \neg G$ 3. $\forall N_j \in \mathcal{N}: \mathcal{A} \wedge N_j \nvdash G$.

Consider any $P_i \in \mathcal{P}$ and any $N_j \in \mathcal{N}$. From 1, we obtain $\vdash (\mathcal{A} \wedge N_j) \rightarrow \neg G$ and thus $\vdash_i \neg\neg((\mathcal{A} \wedge N_j) \rightarrow \neg G)$. Since $\vdash_i \neg\neg\neg\alpha \rightarrow \neg\alpha$, we obtain successively $\vdash_i (\mathcal{A} \wedge N_j) \rightarrow \neg\neg\neg G$, $\vdash_i (\mathcal{A} \wedge N_j) \rightarrow \neg G$ and $\mathcal{A} \wedge N_j \vdash_i \neg G$. From 2 and 3, we derive $\mathcal{A} \wedge P_i \nvdash_i \neg G$ and $\mathcal{A} \wedge N_j \nvdash_i G$, respectively. Thus, $G \in \mathcal{G}_1'$.

The proof of $\mathcal{G}_2 \subseteq_i \mathcal{G}_2'$ (resp. $\mathcal{G}_3 \subseteq_i \mathcal{G}_3'$) is obtained directly, using $\mathcal{G}_1 \subseteq_i \mathcal{G}_1'$ with $\vee P_i$ and $\vee N_j$ (resp. $\wedge P_i$ and $\wedge N_j$) as unique element of $\mathcal{P}$ and $\mathcal{N}$, respectively.

Theorem 16

$\Rightarrow$

Assume $\mathcal{G}_1' = \emptyset$.

Case 1.

Consider any P_h of $\mathcal{P}$. Clearly, $\mathcal{A} \wedge P_h \vdash_i \vee P_i$. Also, $\mathcal{A} \wedge P_h \nvdash_i \neg(\vee P_i)$, otherwise it would result that $\mathcal{A} \wedge P_h \vdash_i \wedge\neg P_i$, and thus that $\mathcal{A} \wedge P_h \vdash_i \neg P_h$, a contradiction with the (intuitionistic) consistency requirement for the positive examples. Since $\mathcal{G}_1' = \emptyset$, this entails that $\exists N_k \in \mathcal{N}: \mathcal{A} \wedge N_k \vdash_i \vee P_i$ (otherwise, $\vee P_i$ would be a generalization).

Case 2.

Consider any N_k of $\mathcal{N}$. Clearly, $\mathcal{A} \wedge N_k \vdash_i \neg\neg(\vee N_j)$. Also, $\mathcal{A} \wedge N_k \nvdash_i \neg(\vee N_j)$, otherwise it would result that $\mathcal{A} \wedge N_k \vdash_i \wedge\neg N_j$, and thus that $\mathcal{A} \wedge N_k \vdash_i \neg N_k$, a contradiction with the (intuitionistic) consistency assumption for the negative examples. Since $\mathcal{G}_1' = \emptyset$, this yields that $\exists P_h \in \mathcal{P}: \mathcal{A} \wedge P_h \vdash_i \neg\neg(\vee N_j)$ must hold (otherwise, $\neg(\vee N_j)$ would be a generalization).

$\Leftarrow$

We have $\exists N_k \in \mathcal{N}: \mathcal{A} \wedge N_k \vdash_i \vee P_i$ and $\exists P_h \in \mathcal{P}: \mathcal{A} \wedge P_h \vdash_i \neg\neg(\vee N_j)$. Assume $\mathcal{G}_1' \neq \emptyset$ and let $G \in \mathcal{G}_1'$.

Case 1.

Suppose $\forall P_i \in \mathcal{P}: \mathcal{A} \wedge P_i \vdash_i G$. Then, $\vdash_i (\mathcal{A} \wedge \vee P_i) \rightarrow G$. Since $\mathcal{A} \wedge N_k \vdash_i \vee P_i$ and thus $\mathcal{A} \wedge N_k \vdash_i \mathcal{A} \wedge \vee P_i$ we have $\mathcal{A} \wedge N_k \vdash_i G$, which contradicts $G \in \mathcal{G}_1'$.

Case 2.

Suppose $\forall N_j \in \mathcal{N}: \mathcal{A} \wedge N_j \vdash_i \neg G$. Then, $\mathcal{A} \vdash_i N_j \rightarrow \neg G$. Accordingly, $\mathcal{A} \vdash_i (\vee N_j) \rightarrow \neg G$. By contraposition, $\mathcal{A} \vdash_i G \rightarrow \neg(\vee N_j)$. By contraposition again, $\mathcal{A} \vdash_i \neg\neg(\vee N_j) \rightarrow \neg G$. Since $\exists P_h \in \mathcal{P}: \mathcal{A} \wedge P_h \vdash_i \neg\neg(\vee N_j)$, we have $\exists P_h \in \mathcal{P}: \mathcal{A} \wedge P_h \vdash_i \neg G$, which contradicts $G \in \mathcal{G}_1$.

Corollary 16.1

Since $\vdash_i$ is monotonic, the condition $(\exists N_k \in \mathcal{N}: N_k \vdash_i \vee P_i$ and $\exists P_h \in \mathcal{P}: P_h \vdash_i \neg\neg\vee N_j)$ entails $(\exists N_k \in \mathcal{N}: \mathcal{A} \wedge N_k \vdash_i \vee P_i$ and $\exists P_h \in \mathcal{P}: \mathcal{A} \wedge P_h \vdash_i \neg\neg\vee N_j)$, which is according to Theorem 16 the necessary and sufficient condition for $\mathcal{G}_1' = \emptyset$.

Corollary 16.2 Direct consequence of Theorem 16.

Corollary 16.3 Direct consequence of Theorem 16.

Theorem 17

Assume $\mathcal{G}_1' = \emptyset$. According to Theorem 16, this is equivalent to $((\exists N_k \in \mathcal{N} : \mathcal{A} \wedge N_k \vdash_i \vee P_i)$ and $(\exists P_h \in \mathcal{P} : \mathcal{A} \wedge P_h \vdash_i \neg\neg \vee N_j))$. Since $\vdash_i$ is monotonic and since for any formula R, T and Q of $\mathcal{L}$, when $R \vdash_i T$ then $R \vdash_i T \vee Q$, this entails $((\exists N_k \in \mathcal{N}_b : \mathcal{A}_b \wedge N_k \vdash_i \vee P_i)$ and $(\exists P_h \in \mathcal{P}_b : \mathcal{A}_b \wedge P_h \vdash_i \neg\neg \vee N_j))$ where $\vee P_i$ and $\vee N_j$ are the disjunctions of all the examples in $\mathcal{P}_b$ and $\mathcal{N}_b$, respectively. This is equivalent to $\mathcal{G}_{1b}' = \emptyset$.

Theorem 18 Apply Theorem 16 with $\mathcal{P} = \{\vee P_i\}$ and $\mathcal{N} = \{\vee N_j\}$.

Corollary 18.1

$\vdash_i \vee N_j \leftrightarrow \vee P_i$ entails $\vee N_j \vdash_i \vee P_i$ and $\vee P_i \vdash_i \vee N_j$. $\vee P_i \vdash_i \vee N_j$ entails $\vee P_i \vdash_i \neg\neg \vee N_j$. Since $\vdash_i$ is monotonic, $\vee N_j \vdash_i \vee P_i$ and $\vee P_i \vdash_i \neg\neg \vee N_j$ entail $\mathcal{A} \wedge \vee N_j \vdash_i \vee P_i$ and $\mathcal{A} \wedge \vee P_i \vdash_i \neg\neg \vee N_j$, and thus, according to Theorem 18, $\mathcal{G}_2' = \emptyset$.

Corollary 18.2

$\Rightarrow$

Let $\mathcal{A} = \emptyset$. According to Theorem 18, if $\mathcal{G}_2' = \emptyset$ then $\vee N_j \vdash_i \vee P_i$ and $\vee P_i \vdash_i \neg\neg \vee N_j$. It follows that $\vee N_j \vdash_i \vee P_i \wedge (\neg\neg \vee N_j \rightarrow \vee N_j)$ and $\vee P_i \wedge (\neg\neg \vee N_j \rightarrow \vee N_j) \vdash_i \vee N_j$.

$\Leftarrow$

If the condition $\vdash_i \vee N_j \leftrightarrow (\vee P_i \wedge (\neg\neg \vee N_j \rightarrow \vee N_j))$ holds then first, $\mathcal{G}_2'{}_+ = \emptyset$. Second, $\neg \vee N_j \vdash_i (\neg\neg \vee N_j \rightarrow \vee N_j)$. Hence, $\vee P_i \wedge \neg \vee N_j \vdash_i \vee P_i \wedge \neg \vee N_j \wedge (\neg\neg \vee N_j \rightarrow \vee N_j)$. Now, $\vee P_i \wedge \neg \vee N_j \wedge (\neg\neg \vee N_j \rightarrow \vee N_j) \vdash_i$ F since $\vee P_i \wedge (\neg\neg \vee N_j \rightarrow \vee N_j) \vdash_i \vee N_j$. Therefore, $\vee P_i \wedge \neg \vee N_j \vdash_i$ F.

Thus, $\vee P_i \vdash_i \neg\neg \vee N_j$. Hence, $\mathcal{G}_2'{}_- = \emptyset$.

Theorem 19 Apply Theorem 16 with $\mathcal{P} = \{\wedge P_i\}$ and $\mathcal{N} = \{\wedge N_j\}$.

Corollary 19.1

$\vdash_i \wedge N_j \leftrightarrow \wedge P_i$ entails $\wedge N_j \vdash_i \wedge P_i$ and $\wedge P_i \vdash_i \wedge N_j$. $\wedge P_i \vdash_i \wedge N_j$ entails $\wedge P_i \vdash_i \neg\neg \wedge N_j$. Since $\vdash_i$ is monotonic, $\wedge N_j \vdash_i \wedge P_i$ and $\wedge P_i \vdash_i \wedge N_j$ entail $\mathcal{A} \wedge \wedge N_j \vdash_i \wedge P_i$ and $\mathcal{A} \wedge \wedge P_i \vdash_i \neg\neg \wedge N_j$, and thus, according to Theorem 19, $\mathcal{G}_3' = \emptyset$.

Corollary 19.2

$\Rightarrow$

Let $\mathcal{A} = \emptyset$. According to Theorem 19, if $\mathcal{G}_3' = \emptyset$ then $\wedge N_j \vdash_i \wedge P_i$ and $\wedge P_i \vdash_i \neg\neg \wedge N_j$. It follows that $\wedge N_j \vdash_i \wedge P_i \wedge (\neg\neg \wedge N_j \rightarrow \wedge N_j)$ and $\wedge P_i \wedge (\neg\neg \wedge N_j \rightarrow \wedge N_j) \vdash_i \wedge N_j$.

$\Leftarrow$

If the condition $\vdash_i \wedge N_j \leftrightarrow (\wedge P_i \wedge (\neg\neg \wedge N_j \rightarrow \wedge N_j))$ holds then first, $\mathcal{G}_3'{}_+ = \emptyset$. Second, $\neg \wedge N_j \vdash_i (\neg\neg \wedge N_j \rightarrow \wedge N_j)$. Hence, $\wedge P_i \wedge \neg \wedge N_j \vdash_i \wedge P_i \wedge \neg \wedge N_j \wedge (\neg\neg \wedge N_j \rightarrow \wedge N_j)$. Now, $\wedge P_i \wedge \neg \wedge N_j \wedge (\neg\neg \wedge N_j \rightarrow \wedge N_j) \vdash_i$ F since $\wedge P_i \wedge (\neg\neg \wedge N_j \rightarrow \wedge N_j) \vdash_i \wedge N_j$. Therefore, $\wedge P_i \wedge \neg \wedge N_j \vdash_i$ F. Thus, $\wedge P_i \vdash_i \neg\neg \wedge N_j$. Hence, $\mathcal{G}_3'{}_- = \emptyset$.

Lemma 20.1

Let $G \in \mathcal{G}_1'$.

$(\forall P_i \in \mathcal{P} : \mathcal{A} \wedge P_i \vdash_i G$ or $\forall N_j \in \mathcal{N} : \mathcal{A} \wedge N_j \vdash_i \neg G)$ entails $(\mathcal{A} \wedge \vee P_i \vdash_i G$ or $\mathcal{A} \wedge \vee N_j \vdash_i \neg G)$. Since $G \in \mathcal{G}_1'$, we have $\forall P_i \in \mathcal{P} : \mathcal{A} \wedge P_i \nvdash_i \neg G$. This entails $\mathcal{A} \wedge \vee P_i \nvdash_i \neg G$. Since we also have $\forall N_j \in \mathcal{N} : \mathcal{A} \wedge N_j \nvdash_i G$, this yields $\mathcal{A} \wedge \vee N_j \nvdash_i G$. Thus, $G \in \mathcal{G}_2'$.

Theorem 20

In Example 1, substitute $\mathcal{P} = \{(P \rightarrow Q) \vee (Q \rightarrow R)\}$, $\mathcal{N} = \{(S \rightarrow T) \vee (T \rightarrow U)\}$ with $\mathcal{P} = \{P \rightarrow P\}$, $\mathcal{N} = \{S \rightarrow S\}$. It is easy to show that $\mathcal{G}_1' = \mathcal{G}_2' = \mathcal{G}_3' = \emptyset$. It is also easy to show that we have an empty set $\mathcal{G}_i$ in one of the Examples 2 to 6, we have that the set $\mathcal{G}_i'$ is empty for the same example, and conversely. Taking Lemma 20.1 into account, this proves the Theorem.

Theorem 21

Assume ($\exists P_h \in \mathcal{P}: \ \vdash_i P_h$) and let $G \in \mathcal{G}_{1+}$. This entails $\mathcal{A} \vdash_i G$ and, since $\vdash_i$ is monotonic, that $\forall N_j \in \mathcal{N} : \mathcal{A} \wedge N_j \vdash_i G$, which contradicts the definition of G. Thus, $\mathcal{G}_1'_+ = \emptyset$.

Assume ($\exists N_k \in \mathcal{N} : \vdash_i \neg\neg N_k$). Then, $\vdash_i \neg\neg(\vee N_j)$ and thus, by the monotonicity of $\vdash_i$, $P_h \vdash_i \neg\neg(\vee N_j)$. Hence, according to Theorem 16, we have $\mathcal{G}_1'_- = \emptyset$.

Theorem 22

Let $G \in \mathcal{G}_1'$.

Case 1.

$\vdash_i \vee P_i$. Assume $G \in \mathcal{G}_1'_+$. Thus, $\forall P_i \in \mathcal{P} : \mathcal{A} \wedge P_i \vdash_i G$, and thus $\mathcal{A} \wedge \vee P_i \vdash_i G$. Since we have $\vdash_i \vee P_i$, this entails $\mathcal{A} \vdash_i G$. Since $\vdash_i$ is monotonic, we have $\forall N_j \in \mathcal{N} : \mathcal{A} \wedge N_j \vdash_i G$, which contradicts the definition of G.

Case 2.

$\vdash_i \neg\neg \vee N_j$. Apply Theorem 16 to $P_h \vdash_i \neg\neg(\vee N_j)$ (obtained by monotonicity of $\vdash_i$).

The above argument also applies for $\mathcal{G}_2'$.

Theorem 23 Similar.

Theorem 24

Direct strenghtening of the right part of Theorem 16, using the fact that $\vdash_i$ is monotonic and that for every formula R, T and Q of $\mathcal{L}$, when $R \vdash_i T$ then $R \vdash_i T \vee Q$.

Theorem 25

Direct consequence of Corollary 18.1, of the fact that for every formula R, T and Q of $\mathcal{L}$, when $R \vdash_i T$ then $R \vdash_i T \vee Q$, and of Lemma 20.1.

Theorem 26 Direct consequence of Theorem 19 and of the fact that $\vdash_i$ is monotonic.

Theorem 27

Assume $\exists N_k \in \mathcal{N}$ such that every literal L in N_k occurs also in $\vee P_i$. Thus $\forall L \in N_k : L \vdash_i \vee P_i$, so $N_k \vdash_i \vee P_i$. According to Corollary 16.1, we have $\mathcal{G}_1'_+ = \emptyset$.

Assume $\exists P_h \in \mathcal{P}$ such that for every literal L in P_h, either L or $\neg\neg L$ occurs also in $\vee N_j$. Thus, $\forall L \in P_h : L \vdash_i \neg\neg(\vee N_j)$, so $P_h \vdash_i \neg\neg(\vee N_j)$. According to Corollary 16.1, we have $\mathcal{G}_1'_- = \emptyset$.

Theorem 28 Direct consequence of Theorem 18.

Theorem 29 Direct consequence of Theorem 19.

Transition Systems and Dynamic Semantics

Tim Fernando
fernando@cwi.nl

CWI, P.O. Box 4079, 1009 AB Amsterdam, The Netherlands

Abstract. Transition systems over first-order models, first-order theories, and families of first-order models are constructed and examined in relation to dynamic semantics (more specifically, DPL, DRT and Update logic). Going the other direction, first-order models are extracted from transition systems, bringing full circle the connection between static and dynamic notions. Only states computationally accessible from an initial state (with minimal information content) are considered, motivating the introduction of an internal notion of proposition on which the concept of an update is analyzed.

Key words and phrases: transition system, bisimulation, dynamic semantics, first-order logic, updates.
Note: This work was funded by the Netherlands Organization for Scientific Research (NWO project NF 102/62-356, 'Structural and Semantic Parallels in Natural Languages and Programming Languages'). The author is gratefully indebted to J. van Benthem, J. van Eijck, C. Gardent, M. Kracht, W. Meyer-Viol and K. Vermeulen for helpful discussions and to CWI for refuge.

As described in van Benthem [4], there is a growing interest in developing a certain broad conception of logic as the processing of information. A particularly important theme to emerge in work on natural language semantics loosely labelled "dynamic semantics" (important examples of which include Kamp [13], Heim [12], Groenendijk and Stokhof [9], and Veltman [20]) is to locate the essence of a proposition in the set of transitions between states (or contexts) that it induces. The present paper analyzes a link uncovered in Groenendijk and Stokhof [9] between these sets of transitions and programs in dynamic logic (see, for example, Harel [11]). The key notion studied is that of a "transition system", the basic thesis being that a good deal of what makes dynamic semantics "dynamic" are constructions that can be fruitfully understood against a background of "static" notions from first-order logic.

Identifying the set of transitions effected by a proposition with a program, a *transition system* is defined, relative to a collection Π of *programs*, to be a triple $\langle S, \{\xrightarrow{\pi}\}_{\pi\in\Pi}, s_0\rangle$ consisting of a non-empty set S of *states*, a family of binary relations $\xrightarrow{\pi} \subseteq S \times S$ on the states (meant as the interpretation of π), and an *initial state* $s_0 \in S$. The "dynamic" shift in perspective on propositions is analyzed below in terms of transition systems built from first-order models, families of first-order models, and first-order theories, in accordance with dynamic logic. In each case, an initial state s_0 with minimal information content is isolated, and the states considered are restricted to those accessible from s_0 by a finite sequence of programs. This simple move is shown to have rather far reaching consequences.

(i) It is exploited in the technical arguments of sections 1 and 3 (some of the details of which are relegated to the Appendix) to exhibit a certain duality between static and dynamic notions.

(ii) It forms the basis for the introduction in section 4 of an internal notion of proposition employed in an account of updates.

The reader familiar with the relevant formal systems will find that (i) relates to DPL (Groenendijk and Stokhof [9]) and DRT (Kamp [13], Heim [12]), while (ii) reconsiders Update logic (Veltman [20]), proposing, in particular, a "syntactic" treatment of might. Along the way, the growth of information between partial states is investigated (section 2), and a view of discourse interpretation as consisting, in part, of the construction of a first-order theory is developed.

1 Transition systems over first-order models

Given a first-order signature (= vocabulary = set of non-logical symbols)[1] L (with equality) and a countable set X of variables, dynamic logic associates with every first-order L-model $M = \langle |M|, \ldots \rangle$ a transition system $[\![M]\!]$ whose states map finitely many variables from X to objects in the universe $|M|$ of M. The collection of programs π involved are given by

$$\pi ::= \varphi? \mid x :=? \mid \pi_1 ; \pi_2 \mid \pi_1 + \pi_2 \mid \pi^* \mid \neg\pi$$

over atomic L-formulas φ with variables x from X. As it turns out, the combination of * and $\neg$ leads to various complications that are best dealt with by confining * to the Appendix, since * has as yet found no applications in the systems of dynamic semantics to be considered below. So, let Π_L be the collection of programs π with*out* an occurrence of *. The initial state of $[\![M]\!]$ is the empty valuation $\emptyset$, from which the remaining states — valuations $\alpha, \beta, \ldots$ from a (non-empty) finite subset of X into $|M|$ — can be obtained through a sequential composition of "random assignments" $x :=?$. More precisely, writing $[\![\pi]\!]_M$ for the relations $\xrightarrow{\pi}$ of $[\![M]\!]$, define

$$\begin{aligned}
\alpha[\![\varphi?]\!]_M\beta &\text{ iff } \alpha = \beta \text{ and } M \models \varphi[\alpha] \\
\alpha[\![x :=?]\!]_M\beta &\text{ iff } \alpha = \beta \text{ except possibly at } x, \text{ and } x \in \text{domain}(\beta) \\
\alpha[\![\pi_1 ; \pi_2]\!]_M\beta &\text{ iff } \alpha[\![\pi_1]\!]_M\gamma \text{ and } \gamma[\![\pi_2]\!]_M\beta \text{ for some } \gamma \\
\alpha[\![\pi_1 + \pi_2]\!]_M\beta &\text{ iff } \alpha[\![\pi_1]\!]_M\beta \text{ or } \alpha[\![\pi_2]\!]_M\beta \\
\alpha[\![\neg\pi]\!]_M\beta &\text{ iff } \alpha = \beta \text{ and there is } no\ \gamma \text{ for which } \alpha[\![\pi]\!]_M\gamma\,.
\end{aligned}$$

Note that if α is related to itself by the test $[\![\varphi?]\!]$, then α must be defined on all the variables in φ. This leads to a certain partiality in that, for instance, it is *not* the case that for some α, $\emptyset[\![x = x?]\!]_M\alpha$ (or, $\emptyset[\![x \neq x?]\!]_M\alpha$). Also, because of closure under negation $\neg\pi$ (written as such in Groenendijk and Stokhof [9], but known more

[1] Certain basic notions from model theory are taken for granted in what follows; definitions can be found in the initial sections of Keisler [16].

traditionally as the test $[\pi]\bot?$), tests for arbitrary first-order L-formulas (on X) as well as modal formulas built from programs can be defined, following

$$(\varphi\&\psi)? = \varphi?;\psi?$$
$$[\pi]\,\varphi? = \neg(\pi;\neg(\varphi?))$$
$$\exists x\varphi? = \neg\neg(x:=?;\varphi?)\ .$$

Our first theorem, however, depends only on the inclusion of atomic tests and random assignments among the programs.

Theorem 1. *For (finite or) countable L-models M and N, $M \cong N$ iff $[\![M]\!] \cong [\![N]\!]$.*

The key to the proof of Theorem 1 (which along with other proofs for this section can be found in the Appendix) lies in the notion of a *partial isomorphism set* from M to N (see, for example, Keisler [16]), which enables an isomorphism between M and N to be built "back and forth". This notion, in turn, corresponds to an equivalence between $[\![M]\!]$ and $[\![N]\!]$, which can be defined more generally for any two transition systems over a set Π of programs as follows. A *bisimulation* (Park [18]) between $\langle S,\{\xrightarrow{\pi}\}_{\pi\in\Pi},s_0\rangle$ and $\langle S',\{\xrightarrow{\pi}{}'\}_{\pi\in\Pi},s'_0\rangle$ is a relation $R\subseteq S\times S'$ such that whenever sRs', then for every $\pi\in\Pi$,

$$\forall t \xleftarrow{\pi} s\ \exists t' \xleftarrow{\pi}{}' s'\ tRt' \text{ and } \forall t' \xleftarrow{\pi}{}' s'\ \exists t \xleftarrow{\pi} s\ tRt'\ .$$

$\langle S,\{\xrightarrow{\pi}\}_{\pi\in\Pi},s_0\rangle$ and $\langle S',\{\xrightarrow{\pi}{}'\}_{\pi\in\Pi},s'_0\rangle$ are *bisimilar*, written

$$\langle S,\{\xrightarrow{\pi}\}_{\pi\in\Pi},s_0\rangle \leftrightarrows \langle S',\{\xrightarrow{\pi}{}'\}_{\pi\in\Pi},s'_0\rangle\ ,$$

if there is a bisimulation between them relating s_0 to s'_0. Now, the predicate

$$[\![M]\!] \leftrightarrows [\![N]\!]$$

can be added to the list of equivalent predicates in Theorem 1.

Theorem 1 states that no information is lost in the passage from a countable first-order model M to its transition system $[\![M]\!]$. A more abstract (i.e., information-decreasing) analysis better suited to bring out the first-order character of M is provided by the following standard construction on transition systems, applied to $[\![M]\!]$. For every $\pi\in\Pi_L$, let

$$s_{\pi,M} = \{\alpha \mid \emptyset[\![\pi]\!]_M\alpha\}\ ,$$

and

$$[\pi]_M = \{(s_{\pi',M},s_{\pi';\pi,M}) \mid \pi'\in\Pi_L,\ s_{\pi',M}\neq\emptyset\neq s_{\pi';\pi,M}\}\ .$$

The states of the new transition system $[M]$ are the *non-empty* sets $s_{\pi,M}$ for some $\pi\in\Pi_L$, the initial state being $\{\emptyset\}$ (i.e., $s_{\neg(x:=?),M}$). The interpretation of π under $[M]$ is $[\pi]_M$, making the transition system *deterministic* in that the transition relations are partial functions.

Example. Suppose L includes two constants 0 and 1, M is an L-model interpreting 0 as 0 and 1 as 1, and $\hat{\pi}$ is the non-deterministic program $x := 0 + x := 1$ (or, to be more precise, $x :=?; (x = 0? + x = 1?)$). Then

$$\emptyset \ [\![\hat{\pi}]\!]_M \ \alpha \text{ iff } \alpha = \{(x,0)\} \text{ or } \alpha = \{(x,1)\}$$
$$\{\emptyset\} \ [\hat{\pi}]_M \ s \text{ iff } s = \{\{(x,0)\},\{(x,1)\}\} \ .$$

Note also that $s_{\pi_0,M} \ [\pi]_M \ s_{\pi_1,M}$ does *not* imply that $\pi_1 = \pi_0;\pi$ (although the converse holds). Take π_0 to be $\neg(x :=?)$, π_1 to be $y :=?$, and π to be $(x = 0)?+y :=?$.

The next theorem should be contrasted with Theorem 1, recalling that over infinite first-order models, isomorphism is typically a much stronger property than elementary equivalence $\equiv_L$ (as a consequence, for instance, of compactness).

Theorem 2. *For L-models M and N, the following are equivalent.*

1. $M \equiv_L N$.
2. $[M] \cong [N]$.
3. $[M] \leftrightarrows [N]$.

Theorem 2 suggests that a transition system be built directly from a first-order theory, rather than a single first-order model. Before proceeding on to such constructions, we pause to consider how information grows between partial states.

2 Partiality of states and information growth

Intuitively, the initial states $\emptyset$ and $\{\emptyset\}$ of $[\![M]\!]$ and $[M]$ respectively have *no* information content — at least when compared with the other states. More formally, $\emptyset$ is the least $[\![M]\!]$-state under the subfunction partial order $\subseteq$, whereas $\{\emptyset\}$ is a minimal $[M]$-state under the so-called "Smyth pre-order" $\leq$ with respect to $\subseteq$

$$s \leq s' \text{ iff } \forall \beta \in s' \ \exists \alpha \in s \ \alpha \subseteq \beta \ .$$

The partial order $\subseteq$ on $[\![M]\!]$-states exposes the "expansive" character of random assignments which more complicated programs can inherit.[2] The move from finite valuations in $[\![M]\!]$ to accessible sets of finite valuations in $[M]$ captures the "eliminative" character of tests $\varphi?$ without discarding the partial order on finite functions. Thus, a program can increase information two ways — expansively and eliminatively. This is not to say, however, that information can never be lost after the execution of a program.

[2] The idea of treating states as partial objects is, of course, hardly novel, going back at least to Goldblatt [8]. A particularly rich structure is introduced in Vermeulen [21], where states range over sequences (recording the values assigned to variables), and an explicit "downdating" program construct is added to provide the possibility of destroying information. The present paper is concerned with a more primitive notion of program variable.

Applied to an $[\![M]\!]$-state already defined on x, the random assignment $x :=?$ destroys information. To protect information while allowing for the possibility of its expansive growth, it is convenient to introduce a *guarded assignment*

$$x := *$$

defined as

$$x = x? + \neg(x = x?); x :=? \ ,$$

the idea being to assign a value to x iff no value has so far been assigned to it. Guarded assignments are examples of *monotone* programs — i.e., programs π such that for every L-model M and $[\![M]\!]$-states α, β $\alpha[\![\pi]\!]_M\beta$ implies $\alpha \subseteq \beta$. Note that every $\pi \in \Pi_L$ can be made monotone by guarding random assignments.

Turning to the second dimension of information growth, call a program π *eliminative* if whenever $\alpha[\![\pi]\!]_M\beta$, it follows that $\alpha = \beta$; whence, for such π, $s[\pi]_M s'$ implies $s' \subseteq s$. Note that tests are eliminative, whereas random (or guarded) assignments are not. Eliminative programs are those programs that can be characterized (statically) by propositions. The point is that every program π has an eliminative approximation $\neg\neg\pi$ that can be constructed from what van Benthem [4] calls *modes* from (static) propositions to (dynamic) procedures, and *projections* going the opposite direction. Assuming a notion of proposition closed under program-labelled modalities (i.e., $\langle\pi\rangle$ and $[\pi]$), compose the projection $\pi \mapsto \langle\pi\rangle\top$ (where $\top$ is, say, $\exists x\ x = x$) with the mode $\varphi \mapsto \varphi?$, where

$$\alpha[\![\varphi?]\!]_M\beta \text{ iff } \alpha = \beta \text{ and } M \models \varphi[\alpha] \ .$$

This yields the map $\pi \mapsto \langle\pi\rangle\top?$, where

$$\begin{aligned} \alpha[\![\langle\pi\rangle\top?]\!]_M\beta \text{ iff } & \alpha = \beta \ \text{ and there is a } \gamma \text{ such that } \alpha[\![\pi]\!]_M\gamma \\ \text{iff } & \alpha[\![\neg\neg\pi]\!]_M\beta \ . \end{aligned}$$

Observe that $\langle\pi\rangle\top?$ (i.e., $\neg\neg\pi$) is semantically equal to π if π is eliminative, while $\langle\varphi?\rangle\top$ is logically equivalent to φ.

3 From semantic evaluation to semantic construction

Two important formalisms associated with dynamic semantics are DPL (Groenendijk and Stokhof [9]) and DRT (Kamp [13], Heim [12]). For our present purposes, it suffices to describe DPL roughly (and a bit incorrectly) as a subsystem of (predicate) dynamic logic subsumed by the *-free programs of section 1, and DRT as a formalism for analyzing the processing of natural language discourse. Hence, whereas the transition systems of section 1 might be associated with the former, it is more natural, for the latter case, to build transition systems out of more partial information — partial information that might be embodied by families of first-order models, or by possibly incomplete theories —, and to extract first-order models from such transition systems. These are presently taken up, in turn.

3.1 Deterministic transition systems induced partially

Given a family $\mathcal{M}$ of first-order L-models, form the transition system $[\mathcal{M}]$ from the transition systems $[M]$, for $M \in \mathcal{M}$, as follows. The states of $[\mathcal{M}]$ are non-empty sets of the form

$$\{(M, s_{\pi,M}) \mid M \in \mathcal{M} \text{ and } s_{\pi,M} \neq \emptyset\}$$

for $\pi \in \Pi_L$. Call the set above, provided it is non-empty, $s_{\pi,\mathcal{M}}$. The initial $[\mathcal{M}]$-state is $s_{\hat{\pi},\mathcal{M}}$ where $\hat{\pi}$ is $\neg x :=?$. The interpretation $[\pi]_{\mathcal{M}}$ of π under $[\mathcal{M}]$ is defined by

$$s_{\pi_0,\mathcal{M}} \; [\pi]_{\mathcal{M}} \; s_{\pi_1,\mathcal{M}} \text{ iff } s_{\pi_1,\mathcal{M}} = s_{\pi_0;\pi,\mathcal{M}} \; .$$

By Theorem 2,

(†) if $\forall M \in \mathcal{M}\ \exists M' \in \mathcal{M}'\ M \equiv_L M'$ and $\forall M' \in \mathcal{M}'\ \exists M \in \mathcal{M}\ M \equiv_L M'$ then $[\mathcal{M}] \,\underline{\leftrightarrow}\, [\mathcal{M}']$.

The converse of (†), however, fails, a counter-example to which will be supplied after presenting an alternative characterization of $[\mathcal{M}]$ in terms of the set $Th\mathcal{M}$ of all L-sentences true in every model of $\mathcal{M}$.

Given a consistent L-theory T, define the following transition system $[T]$. Let s_π be

$$\{(X_0, \psi) \mid X_0 \text{ is the set of variables} \in X \text{ occurring in } \pi, \text{ and for every } L\text{-model } N \text{ and } \beta : X_0 \to |N|,\ \emptyset[\![\pi]\!]_N\beta \text{ iff } N \models \psi[\beta]\}.$$

(It is possible to define s_π without referring to, or quantifying over, L-models N; see Lemma A$^\neg$ of the Appendix.) $[T]$-states are non-empty sets of the form

$$\{(X_0, \psi) \in s_\pi \mid \psi \text{ is consistent with } T\} \; ,$$

for $\pi \in \Pi_L$. Call the set above, provided it is non-empty, $s_{\pi,T}$, and let the initial $[T]$-state $s_{\hat{\pi},T}$ be that induced by $\hat{\pi} = \neg(x :=?)$. The interpretation $[\pi]_T$ of π under $[T]$ is defined by

$$s_{\pi_0,T} \; [\pi]_T \; s_{\pi_1,T} \text{ iff } s_{\pi_1,T} = s_{\pi_0;\pi,T} \; .$$

The definition above was formulated carefully in order to accomodate the possibility of non-monotonicity (i.e., revision in X) as well as partiality (again, due to X). (Readers familiar with DRT might compare $[T]$ with DRS's.) Without requiring any essentially new ideas for its proof,[3] Theorem 2 admits the following generalization.

Theorem 3. *For families $\mathcal{M}$ and $\mathcal{M}'$ of L-models, the following are equivalent.*

1. $Th\mathcal{M} = Th\mathcal{M}'$.
2. $[Th\mathcal{M}] \cong [\mathcal{M}']$.
3. $[\mathcal{M}] \cong [\mathcal{M}']$.
4. $[\mathcal{M}] \,\underline{\leftrightarrow}\, [\mathcal{M}']$.

[3] More specifically, appeal first to Lemma A$^\neg$ to prove $[Th\mathcal{M}] \cong [\mathcal{M}]$, and then show, as in Proposition B$^\neg$, that the only bisimulation on $[\mathcal{M}]$ is equality.

5. $[Th\mathcal{M}] \underline{\leftrightarrow} [\mathcal{M}']$.

Returning now to (†), a counter-example is provided by taking L to consist of a single binary relation, $\mathcal{M}$ to be the family of all finite linear orders, and $\mathcal{M}'$ to be $\mathcal{M}$ together with an infinite linear order constructed, by compactness, to satisfy $Th\mathcal{M}$.

3.2 First-order models from transition systems

To complete the duality between static notions (first-order models, L-formulas φ, and $\equiv_L$) and dynamic notions (transition systems, programs π, and $\underline{\leftrightarrow}$), first-order models are extracted from transition systems below. Thus, transition systems are shown to serve a dual role: a semantic one related to truth (building *on* first-order models), and also a constructive one (*building* first-order models).

Recall (or consult, for example, Keisler [16]) that the usual proof for the completeness theorem of first-order logic consists of the steps

(1) expand the language with (Henkin) witnesses, and
(2) complete a consistent Henkin theory (Lindenbaum's theorem),

followed by a quotient construction over constants. The first step corresponds to the expansive growth of information in a state, and the second, to eliminative growth.

Proposition 4. *There is a sequence $\{\pi_i\}_{i<\omega}$ of monotone programs such that for every L-theory T and countable L-model N, the following are equivalent*

1. *There is a map f from X onto $|N|$ such that for every $n < \omega$, there is a $(X_0, \psi) \in s_{\pi_0;\ldots;\pi_n,T}$ for which ψ is true in N, relative to f.*
2. *N is a model of T.*

Furthermore, given a countable L-model M, there is a sequence $\{\pi_i^M\}_{i<\omega}$ of monotone programs such that for every L-model N, the following are equivalent.

1'. *There is a map f from X onto $|N|$ such that for every $n < \omega$, there is an $\alpha \subset f$ such that $\emptyset[\![\pi_0^M; \ldots; \pi_n^M]\!]_N\alpha$.*
2'. *$N \cong M$.*

Proof. Fix a well-ordering on X, and an enumeration $\{\psi_i\}_{i\in\omega}$ of all L-formulas with variables from X. For every such L-formula φ, let π_φ be

$$x_0 := *; \ldots; x_k := *; \varphi?$$

where $x_0, \ldots, x_k$ are φ's free variables (enumerated according to the fixed ordering). For every $i < \omega$, define a program $\hat{\pi}_i$ inductively as follows.

Let x be the least variable in X not occuring in $\hat{\pi}_j$ for $j < i$. If $\psi_i = \exists y \psi$, then for the least variable x' different from x or any variable occuring in ψ or $\hat{\pi}_j$ for $j < i$, define $\hat{\pi}_i$ to be

$$x := * \; ; \; \pi_{\psi[x'/y]} \; .$$

Otherwise (i.e., ψ_i is not existential), let $\hat{\pi}_i$ be

$$x := * \; ; \; \pi_{\psi_i} \; .$$

(The assignments $x := *$ are inserted so that $|N| = \{f(x) \mid x \in X\}$.) Now, for $i < \omega$, let π_i be the program

$$\hat{\pi}_i + \neg\hat{\pi}_i ,$$

which not only decides the truth of ψ_i but also specifies a witness if ψ_i is existential and tests successfully. That is, running the programs π_i (in sequence $i = 0, 1, \ldots$) over a countable L-model N yields (at the limit) a consistent, complete theory with witness set X, where each element of $|N|$ is the value of some variable in X.

Constructing $\{\pi_i^M\}_{i\in\omega}$ is similar, except that it is carried out relative to an enumeration $\{m_i\}_{i\in\omega}$ of $|M|$ so that the choice between $\hat{\pi}_i$ and $\neg\hat{\pi}_i$ is determined instead by the truth or falsehood of ψ_i relative to the finite part of f built at stage i. ⊣

4 Updates and an internal notion of proposition

As an analysis of the notion of an update at the propositional level, Veltman [20] presents the following system of Update logic, denoted U below. Fix a set $\mathcal{A}$ of propositional variables (written $p, \ldots$). A *world* w is a subset of $\mathcal{A}$, and an *information state* σ is a set of worlds. An *update* A is generated according to

$$A ::= p \mid \sim A \mid A \wedge B \mid A \vee B \mid \mathsf{might}\ A \mid A; B$$

and induces a total function on information states as follows

$$\begin{aligned}
\sigma[p] &= \{w \in \sigma \mid p \in w\} \quad \text{for } p \in \mathcal{A} && (1)\\
\sigma[\sim A] &= \sigma - \sigma[A] && (2)\\
\sigma[A \wedge B] &= \sigma[A] \cap \sigma[B] &&\\
\sigma[A \vee B] &= \sigma[A] \cup \sigma[B] &&\\
\sigma[\mathsf{might}\ A] &= \sigma \quad \text{if } \sigma[A] \neq \emptyset && (3)\\
&= \emptyset \quad \text{otherwise} && (4)\\
\sigma[A; B] &= (\sigma[A])[B] . &&
\end{aligned}$$

Note that for every σ and A, $\sigma[A] \subseteq \sigma$, and that $\emptyset$ is an "absurd" information state from which there is no chance of recovery. Hence, instead of interpreting updates as total functions on a set of information states that includes an absurd one, updates can be taken to be partial functions on a set of non-empty (i.e., non-absurd) information states. From this point of view, might A is "static" in that it can only relate identical states, whereas other programs may relate distinct states.

But what is the intuition behind the notion of an information state above? An information state is a set of worlds, which in turn can be regarded as functions from $\mathcal{A}$ to a set of two truth values. In fact, *total* valuations — in view of clauses (1) and (2) and the fact that the powerset $2^{\mathcal{A}}$ of $\mathcal{A}$ corresponds to the function space from $\mathcal{A}$ to a two-element set. Observe that the defined transitions are "static" in that the worlds in the states do not change: they either persist unaltered through the transition or drop out. Ignoring the might-clauses (3) and (4) for the moment, the updates can be understood as perfectly ordinary eliminative tests, assuming, that is, that the worlds they act on are total. Take away this assumption — i.e., make the worlds (= valuations) finite —, and the question arises as to how an update behaves on a partial world [sic] in which the truth of the proposition (being updated) is undecided. It is important to address this question because, from the point of view of computational practice, total valuations certainly have *no* priority over finite ones.[4] An account based on total valuations must demonstrate that it faithfully models the reality of finite valuations. And as far as the reality of finite valuations is concerned, an update must not only be able to test, but (in the absence of information about the proposition being updated) also stipulate (i.e., establish). That is, we are led to a different conception of updates, which, as we will however see, strips away some of the mystery in might.

4.1 A "dynamic" reconstruction of propositional updates

An account of U within the framework of section 1 can be givn based on the Boolean algebra 2 over $\{0,1\}$, with $L = \{0,1\}$ and the set X of variables equal to the set $\mathcal{A}$ of propositional variables of U. A map $\cdot^u$ from updates A to programs $A^u \in \Pi_L$ is defined as follows. For $p \in \mathcal{A}$, the corresponding program p^u must test if p is true, asserting it to be the case if its truth has not been previously determined. That is, define

$$p^u = p := * \; ; \; p = 1? \; .$$

Extend the translation to might by defining

$$(\mathsf{might}\ A)^u = \neg\neg(A^u) \qquad (= \langle A^u \rangle \top ?) \; ,$$

which is to say that $(\mathsf{might}\ A)^u$ checks that a transition through A^u is possible, without actually making such a transition.[5] In particular, $(\mathsf{might}\ p)^u$ does not assert that p is true, only that p has not been determined to be false. Next, rather than setting $(A \vee B)^u$ to $(\langle A^u \rangle \top \ \vee \ \langle B^u \rangle \top)?$ (which yields a static update), define

$$(A \vee B)^u = A^u + B^u \; ,$$

and, to preserve the commutativity of $\wedge$, let

$$(A \wedge B)^u = A^u ; B^u \ + \ B^u ; A^u$$

[4] Total valuations arise as convenient abstractions from finite valuations. The idea of modelling a finite function as the set of its total extensions not only tends to confuse eliminative with expansive information growth, but also replaces a perfectly finite object with one that is two times infinite: an infinite set of infinite objects. It is perhaps naive to expect that the foundational complications involved in this "reduction" will forever remain buried, even as extensions to richer contexts (for example, predicate formalisms) are considered. At the propositional level, however, it is quite understandable that the expansive growth of information should be overlooked, given the absence of the notion of a program assignment.

[5] Recall that $\neg$ applies to programs, and should be distinguished from the negation $\sim$ in (2) that operates on propositions.

instead of simply $A^u; B^u$, which we reserve for $(A; B)^u$

$$(A; B)^u = A^u; B^u \ .$$

As for $(\sim A)^u$, a treatment of falsehood F symmetric with truth T suggests defining (simultaneously with the above) the "negative" translation A_u as follows

$$\begin{aligned} p_u &= p := * \ ; \ p = 0? \qquad \text{for } p \in \mathcal{A} \\ (A \wedge B)_u &= A_u + B_u \\ (A \vee B)_u &= A_u; B_u \ + \ B_u; A_u \ , \end{aligned}$$

the point being to take

$$\begin{aligned} (\sim A)_u &= A^u \\ (\sim A)^u &= A_u \ . \end{aligned}$$

The laws of double negation and de Morgan then hold (as in U)

$$\begin{aligned} (\sim\sim A)^u &= A^u \\ (\sim(A \wedge B))^u &= (\sim A \vee \sim B)^u \\ (\sim(A \vee B))^u &= (\sim A \wedge \sim B)^u \ . \end{aligned}$$

For completeness, set

$$(\mathsf{might}\ A)_u = \neg(A^u) \quad (= \neg((\mathsf{might}\ A)^u)) \tag{5}$$

$$(A; B)_u = A_u + (\mathsf{might}\ A)^u; B_u \ . \tag{6}$$

Equation (6) respects the sequential order in ";", as well as a principle of minimal change exemplified in the definition above of $(A \vee B)^u$ as $A^u + B^u$, rather than $A^u + B^u + A^u; B^u + B^u; A^u$. The treatment by (5) of might is static, although it is plausible to equate $(\mathsf{might}\ A)_u$ with the dynamic assertion A_u. That is, the update "it is not the case that it might rain" can be construed not only as a test but as possibly establishing the truth of "it does not rain".[6] This would, however, run counter to the fact (in U) that $2^{\mathcal{A}}[\sim(\mathsf{might}\ p)] = \emptyset$. As defined, the translation $\cdot^u$ is faithful to U in a sense to be described shortly. Note that implicit in (5) is a dual must for might given by

$$\begin{aligned} (\mathsf{must}\ A)^u &= \neg(A_u) \\ (\mathsf{must}\ A)_u &= \neg\neg(A_u) \ . \end{aligned}$$

Let Π^u be the subset $\{A^u \mid A \text{ is an update}\}$ of Π_L, and, viewing 2 as an L-model, form, relative to Π^u, the transition systems $[\![2]\!]$ and $[2]$ described in section 1. (So, $[\![A^u]\!]_2$ is the interpretation under $[\![2]\!]$ of A^u; $[2]$-states are non-empty sets of finite valuations from $\mathcal{A}$ into $\{0,1\}$ accessible from $\{\emptyset\}$ by some A^u; and $[A^u]_2$ is

[6] Indeed, replacing the Boolean algebra 2 by a structure with at least 3 elements (see section 4.2), K. Vermeluen has suggested a dynamic definition of $(\mathsf{might}\ A)^u$ analogous to A^u and A_u, but neither testing that A is true nor testing that A is false.

the interpretation under [2] of A^u.) Also, associate with U the transition system U relative to Π^u whose state set is

$$\{2^{\mathcal{A}}\} \cup \{\sigma \mid \sigma \neq \emptyset \text{ and } 2^{\mathcal{A}}[A] = \sigma \text{ for some update } A\} ,$$

and whose initial state is $2^{\mathcal{A}}$. (Recall that $2^{\mathcal{A}}$ is the information state with the least information content.) Define the interpretation $\xrightarrow{A}_U$ of A^u in U by

$$\sigma \xrightarrow{A}_U \sigma' \text{ iff } \sigma[A] = \sigma'$$

for all U-states σ and σ'. While an isomorphism between [2] and U is out of the question — contrast the effects of the update $p \vee \sim p$ on $\{\emptyset\}$ and $2^{\mathcal{A}}$ —, a bisimulation is not. Towards this end, associate with every $[\![2]\!]$-state α the information state $\sigma_\alpha = \{w \mid \alpha \subset w\}$, where a world w (in the sense of U) is identified with its characteristic function from $\mathcal{A}$ to 2.

Proposition 5. *For every update A and every $[\![2]\!]$-state α, there is a finite number of $[\![2]\!]$-states $\beta_1, \ldots, \beta_n$ such that for every $[\![2]\!]$-state β and every world w,*

$$\alpha [\![A^u]\!]_2 \beta \text{ iff } \beta = \beta_i \text{ for some } i \in \{1, \ldots, n\}$$
$$w \in \sigma_\alpha[A] \text{ iff } w \supseteq \beta_i \text{ for some } i \in \{1, \ldots, n\} .$$

Proof. Proceed by induction on A simultaneously with a version for A_u (i.e., $(\sim A)^u$). ⊣

Observe that $\emptyset [\![A^u]\!]_2 \alpha$ does *not* follow from $\forall w \supseteq \alpha \; w \in 2^{\mathcal{A}}[A]$; a counter-example is provided by $A = p \vee \sim p$. On the other hand, Proposition 5 implies

Corollary 6. *For every $[\![2]\!]$-state α, and every update A,*

$$\exists \beta \; \alpha [\![A^u]\!]_2 \beta \text{ iff } \sigma_\alpha[A] \neq \emptyset .$$

In particular, the Π^u-transition systems [2] *and U are bisimilar — i.e.,* $[2] \leftrightarrow U$ *via*

$$\{(\{\emptyset\}, 2^{\mathcal{A}})\} \cup \{(s, \sigma) \mid \exists \text{ update } A \text{ s.t. } \{\emptyset\}[A^u]_2 s \text{ and } 2^{\mathcal{A}} \xrightarrow{A}_U \sigma\} .$$

A distinction is drawn in van Eijck and de Vries [5] between an update A being "acceptable" and "accepted" in an information state σ. A is *acceptable in* σ if $\sigma[A] \neq \emptyset$ and *accepted* if $\sigma[A] = \sigma$. Corollary 6 relates directly to the former notion. As for the latter, the possibility that a program A^u is not eliminative suggests weakening the relation of equality (used in the definition of "accepted") to the so-called "Egli-Milner pre-order" $\sqsubseteq$ on the subset relation $\subseteq$

$$s \sqsubseteq s' \text{ iff } \forall \alpha \in s \; \exists \beta \in s' \; \alpha \subseteq \beta \;\; \wedge \;\; \forall \beta \in s' \; \exists \alpha \in s \; \alpha \subseteq \beta .$$

4.2 Prospects: from truth values to propositions

The analysis of propositions above can be generalized to structures other than the Boolean algebra 2, working with a language L consisting of the unary relation symbols T, F and D (for truth, falsehood and determinateness of truth, respectively). The idea is to replace $p = 1$ by $\mathsf{T}(p)$, $p = 0$ by $\mathsf{F}(p)$, and $p = p$ by $\mathsf{D}(p)$, redefining $p := *$ to

$$\mathsf{D}(p)? + \neg(\mathsf{D}(p)?); p :=? \; .$$

The L-theory with the axiom

$$\mathsf{T}(x) \vee \mathsf{F}(x) \supset \mathsf{D}(x)$$

can, according to the reader's taste, be expanded by various symbols, and extended by suitable axioms. For example, constants 0 and 1, a unary function symbol $\dot{\sim}$, and binary function symbols $\dot{\wedge}, \dot{\vee}$ might be added together with the axioms

$$\begin{array}{ll} \mathsf{T}(\dot{\sim}x) \equiv \mathsf{F}(x) & \mathsf{F}(\dot{\sim}x) \equiv \mathsf{T}(x) \\ \mathsf{T}(x\dot{\vee}y) \equiv \mathsf{T}(x) \vee \mathsf{T}(y) & \mathsf{F}(x\dot{\vee}y) \equiv \mathsf{F}(x) \wedge \mathsf{F}(y) \\ \mathsf{T}(x\dot{\wedge}y) \equiv \mathsf{T}(x) \wedge \mathsf{T}(y) & \mathsf{F}(x\dot{\wedge}y) \equiv \mathsf{F}(x) \vee \mathsf{F}(y) \\ \mathsf{T}(1) & \dot{\sim}0 = 1 \end{array}$$

Rather than adding the axiom $x = 0 \vee x = 1$ for bivalence, however, the move that is being suggested here is one from internalizing truth values in the L-model M to internalizing propositions. Indeed, the rudimentary L-theory above must, for the predicate case, be enriched to provide an account of the formation of a propositional object from a relation plus an appropriate tuple of arguments, where a relation need not be identified with the set of tuples satisfying it. (Note that in this case, propositions would live along side other objects in the first-order model. In other words, DPL and Update logic can be accomodated within the *-free fragment of dynamic logic, provided the language L is expanded to describe an internal notion of propositions.) There is something unmistakably intensional (or syntactic) about these internal propositions. Furthermore, their introduction into the object level is somewhat disturbing, given that an external notion of proposition is already at hand (at the meta-level). These two notions cannot, in general, be expected to coincide, in view of Tarski's theorem on the undefinability of truth (see also Montague [17], Aczel [1] and Turner [19]). The question arises as to which notion of proposition to use for analyzing updates. Appealing again to the finiteness in the computational picture underlying dynamic logic, the author is inclined to choose the internal notion not only because, being internal, it is more managable but also because the external notion (given by a first-order model) is total and, furthermore, extensional. Various reasons have been given (see, for instance, chapter 4 of Barwise [2]) to resist the identification of a proposition with the collection of (total) worlds in which it is true.[7] A somewhat persuasive case against the external notion of proposition may

[7] Returning to the matter of the undefinability of truth, an argument against the totality of worlds can be based on the Liar's paradox, as analyzed in Barwise and Etchemendy [3], and studied from a "dynamic" perspective in Groeneveld [10].

also be based on the logical intractability of a predicate version of U, in which a world is a model of some first-order theory T. In terms of transition systems, the idea is that U, lifted to first-order logic and relativized to a first-order theory T, is essentially the transition system $[T]$ described in section 3.1 above (and identified in Theorem 3 with $[\mathcal{M}]$ for a set $\mathcal{M}$ of models whose common theory is T), extended with programs might φ so that for the initial $[T]$-state s_0,

$$
\begin{aligned}
s_0\ [\mathsf{might}\ \varphi]_T\ s \text{ iff } & s = s_0 \text{ and } \exists s'\ s_0\ [\varphi?]_T\ s' \\
\text{iff } & s = s_0 \text{ and } \varphi \text{ is consistent with } T\ .
\end{aligned}
$$

The problem is that most interesting first-order theories T are undecidable, whence, in contrast to the propositional case U, the predicate "φ is consistent with T" is, in general, not r.e. (in φ). Hence, neither the accepted nor the acceptable updates of this predicate system are axiomatizable.

5 Discussion

The duality between static and dynamic notions established above rests on a careful analysis of the growth of information between partial states. (Observe, for instance, that all of Theorems 1, 2, and 3, Propositions 4 and 5, and Corollary 6 use the finiteness of valuations heavily.) This is not merely an incidental feature of the present paper, but is its fundamental premise: *the computational processes necessitated by the partial character of information are central to the dynamic conception of logic* described in van Benthem [4]. (The reader is asked not to fret over the restrictions to the dynamic conception that the term "computation" might suggest, to construe the term broadly if wished, but to keep in mind that logically tractable conceptions are usually based on some coherent notion of construction.) Thus, as tempting as it may be, for instance, to "simplify matters" by passing from the finite valuations α of $[\![M]\!]$ to the total valuations used in DPL (as well as Harel [11]), the claim that the choice between finite and total valuations is immaterial to *the logic* involved is not only, under a trivial reading, false (— compare the effect of the test $x = x?$ on finite valuations versus total valuations —)[8], but assumes an understanding of "the logic" that we simply do not have, while ignoring a feature of reality (i.e., finiteness) to which we are bound. The distinction between the finite and the infinite is fundamental to foundational studies; and as formalisms are extended, the foundations on which our ideas lie must bear greater weight. To what extent the semantic picture

[8] It is true enough that programs might be restricted to those of the form $\overline{\pi}$, where $\overline{\pi}$ is π preceded by guarded assignments to variables in X occuring in π. (For example, $\overline{x =?}$ is $x := *; x = x?$.) With respect to such programs, transition systems built from finite valuations are, the author expects, bisimilar to corresponding transition systems based on total valuations (generalizing Corollary 6). It would remain then to show that the map from π to $\overline{\pi}$ provides an interpretation of the "logic" of total valuations within the "logic" of finite ones. (Similarly, replacing random assignments by guarded ones is a plausible first step towards an interpretation of first-order intuitionistic logic within the formalism of section 1.) This interpretation would reinforce the feeling (suggested already by the reality of mechanical computation) that the latter is foundationally prior, but *so what?* In reply, the reader is referred back to section 4.

of update logic can, for instance, be supported or developed in its predicate form is called to question in section 4. Looking at the matter theoretically, the expansive growth of information exposed by finite valuations allows the range of possibilities reduced by eliminative programs to be widened, while clearly staying within the scope of first-order logic, the formulas of which have an arbitrary finite number of variables.

A notable omission in the analysis above concerns the concept of inference, typically codified in a logical formalism. An analysis of a particular notion of inference presupposes a certain minimal familiarity with the underlying semantic structures. Just what this "minimal familiarity" is is a somewhat *personal* matter that is rather difficult to spell out, and should perhaps not be imposed universally. Suffice it to say that the work above was carried out with a view towards acquiring (for the author) just that familiarity, the assumption being that transition systems grounded in computation are "the underlying semantic structures." A particular outcome of this work that the author plans to develop and study further is the translation from propositions to programs described in section 4. How the various notions of inference surveyed in van Benthem [4] look under this translation and how the logical formalisms worked out in van Eijck and de Vries [6], [5] can be adapted for this purpose are two natural questions to consider. Intuitively, what is different about this translation from, say, Groenendijk and Stokhof [9], and Veltman [20] is that the program to which a proposition A translates can do one of two things: test that A is true (semantic evaluation), or establish that A is true (semantic construction). (This dual function is not a new idea, but occurs in the analysis of discourse referents in, for instance, Karttunen [14] and Heim [12].) A closer study of the transition systems $[T]$ and $[\mathcal{M}]$ restricted to the translations of propositions may be worthwhile, especially in relation to DRT and Update logic.

Lastly, notice that a basic limitation of the approaches above is that the states considered are "static." Visser [22] disputes the idea that "a state is something static", suggesting instead that "states find their natural home within the saturated-unsaturated distinction" (p. 4). Indeed, the notions of a transition system and a bisimulation are commonly applied in computer science to states that are decidedly dynamic. The interested reader is referred to Fernando [7] for relations between that tradition and dynamic logic (the basis of dynamic semantics, as understood above).

References

1. Peter Aczel. Frege structures and the notions of proposition, truth and set. In *The Kleene symposium*. North-Holland, Amsterdam, 1980.
2. Jon Barwise. *The situation in logic*. CSLI Lecture Notes Number 17, Stanford, 1989.
3. Jon Barwise and John Etchemendy. *The liar: an essay on truth and circularity*. Oxford University Press, Oxford, 1987.
4. Johan van Benthem. Logic and the flow of information. In *Proc. 9th International Congress of Logic, Methodology and Philosophy of Science*. North-Holland, Amsterdam, to appear.
5. J. van Eijck and F.J. de Vries. A sound and complete calculus for update logic. Technical Report CS-R9155, Centre for Mathematics and Computer Science, 1991.

6. J. van Eijck and F.J. de Vries. Dynamic interpretation and Hoare deduction. *Journal of Logic, Language and Information*, 1, 1992.
7. Tim Fernando. Comparative transition system semantics. Manuscript, 1992.
8. Robert Goldblatt. *Axiomatising the logic of computer programming*. LNCS 130. Springer-Verlag, Berlin, 1982.
9. J. Groenendijk and M. Stokhof. Dynamic predicate logic. *Linguistics and Philosophy*, 14, 1991.
10. Willem Groeneveld. Dynamic semantics and circular propositions. University of Amsterdam, ITLI Prepublication Series, LP-91-03, 1991.
11. David Harel. Dynamic logic. In Gabbay et al, editor, *Handbook of Philosophical Logic, Volume 2*. D. Reidel, 1984.
12. Irene Heim. The semantics of definite and indefinite noun phrases. Dissertation, University of Massachusetts, Amherst, 1982.
13. J.A.W. Kamp. A theory of truth and semantic representation. In *Formal methods in the study of language*. Mathematical Centre Tracts 135, Amsterdam, 1981.
14. Lauri Karttunen. Discourse referents. In J. McCawley, editor, *Notes from the Linguistic Underground, Syntax and Semantics 7*. Academic Press, New York, 1976.
15. H. Jerome Keisler. Forcing and the omitting types theorem. In M. Morley, editor, *Studies in model theory*. The Mathematical Association of America, 1973.
16. H. Jerome Keisler. Fundamentals of model theory. In J. Barwise, editor, *Handbook of Mathematical Logic*. North-Holland, Amsterdam, 1977.
17. Richard Montague. Syntactical treatments of modality, with corollaries on reflexion principles and finite axiomatizability. *Acta Phil. Fennica*, 16, 1963.
18. David Park. Concurrency and automata on infinite sequences. In P. Deussen, editor, *Proc. 5th GI Conference*, LNCS 104. Springer-Verlag, Berlin, 1981.
19. Raymond Turner. *Truth and Modality for Knowledge Representation*. Pitman, London, 1990.
20. F. Veltman. Defaults in update semantics. In H. Kamp, editor, *Conditionals, Defaults and Belief Revision*. Edinburgh, Dyana deliverable R2.5.A, 1990.
21. C.F.M. Vermeulen. Sequence semantics for dynamic logic. Technical report, Philosophy Department, Utrecht, 1991.
22. Albert Visser. Actions under presuppositions. Technical report, Philosophy Department, Utrecht, 1992.

Appendix

This appendix presents the technical details behind section 1 (and a bit more), building on the notation introduced there.

Theorem 1. *For countable L-models M and N, the following are equivalent.*

1. $M \cong N$.
2. $[\![M]\!] \cong [\![N]\!]$.
3. $[\![M]\!] \leftrightarrows [\![N]\!]$.

Proof. Only 3 $\Rightarrow$ 1 is not clear. But the definition of $[\![\pi]\!]_M$ has been arranged so that a bisimulation between $[\![M]\!]$ and $[\![N]\!]$ is a partial isomorphism from M to N (see, for example, Keisler [16]): $\emptyset R \emptyset$, and whenever $\alpha R \beta$,

(i) domain(α) = domain(β) (using tests "$x = x?$"),

(ii) (M, α) and (N, β) satisfy the same atomic formulas (using tests for all atomic L-formulas on X), and

(iii) for $x \in X - \text{domain}(\alpha)$,

$$\begin{aligned} &\forall m \in |M|\ \exists n \in |N| \quad \alpha \cup \{(x,m)\}\ R\ \beta \cup \{(x,n)\} \\ \text{and } &\forall n \in |N|\ \exists m \in |M| \quad \alpha \cup \{(x,m)\}\ R\ \beta \cup \{(x,n)\} \end{aligned}$$

(using random assignments $x :=?$).

Thus, an isomorphism can be built "back and forth" between enumerations of $|M|$ and $|N|$. ⊣

Given an L-model M, it is natural to ask whether $[\![M]\!]$ is a "reduced" transition system representation of M in the sense that the only bisimulation on $[\![M]\!]$ is equality. This is the case for L-models M all of whose elements are definable (e.g., $\langle\{0, 1, \ldots\}, +, \times, 0, 1\rangle$). On the other hand, countable counter-examples are also easy to find; take M, for instance, to be the rationals under their usual ordering. Then for any $x \in X$, any two $[\![M]\!]$-states with (the same) domain $\{x\}$ are related by a bisimulation. In particular, it follows from Theorem 1 that (for this instance of M) there is no L-model N such that $[\![N]\!]$ is the reduced form of $[\![M]\!]$.

Before turning to Theorem 2 and *-free programs, let us note that * has a certain computational basis that $\neg$ lacks, and that the preference for $\neg$ over * in the present paper is due solely to the fact that * has so far not found applications to natural language. From the point of view of dynamic logic, however, it is certainly interesting to consider the set Π_* of programs π given by

$$\pi ::= \varphi? \mid x :=? \mid \pi_1 ; \pi_2 \mid \pi_1 + \pi_2 \mid \pi^* \ .$$

Without program negation $\neg$, it becomes convenient to add more tests, and to let φ above range over a set $\Phi(L, X)$ of L-formulas (with free variables from X) that includes all atomic L-formulas, and is closed under conjunction, renaming of variables, and existential quantification. For completeness, let us state the definition

$$\alpha[\![\pi^*]\!]_M\beta \text{ iff } \alpha[\![\pi^k]\!]_M\beta \quad \text{for some } k \geq 0 \ ,$$

where π^0 is $\exists x\ x = x?$, and π^{k+1} is $\pi^k ; \pi$.

Lemma A. *Let M be an L-model, π be a program in Π_*, and α be an $[\![M]\!]$-state with domain X_0 such that $\emptyset[\![\pi]\!]_M\alpha$. Then there is a formula $\psi_{M,\alpha,\pi} \in \Phi(L, X_0)$ satisfied by M, α such that for all L-models N and valuations $\beta : X_0 \to |N|$ satisfying $\psi_{M,\alpha,\pi}$, $\emptyset[\![\pi]\!]_N\beta$.*

Proof. Using the semantic clauses for $+$ and $\cdot^*$, extract a finite sequence $\pi_1 ; \pi_2 ; \ldots ; \pi_n$ of tests and assignments that

(i) is given by ($\Rightarrow$ of) the semantic derivation of $\emptyset[\![\pi]\!]_M\alpha$, and

(ii) satisfies the following condition (by $\Leftarrow$): for every L-model N and $[\![N]\!]$-state β,

$$\emptyset[\![\pi_1 ; \ldots ; \pi_n]\!]_N\beta \text{ implies } \emptyset[\![\pi]\!]_N\beta \ .$$

Then, for $i = 1, \ldots n$, let I_i be the set $\{x_1, \ldots, x_{k_i}\}$ of variables in X_0 mentioned in $\pi_1; \ldots; \pi_i$, and let $x^i_1, \ldots, x^i_{k_i}$ be fresh variables, the intuition being that x^i_j represents the value of x_j after π_i is executed. Now, appealing to the closure properties of $\Phi(L, X)$, let $\psi_{M,\alpha,\pi}$ be

$$\exists x^1_1 \cdots \exists x^1_{k_1} \cdots \exists x^n_1 \cdots \exists x^n_{k_n} \bigwedge_{1 \leq j \leq k_n} x_j = x^n_j \ \& \ \bigwedge_{1 < i \leq n} \varphi_i \ ,$$

where for $1 < i \leq n$, φ_i relates $x^{i-1}_1, \ldots, x^{i-1}_{k_{i-1}}$ to $x^i_1, \ldots, x^i_{k_i}$ after the execution of π_i. ⊣

Call formulas satisfying the requirements for $\psi_{M,\alpha,\pi}$ in Lemma A *(π, M)-records of α*. Given L-models M and N, write $M \equiv_{\Phi(L)} N$ when they satisfy the same sentences in $\Phi(L, \emptyset)$. Now, form the transition system $[M]_*$ from $[\![M]\!]$ as $[M]$ was, but with Π_L replaced by Π_*.

Lemma B. *Assume $M \equiv_{\Phi(L)} N$, and $\{\emptyset\}[\pi]_M s$, $\{\emptyset\}[\pi]_N s'$, where $\pi \in \Pi_*$. Then for every $\pi' \in \Pi_*$,*

$$\{\emptyset\}[\pi']_M s \text{ iff } \{\emptyset\}[\pi']_N s' \ .$$

Proof. By symmetry, it is enough to prove one direction of the bi-conditional. In fact, it suffices, again through an appeal to symmetry, to show (assuming $M \equiv_{\Phi(L)} N$, $\{\emptyset\}[\pi]_M s$ and $\{\emptyset\}[\pi']_M s$) that whenever $\emptyset[\![\pi]\!]_N \beta$, then $\emptyset[\![\pi']\!]_N \beta$.[9] So suppose $\emptyset[\![\pi]\!]_N \beta$. Call domain($\beta$) X_0, and let

$$\begin{aligned} \Psi(X_0, \pi') &:= \{\psi \in \Phi(L, X_0) \mid \exists M' \exists \alpha : X_0 \to |M'| \ \ \psi \text{ is a } (\pi', M')\text{-record of } \alpha\} \\ \Theta(N, \beta) &:= \{\theta \mid N \models \theta[\beta] \text{ and } \theta \text{ or } \neg\theta \in \Phi(L, X_0)\} \ . \end{aligned}$$

If we can demonstrate the consistency of

$$\Theta(N, \beta) \cup \{\bigvee \Psi(X_0, \pi')\} \ ,$$

then we can conclude that $\emptyset[\![\pi']\!]_N \beta$ through an application of Lemma A. But by the General Omitting Types Theorem given in p. 108 of Keisler [15], we need only observe that for every finite piece $\Delta \subset \Theta(N, \beta)$, the conjunction

$$\psi_{N,\beta,\pi} \ \& \ \bigwedge \Delta$$

is satisfied by M relative to some $\alpha : X_0 \to |M|$ (since $M \equiv_{\Phi(L)} N$), which in turn satisfies some $\psi \in \Psi(X_0, \pi')$ (by Lemma A) since $\{\emptyset\}[\pi]_M s \ni \alpha$ and $\{\emptyset\}[\pi']_M s$. ⊣

Theorem C. *For L-models M and N, the following are equivalent.*

1. $M \equiv_{\Phi(L)} N$.
2. $[M]_* \cong [N]_*$.
3. $[M]_* \leftrightarrow [N]_*$.

[9] The point is that the (asymmetric) condition $\{\emptyset\}[\pi]_N s'$ is used only after establishing $\emptyset[\![\pi]\!]_N \beta$ iff $\emptyset[\![\pi']\!]_N \beta$.

Proof. 3 ⇒ 1 follows from the inclusion in Π of tests for all of $\Phi(L)$. 2 ⇒ 3 is trivial (since $\cong$ always implies $\underline{\leftrightarrow}$). All that remains is 1 ⇒ 2. Let $R_{M,N}$ be the relation

$$\{(s, s') \mid \exists\pi\ \{\emptyset\}[\pi]_M s \text{ and } \{\emptyset\}[\pi]_N s'\}\ .$$

It suffices to show that $R_{M,N} : [M]_* \cong [N]_*$, assuming $M \equiv_{\Phi(L)} N$. Proceed as follows.

1. From Lemma A conclude that $R_{M,N}$ relates $\{\emptyset\}$ to $\{\emptyset\}$. (For example, if $\emptyset[\![\pi]\!]_M\alpha$, then as $M \equiv_{\Phi(L)} N$, $N \models \exists\bar{x}\psi_{M,\alpha,\pi}$ and so for some β, $\emptyset[\![\pi]\!]_N\beta$.)
2. Moreover, by Lemma B, $R_{M,N}$ defines a function from $[M]_*$-states to $[N]_*$-states.
3. Writing f for that function, the bi-conditional

$$s_0[\pi]_M s_1 \text{ iff } f(s_0)[\pi]_N f(s_1)$$

follows from the (easily established) fact that

$$s_0[\pi;\pi']_M s_1 \text{ iff } \exists s_2\ \ s_0[\pi]_M s_2 \text{ and } s_2[\pi']_M s_1\ .$$

⊣

Two defects with the duality given in Theorem C are worth pointing out. First, the requirement that $\Phi(L, X)$ be closed under existential quantification cannot simply be dropped by modelling the effect of existential quantification via random assignment (i.e., defining $\exists x\varphi?$ as "$x :=?; \varphi?$") since tests have no side-effects, whereas random assignments affect the assignment of a value to a variable. Second, the Π_*-transition system $[M]_*$ is never "reduced": the $[M]_*$-state generated by $x :=?$ is different from that generated by

$$x :=? + (\exists x\ x = x\ ?)\ ,$$

even though they are related by a Π_*-bisimulation. Both defects can be corrected by closing either $\Phi(L, X)$ under modalities labelled by programs (i.e., forming formulas $[\pi]\varphi$), or (equivalently) the collection Π_* of programs under negation $\neg$. But then Lemma A breaks down (as does Theorem C) — the program

$$x :=?\ ; ([\textbf{while } x > 0 \textbf{ do } x := x - 1]0 = 1)?$$

discriminates between standard and non-standard models of arithmetic. (As suggested by the use of an omitting types argument in Lemma B, infinitary logic is not far behind.)

Thus, to stay within first-order logic, $\neg$ must be introduced only if * is dropped. Replacing Π_* by Π_L, Lemma A can be strengthened (using the finite bound on the non-determinism of programs) to

Lemma A$^\neg$. *Given a program $\pi \in \Pi_L$, and a finite subset X_0 of X, there is an L-formula $\psi_{X_0,\pi}$ with free variables in X_0 such that for every L-model N and $\beta : X_0 \to |N|$,*

$$\emptyset[\![\pi]\!]_N\beta \textit{ iff } N \models \psi_{X_0,\pi}[\beta]\ .$$

Proof. Rather than reducing a program π to a sequence of primitive steps (as in the Kleene normal form theorem), a simpler reduction (in the manner of Tarski's inductive definition of satisfaction) is possible, because (in contrast to Lemma A) the bi-conditional holds. Sequential composition forces us, however, to work with an arbitrary initial state α (instead of simply $\emptyset$). Accordingly, strengthen the induction hypothesis on π as follows: for every finite subset X_0 of X and every $X' \subseteq X_0$, there is an L-formula $\chi_{X',X_0,\pi}$ with free variables from the disjoint union $X' + X_0$ of X' and X_0, and an L-formula $\theta_{X',\pi}$ with free variables from X' such that for every L-model N, $\alpha : X' \to |N|$, and $\beta : X_0 \to |N|$,

$$\alpha[\![\pi]\!]_N\beta \text{ iff } N \models \chi_{X',X_0,\pi}[\alpha + \beta]$$
$$\text{there is } no \ \gamma \text{ such that } \alpha[\![\pi]\!]_N\gamma \text{ iff } N \models \theta_{X',\pi}[\alpha] \ .$$

For the first bi-conditional, the trick is to record an un-corrupted copy of X' in $X' + X_0$ (so as to be able to store relevant conditions on the initial state α) before executing π. Note that for every π, there is a finite set X_π such that if $\alpha[\![\pi]\!]_N\gamma$ then domain(γ) − domain$(\alpha) \subseteq X_\pi$. This pushes through the argument not only for sequential composition, but also for the second bi-conditional. That is, $\theta_{X',\pi}$ can be constructed by taking a conjunction of formulas of the form $\neg\exists\overline{x}\ \chi_{X',X_1,\pi}$ for finitely many X_1's. ⊣

We might now proceed as before, proving $^\neg$-versions of Lemma B and Theorem C (where $\Phi(L, X)$ is the set of all L-formulas on X). Instead, however, let us observe that under Π_L, the states generated by x :=? and by x :=? + $\exists x\ x = x$? are no longer bisimilar since $\neg(x = x?) \in \Pi_L$. In fact,

Proposition B$^\neg$. *If R is a bisimulation on $[M]$ and sRs' then $s = s'$.*

Proof. Assume R is a bisimulation on $[M]$ and sRs'. Choose π and π' that give s and s', respectively (i.e., $\{\emptyset\}[\pi]_M s$ and $\{\emptyset\}[\pi']_M s'$), and let X_1 be a finite subset of X containing all variables of X occuring in π or π'. For every subset X_0 of X_1, let π_{X_0} be the program

$$\bigwedge_{x \in X_0} x = x \ ? \ ; \neg(x_1 = x_1?) \ ; \ \neg(x_2 = x_2?) \ ; \ \ldots \ ; \ \neg(x_k = x_k?) \ ,$$

where $X_1 - X_0 = \{x_1, \ldots, x_k\}$, and fix L-formulas $\psi_{X_0,\pi}$ and $\psi_{X_0,\pi'}$ given by Lemma A$^\neg$, with free variables from X_0. Observe that since sRs' (where R is a bisimulation), it follows from Lemma A$^\neg$ that for every $X_0 \subseteq X_1$,

$$\neg\exists t \ \ s \ [\pi_{X_0} \ ; \ \psi_{X_0,\pi} \& \neg\psi_{X_0,\pi'} \ ?]_M \ t$$

and

$$\neg\exists t' \ \ s' \ [\pi_{X_0} \ ; \ \psi_{X_0,\pi'} \& \neg\psi_{X_0,\pi} \ ?]_M \ t' \ .$$

That is, for every $X_0 \subseteq X_1$, $\psi_{X_0,\pi'}$ and $\psi_{X_0,\pi}$ are satisfied by the same L-models M and functions from X_0 to $|M|$, whence $s = s'$ (again by Lemma A$^\neg$). ⊣

Hence, a bisimulation R between $[M]$ and $[N]$ must be an isomorphism (since $R \circ R^{-1}$ is a bisimulation on $[M]$), and a $^\neg$-version of Theorem C can be proved by showing

$3 \Rightarrow 2 \Rightarrow 1 \Rightarrow 3$ (rather than following the direction $3 \Rightarrow 1 \Rightarrow 2 \Rightarrow 3$ in the proof above of Theorem C).

Theorem 2. *For L-models M and N, the following are equivalent.*

1. $M \equiv_L N$.
2. $[M] \cong [N]$.
3. $[M] \leftrightarroweq [N]$.

Remark (for readers familiar with the model-theoretic notions involved). Prof. van Benthem has suggested replacing in Theorem 1 $M \cong N$ by M is partially isomorphic, or $L_{\infty\omega}$-equivalent to N. Under this modification, the assumption that M and N are countable can be dropped. Futhermore, the parallel with Theorem 2 can then be strengthened by bringing in the algebraic characterization of $\equiv_L$ as being so-called "finitely isomorphic" — a weakened version of partially isomorphic, also having a "back-and-forth" clause, that yields a finite variable hierarchy described, for example, in section 6 of van Benthem [4]. On the other hand, some may find the countability assumption in Theorem 1 (occurring also in Proposition 4) harmless, and a reasonable price for securing an isomorphism. Moreover, the very process of constructing (back-and-forth) an isomorphism is instructive, albeit standard.

This article was processed using the LaTeX macro package with LLNCS style

Declarative Semantics for Inconsistent Database Programs

Marion Sarkis Mircheva

Fac. of Mathematics and Computer Science
Dep. of Logic, Sofia University
James Bouichier str., 5, Sofia 1126

Abstract. In this paper an ongoing research on inconsistent database programs is presented, which can be viewed as an extension of logic programming. By *Literal Database Program* (LDP) we mean a finite set of universally closed clauses of the form $L_0 \leftarrow L_1, \ldots, L_m$ where $m \geq 0$, L_i's are literals.
We define a set of intended models $MOD(P)$ for a given program P with the help of interpretations defined on an extension of the Herbrand base. Instead of resolving inconsistency at all costs, our system is aimed to supply an inference control after the appearance of some *primitive* inconsistent units. It keeps the *good* conflict knowledge, namely those, that come from different, equally trusted sources and proposes mechanism telling one how to act when the primitive inconsistency arises.
A characterization of the semantics for the proposed literal database programs is given by means of general properties of non-monotonic inference operations. Some points for future considerations are also announced.

1 Introduction

The formalization of a reasoning process based upon contradictory databases is a critical issue, as most large knowledge bases are suspected to be inconsistent. We consider an inconsistent database where all data in the base are equally trusted. Such systems are useful for multi-expert databases, where knowledge of several experts is pooled in a single program. We allow for Logic programs to have rules with both A and $\neg A$ in heads and bodies and have any number of independent advisors. The incoming information could be provided from different, equally reliable sources and inconsistency may appear as a result of merging this information. Then the system have to consider what happens when several advisors (expert rules or facts) appear to be in conflict. Generally, the contradictions may involve several steps of reasoning, produce new contradictions or nested themselves in complicated deductions. Either we may have several simple data items or some formal reasonable rules, which together give the contradictory answers, but no single item is to blame. How do we cope with our system in this case?
The existing proposals concerning the problem of reasoning in inconsistent databases can be classified according to the methods and underlying philosophy of treating contradictory knowledge. We separate systems, that define new inference machinery, mainly by modifying the semantics of implication and the other logical connectives. These types of methods are based on some relevant, many-valued or paraconsistent logics. The paper of Lin [10] describes computational mechanisms for defining relevant conclusions from a set of formulae and specify the correct answers that do not depend on "erroneous data". The relevance character of the implication is defined in a syntactic way with the help of the restriction only from the clauses which are used explicitly in a refutation, when a resolution strategy is applied. Some advantages in using paraconsistent logic for knowledge representation in the presence of contradictions are discussed in [2].
The second type of methods offer the possibility of supporting conflicting conclusions

by isolating them in multiple extensions. These methods share the popular idea of creating consistent contexts in the inconsistent database. The classical example is default logic [16] and its variations. Although we can consider default rules as clauses with appropriate semantics, such systems are well adapted to the representation of rules with exceptions, rather then general productions. The paper of Nute [14] is an example of the approach based on using the specialized rules in order to resolve the inconsistency. The work of Martins and Shapiro [13] offers a relevant approach for dealing with inconsistency database where any formula of the knowledge base is associated with the consistent set of axioms from which one can derive it. Similar ideas for management with different consistent subbases of the inconsistent database are presented in the paper of Cholvy [3]. His formalism is supposed to limit the impact of errors to information "related" to erroneous information and is aimed to give the answers to queries by distinguishing reliable information and suspicious ones. In that case suspicious means produced by the union of special minimal inconsistent subsets of the given ones. As in the majority of clausal form reasoning systems, the database consists of definite clauses and a set of integrity constrains presented as a disjunction of negative literals. The presentation of that kind (even without allowing normal programs) is too restrictive to cope with contradictions in the deduction.

A special place among systems devoted to resolve consistency in the inconsistent contexts belongs to the Truth Maintenance System of Doyle [5] and ATMS of De Kleer [4]. Strictly speaking their systems do not perform an inference but are designed with the aim of maintaining the consistency of a set of deductions. They record inferences transmitted to them from an external deduction system and they are in charge of the maintenance of the consistency of a set of assertions. They allow for the management of assumptions, default values and contradictions. According to the incoming values they propose one set of deducible information units. Similarly to "multiple extension" approach the underlying philosophy of TMS is to eradicate inconsistency at all costs. Gabbay and Hunter [8] propose some form of truth maintenance systems to be used in restoring consistency as a result of some *action*, provided by an object-level database. Hence TMS could serve as a supplementary tool for debugging an inconsistent database. The papers [7,8] suggest a general framework for treating inconsistency as a *good* thing, especially as "the norm we should feel happy to be able to formalize". They suggest an approach quite different from the existing at present philosophy in dealing with inconsistency in the deductive databases. According to them, inconsistency in logical systems should provide the effect of reasoning more similar to that of human reasoning, namely:

$$\textit{inconsistency} \longrightarrow \textit{action}$$

In the paper [8] inconsistency is classified according to the appropriate action that is required and might be adopted in a supplementary to original database. The work [6] is an example how inconsistency could be studied as a Labelled Deductive System, where the basic units of information are labelled formulae and the deduction is defined on labelled formulae. Some special cases of deductions can serve as a trigger for performing an action. The idea of taking an appropriate control over the derivation exists in different formalisms dealing with inconsistent knowledge base [15], but in the case of labelled formulae, the labels are necessary in deciding which action to call.

Reasoning in inconsistent contexts can be naturally considered as an integral part of non-monotonic systems. In [9] a certain number of criteria of a technical nature have been proposed for purposes of comparison.

The approach we took, differs from all mentioned above by the method of knowledge representation and by the treatment of contradictions which are deductible in the course of the inference process. We define a Literal Database Program (LDP) as a finite set of clauses of the form:

$$L_0 \longleftarrow L_1, \ldots, L_n$$

where L_0, L_1, ..., L_n are arbitrary literals. The intuitive backgrounds we chose consist of:

1. All the deducible units have "local" consistent support. In other words if Σ is the set of all literals that essentially participate in deducibility of A, then $\Sigma \cup A$ must be consistent.

2. We admit deduction from independent contradictory items. But we have to be sure that if A and $\neg A$ are deductible, then all the literals attending in their deduction are not primitive contradictions. Hence the contradictory knowledge should not be used in the derivation of some other conflict units. In other words nested conflicts are not sanctioned.

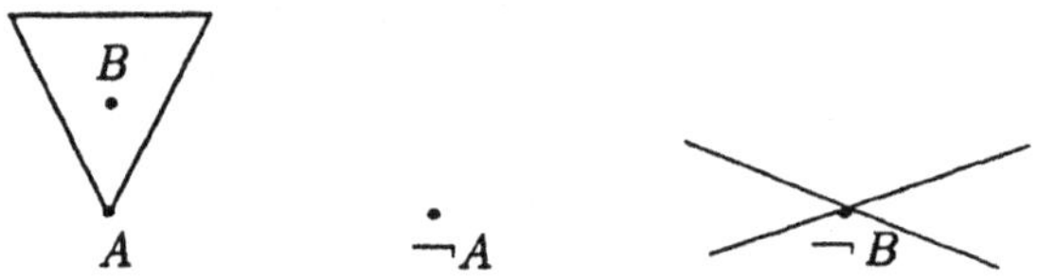

3. Contradictory literals are allowed to support some "soft" B for which $\neg B$ is not deducible. Knowledge, that are produced on the bases of some contradictory units have less informative power then those inferred from a "safety" information.

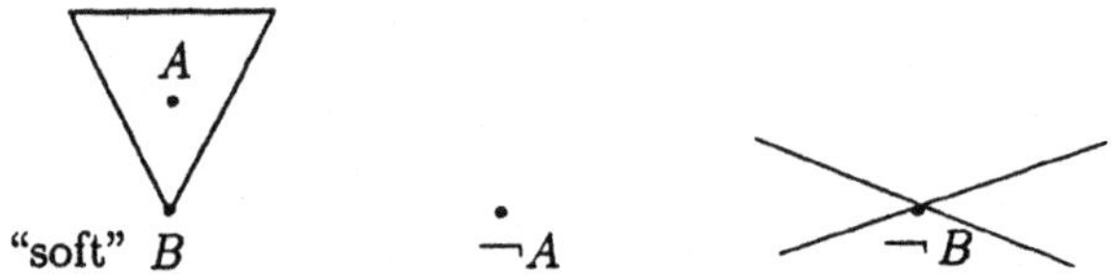

It turns out that all the contradictions deducible from the database are "safety", namely they are not related to some "suspicious" items. The later have got support from the premises that include some A, such that $\neg A$ has independent and safety support too. Thus, opposite to the majority of the approaches that care about the units that generate a contradiction, our system is aimed to control the inference after the appearance of the independent contradictions by deactivation of some of the rules. Instead of resolving inconsistency, the system is adapted to restrict the propagation of inconsistency over nested or dependent conflicts.
In the next section we define the *Literal Database Program* and we describe its translation into the definite program database for the purposes of better semantical explanation. The intuitive backgrounds are considered with the help of some simple examples. In section 3 we define a set of models $MOD(P)$ for a given LPD. The next after it considers some properties of the inference mechanism induced by our semantics from the standpoint of structural theory of non-monotonic reasoning. Section 5 contains concluding remarks.

2 Informal Representation. Some Examples

We begin our study with two motivating examples. By a *literal database program* we mean a finite set of universally closed *clauses* of the form $L_0 \longleftarrow L_1, \ldots, L_n$ where $n \geq 0$, and L_i's are literals. If P is a program then, unless stated otherwise, we will assume that P is instantiated. Conforming to a standard convention, by a ground instantiation of a program P we mean the (possibly infinite) theory consisting of all ground instances of clauses from P.

Example 2.1

$$\begin{aligned} M &\longleftarrow \\ \neg A &\longleftarrow \\ B &\longleftarrow \\ A &\longleftarrow B \\ \neg B &\longleftarrow M \end{aligned}$$

It is obvious, that the rules are in conflict and by ordinary deduction, A and $\neg B$ are inferred in one step from the facts B and M respectively. Notice that all heads of the clauses have consistent derivation trees. If we deduce A and $\neg A$ (they indeed have independent support) we are prevented to deduce at the same time B and $\neg B$, because in that case A would turned out to be supported by an inconsistent unit, namely B. In this case A and $\neg A$ are supposed to be true contradictions, independent from some other conflict literals. Following this intuition, we can admit in the model both A and $\neg A$, also B, M, as maximal possible choice under reasonable deduction. In that case we could call all the units of the model "true" or "sure", because there are not elements of the model inferred from A or $\neg A$. Thus we got the model $\mathcal{M}_1 = \langle\{A, \neg A, M, B\}, \{\emptyset\}\rangle$. $\mathcal{M}_1$ is counted as a model of P because it is a maximal set of literals that satisfy the claims referred in section 1 as intuitive backgrounds for the proposed semantics. In general, each model consists of two disjoint sets of true and suspicious knowledge. The literals from the later set depend on units which opposite literals belong also to the knowledge base. $\mathcal{M}_1$ does not contain any suspicious units, namely those inferred from A or from $\neg A$. Notice that the activation of the rule $\neg B \longleftarrow M$ would produce nested conflict units A, $\neg A$ and B, $\neg B$, because A depends from B. We could describe another two different models for the same program. If B and $\neg B$ are referred as primitive contradiction we get the models $\langle\{B, \neg B, M, \neg A\}, \{\emptyset\}\rangle$ and $\langle\{B, \neg B, M\}, \{A\}\rangle$. When the conflict literals B and $\neg B$ appeared in the knowledge base, they trigger the inference mechanism to control the deducibility of some other conflict literals. Considering all maximal sets of selections we obtain three models for the given program database:

$$\begin{aligned} \mathcal{M}_1 &= \langle\{A, \neg A, M, B\}, \{\emptyset\}\rangle \\ \mathcal{M}_2 &= \langle\{B, \neg B, M, \neg A\}, \{\emptyset\}\rangle \\ \mathcal{M}_3 &= \langle\{B, \neg B, M\}, \{A\}\rangle \end{aligned}$$

Notice that $\mathcal{M}_3$ contains the unit A as "suspicious" knowledge, because it comes from the "independent conflict knowledge" B. Both B and $\neg B$ are true in this situation.

Example 2.2 Let we consider an initial database with a few simple clauses and without nested conflict units (this example is partly considered in [8].

$r_1 : \forall x(fly(x) \longleftarrow bird(x))$
$r_2 : \forall x(\neg fly(x) \longleftarrow bird(x) \wedge big(x))$
$r_3 : bird(a)$
$r_4 : big(a)$

What will be the answer of the query $?bird(b)$. Classical and many non-classical systems (intuitionistic, modal and other systems, that include the axiom of absurdity) will deduce the goal. As we have mentioned in the introduction, systems based on relevant approach will recognize that the contradiction taken from the rules in-

volves another part of the database, namely the part dealing with a and will not infer the goal. A paraconsistent logic will allow the inference of both conflict units, as well as some systems will infer them as "suspicious" information units. A truth maintenance system will recognize the inconsistency and take out some of the date and arrive at a new consistent database. In the case of defeasible logic an preferable ordering between rules is used for the resolution of the contradiction.
If we add the following rule,

$r_5 : \neg\, big(a) \longleftarrow bird(a) \wedge fly(a)$

then not only the rules are in conflict, but the conflict units are interconnected. None of the existing till now approaches pay enough attention to the internal deductive structure of the information units. The problem is, that one has to distinguish the "independent" sources and their consequences, from the units that are related to the contradictions and their consequences. We consider this problem both conceptually and technically. First, in the case of the database $r_1, \ldots, r_4$, we allow for both $fly(a)$ and $\neg\, fly(a)$ to be derivable, but not the goal $?bird(b)$. We also need to know the possible conclusions from the contradictory knowledge. In this simple case we have not any consequences related to $fly(a)$ and $\neg\, fly(a)$ and we get only one extension, which contains equally trustful knowledge, namely the model $\mathcal{M} = \langle\{bird(a),\ big(a),\ fly(a),\ \neg\, fly(a)\},\ \{\varnothing\}\rangle$. If we take the database program $r_1, \ldots, r_5$, we get the models:

$\mathcal{M}_1 = \langle\{bird(a),\ big(a),\ fly(a),\ \neg\, fly(a)\},\ \{\varnothing\}\rangle$ and

$\mathcal{M}_2 = \langle\{bird(a),\ big(a),\ fly(a),\ \neg\, big(a)\},\ \{\varnothing\}\rangle$.

The later model contains the conflict units $big(a)$ and $\neg\, big(a)$ as they both have consistent support, but doesn't contain $\neg\, fly(a)$, because it's deduction depends on the true literal $big(a)$.

The informal semantics we describe have positive informational value about the knowledge base under consideration. It simply says, that some units are trustful and some other are supposed to be considered on the basis of the existing consensus. All points that are not mentioned into the model remain unknown.
As the example 2.1 shows, not any fact of a program P belongs to each extension. The situation will not seem obscure if we remember that there is not any ordering or preference relation between the rules or preference, referring the complexity of the derivations.
One possible modification of the intended models consists in taking only those models which are maximal according to the content of their *sure* knowledge. In this (example 2.1) case we have to admit only first two models as intended models for the program under consideration. But even this stipulation does not guarantee the presence of the facts in each model. In particular, it is of greater relevance to consider what those "models" are and what holds in them, than it is to determine what holds in all of them. The effect of multiple extensions will appear whenever the program contains nested or interconnected conflict units. If there is only one possible choice for selection of primitive contradictions, then obviously only one model will exist. The empty set is the only extension for the programs without any facts.

3 Semantics

Let P be a literal database program. We may look at P as a positive logic program by treating all the negated literals as new items and consider them independently from their semantic connection with the opposite ones. Thus we obtain both a copy of the initial program P and a well defined (*Van Emden and Kowalski*, see[11]) semantics for it. Notice, that in such a case some inference trees should contain conflict literals and many other unacceptable from semantic point of view faults. Let us focus

our attention to the space of *Van Emden and Kowalski* (triggered by any positive program). It simply contains trees. We give now some definitions, based on the structure of the least fixed point of the VEK (*Van Emden and Kowalski*) operator. Each positive program P triggers the couples (τ, A), where τ is an inference tree of A according to the ordinary VEK semantics. The set of all such trees we call VEK space. The notion of inconsistent set of literals is used in ordinary sense.
Given a program P, we consider the set of pairs $\langle \tau, A \rangle$, where τ is an inference tree of A in the VEK space. We consider a correct subset of this one according to the program P.

Definition 3.1 $S_P = \{\langle \tau, A \rangle \mid \tau \cup A$ is consistent set of units $\}$.
Now, S_P is that nice VEK space, we are going to work on it.

Definition 3.2 The unit A has a derivation in $VEK(P)$ iff there exists a set τ of literals such that $\langle \tau, A \rangle \in S_P$.

Definition 3.3 $Concl(P) = \{A \mid A$ has a derivation in $VEK(P)\}$

Definition 3.4 We define the set $MOD(P)$ of all models for a given LDP P.
Let $M \subseteq Concl(P)$ and M satisfies the following conditions:

1) If A and $\neg A$ belong to M, then there exist τ_1 and τ_2 such that $\langle \tau_1, A \rangle \in S_P$, $\langle \tau_2, \neg A \rangle \in S_P$ and for any $X \in \tau_1 \cup \tau_2$, $X' \notin M$ ($X' = X$ if X is negated literal and $X' = \neg X$ otherwise). According to this point, the set M contains only primitive contradictions, namely those which inference trees does not depend on some other conflict units.

2) if $A \in M$, then there exists a derivation τ of A in P, such that for any X, if $X \in \tau$, then $X \in M$.
This condition shows that elements of M have their justifications in M. Notice, that this assumption is weaker than the closure under deduction according to the correct trees in P. It is possible to happen that A has a correct tree in P, whose elements are from M, but A does not belong to M.

3) M is a maximal set in $Concl(P)$ according to the ordinary set theoretical inclusion $\subseteq$.
$MOD(P) = \{M \mid M$ satisfies the conditions from the def.3.4$\}$

Let M be a model of the program P. Then $M = \langle \{T\}, \{S\} \rangle$, where $T \cap S = \emptyset$, $T \cup S = M$ and $T = \{A \mid A \in M$ and A has a derivation τ in $VEK(P)$, such that if $X \in \tau$, then $\neg X \notin M\}$. In other words the set T consists of trustful units, whose justifications do not depend on any primitive contradictions. The set T contains also all the trustful contradictions of M. The units of S depend on knowledge, acceptable under consensus agreement (primitive contradictions). We could be sure that the derivation of the elements of S is based on some unit X for which $\neg X$ is true too.
Let us remind again that we insist upon the local inconsistency, i.e. any particular derivation tree forms a correct set of units. The proposed declarative semantics for a literal logic program P, seems to properly reflect the intended meaning of inconsistent logic programs without preference order between rules and data structure. The motivation behind this approach is based on the idea, that we should minimize all information as much as possible, limiting it to the facts explicitly implied by P and making everything else unknown. We have to keep the information about primitive contradictions (called consensus set) and hold our mind open about consequences getting from the chosen consensus set. The question is, that, if we have some nested contradictions, we should separate different sets of primitive conflict units and get different extensions. In particular it is of greater relevance to discover what these

"models" are and what holds in them, then it is to determine what holds in all of them.

Instead of defining explicitly the set of models for a given program, we can define the set of models of a given program P in a strict model-theoretical point of view by using the notion of *Extended Herbrand Model* (EHM). The idea is that such models could contain conflict literals if they have consistent premises, and moreover, those models consist of only these units, which are explicitly mentioned in the database. After taking out those literals which have not got any consistent support, we obtain the EHM. Let us consider a program P and look at it as a positive database (say rename its negative atoms). Let P has a unique minimal Herbrand model $\mathbf{M}$.

Definition 3.5 $\mathbf{M}_\varepsilon = \{X \mid X \in \mathbf{M}$ and there exists τ such that $(\tau, X) \in S_P\}$
$\mathbf{M}_\varepsilon$ is called the *extended Herbrand model* of the program P.
$MOD(P) = \{M \mid M$ is a maximal subset of $\mathbf{M}_\varepsilon$ and M satisfies (*) $\}$
(*) $M = \{A \mid A \in \mathbf{M}_\varepsilon$ and there exists $\tau \subseteq M$, such that $(\tau, A) \in S_P$ and if $A' \in M$ then $\{X \mid X' \in \tau\} \cap M = \emptyset\}$
(remember that X' is the opposite literal of X)

If $M \in MOD(P)$, then $M = \langle\{T\}, \{S\}\rangle$ where $T = \{X \mid X \in M$ and if X has a derivation τ in $VEK(P)$, then for any $X \in \tau$, $X' \notin M\}$,
$S = M \setminus T$.

It is important to point out the essence of the EHM by run of some examples.
Let T be a system involving two blocks A and B, and a table C. Assume that B is on C. We get the following set of premises:
$T = \{block(A),\ block(B),\ table(C),\ on(B, C)\}$
T has a unique minimal Herbrand model
$M = \{block(A),\ block(B),\ table(C),\ on(B, C),\ \neg table(B),\ \neg table(A),\ \neg on(A, C),\ \neg on(C, A),\ \neg on(B, A),\ \neg on(C, B),\ \neg on(A, B)\}$
According to our intended semantics and correspondingly to the definition of EHM, $\mathbf{M}_\varepsilon = \{block(A),\ block(B),\ table(C),\ on(B, C)\}$. In this case $\mathbf{M}_\varepsilon$ is consistent and T has the unique model that coincides with $\mathbf{M}_\varepsilon$.

If we consider an inconsistent database P, than its unique EHM would be inconsistent. Moreover, we can formally define the inconsistent databases as systems that have inconsistent EHM. Intuitively, an EHM contains those units that have *local* consistent justification set and are explicitly given in the database. Possibly they could not have *independent* derivations and that is why we have to select in appropriate way the set $MOD(P)$ of a given program P from its EHM. What has to be done, is to resolve the nested contradictory units by unbinding them and separate all maximal sets of choices. Thus for the program P from example 3.1 we obtain $\mathbf{M}_\varepsilon = \{B, \neg B, M, \neg A, A\}$ $\quad MOD(P) = \{\mathcal{M}_1, \mathcal{M}_2, \mathcal{M}_3\}$ and for the program $r_1, \ldots, r_5$ from the example 3.2 we get EHM
$\mathbf{M}_\varepsilon = \{bird(a),\ big(a),\ \neg big(a),\ fly(a),\ \neg fly(a)\}$ and $MOD(P) = \{\mathcal{M}_1, \mathcal{M}_2\}$.

As we pointed out at the end of section 2, we could get another intuitive acceptable notion of model and respectively alternative semantics $MOD_a(P)$, if we take the maximal interpretations according to the set T ($\langle T_1, S_1\rangle \prec \langle T_2, S_2\rangle$ iff $T_1 \subseteq T_2$). Then, given a program P, the alternative semantics consists of:
$MOD_a(P) = \{\langle T, S_T\rangle \mid T$ is a maximal subset of $\mathbf{M}_\varepsilon$ and $\langle T, S_T\rangle$ satisfies $(+)\}$
$(+)$ $\quad T = \{A \mid A \in \mathbf{M}_\varepsilon$ and there exists $\tau \subseteq M$, such that $(\tau, A) \in S_P$ and $\{X \mid X' \in \tau\} \cap T = \emptyset\}$
$\quad S_T = \{A \mid A \longleftarrow L_1, \ldots, L_m \in P,\ \{L_1, \ldots, L_m\} \subseteq T \cup S_T$ and at least for one L_i, (L_i and L_i' belong to T) or ($L_i \in S_T$)$\}$

It is easy to check, that for the program P from example 3.1 $MOD_a(P) = \{\mathcal{M}_1, \mathcal{M}_2\}$ and, for the program P from example 3.2 $MOD_a(P) = \{\mathcal{M}_1, \mathcal{M}_2\}$.

Next propositions refer to the semantics $MOD(P)$.

Proposition 3.1 *If $M = \langle\{T\}, \{S\}\rangle$ is a model of a given program P, then there exists a set $\Sigma = \{A_i, \neg A_i \mid i \in I\}$ such that*
$T = \{A_i, \neg A_i \mid i \in I\} \cup \{X \mid X \longleftarrow L_1, \ldots, L_n \in P$ *and* $L_1, \ldots, L_n \in T \setminus \Sigma$ *and* $X' \notin M\}$
$S = \{X \mid X \notin T$ *and* $X \in LAB(P)\}$ *where the set $LAB(P)$ — labeled literals from the program P — is inductively defined as follows:*

1) *if $X \longleftarrow L_1, \ldots, L_n \in P$ and $L_i \in T \cup \Sigma$ and $\sigma_x = \{\cup L_i\} \cap \Sigma \neq \emptyset$ and σ_x is consistent then $X \in LAB(P)$.*

2) *if $X \longleftarrow L_1, \ldots, L_n$ and $L_1, \ldots, L_n \in LAB(P)$ and $\sigma_x = \cup \sigma_{L_i} \notin \emptyset$ and σ_x is consistent then $X \in LAB(P)$.*

Proposition 3.2 *If $M_i \notin M_j$, then $\Sigma_i \notin \Sigma_j$.)*

Proposition 3.3 $\cup M_i = S_P$ where $M_i \in MOD(P)$

Proposition 3.4 *The set of formulae implied by our semantics is recursively enumerable.*

4 Consequence Presentation of Literal Database Program

The proposed semantics for literal database programs induce nonmonotonic entailment relations. We ask for the properties they satisfy in order to gain a better understanding of various problems with inconsistency in the logic programs. General framework for studying nonmonotonic reasoning systems through their consequence relation could be find in [9, 7]. In the later, Gabbay presents three minimal conditions which every non-monotonic consequence $\mid\sim$ should satisfy: *Reflexivity, Restricted Monotonicity* and *Cut*. This framework was further developed in the literature by several authors, thus creating the new area of axiomatic nonmonotonic reasoning. As we need to deal with inconsistent databases, we have to specify more precisely the definition for the consequence relation. A concept of inference operator according to some consensus set (a set of atoms and their negations), is to be defined in order to describe what could happen in each of the models for a given database program. Formally we need to make sense of the symbol $(P, \Sigma) \mid\sim A$, reading, A follows from the database program P on the base of consensus Σ. We can now define the conditions which we call Σ-*reflexivity*, Σ-*restricted monotonicity* and Σ-*cut*.

If $A \in P$ and $A \in \Sigma$ then $(P, \Sigma) \mid\sim A$. (Σ-*reflexivity*)
(in particular $(A, \emptyset) \mid\sim A$)

If $(P, \Sigma) \mid\sim A$ and $(P, \Sigma) \mid\sim B$ then $(P \cup A, \Sigma) \mid\sim B$. ($\Sigma$-*restricted monotonicity*)

If $(P_1, \Sigma) \mid\sim A$ and $(P \cup A, \Sigma) \mid\sim B$, then $(P \cup P_1, \Sigma) \mid\sim B$. ($\Sigma$-*cut*)

Definition 4.1 (Consequence presentation of LDL) Let $\mid\sim \subseteq (P \times \Sigma_P) \times L$ where $P \times \Sigma_P = \{(P, \Sigma) \mid P$ is a database program and there exists a model M for P such that Σ is the set of all conflict units of $M\}$ and L is the literal language.
We say that $(P, \Sigma) \mid\sim A$ iff $A \in M$, where M is that model of P whose contradictory set is Σ.

Clearly, the relation $\mid\sim$ is not defined for arbitrary couple (P, Σ). Also $\mid\sim$ is not monotonic in the sense that we could have $(P, \Sigma) \mid\sim A$, but not $(P \cup X, \Sigma) \mid\sim A$. For example, if we take $(A, \emptyset) \mid\sim A$, then the relation $\mid\sim$ is not even defined on $(A \cup \neg A, \emptyset)$.

Proposition 4.1 *The relation $\mid\sim$ satisfies the conditions Σ-reflexivity, Σ-restricted monotonicity and Σ-cut.*

Many interesting problems in the direction of structural investigation of the LDP consequence relation remained outside of this paper, but they constitute an important point of our current investigations.

5 Perspective and Concluding Remarks

I have proposed in this paper an extension of logic programming for inconsistent knowledge base. I have touched upon some of the semantic issues concerning these logic programs, arguing for a tolerative approach toward inconsistent program databases. My ongoing results are connected with investigation of fixed point semantics on some operators Ψ_Σ acting on ordered sets of interpretations. These operators depend not only on the interpretations, but on the content of the chosen consensus set Σ. Each of them are supposed to describe the inference situation in any separate extension.
The semantics of the clauses of a LDP differs from the standard explication of logical validity, which runs as follows: an argument is valid if its premises cannot be true without its conclusion being true as well. For a clause $L_0 \longleftarrow L_1, \ldots, L_m$, it is possible that $L_1, \ldots, L_m$ belong to a model M, but $L_0 \notin M$. In that case we could be sure that $\neg A \in M$ and A is based on some unit B, such that both B and $\neg B$ are in the model also.
It is obvious that the semantics of the clauses of a LDP modify ordinary default presentation by admitting activation of the clause rules of a given program when some weaker conditions (instead of closed world assumption) are fulfilled. These stipulations refer to the choice of the consensus set under which the calculation is implemented. Our default assumption consists in 1) infer the unit A if it has *local consistent* support in the knowledge base and if $\neg A$ is not in the knowledge base; or 2) one could infer A even if $\neg A$ is in the knowledge base, but one has to be sure that they both have correct justifications, independent from some other conflict units. We consider Net-clause language [12] as a promising tool for implementation of a method for answering queries addressed to an inconsistent databases, presented in the form of LDP. This language is capable to implement a data-driven inference, combined with a kind of default mechanism using a form of resolution and unification as a basic processing paradigm. Some running session of the prototype are under consideration.

References

1. A. Bauval and L. Cholvy: Automated reasoning in case of inconsistency. In: M. De Glas, D. Gabbay(eds) Proc. of WOCFAI'91, 1-5 july 1991, Paris, France, pp. 81-93
2. Ph. Besnard: Logic for automated reasoning in the presence of contradictions. In: North-Holland, Ph. Jorrand, V. Sgurev (eds.): Artificial Intelligence IV Methodology, Systems, Applications. Proc. of AIMSA'90, Albena, Bulgaria, 19-22 September, 1990, pp. 33-43
3. L. Cholvy: Querying an inconsistent database. In: North-Holland, Ph. Jorrand, V. Sgurev (eds.): Artificial Intelligence IV Methodology, Systems, Applications. Proc. of AIMSA'90, Albena, Bulgaria, 19-22 September, 1990, pp. 81-93
4. J. De Kleer: An assumption-based truth maintenance system. Artificial Intelligence 28(1), 127-162 (1986)
5. J. Doyle: A truth maintenance system. Artificial Intelligence, 12(3), 231-272(1979)
6. D. M. Gabbay: Labelled Deductive Systems, draft September 1989.

7. D. M. Gabbay: Theoretical foundations for nonmonotonic reasoning, Part 2: Structured nonmonotonic theories. In: IOS, B. Mayoh (ed.): Odin's ravens. Proc. of the SCAI'91, Roskilde, Denmark, 21–24 May 1991, pp. 19–40
8. D. M. Gabbay and A. Hunter: Making inconsistency respectable: a logical framework for inconsistency in reasoning. In: Proc. of the FAIR'91, September 8–12,19991, Smolenice, Czechoslovakia. Lecture Notes in Computer Science 535. Berlin: Springer 1991, pp. 19–32
9. S. Kraus, D. Lehman and M. Magidor: Non monotonic reasoning, preferential models and cumulative logics. Artificial Intelligence 44, 167–207 (1990)
10. F. Lin: Reasoning in the presence of inconsistency. In: Automated Reasoning, 1988.
11. J. Lloyd: Foundations of Logic Programming. Berlin: Springer-Verlag 1987.
12. Z. Markov, C. Dichev: Logical inference in a network environment. In: North-Holland, Ph. Jorrand, V. Sgurev (eds.): Artificial Intelligence IV Methodology, Systems, Applications. Proc. of AIMSA'90, Albena, Bulgaria, 19–22 September, 1990, pp. 169–179
13. J. Martins and S. Shapiro: A model for belief revision. Artificial Intelligence 35(1), 25–79 (1988)
14. D. Nute: Defeasible reasoning and decision support systems. Decision Support System 4, 97–110 (1988)
15. I. Popchev, N. Zlatareva, M. Mircheva: A truth maintenance theory: an alternative approach. In: Proc. of ECAI'90, Stockholm, Sweden, pp. 509–515
16. R. Reiter: A logic for default reasoning. Artificial Intelligence 13(1, 2), 81–132 (1980)

Tableau-Based Theorem Proving and Synthesis of λ-Terms in the Intuitionistic Logic

Oliver Bittel
Fachhochschule Konstanz
Postfach 10 05 43
D-7750 Konstanz
Germany
E-mail: bittel@fh-konstanz.de

Abstract

Because of its constructive aspect, the intuitionistic logic plays an important role in the context of the programming paradigm "programming by proving". Programs are expressed by λ-terms which can be seen as compact representations of natural deduction proofs. We are presenting a tableau calculus for the first-order intuitionistic logic which allows to synthesize λ-terms. The calculus is obtained from the tableau calculus for the classical logic by extending its rules by λ-terms. In each rule application and closing of tableau branches, λ-terms are synthesized by unification. Particularly, a new λ-term construct (implicit case analysis) is introduced for the the disjunction rules.

In contrary to existing approachs based on the natural deduction calculus, our calculus is very appropriate for automatic reasoning. We implemented the calculus in Prolog. A strategy which is similar to model elimination has been built in. Several formulas (including program synthesis problems) have been proven automatically.

Key words: Intuitionistic Logic, Automatic Theorem Proving, Program Synthesis, Typed λ-Calculus

1 Introduction

In many papers, the intuitionistic logic has been proposed as a logic for program synthesis (see e.g. [Mar82], [Con86], [BSH90]). A formula of the form $\forall x.\exists y.P(x,y)$ specifies a program, which takes an input x and yields an output y such that the condition $P(x,y)$ holds. A constructive (intuitionistic) proof of such a specification can be seen as a program satisfying this specification. E.g. the following formula specifies the integer square root function:

$$\forall n \in Nat.\, \exists r \in Nat.\, r^2 \leq n < (r+1)^2$$

A proof of this formula consists of a function f mapping a natural number n to a pair $\langle r, M\rangle$, whereby r is the integer square root of n and M is a proof for that.

The main characteristic of this program synthesis paradigm is that proofs are performed in the natural deduction calculus. Instead of taking a natural deduction proof as program, it is more appropriate to use a more compact represention, usually a λ-term. [1] There is a direct correspondence between proof rules and program constructs: case analysis in proofs and λ-terms, inductions and recursions, and finally lemmas and subroutines.

In order to solve realistic problems, it is important to automate at least the trivial parts of a proof. For that, we need an automatic theorem prover for the intuitionistic logic which is efficient and yields λ-terms if formulas are valid. Two approaches for automatic theorem proving in intuitionistic logic are mainly used:

1. Systems such as NUPRL [Con86] and OYSTER [BSH90] are based on the natural deduction calculus. They provide tactics and tacticals for conducting backward proofs, i.e. we start with the formula to be proven and apply rules in a backward manner until axioms are reached. In each proof step the desired λ-term is refined. As soon as the proof of a formula is found, we also get a λ-term. For several reasons, the natural deduction calculus is not quite appropriate for an efficient automatic theorem prover. The main problem with this calculus is the strong order dependence on the rules. E.g., in order to prove a disjunction $\alpha \vee \beta$ under a given set of assumptions Σ, one of the following rules can be applied in a backward manner:

   $$\text{(VI)} \quad \frac{\Sigma \vdash \alpha}{\Sigma \vdash \alpha \vee \beta} \qquad \frac{\Sigma \vdash \beta}{\Sigma \vdash \alpha \vee \beta}$$ [2]

   Both rule applications might lead to a deadlock, since it could be necessary to decompose a formula of Σ first. [3]

2. In the last years, several proof methods for modal and intuitionistic logics have been proposed,[4] among them matrix-based proof methods [Wal88] and resolution calculi [Ohl88]. Both approaches are based on Kripke-style semantics and refutation theorem proving. The main idea is that the formula to be proven is translated to a matrix [Wal88] or a set of clauses [Ohl88] such that it can be handled like in the classical logic. The Kripke-style semantics is taken into account by introducing so-called world paths in predicates and functions and to use a specific unification procedure for the world paths.

   These proof procedures work well if only the question of validity of a formula is interesting. However, in order to integrate an automatic theorem prover in the program synthesis approach discussed here it is mandatory to get a λ-term (or proof in the natural deduction calculus) in case a formula is proven valid. At this point it should be mentioned that in classical logic there are some methods which

[1] A λ-term can also be seen as a functional program.

[2] We prefer sequents for formalizing the natural deduction calculus.

[3] The problem of order-dependance can not be overcome even if the calculus is restricted to normal form derivations or the sequent calculus is taken. The same is true for the standard tableau calculus for the intuitionistic logic [Fitt83], where the order-dependance is due to the branch modification rule.

[4] There is a strong correspondence between the modal logic S4 and the intuitionistic logic [Fit83].

allow the transformation from matrix and resolution proofs to natural deduction proofs [And80], [Lin89] (see also [Wos90]). However, these translations generate non-constructive proofs in some cases. We suppose that it is difficult to get this methods work for the intutionistic logic.

In this paper we suggest a tableau-based calculus for the first-order intuitionistic logic which is efficient and yields a λ-term if the formula to be proven is valid. In our approach we start with the tableau calculus for the classical logic as base[5] and extend its rules by λ-terms. During the construction of a tableau proof, λ-terms are synthesized by unification in each rule application and closing of tableau branches. By some variable restriction on the λ-terms, the tableau proofs are forced to be constructive. Particularly, a new λ-term construct (implicit case analysis) is introduced in order to treat the disjunction rules.

After defining some preliminary notations we first present the tableau calculus in a ground version. We call this version test calculus, because the desired λ-term has to be guessed and than tested that it really represents a correct proof of the considered formula. In the next section, we lift the test calculus to a synthesis calculus by using meta-variables for λ-terms which are instantiated by unification. Soundness and completeness results will be given.[6]

2 First-Order Intuitionistic Logic

In this section, a first-order language, λ-terms and the calculus of natural deduction with λ-terms are defined. This calculus can be seen as a subset of Martin-Löf's intuitionistic type theory [Mar82]. The definition is very similar to the one found in [Coq90].

We assume a first-order language in the usual way. Let $\mathcal{V}$ be a set of variables. An *individual term* is either a variable or of the form $f(t_1, ..., t_n)$, where f is a function symbol with arity $n \geq 0$ and t_i are individual terms. The set of all individual terms are named by $\mathcal{T}$. *Fomulas* are defined in the usual way by using relation symbols, individual terms and the logical connectives $\wedge$ (and), $\vee$ (or), $\rightarrow$ (implies), $\perp$ (falsity), quantifier symbols $\forall$ and $\exists$. $\neg\alpha$ is defined as $\alpha \rightarrow \perp$. For a formula α, a variable x and a term t, $\alpha[x/t]$ denotes the result of substituting each free occurrence of x in α by t.[7] $FV(\alpha)$ is defined to be the set of all free variables in α.

The set Λ of all λ-*terms* is defined by the following inductive definition. Let $x \in \mathcal{V}$, $t \in \mathcal{T}$ and M, N, N' λ-terms. Then the following expressions are also λ-terms:

x	*variable*
t	*individual term*
MN	*application*
$\lambda x.M$	*abstraction*
$\langle M, N \rangle$	*pairing*
$\mathsf{fst}(M)$, $\mathsf{snd}(M)$	*projections*
$\mathsf{split}(M, N)$	*pair application*

[5] Therefore, we profit by the fact that no branch modification rules are necessary.
[6] Complete proofs can be found in [Bit91].
[7] We assume that renaming of bounded variables is done in order to avoid name clashes.

inl(M), **inr**(M)	*injections*
when(M, N, N')	*case analysis*
absurd(M)	

For a λ-term M, a variable x and a λ-term N, $M[x/N]$ denotes the result of substituting each free occurrence of x in M by N.[7] $FV(M)$ means the set of all free variables in M. There are also some conversion rules, e.g. [8]

$$(\lambda x.M)N \rhd M[x/N]$$
$$\mathbf{when}(\mathbf{inl}(N), \lambda x.M, O) \rhd M[x/N]$$
$$\mathbf{split}(\langle N_1, N_2\rangle, \lambda x_1.\lambda x_2.M) \rhd M[x_1/N_1][x_2/N_2]$$

A *context* is a finite set $\{x_1 : \alpha_1, ..., x_n : \alpha_n\}$, where $\alpha_1, ..., \alpha_n$ are formulas and $x_1, ..., x_n$ are pairwise distinct variables which must not occur in $\alpha_1,..., \alpha_n$. An expressions like $M : \alpha$ is also called a *type relation.*[9] For any context Σ, the set of all free variables occuring in Σ is denoted by $FV(\Sigma)$.

For any context Σ, λ-term M and formula α, we define the relation $\Sigma \vdash M : \alpha$ by the rules shown in table 1. If $\Sigma \vdash M : \alpha$ holds we also say that α is *(intuitionistically) valid under the context* Σ *and the* λ*-term* M *is a proof for that.* If $\emptyset \vdash M : \alpha$, then α is *(intuitionistically) valid.*

Remark: If the λ-terms in the rules are omitted we exactly get the usual natural deduction calculus for intuitionistic logic. However, the calculus with λ-terms offers the possibilty not only to check the validity of a formula α but also to construct a λ-term which can be seen as the desired program satisfying α.

3 The Test Calculus

In this section a calculus is presented which tests whether $\Sigma \vdash M : \alpha$ holds. The test calculus is obtained from the tableau calculus for the classical logic by extending its rules by λ-terms. For that, we introduce projections for existentially bounded formulas and a new λ-term construct which we call implicit case analysis. With the help of a small transformation procedure, projections and implicit case analysis can be replaced by pair-application (**split**-construct) and explicit case analysis (**when**-construct), respectively.

3.1 Informal Description of the Calculus

The calculus works on tableaux which are trees labeled with signed type relations of the form $+M : \alpha$ and $-M : \alpha$. The intuitive meaning is the following:

occurs in a tableau branch π	meaning
$+M : \alpha$	$\Sigma_\pi \vdash M : \alpha$ [11]
$-M : \alpha$	$\Sigma_\pi \not\vdash M : \alpha$

[8] Particularly, we have omitted the so-called commuting conversions. For a complete definition see e.g. [Gir89], [Bit91].

[9] I.e. M is of type α. It is well-known from literature, that there is an isomorphism between types and formulas (see e.g. [How80]).

[10] We write $\Sigma, x : \alpha$ instead of $\Sigma \cup \{x : \alpha\}$.

(Axiom) $\Sigma, x : \alpha \vdash x : \alpha$ [10]

$$(\wedge I)\ \frac{\Sigma \vdash M : \alpha \qquad \Sigma \vdash N : \beta}{\Sigma \vdash \langle M, N\rangle : \alpha \wedge \beta} \qquad (\wedge E)\ \frac{\Sigma \vdash M : \alpha \wedge \beta}{\Sigma \vdash \mathbf{fst}(M) : \alpha} \quad \frac{\Sigma \vdash M : \alpha \wedge \beta}{\Sigma \vdash \mathbf{snd}(M) : \beta}$$

$$(\vee I)\ \frac{\Sigma \vdash M : \alpha}{\Sigma \vdash \mathbf{inl}(M) : \alpha \vee \beta} \quad \frac{\Sigma \vdash M : \beta}{\Sigma \vdash \mathbf{inr}(M) : \alpha \vee \beta}$$

$$(\vee E)\ \frac{\Sigma \vdash M : \alpha \vee \beta \qquad \Sigma, x_1 : \alpha \vdash N_1 : \gamma \qquad \Sigma, x_2 : \beta \vdash N_2 : \gamma}{\Sigma \vdash \mathbf{when}(M, \lambda x_1.N_1, \lambda x_2.N_2) : \gamma}$$

$$(\rightarrow I)\ \frac{\Sigma, x : \alpha \vdash M : \beta}{\Sigma \vdash \lambda x.M : \alpha \rightarrow \beta} \qquad (\rightarrow E)\ \frac{\Sigma \vdash M : \alpha \rightarrow \beta \qquad \Sigma \vdash N : \alpha}{\Sigma \vdash MN : \beta}$$

$$(\bot E)\ \frac{\Sigma \vdash M : \bot}{\Sigma \vdash \mathbf{absurd}(M) : \alpha}$$

$$(\forall I)\ \frac{\Sigma \vdash M : \alpha}{\Sigma \vdash \lambda x.M : \forall x.\alpha}\ x \notin FV(\Sigma) \qquad (\forall E)\ \frac{\Sigma \vdash M : \forall x.\alpha}{\Sigma \vdash Mt : \alpha[x/t]}$$

$$(\exists I)\ \frac{\Sigma \vdash M : \alpha[x/t]}{\Sigma \vdash \langle t, M\rangle : \exists x.\alpha}$$

$$(\exists E)\ \frac{\Sigma \vdash M : \exists x.\alpha \qquad \Sigma, y : \alpha \vdash N : \beta}{\Sigma \vdash \mathbf{split}(M, \lambda x.\lambda y.N) : \beta}\ x \notin FV(\Sigma \cup \{\beta\})$$

Table 1: Natural deduction calculus with λ-terms

In order to check $\{x_1 : \gamma_1, ..., x_n : \gamma_n\} \vdash M : \alpha$, we start with the following intitial tableau which consists of one branch with $n+1$ nodes:

$$\begin{array}{c} +x_1 : \gamma_1 \\ \vdots \\ +x_n : \gamma_n \\ -M : \alpha \end{array}$$

To this initial tableau several tableau rules (shown in table 2) are applied until a closed tableau is reached, i.e. a tableau where each branch contains a complementary pair $+N : \gamma$ and $-N : \gamma$.

Most of the tableau rules are straightforward translations of the natural deduction rules with λ-terms. We want to motivate some of them. If $\lambda x.M$ is not a proof for $\alpha \rightarrow \beta$ (i.e. $-\lambda x.M : \alpha \rightarrow \beta$) then by the $(- \rightarrow)$-rule, M can not be a proof for β (i.e. $-M : \beta$) under the assumption that x is a proof for α (i.e. $+x : \alpha$). If we know M to be a proof for $\alpha \rightarrow \beta$ (i.e. $+M : \alpha \rightarrow \beta$) we may introduce a branching using the $(+ \rightarrow)$-rule. For an arbitrary λ-term N, we may assume that either N is not a proof

[11] Σ_π is the context that corresponds to the branch π, i.e. $\Sigma_\pi = \{x : \alpha \mid +x : \alpha \in \pi \text{ and } x \in \mathcal{V}\}$.

for α (i.e. $-N : \alpha$; left branch) or N is a proof for α. In the latter case, we infer that MN is a proof for β (i.e. $+MN : \beta$; right branch).

The most interesting rules are the disjunction rules and the rules for the existential quantifier.[12]

If we have a proof M for $\alpha \vee \beta$, then M represents either a proof for α or one for β. But we have no linguistic means (λ-term constructs) to express this. For that, we introduce two new constructors **l** and **r**. $\mathbf{l}(M)$ is only defined if M represents a proof for the left part of a disjunction. Thus, $\mathbf{l}(\mathbf{inl}(N))$ is defined whereas $\mathbf{l}(\mathbf{inr}(N))$ is not. This property is expressed by the following conversion rules:

$$\mathbf{l}(\mathbf{inl}(N)) \rhd N$$
$$\mathbf{l}(\mathbf{inr}(N)) \rhd \uparrow \text{ (undefined)}$$

$\mathbf{r}(M)$ can be considered analogously. Thus, $\mathbf{l}(M)$ and $\mathbf{r}(M)$ are partially defined λ-terms. In some sense, **l** and **r** is inverse to **inl** and **inr**, respectively. With the new constructors, the formulation of $(+\vee)$-rule becomes straightforward.

Next, we introduce the implicit case analysis, which is a list of partially or totally defined λ-terms:

$$[\![M_1, M_2, ..., M_n]\!]$$ [13]

Each M_i represents one case. At least one of the cases must be defined.

$$[\![M_1, M_2, ..., M_n]\!] \rhd M_i, \text{ if not } M_i \rhd \uparrow$$

The implicit case analysis is very similar to a guarded command, where the conditions are left implicit. The usual (explicit) case analysis

$$\mathbf{when}(M, \lambda p_1.N_1, \lambda p_2.N_2)$$

can be expressed by an implicit case analysis

$$[\![N_1[p_1/\mathbf{l}(M)], N_2[p_2/\mathbf{r}(M)]]\!]$$

The $(-\vee)$-rule can now be read in the following way: if $[\![\mathbf{inl}(M), \mathbf{inr}(N)]\!]$ is not a proof for $\alpha \vee \beta$, then both M is not a proof for α and N is not a proof for β.

Since a proof for an arbitrary formula may consist of a case analysis the rule $(-[\![\,]\!])$ is also needed. As we will see in the following section λ-terms with implicit case analysis can easily be transformed to those with the usual explicit case analysis.

At last, the $(+\exists)$-rule shall be considered. If M is a proof for $\exists x.\alpha$, then M represents a pair, consisting of a term t and a proof for $\alpha[x/t]$. We use the projection functions $\mathbf{fst}_\exists$ and $\mathbf{snd}_\exists$ to access to the pair components. Thus, the formulation of the $(+\exists)$-rule becomes straightforward. The index $\exists$ in the projection functions is used to distinguish them from the projections used for conjunctions. Now, λ-terms of the form $\mathbf{fst}_\exists(M)$ which denote in some way constants are also allowed to occur in formulas. Those terms together with individual terms will be called extended individual terms. t in the rules $(-\exists)$ and $(+\forall)$ is restricted to be an extended individual term.

[12] Girard writes in his book [Gir89]: "Yet disjunction and existence are the two most typically intuitionistic connectors!"

[13] $[\![M_1, M_2, ..., M_n]\!]$ is an abbreviation for $cons(M_1, cons(M_2, ...cons(M_n, nil)...))$.

[14] x must not occur free in the branch where the rule is applied.

$$(-\wedge)\ \frac{-\langle M,N\rangle : \alpha \wedge \beta}{-M : \alpha \mid -N : \beta} \qquad (+\wedge)\ \frac{+M : \alpha \wedge \beta}{\begin{array}{c}+\mathbf{fst}(M) : \alpha \\ +\mathbf{snd}(M) : \beta\end{array}}$$

$$(-\vee)\ \frac{-[\![\mathbf{inl}(M), \mathbf{inr}(N)]\!] : \alpha \vee \beta}{\begin{array}{c}-M : \alpha \\ -N : \beta\end{array}} \qquad (+\vee)\ \frac{+M : \alpha \vee \beta}{+\mathbf{l}(M) : \alpha \mid +\mathbf{r}(M) : \beta}$$

$$(-\rightarrow)\ \frac{-\lambda x.M : \alpha \rightarrow \beta}{\begin{array}{c}+x : \alpha \\ -M : \beta\end{array}}\ ^{14} \qquad (+\rightarrow)\ \frac{+M : \alpha \rightarrow \beta}{-N : \alpha \mid +MN : \beta}$$

$$(-\forall)\ \frac{-\lambda x.M : \forall x.\alpha}{-M : \alpha}\ ^{14} \qquad (+\forall)\ \frac{+M : \forall x.\alpha}{+Mt : \alpha[x/t]}$$

$$(-\exists)\ \frac{-\langle t, M\rangle : \exists x.\alpha}{-M : \alpha[x/t]} \qquad (+\exists)\ \frac{+M : \exists x.\alpha}{+\mathbf{snd}_\exists(M) : \alpha[x/\mathbf{fst}_\exists(\mathrm{M})]}$$

$$(-[\![\,]\!])\ \frac{-[\![M_1, M_2, ..., M_n]\!] : \alpha}{\begin{array}{c}-M_1 : \alpha \\ -M_2 : \alpha \\ \vdots \\ -M_n : \alpha\end{array}}$$

Table 2: Tableau calculus with λ-terms

3.2 Formal Definition of the Calculus

Definition 1 (Extended λ-terms) *The set Λ_E of all extended λ-terms is defined inductively as follows. Let f be a function symbol with arity $n \geq 0$, $x \in \mathcal{V}$ and $M, N, M_1, ..., M_n$ extended λ-terms. Then the following expressions are also extended λ-terms:*

x	*variable*
$f(M_1, ..., M_n)$	*individual term*
MN	*application*
$\lambda x.M$	*abstraction*
$\langle M, N\rangle$	*pairing*
$\mathbf{fst}(M)$, $\mathbf{snd}(M)$	*projections for terms of conjunctive type*
$\mathbf{fst}_\exists(M)$, $\mathbf{snd}_\exists(M)$	*projections for terms of existential type*
$\mathbf{inl}(M)$, $\mathbf{inr}(M)$	*injections*
$\mathbf{l}(M)$, $\mathbf{r}(M)$	*inverses to the injections*
$[\![M_1, ..., M_n]\!], n \geq 0$	*implicit case analysis*
$\mathbf{absurd}(M)$	

Definition 2 (Extended individual terms) *The set $\mathcal{T}_E$ of all extended individual terms is defined inductively as follows. Let $x \in \mathcal{V}$, $M \in \Lambda_E$, f a function symbol with*

arity $n \geq 0$ *and* $t_1, ..., t_n$ *extended individual terms. Then* $x, f(t_1, ..., t_n), \mathbf{fst}_\exists(M)$ *are extended individual terms, too.*

Definition 3 (Extended formula) *The set of all extended formulas is defined inductively. Each formula is also an extended formula. If* α *is an extended formula and* $t \in \mathcal{T}_E$*, then* $\alpha[x/t]$ *is an extended formula.*

Definition 4 (Tableau) *A tableau is a binary tree which is labeled with signed type relations of the form* $+M : \alpha$ *and* $-M : \alpha$ *where* $M \in \Lambda_E$ *and* α *is an extended formula. Let* Γ *be a set of signed type relations. An initial tableau for* Γ *consists of exactly one branch* π *such that* Γ *is the set of all labels in* π*. If no confusion is possible, we identify* π *with some initial tableau for* π*. A branch* π *is called closed if both* $+M : \alpha$ *and* $-M : \alpha$ *or if both* $+M : \bot$ *and* $-\mathbf{absurd}(M) : \alpha$ *(complementary pairs) occur in* π *for some* $M \in \Lambda_E$ *and some extended formula* α*. A tableau is closed if all its branches are closed.*

Definition 5 (Tableau extension) *Let* T, T' *be tableaux and* r *one of the tableau rules. We define the relation* $T \xrightarrow{r} T'$ *(*T *is extended to* T' *by rule* r*). In general,* r *is either a branching or a non-branching rule of the following form:*

$$(1) \quad \frac{A}{A_1 \mid A_2} \qquad\qquad (2) \quad \frac{A}{\begin{array}{c} A_1 \\ \vdots \\ A_n \end{array}} \quad n \geq 1$$

$T \xrightarrow{r} T'$ *holds if there exists some branch* π *in* T *such that* A *(the upper part of* r*) occurs in* π *and* T' *is obtained from* T *by extending* π *in the following way:*

- *If* r *is a branching rule of the form (1), then* π *is extended by a branching with one left and right node labeled with* A_1 *and* A_2*, respectively.*
- *If* r *is a non-branching rule of the form (2), then* π *is extended by* n *nodes which are labeled with* $A_1, ..., A_n$*.*

The proviso of the rules $(- \rightarrow)$ *and* $(-\forall)$ *must be taken into account. We write* $T \xrightarrow{TC} T'$ *if there exists a rule* r *such that* $T \xrightarrow{r} T'$*.* $\xrightarrow{TC*}$ *is defined to be the reflexive and transitive closure of* $\xrightarrow{TC}$*.*

Example: Let $\Sigma = \{y_0 : a \wedge b \rightarrow g, y_1 : c \rightarrow g, y_2 : b \vee c\}$ be a context. The figure 1 shows a closed tableau T where

$$+\Sigma \cup \{-\lambda x.[\![y_1 \mathrm{r}(y_2), y_0\langle x, \mathrm{l}(y_2)\rangle]\!] : a \rightarrow g\} \xrightarrow{TC*} T$$ [15]

The nodes (5) and (6) result from node (4) by a rule application. Since (4) is a negatively signed implication, that extension was done by rule $(- \rightarrow)$. Note, that the pairs of nodes (9), (10) and (15), (16) result from the application of the $(+ \rightarrow)$ rule. For that, the λ-terms in node (9) and (15) were guessed. ∎

[15] $+\Sigma = \{+M : \alpha \mid M : \alpha \in \Sigma\}$.

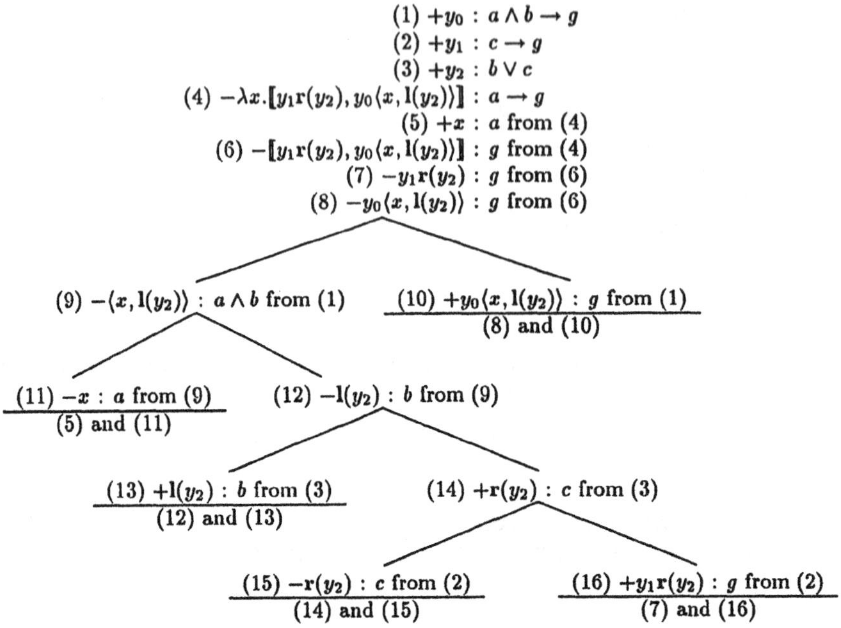

Figure 1: Closed tableau

Comparison between tableau calculus and natural deduction calculus: Suppose we want to find a proof in the natural deduction calculus for $\Sigma \vdash a \rightarrow g$.[16] For that, it is appropriate to start with the goal sequent $\Sigma \vdash a \rightarrow g$ and apply rules in a backward manner. Using the ($\rightarrow$I) rule we obtain the subgoal $\Sigma, a \vdash g$. Now, there are several possibilities to continue. If we decide to decompose $a \wedge b \rightarrow g \in \Sigma$ with the ($\rightarrow$E)-rule, we get the two subgoals

$$\Sigma, a \vdash a \wedge b \rightarrow g \text{ and } \Sigma, a \vdash a \wedge b.$$

The first sequent is trivially true but the second sequent can not be derived. Thus, we went into a deadlock. The right way would be first to carry out a case analysis using the assumption $b \vee c$ and the ($\vee$E)-rule. It is clear that this property of the natural deduction calculus becomes worse if Σ increases in size or nested case analysis are required.

On the contrary, the tableau calculus is deadlock-free. It does not matter in which order rules are applied. As we can see in the example, first $+a \wedge b \rightarrow g$ was decomposed before we introduce a case analysis by decomposing $+b \vee c$. The tableau would also be closed if the rule order would be changed. The order of the rule applications in the example comes up if a strategy is used which requires branches to be closed as soon as possible.[17] E.g. the decomposition of $+a \wedge b \rightarrow g$ leads to two closed branches (see node (11) and (10)). ■

[16] We choose Σ as in the example above. For simplicity, λ-terms are omitted.

[17] Actually, this strategy is taken form [Sch85] and is similar to model elimination.

In order to formally define validity wrt. tableaux global constraints on tableaux are necessary. As the example above shows, the tableau would remain closed if the implicit case analysis in node (4) and (6) would be extented by a further arbitrary λ-term N.[18] Since we want to transform λ-terms of Λ_E to usual λ-terms of Λ, such an additional arbitrary λ-term N would have a disturbing effect. Thus, we require each negatively signed type relation to be participated either in a rule application or in a complementary pair. Another problem comes from the fact that in the $(+\forall)$-rule t may contain a λ-term $\mathbf{fst}_\exists(N)$, where N can be a completely senseless term. However, if a node labeled with $+N : \exists x.\alpha$ appears in the tableau, the term $\mathbf{fst}_\exists(N)$ would make sense. We comprise these global constraints in the following minimality condition.

Definition 6 (Minimal tableaux) *A tableau T is called minimal iff two conditions hold:*

(i) For each $-M : \alpha$ occuring in T one of the following situations are true.

- *$M = [\![\,]\!]$ (empty case analysis), or*
- *a tableau rule is applied to $-M : \alpha$, or*
- *there exists some $+M : \alpha$, such that both $-M : \alpha$ and $+M : \alpha$ occur in the same branch (complementary pair), or*
- *there exists some $+N : \bot$, such that $M = \mathbf{absurd}(N)$ and both $-M : \alpha$ and $+N : \bot$ occur in the same branch (complementary pair).*

(ii) For each $\mathbf{fst}_\exists(N)$ occuring in T there exists some formula $\exists x.\alpha$ and some node in T which is labeled with $+N : \exists x.\alpha$.

Definition 7 (TC-Validity) *Let Σ be any context, $M \in \Lambda_E$ and α any formula. We define $\Sigma \vdash_{TC} M : \alpha$ if there exists some closed and minimal tableau T such that $+\Sigma \cup \{-M : \alpha\} \xrightarrow{TC*} T$.*

3.3 Soundness

As we have seen in the previous section extended λ-terms are an appropriate means for representing tableau proofs. Extended λ-terms can also be seen as programs which can be evaluated by some special conversion rules. However for efficiency, λ-terms with explicit case analysis are more suitable for programs. Therefore, we give a procedure which transfoms extended λ-terms to usual λ-terms with explicit case analysis (instead of implicit) and pair applications (instead of projections). This transformation is also used to establish a soundness result.

Transformation

The idea is simple: the λ-term is visited from outermost to innermost. As far as λ-terms $\mathrm{l}(N)$ and $\mathrm{r}(N)$ are detected, where the transformation of $N \in \Lambda_E$ leads to a λ-term $N' \in \Lambda$, we replace $\mathrm{l}(N)$ and $\mathrm{r}(N)$ by introducing a **when**-construct with N' as first argument. Similarly, projections are handled.

[18] An extension by N would lead to $[\![y_1\mathrm{r}(y_2), y_0\langle x, \mathrm{l}(y_2)\rangle, N]\!]$.

Definition 8 (*trans*)

$$\begin{aligned}
trans(M) &= \mathrm{when}(trans(N), \lambda x_1.trans(M[\mathrm{l}(N)/x_1]), \lambda x_2.trans(M[\mathrm{r}(N)/x_2])) \\
&\qquad if\ S_{lr}(M) \neq \emptyset,\ N = min(S_{lr}(M)) = min(S(M))^{19} \\
trans(M) &= \mathrm{split}(trans(N), \lambda x_1.\lambda x_2.trans(M[\mathrm{fst}_\exists(N)/x_1][\mathrm{snd}_\exists(N)/x_2])) \\
&\qquad if\ S_{fstsnd}(M) \neq \emptyset,\ N = min(S_{fstsnd}(M)) = min(S(M))^{19} \\
trans(M) &= trans_1(M),\ if\ S(M) = \emptyset.
\end{aligned}$$

$$\begin{aligned}
S_{lr}(M) &= \{P \mid \mathrm{l}(P) \in F(M), \mathrm{r}(P) \in F(M)\ and\ trans(P) \in \Lambda\}^{20} \\
S_{fstsnd}(M) &= \{P \mid \mathrm{fst}_\exists(P) \in F(M)\ and\ trans(P) \in \Lambda\} \cup \\
&\quad \{P \mid \mathrm{snd}_\exists(P) \in F(M)\ and\ trans(P) \in \Lambda\} \\
S(M) &= S_{lr}(M) \cup S_{fstsnd}(M)
\end{aligned}$$

$$\begin{aligned}
& trans_1(\lambda x.M) = \lambda x.trans(M) \\
& trans_1(MN) = trans(M)trans(N) \\
& trans_1(\langle M, N\rangle) = \langle trans(M), trans(N)\rangle \\
& trans_1(\mathrm{inl}(M)) = \mathrm{inl}(trans(M)) \\
& trans_1(\mathrm{inr}(M)) = \mathrm{inr}(trans(M)) \\
& trans_1(\mathbf{absurd}(M)) = \mathbf{absurd}(trans(M)) \\
& trans_1([\![M_1, M_2, ..., M_n]\!]) = min\{trans(M_i) \mid 1 \leq i \leq n\ and\ trans(M_i) \in \Lambda\} \\
& trans_1(f(M_1, M_2, ..., M_n)) = f(trans(M_1), trans(M_2), ..., trans(M_n)), \\
& trans_1(\mathrm{fst}(M)) = \mathrm{fst}(trans(M)) \\
& trans_1(\mathrm{snd}(M)) = \mathrm{snd}(trans(M)) \\
& trans_1(\mathrm{fst}_\exists(M)) = c,\ where\ c\ is\ any\ constant^{21} \\
& trans_1(x) = x
\end{aligned}$$

The function *trans* is partial. E.g. $trans(\lambda x.\mathrm{l}(x))$ is not defined.

Theorem 9 (Soundness) *If* $\Sigma \vdash_{TC} M : \alpha$ *then* $\Sigma \vdash trans(M) : \alpha$.

Proof: An inductive proof on closed and minimal tableaux seems to be difficult because of the global constraint of minimality. So, we decided to divide the proof in four parts.

Normal form lemma: It follows directly from the tableau rules that λ-terms occuring in closed and minimal tableaux are of a certain normal form, which is called *S-normal form.* Roughly speaking, S-normal forms are built up from *selector terms* using abstractions, pairings, injections, and implicit case analysis. *Selector terms* are built up from variables, applications, projections and injection inverses.

Weak soundness: We introduce an intermediate calculus which allows the derivation of sequents $\Sigma \vdash_I M : \alpha$, where $M \in \Lambda_E$. This calculus is very similar to the natural deduction calculus. In contrary to the tableau calculus, this calculus is weaker since it admits to give types to partial defined terms. E.g. $\emptyset \vdash_I \lambda x.\mathrm{l}(x) : a \vee b \rightarrow a$, whereas $\emptyset \not\vdash_{TC} \lambda x.\mathrm{l}(x) : a \vee b \rightarrow a$. It can be shown inductively on M, that if $zM : \alpha$, $z \in \{+, -\}$, occurs at node k in a minimal tableau (minimality is essential), then we get $\Sigma_k \vdash_I M : \alpha$. Hence, from $\Sigma \vdash_{TC} M : \alpha$ it follows $\Sigma \vdash_I M : \alpha$.

[19] x_1, x_2 are new. *min* is defined wrt. any given term ordering.

[20] $F(M) = \{N \mid N$ occurrs in M and no free variable in N is bound in $M\}$. E.g., $F(\lambda x.x(yz)) = \{y, z, yz, \lambda x.x(yz)\}$.

[21] This case is helpful in the proof of soundness.

Definedness of trans: If there exists a closed tableau for $+\Sigma \cup \{-M_1 : \alpha_1, ..., -M_n : \alpha_n\}$, then there exists at least one M_i such that $trans(M_i)$ is defined. Note, that no minimality condition is assumed. This makes the proof easier which is conducted inductively on the number of constructors and selectors in $M_1, ..., M_n$.

Finally, we can prove inductively on S-normal forms M, that if $\Sigma \vdash_I M : \alpha$ and $trans(M)$ is defined, it follows $\Sigma \vdash trans(M) : \alpha$. ■

3.4 Completeness

The main key to prove completeness is the normalization theorem (see e.g. [Gir89]), which says that each typed λ-term $M \in \Lambda$ (i.e. $\Sigma \vdash M : \alpha$, for some Σ and α), can be brought into a normal form M' such that $M \rhd^* M'$ and $\Sigma \vdash M' : \alpha$. Then, by using a function $trans' : \Lambda \longrightarrow \Lambda_E$, we can show that from $\Sigma \vdash M : \alpha$ it follows $\Sigma \vdash_{TC} trans'(M) : \alpha$ for each normal form M.

Definition 10 ($trans'$)

$$
\begin{aligned}
&trans'(\lambda x.M) = \lambda x.trans'(M)\\
&trans'(MN) = trans'(M)trans'(N)\\
&trans'(\langle M, N\rangle) = \langle trans'(M), trans'(N)\rangle\\
&trans'(\mathrm{inl}(M)) = [\![\mathrm{inl}(trans'(M)), \mathrm{inr}([\![\,]\!])]\!]\\
&trans'(\mathrm{inr}(M)) = [\![\mathrm{inl}([\![\,]\!]), \mathrm{inr}(trans'(M))]\!]\\
&trans'(\mathrm{fst}(M)) = \mathrm{fst}(trans'(M))\\
&trans'(\mathrm{snd}(M)) = \mathrm{snd}(trans'(M))\\
&trans'(\mathrm{absurd}(M)) = \mathrm{absurd}(trans'(M))\\
&trans'(\mathrm{split}(M, \lambda x_1.\lambda x_2.N))\\
&\qquad = trans'(N)[x_1/\mathrm{fst}_\exists(trans'(M))][x_2/\mathrm{snd}_\exists(trans'(M))]\\
&trans'(\mathrm{when}(M, \lambda x_1.N_1, \lambda x_2.N_2)) = [\![\ trans'(N_1)[x_1/\mathrm{l}(trans'(M))],\\
&\qquad\qquad trans'(N_2)[x_2/\mathrm{r}(trans'(M))]\ ]\!]\\
&trans'(f(M_1, M_2, ..., M_n)) = f(trans'(M_1), trans'(M_2), ..., trans'(M_n))\\
&trans'(x) = x
\end{aligned}
$$

Lemma 11 *If $\Sigma \vdash M : \alpha$ then $\Sigma \vdash_{TC} trans'(M) : \alpha$, for any $M \in \Lambda$ in normal form (wrt. $\rhd$).*

Proof: By induction on M. ■

From this, the completeness result is obtained by the normalization theorem.

Theorem 12 (Completeness) *If $\Sigma \vdash M : \alpha$ then there exists some $M' \in \Lambda_E$ such that $\Sigma \vdash_{TC} M' : \alpha$.*

4 The Synthesis Calculus

In contrast to the test calculus, the synthesis calculus avoids guessing λ-terms by using metavariables, which may be instantiated by λ-terms during tableau extensions. The instantiations result from applications of tableau rules and closing of branches. This technique is also known as lifting technique. Actually, the synthesis calculus is not a direct lifted version of the test calculus, because the ($-[\![\,]\!]$)-rule has been made implicit for some reasons which shall be discussed later.

4.1 Informal Description of the Calculus

The rules e.g. for conjunction are as follows

$$(+\wedge)\quad \frac{+M : \alpha \wedge \beta}{\begin{array}{c}+\mathbf{fst}(M) : \alpha \\ +\mathbf{snd}(M) : \beta\end{array}} \qquad\qquad (-\wedge)\quad \frac{-\langle X, Y\rangle : \alpha \wedge \beta}{-X : \alpha \mid -Y : \beta}$$

where X and Y are metavariables. The rules for positively signed formulas work as before whereas applications of rules for negatively signed formulas lead to instantiations. In order to synthesize a λ-term M such that $x : a \wedge b \vdash_{TC} M : b \wedge a$ we start with the initial tableau T_1 (shown in figure 2), where X_1 is a metavariable. By application of $(+\wedge)$ to (1) we get tableau T_2. Applying $(-\wedge)$ to (2), the tableau is extended to T_3 and X_1 is instantiated by $\langle X_2, X_3\rangle$, where X_2, X_3 are new metavariables. Both branches can be closed by the substitutions $[X_2/\mathbf{snd}(x)]$ and $[X_3/\mathbf{fst}(x)]$ (see T_4). They can be found out by unification. The desired λ-term $\langle\mathbf{snd}(x), \mathbf{fst}(x)\rangle$ can now be picked up at node (2).

$$\begin{array}{c}(1)\ +x : a \wedge b \\ (2)\ -X_1 : b \wedge a \\ \\ T_1\end{array}
\qquad
\begin{array}{c}(1)\ +x : a \wedge b \\ (2)\ -X_1 : b \wedge a \\ (3)\ +\mathbf{fst}(x) : a \\ (4)\ +\mathbf{snd}(x) : b \\ \\ T_2\end{array}
\qquad
\begin{array}{c}(1)\ +x : a \wedge b \\ (2)\ -\langle X_2, X_3\rangle : b \wedge a \\ (3)\ +\mathbf{fst}(x) : a \\ (4)\ +\mathbf{snd}(x) : b \\ \diagup \qquad \diagdown \\ (5)\ -X_2 : b \quad (6)\ -X_3 : a \\ \\ T_3\end{array}
\qquad
\begin{array}{c}(1)\ +x : a \wedge b \\ (2)\ -\langle\mathbf{snd}(x), \mathbf{fst}(x)\rangle : b \wedge a \\ (3)\ +\mathbf{fst}(x) : a \\ (4)\ +\mathbf{snd}(x) : b \\ \diagup \qquad \diagdown \\ \underline{(5)\ -\mathbf{snd}(x) : b} \ \underline{(6)\ -\mathbf{fst}(x) : a} \\ \\ T_4\end{array}$$

Figure 2: Some tableau extensions with the synthesis calculus

In order to fulfill the proviso for the rules $(-\rightarrow)$ and $(-\forall)$, we fall back upon the Skolemization technique.

$$(-\rightarrow)\quad \frac{-\mathbf{lambda}(f(VL), L) : \alpha \rightarrow \beta}{\begin{array}{c}+f(VL) : \alpha \\ -L : \beta\end{array}}$$

Here, f is a new Skolem function, and VL is the set of all metavariables occuring in the branch where the rule is applied. So, the Skolem expression $f(VL)$ can be understood as a new variable. In order to avoid conflicts with substitutions, we use a non-binding abstraction $\mathbf{lambda}(N, M)$. As far as tableaux are closed, non-binding abstractions $\mathbf{lambda}(N, M)$ are replaced by usual abstractions $\lambda x.M[N/x]$.

The chief difficulty arises from the implicit case analysis. If the ($[\![\,]\!]$)-rule would be applied the number of cases n would have to be guessed. For that reason, this rule is omitted and we presume that the λ-term in each negatively signed type relation is a priori an implicit case analysis. In general, a negatively signed type relation is of the following form:

$$-[\![M_1, M_2, ..., M_n|L]\!] : \alpha, \text{ [22] where } n \geq 0. \qquad (1)$$

L is a metavariable which stands for a list of further cases. L can be refined by rule applications and closing of branches. If α in (1) would be $\beta \wedge \gamma$, then $(-\wedge)$ could be applied. That would lead to an extension of the tableau and a refinement of (1) by a further case:

$$-[\![M_1, M_2, ..., M_n, \langle L_1, L_2\rangle | L_3]\!] : \beta \wedge \gamma$$

The tableau would be extended by a branching with two new nodes labeled with

$$-L_1 : \beta \text{ and } -L_2 : \gamma$$

Now, L_1, L_2, L_3 are metavariables which may be refined later.

Similarly, L would have been refined if (1) would have been involved in closing a branch.

4.2 Formal Definition of the Calculus

For the sequel, we assume that metavariables, Skolem expressions and **lambda**-expressions may appear in λ-terms, formulas and tableaux. The tableau rules are defined as described in the previous subsection. Rules for positively signed type relations remain as before (see table 2). The only exception is for the rules $(+\rightarrow)$ and $(+\forall)$, where metavariables L and X are used instead of N and t, respectively. Rules for negatively signed type relations are modified by introducing metavariables and Skolem expressions (in the rules $(-\rightarrow)$ and $(-\forall)$). The rule $(-[\![\,]\!])$ is omitted.

Definition 13 (Tableau extension) *Let T, T' be tableaux and r one of the tableau rules. In the case that r is a rule for positively signed type relations, $T \xrightarrow{r}_S T'$ (T is extended to T' by rule r) is defined like $T \xrightarrow{r} T'$ (see Definition 5). In the case r has one of the following form*

$$(1)\quad \frac{-M : \alpha}{A_1 \mid A_2} \qquad\qquad (2)\quad \frac{-M : \alpha}{\begin{array}{c} A_1 \\ \vdots \\ A_n \end{array}} \qquad n = 1, 2$$

then $T \xrightarrow{r}_S T'$ holds if there exists some branch π in T such that $-[\![N_1, ..., N_m|L]\!] : \alpha$ occurs in π and T' is obtained from T in the following way: First, T is instantiated by $\sigma = [L/[\![M|L']\!]]$, where L' is a new variable.[23] *Then $\pi\sigma$ is extended either by a branching with the nodes A_1 and A_2 or by n further nodes labeled with $A_1, ..., A_n$. We write $T \xrightarrow{SC} T'$ if there exists a rule r such that $T \xrightarrow{r}_S T'$.*

[22] $[\![M_1, M_2, ..., M_n|L]\!]$ is an abbreviation for $cons(M_1, cons(M_2, ... cons(M_n, L)...))$.

[23] This instantiation leads to a refinement by one further case. $\pi\sigma$ now contains some node labeled with $-[\![N_1, ..., N_m, M|L']\!]$.

Definition 14 (Closing of a branch) *Let T, T' be tableaux. $T \xrightarrow{c}_S T'$ (T is transformed to T' by closing a branch) holds if there exists some branch π in T such that both $+M : \alpha$ and $-[\![N_1, ..., N_m | L]\!] : \alpha'$ occur in π for some $N_1, ..., N_m, M, L, \alpha, \alpha'$ holding the following conditions:*

(i) If $\alpha \neq \bot$ then α and α' has to be unifiable.[24] *Let σ the most common unificator. Then $T' = (T[L/[\![M|L']\!]])\sigma$, where L' is a new variable.*

(ii) If $\alpha = \bot$ then $T' = T[L/[\![\mathbf{absurd}(M)|L']\!]]$, where L' is a new variable.

$\xrightarrow{SC*}$ is defined to be the reflexive and transitive closure of $\xrightarrow{SC}$ and $\xrightarrow{c}_S$.

Definition 15 (Closed Tableau) *A branch π is called closed if both $+M : \alpha$ and $-[\![N_1, ..., N_m]\!] : \alpha'$ occur in π for some $M, N_1, ..., N_m, \alpha, \alpha'$ such that either $\alpha = \alpha'$ and $M = N_i$ for some i, $1 \leq i \leq n$, or $\alpha = \bot$ and $\mathbf{absurd}(M) = N_i$ for some i, $1 \leq i \leq n$. A tableau is closed if all its branches are closed.*

Definition 16 (Transformation to a ground tableau) *Let T, T' be tableaux. We define $T \xrightarrow{g}_S T'$ (T is transformed to a ground tableau T'), if T' is obtained from T by the following two steps:*

1. Each implicit case analysis $[\![N_1, ..., N_n | L]\!]$ is replaced by $[\![N_1, ..., N_n]\!]$. Each remaining metavariable X is substituted by some constant c.

2. Each expression of the form $\mathbf{lambda}(f(t_1, ..., t_n), M)$ is replaced by $\lambda x.(M[f(t_1, ..., t_n)/x])$, where $x \in \mathcal{V}$ is new.

Definition 17 (SC-Validity) *Let Σ be any context, $M \in \Lambda_E$ and α any formula. We define $\Sigma \vdash_{SC} M : \alpha$ if there exists some closed tableau T such that*

$$+\Sigma \cup \{-L : \alpha\} \xrightarrow{SC*} T' \xrightarrow{g} T,$$

where L is some metavariable and T' some tableau, and L is instantiated by M (i.e. T contains in its initial part a node labeled with $-M : \alpha$).

Theorem 18 (Soundness) *If $\Sigma \vdash_{SC} M : \alpha$ then $\Sigma \vdash_{TC} M : \alpha$.*

Proof: Inductively on the number of rule applications. Let be given a closed ground tableau T produced by the synthesis calculus. By inserting applications of the $(-[\![\,]\!])$-rule to each node in T, where it is possible, we get a closed and minimal tableau, which can be produced by the test calculus. ■

Theorem 19 (Completeness) *If $\Sigma \vdash_{TC} M : \alpha$ then there exists some $M \in \lambda_E$ such that $\Sigma \vdash_{SC} M' : \alpha$.*

Proof: It can be shown that each derivation of a closed and minimal tableau in the test calculus can be lifted to a closed one in the synthesis calculus modulo some modifications. For that, each application of the $(-[\![\,]\!])$-rule has to be cut off. In addition, some brackets of the form $[\![$ and $]\!]$ must be inserted, since in tableaux produced by the synthesis calculus each λ-terms occuring at a negatively signed formula is an implicit case analysis (even if it contains only one case). ■

[24]Unification is restricted to metavariables.

5 Conclusions

There are two worlds of proof methods for the intuitonistic first-order logic: proof-theoretic calculi and semantic-based calculi. The natural deduction calculus and the related sequent calculus come from the proof theory area. Because of the order-dependence on the rules, they are not very appropriate for automatic reasoning. However, λ-terms are given easily, when a formula is proven valid. Resolution and matrix method are semantic-based proof methods, dedicated to automatic reasoning. However, synthesis of λ-terms is not considered.

The tableau calculus presented here subsumes the advantages of both worlds: it is automatic and efficient (deadlock-free) and allows the synthesis of λ-terms. One of the key notions is case analysis. In natural deduction proofs case analysis (($\vee$E)-rule) has often to be done before other rule applications, while in tableau proofs case analysis (($+\vee_i$)-rule) can be arbitrarily delayed. This property of the tableau calculus is reflected by a new λ-term construct, which we call implicit case analysis. Therefore and because of the use of metavariables and unification, the order of rule applications does not play any role.

Since the new calculus is based on the one for classical logic, all the techniques developed in the area of automated tableau-based theorem proving for classical logic (e.g. [Sch85], [OpSu88], [Fit90]) can be carried over. We have implemented the calculus in Prolog. A strategy [Sch85] which is similar to model elimination has been built in. Several formulas like the maximum of two and of three numbers and integer square root were proven automatically.[25] It is clear that this theorem prover can easily be integrated as automatic tool in a more interactive system like NUPRL and OYSTER.

There is an interesting relation to the deductive program synthesis approach of Manna and Waldinger [MaWa80]. Their approach is also based on the classical tableau calculus, but branching (splitting) is not allowed. E.g. a goal $\alpha \wedge \beta$ (in our calculus $-\alpha \wedge \beta$) is not allowed to be decomposed. This property lies inherently in their calculus. In contrary, in our calculus branching is allowed and therefore decompostion of all formula kinds is possible. The advantage is more flexibility in integrating different strategies.

Acknowledgement. I would like to thank Jochen Burghardt, Viktor Friesen and Birgit Heinz for reading an earlier draft of this paper.

References

[And80] Andrews, P. B., *Transforming Matings into Natural Deduction Proofs*, 5. CADE, LNCS 87, 1980.

[Bit91] Bittel, O., *Ein tableaubasierter Theorembeweiser für die intuitionistische Logik*, Ph.D. thesis, GMD-Report 198, Oldenbourg-Verlag, 1991.

[BSH90] Bundy, A., Smaill, A. Hesketh, J., *Turning Eureka Steps into Calculations in Automatic Programm Syntesis*, in Proceedings of the First Workshop on Logical Frameworks, Antibes, 1990.

[25] For the last example induction on natural numbers is necessary. The induction step was proven fully automatic.

[Con86] Constable, R. L. et al., *Implementing Mathematics with the Nuprl Proof Development System,* Prentice-Hall, 1986.

[Coq90] Coquand, T., *On the Analogy Between Propositions and Types*, in Logical Foundations of Functional Programming, G. Huet (ed.), Addison-Wesley, 1990.

[Fit83] Fitting, M., *Proof Methods for Modal and Intuitionistic Logics,* Holland, 1983.

[Fit90] Fitting, M., *First-Order Logic and Automated Theorem Proving*, Springer-Verlag, 1990.

[Gen35] Gentzen, G., *Untersuchungen über das logische Schlie"sen,* Mathematische Zeitschrift, 39:176-210, 405-431, 1935.

[Gir89] Girard, J.-Y., Lafont, Y., Taylor, P., *Proofs and Types,* Cambridge Tracts in Theoretical Computer Science 7, 1989.

[How80] Howard, W. A., *The Formulae as Types Notion of Construction,* in H. B. Curry – essays on Combinatory Logic, λ -calculus and Formalism, Seldin/Hindley (Eds.), 579-606, Academic Press, 1980.

[Lin89] Lingenfelder, C., *Structuring Computer Generated Proofs*, IJCAI, 1989.

[Mar82] Martin-Löf, P., *Constructive Mathematics and Computer Programming,* in I. J. Cohen, J. Los, H. Pfeiffer and K. D. Podewski (Eds.), Logic, Methodology and Phylosophy of Science VI, 153-179, North-Holland, 1982.

[MaWa80] Manna, Z., Waldinger, R., *A Deductive Approach to Program Synthesis,* ACM TOPLAS Vol. 2, No. 1, 90-121, 1980.

[Mil84] Miller, D. A., *Expansion Tree Proofs and their Conversion to Natural Deduction Proofs*, 7. CADE, LNCS 170, 1984.

[Ohl88] Ohlbach, H. J., *A Resolution Calculus for Modal Logics*, 9. CADE, LNCS 310, 1988.

[OpSu88] Oppacher, F., Suen, E., *HARP: A Tableau-Based Theorem Prover*, Journal of Automated Reasoning, Vol. 4, 69 - 100, 1988.

[Pra65] Prawitz, D., *Natural Deduction*, Almquist & Wiksell, Stockholm, 1965.

[Sch85] Schönfeld, W., *Prolog Extensions Based on Tableau Calculus*, IJCAI, 1985.

[Wal88] Wallen, L. A., *Matrix Proof Methods for Modal Logics*, IJCAI, 1988.

[Wos90] Wos, L., *The Problem of Finding a Mapping between Clause Representation and Natural-Deduction Representation,* Journal of Automated Reasoning 6, 211-212, 1990.

A Constructive Type System Based on Data Terms

Hans-Joachim Goltz
German National Research Center for Computer Science
GMD–FIRST, Rudower Chaussee 5
D – 1199 Berlin, Germany

Abstract

The syntax and semantics of a new kind of a type system are described and type unification is discussed. Special algebraic structures are used for the semantics. The universe of such a structure consists of ground data terms (for the representation of the ground objects) and of ground type terms (for the representation of the types). A binary function ":" is used to determine the elements belonging to a type. This function is defined by special rewrite rules called type rules. The main aim of this approach is computing with types. The elements of a type can be generated by using the type rules. The type unification can generate the elements belonging to the intersection of two types.

1 Introduction

The syntax and semantics of a new kind of type system are described in the present paper, a unification algorithm for the unification of typed terms is given, and properties of this algorithm are discussed. In the suggested type system, types are subsets of the universe of discourse and are defined by type rules. The elements belonging to a type can be generated by using these type rules.

In general, there are several objectives for introducing types, e.g. type checking (avoiding useless inferences, avoiding program errors), optimized compilation, definition of data structures, computing with types (as an extension of type inference). In our approach, the main aim is computing with types. Especially, data terms can be generated during the inferences. The other objectives mentioned above are supported, too.

Structuring the universe of discourse is a successful method in the field of representing knowledge and automated reasoning. The universe is subdivided into a certain number of subuniverses. Subuniverses may contain one another, may partially overlap, or may be disjoint. Relations between subuniverses form a special kind of information, sometimes called taxonomic information, which helps to avoid meaning-less or wrong conclusions. The benefit of using such information is widely recognized within the field of knowledge representation and automated theorem proving (see e.g. [21],[26]). The introduction of types and sorts provides an appropiate means for such a structuring of the universe. Especially, the incorporation of types or sorts into logic programming is of great interest. Many research activities have been devoted to this task and several proposals for various type systems have been made. The majority of the approaches are based on many-sorted logic or order-sorted logic. In these approaches either a common universe of all ground elements is assumed (then types are interpreted by subsets of this domain) or there are different domains for different types.

In the following some important papers on type systems in logic programming are shortly reviewed. Mycroft and O'Keefe ([22]) have proposed a polymorphic type system for Prolog based on many-sorted logic. A formal semantics of this approach is discussed in [18]. Different domains are associated with distinct types and type declarations form an integral part of programs. Subset relations between types are not allowed. A typed functional extension of Logic programming is defined in [23]. This language is essentially based on Horn clause logic with equality and a polymorphic type system that is an extension of Mycroft and O'Keefe's system.

In order-sorted type systems different types may be related by an inclusion relation ([24]). OBJ is a logic programming language based on order-sorted equational logic ([17]). Programs are order-sorted equational specifications and computation is a form of equational deduction by rewriting. Logic Programming with polymorphically order-sorted types is proposed in [25]. Type functions must be monotonic in their arguments with respect to inclusion order. In [13] the subtype order is defined by Horn clauses and the monotony of type functions is not required. PROTOS-L ([2]) is a logic programming language based on the foundations given in [25].

In [1] another direction of an approach for types in logic programming is given. Ordinary first-order terms are replaced by record structures and a special type unification is used. A further direction is proposed in [6] and [5]. Types are defined by functions on terms (by means of equations) and it is shown that order-sortedness can easily be expressed within the framework of equational logic programming. Equations are also used for the type system in [14], but there is a clear distinction between types and functions (and predicates). The subsort relationship is described by equations and the type structure is specified by a many-sorted signature with equational axioms.

The basic idea of a type system discussed in this paper is very similar to the approach given in [6] and [5], respectively. Types can be defined by special rewrite rules (called type rules). In contrast to [6] and [5], a special semantics and a special type unification are defined. Our main aim is the definition of subsets of the universe by means of type rules such that the elements belonging to a subset can be generated by these rewrite rules, and computing with such subsets. These subsets are regarded as types.

In the following we discuss how rewrite rules can be used for the definition of types. For instance, the type of *natural numbers* can be defined by the following rewrite rules:

$$nat(0) => 0$$
$$nat(s(x)) => s(nat(x))$$

If we write "*term* : *type*" instead of "*type*(*term*)" (in this case, ":" can be regarded as the binary function *apply*), then we get

$$0 : nat => 0$$
$$s(x) : nat => s(x : nat)$$

The functional evaluation (reduction) of $s(s(0)) : nat$ gives $s(s(0))$ and if t is not a natural number, then $t : nat$ can not be reduced. For type checking and computation with types (type inference) an extended unification (e.g. narrowing) can be used. If the extended unification includes the syntactical unification, then the unification of the terms $x : nat$ and $t : nat$ yields the substitution $x = t$ for any term t (this is the case if the wellknown narrowing is used, also in [6] and [5] this substitution would be infered), but this is not desirable for the computation with types. In our framework given in [10], terms can be equal only if they are defined terms. Thus, the unification of $x : nat$ and $t : nat$ is not succesful, if t is not a natural number.

In this paper we assume that the universe of discourse can be represented by ground data terms. The elements of the universe are also called ground objects. Types are subsets of the universe and are represented by certain ground terms called *ground type terms*. We use the special binary function ":" in order to determine the elements belonging to a type. If r is a notation of a type (i.e. r is a ground type term) and t is a ground data term (i.e. t denotes a ground object), then $t : r$ yields t if t belongs to r and $t : r$ is not defined in the other cases. The function ":" is defined by rewrite rules of the form: $t : r => t$, where t is a term and r a notation for a type. Thus, the second rule of the example given above is to be modified:

$$s(x : nat) : nat => s(x : nat)$$

Since in the rule $t : r => t$ the term t occurs twice, we write $r :=> t$ for this rule and denote such rules as *type rules*. Hence, the type of natural numbers can be defined by the following type rules:

$$nat :=> 0$$
$$nat :=> s(x : nat)$$

Type rules can also be regarded as rules of construction for the generation of the elements belonging to the corresponding type. A subset relation between types can be represented by type rules, too. For example, the type rule $int :=> x : nat$ means that the type "*nat*" is a subset of the type "*int*".

Furthermore, polymorphic types can also be defined. Variables for types are marked by a special unary function "*type*". In contrast to other type systems, the types of the suggested type system can depend on further parameters (where these parameters are ground objects). For example, the polymorhic type of lists with a given number of elements can be defined by

$$list(0, type(x)) :=> []$$
$$list(s(y : nat), type(x)) :=> [z_1 : type(x) \mid z_2 : list(y, x)]$$

Thus, the type term $list(s(s(s(0))), nat)$ characterizes the set of lists with three natural numbers. In the rules of this example, there are two kinds of variables. x is a variable for a type and the others are variables for ground objects. This distinction of the variables may also be achieved by introducing various sorts. But, we prefer an approach based on a one-sorted logic. In the term on the right of the second rule we can write $list(y, x)$ instead of $list(y, type(x))$ since the kind of the variable x is determined in the term on the left-hand side. But, for technical reasons, we have to write $t:type(x)$ instead of $t:x$.

Special algebraic structures are used for the semantics. The universe of such a structure consists of ground data terms (for the representation of the ground objects) and of ground type terms (for the representation of the types). A special element *"undef"* is added to the universe in order to avoid the use of partial functions. Let u be a ground data term and r be a ground type term. The interpretation of $u : r$ is identical to u, if "u belongs to r", and is $undef$ in the other case. The computing with types including type inference and type checking is based on a type unification and type unification is a special kind of E-unification. Some advantages of the suggested type system are:

- elements belonging to a type can be generated by type inference;
- the type inference can generate the elements of the intersection of two types (this intersection has not to be a type);
- it is not necessary to declare the order between the types explicitly;
- usual type checking is included.

This paper is organized as follows. In the next section we briefly quote some basic definitions and notations. Section 3 is devoted to the syntax and some syntactical relations. Furthermore, an axiom system used for the semantics is defined in this section. The semantics of the type system is defined in Section 4. In the fifth section an algorithm for the type unification is described, correctness and completeness are discussed.

We assume the reader to be familiar with the usual notations and basic results of the foundations of logic programming (see e.g. [19, 9]).

2 Basic Definitions and Notations

A *signature* is determined by a set $\mathcal{F}$ of function symbols and a set $\mathcal{R}$ of relation symbols. Each function symbol and relation symbol is associated with an arity, where the arity is a natural number. We use the notation f/n if we want to express that a function symbol f is associated with the arity n. If $n \neq m$, then f/n and f/m are different function symbols.

For our investigations $\mathcal{F}$ is the union of a set $\mathcal{F}_c$ of constructor symbols, a set $\mathcal{F}_t$ of symbols for type functions, and a set of special function symbols $\{:, type\}$, where $:$ is a binary and $type$ is a unary function symbol, $\mathcal{F}_c \cap \mathcal{F}_t = \emptyset$, $\mathcal{F} \cap \mathcal{R} = \emptyset$, and $(\mathcal{F}_c \cup \mathcal{F}_t) \cap \{:, type\} = \emptyset$. We assume that $\mathcal{F}_c$ contains at least one function symbol of arity 0 (a constant). Elements of $\mathcal{F}_t$ are also called type symbols.

In the following we suppose that $\mathcal{F}$ is given and that $\mathcal{R}$ contains the binary relations $=$, $\cong$, and $:=>$, only. Furthermore, let $\mathcal{V} = \{x, y, z, x_1, y_1, z_1, x_2, \ldots\}$ be a denumerable set of variables (throughout this paper we use this notation for variables). Two definitions for the generation of terms are given as follows.

Definition 1 Given a set M and a set F of function symbols ($F \subseteq \mathcal{F}$), $Tm(M, F)$ is the least set satisfying the following conditions:

- $M \subseteq Tm(M, F)$;
- if $f \in F$, 0-ary, then f belongs to $Tm(M, F)$;
- if $f/n \in F$ $(n > 0)$ and $t_1, \ldots, t_n \in Tm(M, F)$, then $f(t_1, \ldots, t_n)$ belongs to $Tm(M, F)$.

Using this definition, the usual *first-order terms* can be defined by $Tm(\mathcal{V}, \mathcal{F})$ and the set of data terms $\mathcal{D}$ by $Tm(\mathcal{V}, \mathcal{F}_c)$.

Definition 2 Let V be a set of variables ($V \subseteq \mathcal{V}$). The set $\mathcal{T}_t(V)$ is the least set satisfying the following conditions:

- if $h \in \mathcal{F}_t$, 0-ary, then h belongs to $\mathcal{T}_t(V)$;
- if $h/n \in \mathcal{F}_t$ $(n > 0)$ and $t_1, \ldots, t_n \in \mathcal{T}_t(V) \cup Tm(V, \mathcal{F}_c)$, then $h(t_1, \ldots, t_n)$ belongs to $\mathcal{T}_t(V)$.

Obviously, $\mathcal{T}_t(\mathcal{V})$ is a subset of $Tm(\mathcal{V}, \mathcal{F})$. The elements of the set $\mathcal{T}_t(\mathcal{V})$ are called *type terms*. We say a term is *ground*, if there is not any variable in it. Consequently, the set of *ground data terms* $\mathcal{D}_0$ can be defined by $Tm(\emptyset, \mathcal{F}_c)$ and the set of *ground type terms* $\mathcal{T}_t$ by $\mathcal{T}_t(\emptyset)$. In the following we define some further subsets of the set of terms $Tm(\mathcal{V}, \mathcal{F})$:

typed variables $\mathcal{V}_t := \{x : s \mid x \in \mathcal{V}, s \in \mathcal{T}_t(\mathcal{V})\}$;

polymorphic typed variables $\mathcal{V}_p := \{x : type(y) \mid x, y \in \mathcal{V}\}$;

simple typed variables $\mathcal{V}_t^0 := \{x : s \mid x \in \mathcal{V}, s \in \mathcal{T}_t\}$;

general data terms $\mathcal{D}_g := Tm(\mathcal{V}_t \cup \mathcal{V} \cup \mathcal{V}_p, \mathcal{F}_c)$.

typed data terms $\mathcal{D}_t := Tm(\mathcal{V}_t, \mathcal{F}_c)$

simple typed data terms $\mathcal{D}_t^0 := Tm(\mathcal{V}_t^0 \cup \mathcal{V}, \mathcal{F}_c)$

Definition 3 (object terms) The set $\mathcal{O}_t$ of *object terms* is the least set satisfying the following conditions:

- $Tm(\mathcal{V}, \mathcal{F}_c) \subseteq \mathcal{O}_t$;
- if $t \in \mathcal{O}_t$ and $r \in \mathcal{T}_t$, then $t : r \in \mathcal{O}_t$ and $t : type(r) \in \mathcal{O}_t$;

- if $f/n \in \mathcal{F}_c$ $(n > 0)$ and $t_1, \ldots, t_n \in \mathcal{O}_t$, then $f(t_1, \ldots, t_n) \in \mathcal{O}_t$.

The extended unification including type inference (see section 5) is considered between object terms, only. We denote the set of ground object terms by $\mathcal{O}_t^0$. Note that $\mathcal{D}_t^0$ is a subset of $\mathcal{O}_t$ and that $\mathcal{O}_t^0 = \mathcal{O}_t \cap Tm(\emptyset, \mathcal{F})$.

The set of variables in a set M of terms is denoted by $var(M)$. If $M = \{t\}$ for a term t, we use $var(t)$ instead of $var(\{t\})$. A variable x of a term t is called *object variable in* t, if there is a subterm $x : r$ of t, and x is called *type variable in* t, if there is a subterm $type(x)$ of t.

A *substitution* σ is a mapping from $\mathcal{V}$ to $Tm(\mathcal{V}, \mathcal{F})$ such that the set $\mathrm{dom}(\sigma)$ defined by $\{x \in \mathcal{V} \mid \sigma(x) \neq x\}$ is finite. We use the notation $\sigma(t)$ to represent the term obtained by replacing the variables of t by their images under σ, i.e. a substitution σ can be regarded as a mapping from $Tm(\mathcal{V}, \mathcal{F})$ to $Tm(\mathcal{V}, \mathcal{F})$. For a substitution σ the set $\{\sigma(x) \mid x \in dom(\sigma)\}$ is denoted by $codom(\sigma)$. A substitution σ is determined by a set $\{x_1/t_1, \ldots, x_n/t_n\}$, i.e. $dom(\sigma) = \{x_1, \ldots, x_n\}$ and $codom(\sigma) = \{t_1, \ldots, t_n\}$. The composition $\sigma \circ \tau$ of two substitutions σ and τ is defined by $\sigma \circ \tau(x) = \tau(\sigma(x))$.

Lists are special data terms. We use the notation $[t_1, \ldots, t_n]$ for a list with the elements $t_1, \ldots, t_n$. Then $[t \mid list]$ is the usual notation of a list with the first element t. The empty list is denoted by $[]$.

3 Syntax and Syntactic Relations

Definition 4 (type rule) The expression $s :=> u$ is called *type rule*, if the following conditions are satisfied (x denotes any variable):

(1) s is a 0-ary type symbol or s is identical to a term $h(s_1, \ldots, s_n)$, where $h \in \mathcal{F}_t$ and for each s_i holds: s_i is a typed data term or s_i is identical to a term $type(x)$;

(2) u is a general data term, but u is not identical to a polymorphic typed variable;

(3) if $x \in var(u)\backslash var(s)$, then there is a subterm $x : r$ of u ;

(4) if $g \in \mathcal{F}_t$ is an m-ary symbol of a type function and $g(u_1, \ldots, u_m)$ is a subterm of u then $var(g(u_1, \ldots, u_m)) \subseteq var(s)$;

(5) there is not any variable which is a type variable and an object variable in $s :=> u$.

A type rule $s :=> u$ can be considered a special rewrite rule $u : s => u$ (see also Section 1). Variables introduced by a type rule $s :=> u$ cannot occur in a type term contained in u (condition 4), and, furthermore, these variables are object variables in u (condition 3). By means of the unary function symbol *type*, variables for type terms can be marked. These variables have to be different from the variables for object terms.

Example 1 The following is a set of type rules :

$$nat :=> 0$$
$$nat :=> s(x : nat)$$
$$int :=> x : nat$$
$$int :=> -s(x : nat)$$
$$list(0, type(x)) :=> []$$
$$list(s(x : nat), type(y)) :=> [z_1 : type(y) \mid z_2 : list(x, y)]$$

□

Definition 5 (ground substitution) A substitution σ is called *ground substitution for a type rule* $s :=> u$, if the following conditions are satisfied:

- $var(s) \cup var(u) \subseteq dom(\sigma)$ and $codom(\sigma) \subseteq \mathcal{D}_0 \cup \mathcal{T}_t$;
- if $x : r$ is a subterm of s or u, then $\sigma(x) \in \mathcal{D}_0$;
- if $type(x)$ is a subterm of s or u, then $\sigma(x) \in \mathcal{T}_t$;
- $\sigma(u)$ is an object term.

Let σ be a ground substitution for some type rule $s :=> u$. Thus $\sigma(s)$ and $\sigma(u)$ are ground terms. Furthermore, if $\sigma(s) = h(s_1, \ldots, s_n)$ for some $h \in \mathcal{F}_t$, then, for each $i \in \{1, \ldots, n\}$, s_i is an object term or s_i is identical to a term $type(r)$ for some $r \in \mathcal{T}_t$.

Since type rules are regarded as special rewrite rules, the usual reduction relation of terms (term rewriting) may be used ([16]). The reduction of terms using type rules means type checking. In the following we define a special kind of term reduction.

Definition 6 (term reduction) Let M be a set of type rules. Suppose that there are not any two type rules $h(s_1, \ldots, s_n) :=> u$ and $h(r_1, \ldots, r_m) :=> v$ of this set M such that, for some $i \in \{1, \ldots, n\}$, the term s_i is the term $type(x)$ for a variable x and r_i is not a term of such a kind. A ground term t *is reducible to* a term u (with respect to M) if one of the following conditions is satisfied:

(1) t is identical to u;

(2) there is a function symbol $f/n \in \mathcal{F}_c$ $(n > 0)$ such that $t \equiv f(t_1, \ldots, t_n)$, $u \equiv f(u_1, \ldots, u_n)$, and t_i is reducible to u_i for each $i \leq n$;

(3) $t \equiv w{:}s$ or $t \equiv w{:}type(s)$, the terms u, w are ground terms, w is reducible to u , $s \in \mathcal{T}_t$, and there are a type rule $r :=> v$ and a ground substitution σ for this type rule such that $\sigma(v)$ is reducible to u and one of the two conditions is true:

- $\sigma(r) = r$ and $r \equiv s$,
- there is a $h/n \in \mathcal{F}_t$ $(n > 0)$ such that $s \equiv h(s_1, \ldots, s_n)$, $\sigma(r) \equiv h(r_1, \ldots, r_n)$, and r_i is reducible to s_i for each $i \in \{1, \ldots, n\}$.

(4) $t \equiv type(u)$ and $u \in \mathcal{T}_t$.

Example 2 We regard the definition of types given in Example 1. The ground term $[0, s(0)] : list(s(s(0)), nat)$ can be reduced to the following terms:
$[0 : nat \mid [s(0)] : list(s(0), nat)]$, $[0 : nat \mid [s(0) : nat \mid [\,] : list(0, nat)]]$,
$[0 : nat, s(0) : nat]$, $[0, s(0)]$, and to some others. □

Definition 7 (relation $\lhd$) The binary relation $\lhd$ between two ground terms is defined as follows:

- for each 0-ary symbol $c \in \mathcal{F}_t \cup \mathcal{F}_c$: $c \lhd c$;
- for each n−ary function symbol $f \in \mathcal{F}_t \cup \mathcal{F}_c$ $(n > 0)$ and terms $t_1, \ldots, t_n$, $u_1, \ldots, u_n$:

$$t_1 \lhd u_1, \ldots, t_n \lhd u_n \quad \text{iff} \quad f(t_1, \ldots, t_n) \lhd f(u_1, \ldots, u_n) ;$$

- $t \lhd t : s$ and $s \lhd type(s)$ for any term t, s .

Informally, if t, u are ground terms, then the relation $t \lhd u$ means that t and u are identically, if a part of type checking is ignored in u. If $t \in \mathcal{D}_0 \cup \mathcal{T}_t$ and $u \lhd t$ for some term u, then the terms t and u are identical. For each ground term t there is a unique term $u \in \mathcal{D}_0 \cup \mathcal{T}_t$ such that $u \lhd t$. The following lemma is a consequence of the corresponding definitions.

Lemma 1 *If a ground term t is reducible to a term $u \in \mathcal{D}_0 \cup \mathcal{T}_t$, then $u \lhd t$.*

Hence, there is at most one $u \in \mathcal{D}_0 \cup \mathcal{T}_t$ for each ground term t such that t is reducible to a term of $\mathcal{D}_0 \cup \mathcal{T}_t$.

Definition 8 (match substitution) Let $s :=> u$ be a type rule and $r \in \mathcal{T}_t$. A substitution σ is called a *match substitution for* $(r, s :=> u)$, if the following conditions are satisfied

- $var(s) \subseteq dom(\sigma)$ and $codom(\sigma) \subseteq \mathcal{D}_0 \cup \mathcal{T}_t$;
- if $x : s_1$ is a subterm of s, then $\sigma(x) \in \mathcal{D}_0$;
- if $type(x)$ is a subterm of u or s, then $\sigma(x) \in \mathcal{T}_t$;
- $\sigma(u)$ is an object term ;
- if s is a 0-ary type symbol, then s is identical to r ;
- if $r \equiv h(r_1, \ldots, r_n)$ for some $h/n \in \mathcal{F}_t$ $(n > 0)$, then $s \equiv h(s_1, \ldots, s_n)$ and $\sigma(s_i)$ is reducible to r_i for each $i \in \{1, \ldots, n\}$.

We need a match substitution for the application of type rules. If $r \in \mathcal{T}_t$ and σ is a match substitution for $(r, s :=> u)$, then we have:

- if $var(u) \subseteq dom(\sigma)$, then σ is a ground substitution;
- if $x \in var(\sigma(u))$, then x is an object variable in $\sigma(u)$;
- if u is a variable, then $\sigma(u) \in \mathcal{D}_0$.

Definition 9 (definition of types) A *definition of types* is a set TD of type rules where the following conditions are satisfied:

- there are not any two type rules $h(t_1, \ldots, t_n) :=> t$ and $h(u_1, \ldots, u_m) :=> u$ of this set such that, for some $i \in \{1, \ldots, n\}$, the term t_i is the term $type(x)$ for a variable x and u_i is not a term of such a kind;
- for each $u : s$ with $u \in \mathcal{D}_0$ and $s \in \mathcal{T}_t$, the reduction is finite (succesful or not);
- if $s :=> v$ is a type rule of TD and if $r \in \mathcal{T}_0$ such that there is a match substitution σ for $(r, s :=> v)$, then there exists a ground term $u \in \mathcal{D}_0$ such that $u : r$ is reducible to u.

In the following we define a set of definite clauses called axiom system AX. This set depends on the given set $\mathcal{F}$ of function symbols. A model of a definition of types is defined by means of this axiom system (see next section). The binary relation $\cong$ characterises the identy for elements of $\mathcal{D}_0$ and $\mathcal{T}_t$.

Definition 10 (axiom system AX) . Let AX be the following set of definite clauses:

(1) $x \cong y \longleftarrow y \cong x$

$x = y \longleftarrow y = x$

$x = y \longleftarrow x \cong y$

(2) for each 0-ary symbol $t \in \mathcal{F}_c \cup \mathcal{F}_t$: $\quad t \cong t$

(3) for each $n-$ary function symbol $f \in \mathcal{F}_c \cup \mathcal{F}_t$ $(n > 0)$:

$$f(x_1, \ldots, x_n) \cong f(y_1, \ldots, y_n) \longleftarrow x_1 \cong y_1, \ldots, x_n \cong y_n$$

(4) for each $n-$ary function symbol $f \in \mathcal{F}_c$ $(n > 0)$:

$$f(x_1, \ldots, x_n) = f(y_1, \ldots, y_n) \longleftarrow x_1 = y_1, \ldots, x_n = y_n$$

(5) for each 0-ary type symbol $s \in \mathcal{F}_t$:

$x : s = y \longleftarrow s :=> z,\ x = y,\ y = z$

$s = type(s)$

$x : type(s) = y \longleftarrow x : s = y$

(6) for each n-ary type function symbol $h \in \mathcal{F}_t$ $(n > 0)$:

$$x : h(x_1, \ldots, x_n) = y \quad \longleftarrow \quad h(z_1, \ldots, z_n) :=> z,\ x_1 = z_1,\ \ldots,\ x_n = z_n,\ h(x_1, \ldots, x_n) \cong h(x_1, \ldots, x_n),\ x = y,\ y = z$$

$$h(x_1, \ldots, x_n) = type(h(x_1, \ldots, x_n)) \quad \longleftarrow \quad h(x_1, \ldots, x_n) \cong h(x_1, \ldots, x_n)$$

$$x : type(h(x_1, \ldots, x_n)) = y \quad \longleftarrow \quad x : h(x_1, \ldots, x_n) = y,\ h(x_1, \ldots, x_n) \cong h(x_1, \ldots, x_n)$$

Given a definition of types TD, the set $\Sigma(TD)$ is defined by
$\{\sigma(s) :=> \sigma(u) \mid s :=> u \in TD\ ,\ \sigma$ is a ground substitution for $s :=> u\ \}$.

Since $\Sigma(TD) \cup AX$ is a set of definite clauses, there exists a least Herbrand model of $\Sigma(TD) \cup AX$ and for the construction of such a model the fixpoint semantics can be used (see [19] or [9]). Some properties of such a Herbrand model is given in the following propositions.

Lemma 2 *Let TD be a definition of types and $\mathcal{H}$ be the least Herbrand model of the set of definite clauses $\Sigma(TD) \cup AX$.*

(1) If $t = u \in \mathcal{H}$ and $u = v \in \mathcal{H}$, then $t = v \in \mathcal{H}$.

(2) Suppose that t and u are ground terms. The terms t and u are identical terms of $\mathcal{D}_0 \cup \mathcal{T}_t$ if and only if, $t \cong u \in \mathcal{H}$.

(3) If $t \in \mathcal{D}_0 \cup \mathcal{T}_t$, then $t = t \in \mathcal{H}$.

(4) If $t : s = u \in \mathcal{H}$, then $s \in \mathcal{T}_t$ or s is identical to a term $type(s^*)$ for some $s^* \in \mathcal{T}_t$.

Proof. Proposition (1) can be proved inductively on the term construction of u for all terms v, t . For example, if $u \equiv u_1 : s$, $u_1 : s = t \in \mathcal{H}$ and $u_1 : s = v \in \mathcal{H}$, then $\mathcal{H}$ contains the equations $u_1 = t$ and $u_1 = r$ (see definition of $\mathcal{H}$). Thus $t = v \in \mathcal{H}$ follows from the supposition of induction.

The if-part of proposition (2) can be proved inductively on the term construction (by using (2) and (3) of Definition 10). Since $\mathcal{H}$ is the least Herbrand model, the other part of this proposition is true.

The propositions (3) and (4) follow from (2) and the definition of $\mathcal{H}$. □

Theorem 1 *Let $\mathcal{H}$ be the least Herbrand model of $\Sigma(TD) \cup AX$, where TD is a given definition of types. If $t_1 = t_2 \in \mathcal{H}$, then there is a term $u \in \mathcal{D}_0 \cup \mathcal{T}_t$ such that $t_1 = u \in \mathcal{H}$ and $t_2 = u \in \mathcal{H}$.*

Proof. Let $M_0, M_1, M_2, \ldots$ be a sequence of sets of ground atomic formulas such that this sequence corresponds with the construction of the least Herbrand model $\mathcal{H}$ by using the fixpoint semantics, i.e. $\mathcal{H} = \cup_{i<\omega} M_i$. We prove inductively on i the following proposition: if $t_1 = t_2 \in M_i$, then there is a $u \in \mathcal{D}_0 \cup \mathcal{T}_t$ such that $t_1 = u \in \mathcal{H}$ and $t_2 = u \in \mathcal{H}$. If the corresponding definitions and Lemma 2 are used, the proof of this proposition is straightforward. □

Theorem 2 . *Let* TD *be a definition of types and let* $\mathcal{H}$ *be the least Herbrand model of the definite clauses* $\Sigma(TD) \cup AX$. *If* t *is a ground term* $u \in \mathcal{D}_0 \cup \mathcal{T}_t$, *then* t *is reducible to* u *if and only if,* $t = u \in \mathcal{H}$.

Proof. Firstly, for each term pair (t, u) with $t \in Tm(\emptyset, \mathcal{F})$ and $u \in \mathcal{D}_0 \cup \mathcal{T}_t$, we define the number $n(t, u)$ of reduction steps, if t is reducibles to u . Corresponding to the cases of Definition 6 (the same notations are chosen) $n(t, u)$ is definied by

(1) $n(t, u) = 0$, if t is identical to u ;

(2) $n(t, u) = 1 + \Sigma_i n(t_i, u_i)$, if $t \equiv f(t_1, \ldots, t_n)$ and $u \equiv f(u_1, \ldots, u_n)$;

(3) $n(t, u) = 1 + n(w, u) + n(\sigma(v), u) + \Sigma_i n(r_i, s_i)$, if $t \equiv w : s$ or $t \equiv w : type(s)$ ($\Sigma_i n(r_i, s_i) = 0$ if $r \equiv s$) ;

(4) $n(t, u) = 1$, if $t = type(u)$.

Then the if-part of this theorem can be proved inductively on the number of reduction steps. Using the corresponding definitions and Lemma 2 , this inductive proof is straightforward.

The other part of the theorem can easily be proved by using the fixpoint construction of $\mathcal{H}$ (inductive proof on this construction). Note, $t \cong u \in \mathcal{H}$ implies that u and t are identical terms. □

4 Semantics

Special algebraic structures called *t-structures* are used for the semantics. The universe of such a structure consists of ground data terms and of ground type terms. We add a special element *"undef"* to the universe in order to avoid the use of partial functions.

Definition 11 (t-structure) A *t-structure* $\mathcal{A}$ is a tuple $(U, (:)^{\mathcal{A}}, type^{\mathcal{A}}, =_{\mathcal{A}}, \varphi)$ such that

(1) the universe U of $\mathcal{A}$ is the set $\mathcal{D}_0 \cup \mathcal{T}_t \cup \{undef\}$;

(2) $(:)^{\mathcal{A}}$ is a binary function $U \times U \rightarrow \mathcal{D}_0 \cup \{undef\}$ such that

- $(:)^{\mathcal{A}}(a, b) = undef$, if $a \in \mathcal{T}_t \cup \{undef\}$ or $b \in \mathcal{D}_0 \cup \{undef\}$,
- $(:)^{\mathcal{A}}(a, b) \in \{a , undef\}$, if $a \in \mathcal{D}_0$ and $b \in \mathcal{T}_t$;

(3) $type^{\mathcal{A}}$ is a unary function $U \rightarrow \mathcal{T}_t \cup \{undef\}$ such that

- $type^{\mathcal{A}}(a) = undef$, if $a \in \mathcal{D}_0 \cup \{undef\}$,
- $type^{\mathcal{A}}(a) = a$, if $a \in \mathcal{T}_t$;

(4) $=_{\mathcal{A}}$ is a binary relation and is defined by the set $\{(t, t) \mid t \in \mathcal{D}_0 \cup \mathcal{T}_t\}$;

(5) φ is a mapping from the set of ground terms to the universe of $\mathcal{A}$ satisfying the following conditions:

- if $t \in \mathcal{D}_0 \cup \mathcal{T}_t$, then $\varphi(t) = t$;
- if $h/n \in \mathcal{F}_t$ $(n > 0)$ and $h(s_1, \ldots, s_n) \notin \mathcal{T}_t$, then $\varphi(h(s_1, \ldots, s_n)) = undef$;
- if $f/n \in \mathcal{F}_c$ $(n > 0)$ and $f(t_1, \ldots, t_n) \notin \mathcal{D}_0$, then
  - $\varphi(f(t_1, \ldots, t_n)) = undef$, if there is a t_i with $\varphi(t_i) = undef$,
  - $\varphi(f(t_1, \ldots, t_n)) = f(\varphi(t_1), \ldots, \varphi(t_n))$ if there is not such a t_i ;
- $\varphi(type(s)) = s$, if $s \in \mathcal{T}_t$;
- $\varphi(type(s)) = undef$, if $s \notin \mathcal{T}_t$;
- $\varphi(t : s) = (:)^{\mathcal{A}}(\varphi(t), \varphi(s))$.

Obviously, a t-structure is unique determined, if the function symbols are given and the function $(:)^{\mathcal{A}}$ is defined (and satisfies the conditions). The following lemma is a consequence of the definition.

Lemma 3 *Let* $\mathcal{A} = (U, (:)^{\mathcal{A}}, type^{\mathcal{A}}, =_{\mathcal{A}}, \varphi)$ *be a t-structure and* t *be a ground term. If* $\varphi(t) \neq undef$, *then* $\varphi(t) \lhd t$ *and* $t \in \mathcal{O}_t^0 \cup \mathcal{T}_t \cup \{type(t) \mid t \in \mathcal{T}_t\}$.

Definition 12 Let $\mathcal{A} = (U, (:)^{\mathcal{A}}, type^{\mathcal{A}}, =_{\mathcal{A}}, \varphi)$ be a t-structure and t, u be ground terms. The equation $t = u$ is *valid* in $\mathcal{A}$ (denoted by $\mathcal{A} \models t = u$), if $(\varphi(t), \varphi(u))$ belongs to $=_{\mathcal{A}}$.

Definition 13 (model) Suppose TD to be a of definition of types. Let $\mathcal{H}$ be the least Herbrand model of $\Sigma(TD) \cup AX$. A t-structure $\mathcal{A}$ is called a *model* of TD, if for any ground terms t and u hold: $\mathcal{A} \models t = u$ iff $t = u \in \mathcal{H}$.

Corollary 4 *If* $\mathcal{A} = (U, (:)^{\mathcal{A}}, type^{\mathcal{A}}, =_{\mathcal{A}}, \varphi)$ *is a* model *of a definition of types* TD *and* $\mathcal{H}$ *is the least Herbrand model of* $\Sigma(TD) \cup AX$, *then for each* $a \in \mathcal{D}_0$ *and each* $b \in \mathcal{T}_t$ *holds:* $(:)^{\mathcal{A}}(a, b) := a$ *iff* $a : b = a \in \mathcal{H}$.

For each $s \in \mathcal{T}_t$ the set $I(s) = \{ u \in \mathcal{D}_0 \mid (:)^{\mathcal{A}}(u, s) = u \}$ can be regarded as the interpretation of the type s (i.e. $\varphi(u : s) = u$ iff $u \in I(s)$). Clearly, $I(s)$ may be empty. But, if there are a type rule $r :=> u$ of a definition of types and a match substitution for $(s, r :=> u)$, then $I(s) \neq \emptyset$.

Lemma 5 *Let* $\mathcal{A} = (U, (:)^{\mathcal{A}}, type^{\mathcal{A}}, =_{\mathcal{A}}, \varphi)$ *be a* model *of a definition of types* TD *and* $\mathcal{H}$ *be the least Herbrand model of* $\Sigma(TD) \cup AX$.

(1) Let t *be ground term and* $v \in \mathcal{D}_0 \cup \mathcal{T}_t$. *Then the following proposition is true:* $\varphi(t) = v$ *if and only if,* $t = v \in \mathcal{H}$.

(2) If $h(s_1, \ldots, s_n) :=> u$ *is an element of* $\Sigma(TD)$, $\varphi(s_i) \neq undef$ *for each* $i \in \{1, \ldots, n\}$, *and* $\varphi(u) \neq undef$, *then*

- $\mathcal{A} \models \varphi(u) : h(\varphi(s_1), \ldots, \varphi(s_n)) = \varphi(u)$ and
- $\mathcal{A} \models u : h(s_1, \ldots, s_n) = u$.

(3) If $t = u \in \mathcal{H}$ *and* $u \in \mathcal{D}_0 \cup \mathcal{T}_t$ *, then* $u \lhd t$.

Proof. The two parts of proposition (1) can be proved inductively on the term construction of t. Here, we show the case that t is identical to $u : s$, only. The proofs of the other cases are straightforward.

Firstly, we suppose that $\varphi(t) = v$, i.e. $\varphi(t) \neq undef$. Since $\varphi(u : s)$ is defined by $(:)^{\mathcal{A}}(\varphi(u), \varphi(s))$, $\varphi(s)$ belongs to $\mathcal{T}_t$ and, therefore, $s \in \mathcal{T}_t$ or $s = type(s^*)$ with $s^* \in \mathcal{T}_t$ (Lemma 2). Assume that $s \in \mathcal{T}_t$ and $s \equiv f(s_1, \ldots, s_n)$. Because of the suppositions, v and $\varphi(u)$ are identical elements of $\mathcal{D}_0$ and $v : s = v \in \mathcal{H}$. Thus there is an element $f(r_1, \ldots, r_n) :=> w$ of $\Sigma(TD)$ and $\mathcal{H}$ contains the atomic formulas: $f(s_1, \ldots, s_n) \cong f(s_1, \ldots, s_n)$, $v = w$, $s_1 = r_1, \ldots, s_n = r_n$. Since $\varphi(u) = u \in \mathcal{H}$ (supposition of the induction), $\mathcal{H}$ contains $u = v$ ($v \equiv \varphi(u)$). Then, the proposition $u : s = v \in \mathcal{H}$ is a consequence the definitions of AX and of $\mathcal{H}$.

For the proof of the other part of the proposition (1), we suppose that the equation $u : s = v$ belongs to $\mathcal{H}$. Then, $s \in \mathcal{T}_t$ or $s = type(s^*)$ with $s^* \in \mathcal{T}_t$ (Lemma 2). Assumed that $s \in \mathcal{T}_t$ and $s \equiv f(s_1, \ldots, s_n)$. Since $u : s = v \in \mathcal{H}$, there is a type rule $f(r_1, \ldots, r_n) :=> w$ of $\Sigma(TD)$ such that $\mathcal{H}$ contains the atomic formulas: $v = w$, $u = v$, $s_1 = r_1, \ldots, s_n = r_n$. Because of the supposition of induction, w is reducible to v, u is reducible to v, and r_i is reducible to s_i for each $i \in \{1, \ldots, n\}$. Thus $u : s$ is reducible to v. The case "$s \equiv type(s^*)$" can be proved anologously.

Now we prove proposition (2). Using (1), we get $u = \varphi(u) \in \mathcal{H}$ and $\varphi(s_i) = s_i \in \mathcal{H}$ for each $i \in \{1, \ldots, n\}$. Futhermore, $h(\varphi(s_1), \ldots, \varphi(s_n)) \cong h(\varphi(s_1), \ldots, \varphi(s_n)) \in \mathcal{H}$ and $\varphi(u) = \varphi(u) \in \mathcal{H}$ (Lemma 2). Then the proposition follows from the definition of $\mathcal{H}$ (see also Definition 10) and the definition of φ.

The proposition (3) follows from (1) and Lemma 3. □

Theorem 3 *There is a unique model for a given definition of types* TD.

Proof. Let $\mathcal{H}$ be the least Herbrand model of $\Sigma(TD) \cup AX$. We define a t-structure $\mathcal{A} = (U, (:)^{\mathcal{A}}, type^{\mathcal{A}}, =_{\mathcal{A}}, \varphi)$ such that for each $a \in \mathcal{D}_0$ and $b \in \mathcal{T}_t$ holds: $(:)^{\mathcal{A}}(a, b) = a$ iff $a : b = a \in \mathcal{H}$. We have to prove that $(\varphi(t), \varphi(u)) \in =_{\mathcal{A}}$ iff $t = u \in \mathcal{H}$.

Firstly, let $(\varphi(t), \varphi(u)) \in =_{\mathcal{A}}$, i.e. $\varphi(t)$ and $\varphi(u)$ are identical terms of $\mathcal{D}_0 \cup \mathcal{T}_t$. Then, $t = u \in \mathcal{H}$ follows from Lemma 5 (1) and from Lemma 2 (1).

Now we suppose that $t = u \in \mathcal{H}$. Because of Theorem 1, there is a term $r \in \mathcal{D}_0 \cup \mathcal{T}_t$ such that $t = r \in \mathcal{H}$ and $u = r \in \mathcal{H}$. Using Lemma 5 (1), we get $\varphi(t) = r$ and $\varphi(u) = r$. Thus $(\varphi(t), \varphi(u)) \in =_{\mathcal{A}}$. □

Definition 14 A ground term t is called *a defined term* with respect to a definition of types TD, if $\mathcal{A} \models t = t$ for the model $\mathcal{A}$ of TD.

Lemma 6 *Let* $\mathcal{A} = (U, (:)^{\mathcal{A}}, type^{\mathcal{A}}, =_{\mathcal{A}}, \varphi)$ *be a* model *of a definition of types* TD *and* $\mathcal{H}$ *be the least Herbrand model of* $\Sigma(TD) \cup AX$.

(1) For each ground term t *holds:* $\varphi(t) \neq undef$ *iff* t *is a defined term.*

(2) If t *is a defined ground term, then there is a term* $u \in \mathcal{T}_t \cup \mathcal{D}_0$ *such that* $t = u \in \mathcal{H}$, $\mathcal{A} \models t = u$ *and* $\varphi(t) = u$.

(3) If $s \in \mathcal{T}_t$ *and* $t : s$, $u : s$ *are defined ground terms, then*

(a) $\mathcal{A} \models t : s = t$ and

(b) $\mathcal{A} \models t : s = u$ *implies* $\mathcal{A} \models t = u : s$.

Proof. The propositions (1) and (3a) follow from the corresponding definitions. The second proposition (2) is a consequence of Theorem 1, Lemma 5 (1), and the Definitions 13 and 14. The last proposition (3b) can be shown by means of (3a). □

Using Lemma 6 (3), Lemma 2 (1), and the corresponding definitions, the following corollary can easily be shown.

Corollary 7 *If* $\mathcal{A} = (U, (:)^{\mathcal{A}}, type^{\mathcal{A}}, =_{\mathcal{A}}, \varphi)$ *is a model of a definition of types* TD, t, u *are ground terms, and* $s \in \mathcal{T}_t$, *then*

- $\mathcal{A} \models t{:}s{:}s = u$ *iff* $\mathcal{A} \models t{:}s = u$;
- $\mathcal{A} \models t{:}s_1{:}s_2 = u$ *iff* $\mathcal{A} \models t{:}s_1 = t{:}s_2$ *and* $\mathcal{A} \models t{:}s_1 = u$.

Lemma 8 *Suppose that* $\mathcal{A}$ *is a model of a definition of types,* t *is a data term,* $\sigma_x = \{x/t\}$ *is a substitution, and* $\mathcal{A} \models \sigma(x) = \sigma(t)$ *for a substitution* σ . *If* u *is a object term,* $\sigma(u)$ *and* $\sigma(\sigma_x(u))$ *are ground terms, and* $\sigma(u)$ *or* $\sigma(\sigma_x(u))$ *is a defined term, then* $\mathcal{A} \models \sigma(\sigma_x(u)) = \sigma(u)$.

Proof. The proposition can be shown inductively on the construction of the term. For $s \in \mathcal{T}_t$ the following equations are true in $\mathcal{A}$: $\sigma(x : s) = \sigma(x) : s = \sigma(t) : s = \sigma(\sigma_x(x)) : s$. Then this inductive proof is straightforward. □

Definition 15 (well-typed ground substitution) Let σ be a substitution such that $codom(\sigma) \subseteq \mathcal{D}_t^0$. A substitution τ is a *well-typed ground substitution for* σ if the following conditions are satisfied:

- $var(codom(\sigma)) \subseteq dom(\tau)$, $codom(\tau) \subseteq \mathcal{D}_0$;
- $\tau(t)$ is a defined term for each $t \in codom(\sigma)$.

Example 3 Some further examples of definition of types are given in the following, where the definition of the type of natural numbers $nat/0$ is supposed (see Introduction or Example 1):

- *set of finite natural numbers* $\{0, s(0), \ldots, x\}$:

$$nat(x : nat) \;{:=>}\; x$$
$$nat(s(x : nat)) \;{:=>}\; y : nat(x)$$

- *type of polymorphic lists* :

$$list(type(x)) \;{:=>}\; []$$
$$list(type(x)) \;{:=>}\; [y : type(x) | z : list(x)]$$

- *type of (polymorphic) lists with a certain number of elements* (see also Example 1)

$$list(0, type(y)) \;{:=>}\; []$$
$$list(s(x : nat), type(y)) \;{:=>}\; [z_1 : type(y) | z_2 : list(x, y)]$$

- *type of (poloymorphic) lists with at most x elements*, where x is a parameter (the definition of the type *list*/2 is used):

$$list_1(x : nat, type(y)) \;{:=>}\; z : list(x, y)$$
$$list_1(s(x : nat), type(y)) \;{:=>}\; z : list_1(x, y)$$

- *type of (polymorphic)* $n \times m$ *matrices* (the definition of the type *list*/2 is used):

$$matrix(x_1 : nat, x_2 : nat, type(y)) \;{:=>}\; z : list(x_1, list(x_2, y))$$

□

5 Type Unification

Computing with types is based on type unification and type unification includes type checking. In the following the notation *"type unification"* is defined and an algorithm for type unification is given. Furthermore, correctness and completeness of this algorithm are discussed.

In this section, we assume that a definition of types TD is given and that $\mathcal{A}$ is a t-structure and the model of TD. Type unification is only regarded for object terms. Note that a variable of an object term can not belong to a subterm $f(t_1, \ldots, t_n)$ with $f \in \mathcal{F}_t$.

Definition 16 (t-unification) Two object terms t and u are *t-unifiable*, if there is a substitution σ (called *t-unifier for t and u*) such that the following conditions are satisfied:

- $codom(\sigma) \subseteq \mathcal{D}_t^0$, $dom(\sigma) \cap var(codom(\sigma)) = \emptyset$;
- there is a well-typed ground substitution τ for σ ;
- if τ is a well-typed ground substitution for σ and $var(\sigma(t)) \cup var(\sigma(u))$ is a subset of $dom(\tau)$, then $\mathcal{A} \models \tau(\sigma(t)) = \tau(\sigma(u))$.

Since $\mathcal{D}_t^0 \subseteq \mathcal{O}_t$ the following proposition is true: if σ is a t-unifier of two object terms t and u, then $\sigma(t)$ and $\sigma(u)$ are object terms.

Example 4 Using the definition of types given in Example 1, a t-unifier of the terms $x : int$ and $y : nat$ is the substitution $\sigma = \{x/y : nat\}$. The equation $\sigma(x : int) = \sigma(y : nat)$ is valid for natural numbers, only (i.e. if y is bound to an element such that the corresponding terms are defined). □

The following proposition is a consequence of the definition of t-unification.

Corollary 9 *If σ is a t-unifier for two object terms t, u and τ is a well-typed ground substitution for σ with $var(\sigma(t)) \cup var(\sigma(u)) \subseteq dom(\tau)$, then $\sigma \circ \tau(t)$ and $\sigma \circ \tau(u)$ are defined terms.*

An algorithm for the t-unification can be specified by means of the method of transformation of terms for solving unification problems. This method was developed in [20]. A unification problem is regarded as a set of term pairs $\{(t_1, u_1), \ldots, (t_n, u_n)\}$. The method of transformation consists in applying simple transformations until a system in solved form is obtained, the solution of which is obvious.

Definition 17 A substitution σ is a *t-unifier* for a set of term pairs M, if the following conditions are satisfied:

(1) $codom(\sigma) \subseteq \mathcal{D}_t^0$, $dom(\sigma) \cap var(codom(\sigma)) = \emptyset$;

(2) there is a well-typed ground substitution τ for σ ;

(3) if τ is a well-typed ground substitution for σ and $var(\sigma(M)) \subseteq dom(\tau)$, then $\mathcal{A} \models \tau(\sigma(t_i)) = \tau(\sigma(u_i))$ for each $(t_i, u_i) \in M$.

A set of term pairs $M = \{(t_1, u_1), \ldots, (t_n, u_n)\}$ is called *a condition for typed variables in solved form*, if the following conditions are satisfied:

- if $(t, u) \in M$, then
  - t and u are variables or simple typed variables ($t, u \in \mathcal{V} \cup \mathcal{V}_t^0$),
  - t or u is a simple typed variable ($t \in \mathcal{V}_t^0$ or $u \in \mathcal{V}_t^0$) ;
- if $(x : s, u) \in M$ or $(u, x : s) \in M$, then for each term t of M (i.e. $(t, t_1) \in M$ or $(t_1, t) \in M$) holds: $x \notin var(t)$ or $t \equiv x : s$ or $t \equiv x$;
- if $(x : s, u) \in M$ or $(u, x : s) \in M$, then $u \equiv y : s$ for some variable y ;
- if $(x, u) \in M$ or $(u, x) \in M$ and the typed variable $x : s$ occurs in M , then $u \equiv y : s$ for some variable y ;
- if $(x : s, u) \in M$ or $(u, x : s) \in M$, then there are a type rule $r :=> v$ of TD and a match substitution σ for $(s, r :=> v)$.

Informally, these conditions means that variables which are identical or are equal (with respect to $=$ in $\mathcal{A}$) have to belong to the same nonempty type (see also Definition 9).

If M is a condition for typed variables in solved form, then we can generate a t-unifier τ for M step by step. We define: $M_0 := M$ and $\tau_0 = \emptyset$. Suppose that $(t, u) \in M_i$, $x \in var(t)$, $y \in var(u)$, and $t \equiv x:s$ or $u \equiv y:s$. Then the set M_{i+1} is defined by $M_{i+1} := M_i \setminus \{(t, u)\}$ and τ_{i+1} is defined by :

- if $x, y \in dom(\tau_i)$, then $\tau_{i+1} := \tau_i$;
- if $x \notin dom(\tau_i)$ and $y/z:s \in \tau_i$, then $\tau_{i+1} := \tau_i \cup \{x/z:s\}$;
- if $y \notin dom(\tau_i)$ and $x/z:s \in \tau_i$, then $\tau_{i+1} := \tau_i \cup \{y/z:s\}$;
- if $x, y \notin dom(\tau_i)$, then $\tau_{i+1} := \tau_i \cup \{x/z:s, y/z:s\}$, where z is a new variable (i.e. $z \notin var(\tau_i) \cup var(M)$) .

If M contains n elements, then $\tau := \tau_n$. Obviously, $codom(\tau) \subseteq \mathcal{D}_t^0$ and $dom(\tau) \cap var(codom(\tau)) = \emptyset$. From the definitions of TD and of M it follows that there is a well-typed ground substitution for τ. Since $\mathcal{A} \models t:s:s = t:s$ for a defined ground term $t:s$, τ satisfies condition (3) of the Definition 17. Thus τ is a t-unifier for M.

Definition 18 A set of term pairs M is said to be *in solved form*, if M can be partitioned into two sets V and D (i.e. $M = V \cup D$ and $V \cap D = \emptyset$) and the following conditions are satisfied (V or D can also be empty):

- V is a condition for typed variables in solved form and satisfies the conditions described above;
- if $(t, u) \in D$, then t is a variable occuring only once in M and u is a data term (i.e. $u \in \mathcal{D}$).

Lemma 10 *If $M = V \cup D$ is a set of term pairs in solved form, V is a condition for typed variables in solved form, $\tau = \{x_1/z_1 : s_1, \ldots, x_m/z_m : s_m\}$ is the generated t-unifier for V, and $D = \{(y_1, t_1), \ldots, (y_n, t_n)\}$ satisfies the conditions described in Definition 18 , then the substitution $\sigma = \{y_1/\tau(t_1), \ldots, y_n/\tau(t_n), x_1/z_1 : s_1, \ldots, x_m/z_m : s_m\}$ is a t-unifier for the set M .*

Proof. Obviously, σ satisfies the condition (1) of the Definition 17. Because of the definition of types, the second condition of this definition is true.

Let τ be a well-typed ground substition for σ. If we replaced terms of the form $t:s:s$ (t a ground term) by $t:s$ in $\sigma \circ \tau(M)$, then we have $t \equiv u$ for each $(t, u) \in M$. Since $\mathcal{A} \models t:s:s = t:s$ (Corollary 7, $t:s$ is a defined term), the last condition of a t-unifier for M is satisfied. □

In the following we define transformation rules on sets of term pairs for an algorithm of t-unification. Let M be a set of term pairs. Formally, the set M is partitioned into two parts U and D, where the set D has to satisfy the second condition of Definition 18. We denote this by $M = U + D$.

(1) Trivial reduction: $(\{(x,x)\} \cup U) + D \implies U + D$

(2) Term decomposition:
$(\{(f(t_1,\ldots,t_n), f(u_1,\ldots,u_n))\} \cup U) + D \implies (\{(t_1,u_1),\ldots,(t_n,u_n)\} \cup U) + D$,
where $f \in \mathcal{F}_c$ (if $n = 0$ this rule is reduced to $(\{(f,f)\} \cup U) + D \implies U + D$).

(3) Variable substitution:
$(\{(x,t)\} \cup U) + D \implies \mu(U) + (\mu(D) \cup \{(x,t)\}))$,
where $t \in \mathcal{D}$, x does not occur in t, and μ is the substitution $\{x/t\}$.

(4) Simplifications:
$(\{(t:s:s,u)\} \cup U) + D \implies (\{(t:s,u)\} \cup U) + D$, where $s \in \mathcal{T}_t$.
$(\{(t:s:r,u)\} \cup U) + D \implies (\{(t:s, t:r),(t:s,u)\} \cup U) + D$, where $s, r \in \mathcal{T}_t$.

(5) Type term test:
$(\{(t : type(s), u)\} \cup U) + D \implies (\{(t : s, u)\} \cup U) + D$, if $s \in \mathcal{T}_t$.

(6) Rewriting:
$(\{(t : s, u)\} \cup U) + D \implies (\{(t, \lambda(v)), (u, \lambda(v))\} \cup U) + D$,
if there are a type rule $r :=> v$ and a match substitution λ for $(s, r :=> v)$, and if transformation rule (4) can not be applied.

(7) Term generation:
$(\{(x, f(t_1,\ldots,t_n))\} \cup U) + D \implies$
$(\{(x_1,t_1),\ldots,(x_n,t_n),(x, f(x_1,\ldots,x_n))\} \cup U) + D$,
if f is a constructor symbol (i.e. $f \in \mathcal{F}_c$), $f(t_1,\ldots,t_n) \notin \mathcal{D}$, x does not occur in $f(t_1,\ldots,t_n)$, and $x_1,\ldots,x_n$ are new variables.

The transformation rules are defined analogously, if the elements of the first term pair are exchanged. In the following, we prove some properties of these transformation rules.

Lemma 11 *Let M_1 be a set of term pairs obtaining by applying one of the transformation rules to a set of term pairs M. If a substitution σ is a t-unifier for M_1, then σ is also a t-unifier for M.*

Proof. Suppose that σ is a t-unifier for M_1 and that τ is a well-typed ground substitution for σ with $var(\sigma(M)) \subseteq dom(\tau)$. We have to show that $\mathcal{A} \models \tau(\sigma(t)) = \tau(\sigma(u))$ for each $(t,u) \in M$. We regard the following cases corresponding to the number of transformation rule which was be applied to M (we choose the notations such as in the corresponding transformation rule):

(1) The proof of this case is trivial.

(2) Since the relation $\models$ is defined by means of the least Herbrand model, the following proposition is true: $\mathcal{A} \models \tau(\sigma(f(t_1,\ldots,t_n))) = \tau(\sigma(f(u_1,\ldots,u_n)))$ iff $\mathcal{A} \models \tau(\sigma(t_i)) = \tau(\sigma(u_i))$ for each $i \in \{1,\ldots,n\}$.

(3) Using Lemma 8, the proof of this case is staightforward.

(4) In this case, the proposition follows from Corollary 7.

(5) $\mathcal{A} \models type(t) = t$ for each $t \in \mathcal{T}_t$.

(6) Let $\mathcal{H}$ be the least Herbrand model of $AX \cup \Sigma(TD)$. We assume that $s \equiv h(s_1, \ldots, s_m)$ and $r \equiv h(r_1, \ldots, r_m)$. From Theorem 2 it follows that $\{s_1 = \lambda(r_1), \ldots, s_m = \lambda(r_m)\} \subseteq \mathcal{H}$. There is a substitution μ such that $var(\lambda(v)) = dom(\mu)$, $codom(\mu) \subseteq \mathcal{D}_0$, and $\mu(\lambda(v)) = \tau(\sigma(\lambda(v))) \in \mathcal{H}$. Then $\lambda(r) :=> \mu(\lambda(v))$ belongs to $\Sigma(TD)$. Using the definition of AX and the definition of $\mathcal{A}$, the assertion can easily be shown.

(7) For this case, the proof is obviously. □

Lemma 12 *Suppose $M = U + D$ to be a set of term pairs which is not in solved form. If there is a term pair $(t, u) \in U$ of M such that there is not any transformation rule which can be applied to this term pair, then there is not any t-unifier for M.*

Proof. If there is not any transformation rule which can be applied to the term pair (t, u), then there are the following possibilities (up to the analogous cases):

- $t = f(t_1, \ldots, t_n)$ and $u = g(u_1, \ldots, u_m)$ with $f \not\equiv g$ or $n \neq m$ $(f, g \in \mathcal{F}_c)$;
- $t = x$ for a variable $x \in \mathcal{V}$, $u = f(u_1, \ldots, u_m)$ with $f \in \mathcal{F}_c$, and x occurs in u ;
- $t = t_1 : s$ with $s \in \mathcal{T}_t$, the transformation rule (4) can not be applied to (t, u), and there are not any type rule $r :=> v$ and match substitution λ for $(s, r :=> v)$.

Obviously, in the first and the second case, a t-unifier can not exist for M. We assume that the last case is true and there is a t-unifier for M. Then there is a substitution σ such that $dom(\sigma) = var(t) \cup var(u)$, $codom(\sigma) \subseteq \mathcal{D}_0$, and $\sigma(t) = \sigma(u) \in \mathcal{H}$. Hence, there is a SLD-refutation of $t = u$ which generates an answer substitution τ such that $\sigma = \tau \circ \lambda$ for a substitution λ (see [19] or [9], for instance). In the first step, this refutation has to use one of the axioms in AX (first axioms of (5) or (6) in Definition 10). Therefore, if s is a 0-ary type symbol, then there is a type rule such that the transformation rule (6) can be applied to (t, u). Now we suppose that $s \equiv h(s_1, \ldots, s_m)$ for some $h/m \in \mathcal{F}_t$ with $m > 0$. Because of the refutation, there is a type rule $r :=> v$ of TD with $r \equiv h(r_1, \ldots, r_m)$ and a ground substitution μ for this type rule such that $\mu(r_i) = s_i \in \mathcal{H}$ for each $i \in \{1, \ldots, m\}$. Using Theorem 2, μ is a match substitution for $(s, r :=> v)$. Thus the transformation rule (6) can be applied to (t, u). This is a contradiction to the assumption. Note that $dom(\mu) \cap dom(\sigma) = \emptyset$ and that $\sigma \circ \mu$ is a t-unifier for M_1, where M_1 is obtained by applying the transformation rule (6) to M. □

Lemma 13 *Let M be a set of term pairs and let σ be a t-unifier for M, where M is not a set of term pairs in solved form and $dom(\sigma) \subseteq var(M)$. Then there is a transformation rule and a t-unifier σ_1 for M_1, where M_1 is obtained by applying this transformation rule to M and*

- $\sigma_1(x) = \sigma(x)$ *for each* $x \in dom(\sigma)$, *if the transformation rule (6) has not been applied;*
- $\sigma_1 = \sigma \circ \mu$ *for some substitution* μ, *if the transformation rule (6) has been applied.*

Proof. Because of Lemma 12 there is a transformation rule which can be applied to M. The definition of σ_1 depends on the applied transformation rule. If σ_1 is defined and τ is a well-typed ground substitution for σ, then we have to show that $\mathcal{A} \models \sigma_1 \circ \tau(t) = \sigma_1 \circ \tau(u)$ for each $(t, u) \in M$. Such as in the proof of Lemma 11 we regard the cases corresponding to the numbers of the applied transformation rule (also, the notations are chosen in the same way). In the cases (1), (2), (3), (4), and (5), we can define $\sigma_1 := \sigma$ and the proposition can easily be shown (see also proof of lemma 11 in these cases). The proposition for case (6) follows from the proof of Lemma 12. We define $\sigma_1 := \sigma \cup \{x_1/\sigma(t_1), \ldots, x_n/\sigma(t_n)\}$ in the case (7). Since $\sigma_1(x) = \sigma(x)$, $\sigma_1(x_i) = \sigma(t_i)$, and $\sigma_1(t_i) = \sigma(t_i)$, the proof is straightforward. □

Now we can state the algorithm for the t-unification. We suppose that two object terms t, u are given. Initially, $U = \{(t, u)\}$ and D is empty, and at the end U must be empty or a condition for typed variables in solved form, if a t-unifier exists. The kernel of this algorithm looks like this:

```
repeat
    choose a term pair (t,u) of U ;
    if    some transformation rule can be applied to (t,u) or to (u,t)
    then  apply this rule to ({(t,u)} ∪ U) + D and
          let U1 + D1 be the result of this transformation ;
    else  stop with failure ;
    U := U1 ; D := D1
until U is empty or U is a condition of typed variables in solved form
      or failure
```

This unification algorithm is non-deterministic with respect to the selection of the term pair as well as to the choice of the type rule to be used if the transformation rule (6) is applied. We can suppose that the application of transformation rule (7) is always followed by the application of rule (3) with respect to the term pair $(x, f(x_1, \ldots, x_n))$ generated by rule (7). If the algorithm stops with a set in solved form, it generates a t-unifier (see Lemma 10). Otherwise, the algorithm will either stop with failure or does not terminate. If the algorithm stops with failure, then there is not any t-unifier for the given terms.

Suppose σ to be a t-unifier of the object terms t and u such that $\sigma(t)$, $\sigma(u)$ are ground terms. Then $\sigma(t) = \sigma(u) \in \mathcal{H}$ and there is a SLD-refutation of $t = u$ which generates an answer substitution τ such that $\sigma = \tau \circ \lambda$ for a substitution λ. Since this SLD-refutation can be used for a control mechanism of the unification algorithm, there is a sequence of transformation rules such that the algorithm terminates for the unification of t and u.

Using Lemma 11 and Lemma 10 the correctness of the given unification algorithm can easily be proved.

Theorem 4 (correctness) *Let t, u be two object terms. If σ is a substitution generated by the given algorithm, then σ is a t-unifier for t and u.*

Since, for a definition of types, we do not require that the elements of a type can be generated by the types rules in a unique way, the given algorithm does not satisfy the general completeness property. This means that, in general, the algorithm can not generate a t-unifier which is equal or more general than a given t-unifier of two terms (see also Lemma 13).

Example 5 We regard the following definition of types:

$$nat :=> 0 \qquad neg :=> -s(x : nat)$$
$$nat :=> s(x : nat)$$

$$int :=> x : nat \qquad r :=> c(y : nat)$$
$$int :=> x : neg \qquad r :=> c(y : neg)$$

A t-unifier for the terms $x : r$ and $c(y : r)$ is the substitution σ defined by $\{x/c(y{:}int)\}$. This t-unifier can not be generated by the unification algorithm. But the algorithm can generate the following t-unifiers: $\sigma_1 = \{x/c(z : nat), y/z : nat\}$ and $\sigma_2 = \{x/c(z : neg), y/z : neg\}$. If the type rules for the definition of the type r is equivalently replaced by $r :=> c(y : int)$, then the unification algorithm can generate the t-unifier σ. □

Using Lemma 13, Lemma 12, and Lemma 11 the following weak compelteness of the given unification algorithm can easily be shown.

Theorem 5 (weak completenes) *Let t and u be two object terms. If σ is a t-unifier for t, u and τ is a well-typed ground substitution for σ, then the given unification algorithm can generate a t-unifier μ for t, u such that $\sigma \circ \tau = \mu \circ \lambda$ for some substitution λ.*

Note that the proposed unification algorithm is given for theoretical investigations. This algorithm has to be extended in order to get an efficient unification algorithm. Some sequences of transformation rules can be summerized to one rule and, for special cases, new transformation rules can be added. For instance, the following transformation rule is useful in some cases:

(8) $(\{(t : s, u)\} \cup U) + D \implies (\{t : \lambda(r_1), u)\} \cup U) + D$,
if there is a type rule $r :=> z : r_1$ and a match substitution λ for $(s, r :=> z : r_1)$, and transformation rule (4) can not be applied.

Furthermore, the control mechanism of the unification algorithm can be improved. The information which data terms belong to the types could be used for the control of this algorithm. For example, the t-unification of the terms $x:s$ and $y:r$ are impossible, if there is not any data term belonging to s and r. A control mechanism using this kind of information will be discussed in further research. This control mechanism is based on the idea given in [11] for E-unification.

The t-unification of a ground term t and a variable can only be succesful, if type checking of t is succesful, i.e. the algorithm for t-unification includes type checking. Note that the algorithm for t-unification can generate the elements belonging to the intersection of two types (this intersection has not to be a defined type).

Example 6 We regard the following part of a definition of types, where c_1, c_2, c_3, c_4, c_5 belong to $\mathcal{F}_c$ and $\mathcal{F}_t = \{a_1, a_2, a_3, b_1, b_2, b_3, b_4, \ldots\}$:

$$\begin{array}{lll} a_1 :=> x:b_1 & a_2 :=> x:b_2 & a_3 :=> x:b_3 \\ a_1 :=> x:b_2 & a_2 :=> x:b_3 & a_3 :=> x:b_4 \end{array}$$

$$\begin{array}{ll} b_2 :=> c_1 & b_3 :=> c_3 \\ b_2 :=> c_2 & b_3 :=> c_4 \\ b_2 :=> c_3 & b_3 :=> c_5 \\ b_2 :=> c_4 & \ldots \\ \ldots & \end{array}$$

Using the transformation rules given above, a succesful sequence of transformations for the t-unification of the terms $x:a_1$ and $y:a_2$ is

$$\{(x:a_1, y:a_2)\} + \emptyset \Longrightarrow_{(6)} \{(x, x_1:b_2), (y:a_2, x_1:b_2)\} + \emptyset$$
$$\Longrightarrow_{(6)} \{(x, x_1:b_2), (y, y_1:b_2), (x_1:b_2, y_1:b_2)\} + \emptyset$$

Thus, the substitution $\{x/z:b_2, y/z:b_2\}$ is a t-unifier for these terms.

If we can also use the additional transformation rule (8), then we get the following sequence of transformations for the same problem:

$$\{(x:a_1, y:a_2)\} + \emptyset \Longrightarrow_{(8)} \{(x:b_2, y:a_2)\} + \emptyset$$
$$\Longrightarrow_{(8)} \{(x:b_2, y:b_2)\} + \emptyset$$

A succesful sequence of transformations for the t-unification of the terms $x:a_1$ and $y:a_3$ is

$$\{(x:a_1, y:a_3)\} + \emptyset \Longrightarrow_{(8)} \{(x:b_2, y:a_3)\} + \emptyset$$
$$\Longrightarrow_{(8)} \{(x:b_2, y:b_3)\} + \emptyset \Longrightarrow_{(6)} \{(x:b_2, c_3), (y, c_3)\} + \emptyset$$
$$\Longrightarrow_{(6)} \{(x, c_3), (c_3, c_3), (y, c_3)\} + \emptyset \Longrightarrow_{(2)} \{(x, c_3), (y, c_3)\} + \emptyset$$
$$\Longrightarrow_{(3)} \{(x, c_3)\} + \{(y, c_3)\} \Longrightarrow_{(3)} \emptyset + \{(x, c_3), (y, c_3)\}$$

Thus one solution of this t-unification is the substitution $\{x/c_3, y/c_3\}$ (another solution is $\{x/c_4, y/c_4\}$). □

Another kind of type unification is given in the next definition.

Definition 19 (strong t-unification) Suppose TD to be a definition of types. Two object terms t and u are *strongly t-unifiable* if there is a substitution σ (called *strong t-unifier for t and u*) such that the following conditions are satisfied:

- $dom(\sigma) \cap var(codom(\sigma)) = \emptyset$ and $codom(\sigma) \subseteq \mathcal{D}$;
- there is a substitution τ such that $var(\sigma(t)) \cup var(\sigma(u)) \subseteq dom(\tau)$, $codom(\tau) \subseteq \mathcal{D}_0$, and $\mathcal{A} \models \tau\circ\sigma(t) = \tau\circ\sigma(u)$;
- if τ is a substitution such that $var(\sigma(t)) \cup var(\sigma(u)) \subseteq dom(\tau)$, and $codom(\tau) \subseteq \mathcal{D}_0$, then $\mathcal{A} \models \tau\circ\sigma(t) = \tau\circ\sigma(u)$.

Obviously, a strong t-unifier for two object terms is a t-unifier for these terms, too. Let s_1 and s_2 be ground type terms, i.e. s_1 and s_2 denote types. If $x : s_1$ and $y : s_2$ are strongly t-unifiable, then the strong t-unifiers for these terms could generate the elements of the intersection of the given types. Assume that the type r is the intersection of s_1 and s_2. The substitution $\{x/z:r,\ y/z:r\}$ is a t-unifier, but not a strong t-unifier for $x : s_1$ and $y : s_2$. Nevertheless, the strong t-unification is useful for some applications.

A modified version of the given unification algorithm can be used for the strong t-unification. In the modified version, U has to be empty, if the algorithm stops succesfully.

6 Concluding Remarks

The use of declarations for type rules simplifies the syntax of type rules and may avoid errors. For example, the type of polymorphic lists with a certain number of elements may be defined by

type *list* : *nat* × *type*
$list(0, y)$:=> []
$list(s(x), y)$:=> $[z_1 : type(y) | z_2 : list(x, y)]$

Furthermore, abbreviations can be introduced for some sets of type rules. For example, the notation

$car := \{ford, opel, mercedes, vw, bmw\}$

may be used instead of the type rules:

car :=> *ford*
car :=> *opel*
car :=> *mercedes*
car :=> *vw*
car :=> *bmw*

The notation

vehicle := *car* ++ *ship* ++ *airplane*

may be used instead of the type rules:

vehicle :=> x : *car*
vehicle :=> x : *ship*
vehicle :=> x : *airplane*

The presented type system can be integrated in deduction systems. Especially, if the elements of the universe of discourse can be represented by data terms, the suggested type system has many advantages (see Introduction). This type system is also a natural and powerful extension of a functional-logic programming language, if the functional component is based on term rewriting and the logical component is based on Horn clauses ([6], [5], [10]). In such a programming language, it is very useful to extend the suggested type system by conditional type rules (for example, $t :=> x \; :- \; p(x)$, where $p/1$ is a predicate defined by Horn clauses). Thus a type may be defined by means of predicates. The logical foundations of a functional-logic programming language with an extension of of the suggested type system is one of our aims of research. Further research includes the extension of the type system in the sense that type inference is also defined for terms included type terms with variables. Note that the idea of the development of the suggested type system has been originated from experiments with a functional extension of Prolog called F-Prolog ([10]). Currently, this type system is being implemented in Prolog.

References

[1] Ait-Kaci,H., and R.Nasr: *LOGIN: A Logic Programming Language with Built-In Inheritance.* J. Logic Programming 3 (1986), 185-215.

[2] Beierle,C.: *Types, Modules and Databases in the Logic Programming Language PROTOS-L.* In [3], 73-110.

[3] Bläsius,K.H., U.Hedtstück, and C.-R. Rollinger (Eds.): *Sorts and Types in Artificial Intelligence.* LNAI 418, Springer-Verlag, 1990.

[4] DeGroot,D., and G.Lindstrom (eds.): *Logic Programming. Functions, Relations and Equations.* Prentice-Hall, Englewood Cliffs (New Jersey) 1986.

[5] Furbach,U., and S.Hölldobler: *Equations, Order-Sortedness and Inheritance in Logic Programming.* Technical Report, FKI–110–89, TU München 1989.

[6] Furbach,U.: *Logische und Funktionale Programmierung. Grundlagen einer Kombination.* Vieweg Verlag, Braunschweig 1991.

[7] Genesereth,M.R., and N.J.Nilsson: *Logical Foundations of Artificial Intelligence.* Morgan Kaufmann Publ., Los Altos 1987.

[8] Goltz,H.-J.: *Functional Data Term Models and Semantic Unification.* In: [12], 158-167.

[9] Goltz,H.-J., and H.Herre: *Grundlagen der logischen Programmierung.* Akademie Verlag, Berlin 1990.

[10] Goltz,H.-J.: *Functional Extension of Logic Programming.* In: U.Geske and D.Koch: Contributions to Artificial Intelligence. Research in Informatics vol. 1, Akademie Verlag, Berlin 1991, 55 – 73.

[11] Goltz,H.-J.: *Ein praktischer Algorithmus für die E-Unifikation.* IWBS-Report 166, IBM Stuttgart 1991.

[12] Grabowski,J., P.Lescanne, and W.Wechler (eds.): *Algebraic and Logic Programming*. Akademie-Verlag Berlin 1988 (also Lecture Notes in Computer Sciences 343, Springer-Verlag, Berlin, Heidelberg, New York, Tokyo 1988).

[13] Hanus,M.: *Parametric Order-Sorted Types in Logic Programming*. In: Proceedings of the TAPSOFT'91, LNCS 494, Springer-Verlag, 1991, 181-200.

[14] Hanus,M.: *Logic Programming with Type Specifications*. To appear in: F.Pfenning (ed.), Types in Logic Programming, MIT Press, 1992.

[15] Hölldobler,S.: *Foundations in Equational Logic Programming*. LNCS 353, Springer-Verlag, 1989.

[16] Huet,G., and D.Oppen: *Equations and Rewrite Rules: A Survey*. In: R.Book (ed.), Formal Language Theory: Perspectives and Open Problems, Academic Press, New York, London 1980, 349- 405.

[17] Jouannaud,J.-P., C.Kirchner, H.Kirchner, and A.Megrelis: *OBJ: Programming with Equalities, Subsorts, Overloading and Parameterization*. In: [12], 41-52.

[18] Lakshman,T.K. and U.S.Reddy: *Typed Prolog: A Semantic Reconstruction of the Mycroft-O'Keefe Type System*. In: V.Saraswat, K.Ueda (eds.), Logic Programming, Proc. Int. Symposium 1991, MIT Press, 1991, 201-217.

[19] Lloyd,J.W.: *Foundations of Logic Programming*. 2.Ed., Springer-Verlag, Berlin, Heidelberg, New York, Tokyo 1987.

[20] Martelli,A., and U.Montanari: *An Efficient Unification Algorithm*. ACM Transactions on Programming Languages and Systems 4 (1982), 258-282.

[21] Montini,G.: *Efficiency Considarations on Built-in Taxonomic Reasoning in Prolog*. In: Proc. Int. Joint Conf. Artificial Intelligence 1987, 68-75.

[22] Mycroft,A., and R.A.O'Keefe: *A Polymorphic Type System for Prolog*. Artificial Intelligence 23 (1984), 295-307.

[23] Shin,D.W., J.H.Nang, S.R.Maeng, and J.W.Cho: *A Typed Functional Extension of Logic Programming*. Journal New Generation Computing 10 (1992), 197-221.

[24] Smolka,S., W.Nutt, J.A.Goguen, and J.Meseguer: *Order-Sorted Equational Computation*. SEKI-Report SR-87-14, FB Informatik, Universität Kaiserslautern, 1987.

[25] Smolka,G.: *Logic Programming with Polymorphicaly Order-Sorted Types*. In: [12], 53-70.

[26] Walther,C.: *A Many-sorted Calculus Based on Resolution and Paramodulation*. Research Notes in Artificial Intelligence, Pitman and Morgan Kaufman Publ., London; Morgan Kaufmann Publ., Los Altos (Calif.) 1987.

An Ordered Resolution and Paramodulation Calculus for Finite Many-Valued Logics*

Nicolas Zabel

LIFIA–IMAG–CNRS
46, Av. Félix Viallet
F 38 031 Grenoble Cedex - FRANCE
e-mail: zabel@lifia.imag.fr(uucp)

Abstract. The paper deals with a *graded equality relation* for first-order finite many-valued logics with function symbols. The class of interpretations compatible with a graded equality is characterized, thus clarifying the assumptions underlying graded equality. We present also a resolution/paramodulation calculus for first-order finite many-valued logics including graded equality, this calculus is refined using complete simplification orderings.

1 Introduction

The need for many-valued logics naturally arises when one wishes to model fuzzy or incomplete knowledge of the world. The additional truth values represent degrees of belief. Our partial knowledge of the world may also affect the individuals: they may be quite "similar", but are nevertheless distinct from each other. Within many-valued logics it is possible to express this by extending the equality relation, essentially bivalent [5], to a *graded equality relation* ranging over the whole set of truth values [6]. Actually we supply the set of individuals with a notion of distance ranging over the truth values [16], and Scott links the graded equality to this distance between two individuals. It seems then quite natural to assume that the closer the individuals are, the more similar they behave with respect to the propositions.

In [6], Morgan discusses in detail the assumptions underlying the similarity relation; in his reading of the graded equality, two terms are the more similar, the more they share some properties.

We define here a Resolution and Paramodulation calculus for such similarity relations. The Resolution and Paramodulation are the most widely used inference rules in automated theorem proving [13, 12, 4], basically because of their efficiency. Therefore, the Resolution Principle (clausal or not) for the classical logic has often been shifted to non-classical logics [9, 10, 18].

In the remainder of this section, we explain briefly our notations and give the semantics of the logics considered. In section 2 we recall the definition of transfinite semantic trees due to Rusinowitch [15] (an alternative presentation of this

* This work has been partially supported by the MEDLAR–BRA Esprit project (CEC $N°3125$) and the PRC - IA (MRT–CNRS, France).

technique and its links to forcing can be found in [11]) and extend this technique to reasoning with similarity-interpretation. The links between the similarity relation and the metrics on the set of individuals are investigated. The extended transfinite trees technique is used in section 3 to prove the completeness of the calculus introduced in that section. We conclude the paper with some directions of future work.

1.1 Some Notations

In the sequel, Σ and Ω denote respectively a signature of terms and of atoms (see for ex. [2]). For any $n \in IN$, Σ_n and Ω_n are the sets of all n-ary symbols in Σ and Ω respectively $\mathcal{V}$ is an infinite set of variables sharing no element with Σ or Ω. $\mathcal{G}(\Sigma)$ is the set of all ground terms (Herbrand Universe) and $\mathcal{T}(\Sigma, \mathcal{V})$ is the set of all terms with variables from $\mathcal{V}$. $\mathcal{A}(\Sigma, \Omega)$ is the set all ground atoms (Herbrand base) and $\mathcal{A}(\Sigma, \Omega, \mathcal{V})$ the set of all atoms with variables from $\mathcal{V}$.

If f and P are n-ary symbols of respectively Σ and Ω, $f\vec{s}$ and $P\vec{s}$ are the objects whose arguments are $s_1, \ldots, s_n$. If p is a tree address (or position) of $f\vec{s}$ and $P\vec{s}$ (see for ex. [2]), $f\vec{s}|_p$ and $P\vec{s}|_p$ is the object at position p. (For atoms we assume tacitly that p is not the top position, hence the object is always a term.) The replacement of that object by the term t is written $f\vec{s}[t]_p$ and $P\vec{s}[t]_p$. The set of all positions of $f\vec{s}$ or $P\vec{s}$ is noted by $\mathcal{P}os(f\vec{s})$ and $\mathcal{P}os(P\vec{s})$. The topmost (rootmost) position is written Λ.

If σ is a substitution i.e. a mapping from $\mathcal{V}$ to $\mathcal{T}(\Sigma, \mathcal{V})$, then its domain (i.e. the finite set of all these variables such that $x\sigma \neq x$), is written $\mathcal{D}om(\sigma)$. $\mathcal{V}Ran(\sigma)$ is the set of these variables which occur in some term $x\sigma$ with $x \in \mathcal{D}om(\sigma)$. ι denotes the substitution with the empty domain i.e. the identity. Sometimes we shall give explicitly the substitution σ by writing $\{x \mapsto x\sigma : x \in \mathcal{D}om(\sigma)\}$. A substitution is ground iff for each $x \in \mathcal{D}om(\sigma)$ $x\sigma$ is a ground term. If $\vec{s}$ and $\vec{t}$ are two n-tuples of terms, and if these terms are unifiable, let $\vec{s} \sqcap \vec{t}$ denote the most general unifier of $\vec{s}$ and $\vec{t}$ (normalized in some unspecified way in order to insure the uniqueness). If the tuples were not unifiable, $\vec{s} \sqcap \vec{t}$ is not defined.

1.2 Semantics of the Logics

The logic used in this paper, is similar to that in [14, 7]. M denotes the number of truth values. $\mathcal{V}al = \{0, \frac{1}{M-1}, \ldots, \frac{M-2}{M-1}, 1\}$ are the truth values. S is a threshold $1 < S \leq M$, that is to say the truth values in $\{0, \ldots, \frac{M-S}{M-1}\}$ are considered as false. Those in $\mathcal{D}es = \{\frac{M-S+1}{M-1}, \ldots, 1\}$ are accepted as true, the latter are called the *designated values*.

The connectives considered here are $\vee, \wedge, \forall, \exists$ and J_k for each $1 \leq k \leq M$. Every atom (built up with symbols from Σ,Ω and $\mathcal{V}$) is a well formed formula (WFF). If F and G are WFFs, then $F \vee G$, $F \wedge G$, $J_k(F)$ are WFFs. If F is a WFF and $x \in \mathcal{V}$, then $\forall x F$ and $\exists x F$ are WFFs.

An *interpretation* $\Im$ with *domain* $\mathcal{I}$ is a mapping which associates to each non logical symbol (i.e. to the elements of $\Sigma, \Omega, \mathcal{V}$) respectively a mapping into $\mathcal{I}$, a mapping from $\mathcal{I}^n$ to $\mathcal{V}al$, and an element of $\mathcal{I}$:

$$\begin{aligned} \Im : \Sigma_n &\rightarrow \mathcal{I}^{\mathcal{I}^n} && \text{for each } n \in I\!N \\ \Omega_n &\rightarrow \mathcal{V}al^{\mathcal{I}^n} && \text{for each } n \in I\!N \\ \mathcal{V} &\rightarrow \mathcal{I} \end{aligned}$$

If $x \in \mathcal{V}$ and $c \in \mathcal{I}$, $\Im_c^x$ is the interpretation that differs only at x from $\Im$ and $\Im_c^x(x) = c$.

For the sake of conciseness, we shall confuse $\Im$ with its homomorphic extension to the terms and the WFFs. The semantics of the logics is given in Figure 1. A sentence F is a *theorem* iff in every interpretation $\Im$, $\Im(F) \in \mathcal{D}es$.

$$\begin{aligned} \Im(f\vec{s}) &= \Im(f)(\Im(s_1), \ldots, \Im(s_n)) \text{ where } f \in \Sigma_n \text{ and } s_i \text{ are terms} \\ \Im(P\vec{s}) &= \Im(P)(\Im(s_1), \ldots, \Im(s_n)) \text{ where } P \in \Omega_n \text{ and } s_i \text{ are terms} \\ \Im(F \vee G) &= \max(\Im(F), \Im(G)) \\ \Im(F \wedge G) &= \min(\Im(F), \Im(G)) \\ \Im(J_k(F)) &= \begin{cases} 1 \text{ if } \Im(F) = k \\ 0 \text{ otherwise} \end{cases} \\ \Im(\forall x F) &= \min_{c \in \mathcal{I}}(\Im_c^x(F)) \\ \Im(\exists x F) &= \max_{c \in \mathcal{I}}(\Im_c^x(F)) \end{aligned}$$

Fig. 1. Semantics of the Logics

Later we shall introduce two interpreted predicates $\doteq$ (for bivalent equality) and $\approx$ (for similarity). The class of interpretations considered then will be given at that moment. We have chosen these many-valued logics, because they are the most intuitive which has a Skolem Normal Form and a Clausal Form where only atomic formulas are in the scope of J_k [7].

2 Semantical Issues and Semantic Trees

This section heavily uses the results of Rusinowitch in [15]. In that work, Rusinowitch investigates in detail the benefit for automated theorem proving in using rewriting techniques. One of its most striking achievements is a technique which allows to prove the completeness of resolution and paramodulation strategies by reasoning on transfinite trees.

Its author states the results for the classical logics; they are extended here to the many-valued case.

2.1 Partial Interpretations

Definition 1. An ordering $<$ is called an *CSO* (*complete simplification ordering*) iff it satisfies the following conditions:

- $<$ is total on $\mathcal{G}(\Sigma)$ and $\mathcal{A}(\Sigma, \Omega)$
- $<$ is well founded
- $<$ is substitution-stable i. e. if $v, w \in \mathcal{G}(\Sigma) \cup \mathcal{A}(\Sigma, \Omega)$ and θ is a substitution and if $v < w$ then $v\theta < w\theta$
- $<$ is monotone i. e. if $s, t \in \mathcal{G}(\Sigma)$ and $w \in \mathcal{G}(\Sigma) \cup \mathcal{A}(\Sigma, \Omega)$ and if $s < t$ than $w[s] < w[t]$.
- $<$ has the subterm property:
  - if s is a strict subterm of u, then $s < u$
  - if s is a strict subterm of w and w is neither an $\doteq$-atom nor an $\approx$-atom, then $s \doteq t < s \approx t$, $s \doteq t < w$ and $s \approx t < w$
  - if s is a subterm of a or b, then $s \doteq t < a \doteq b$ and $s \approx t < a \approx b$

Definition 2. Let $<$ be a *CSO* on the Herbrand base $\mathcal{A}(\Sigma, \Omega)$ whose sequence of ground atoms is $\{A_i\}_{i<\lambda}$, where λ is its ordinality. Given A_α, let us note W_α the initial segment $\{A_i : i < \alpha\}$. An application $\Im$ from W_α into $\mathcal{V}al$, a *partial E-interpretation on* W_α, iff the following conditions are fulfilled:

- (*bivalence of equality*) for each term a and b, if $(a \doteq b) \in W_\alpha$, either $\Im(a \doteq b) = 1$ or $\Im(a \doteq b) = 0$
- (*reflexivity of equality*) for each term a, if $(a \doteq a) \in W_\alpha$, $\Im(a \doteq a) = 1$
- (*substitutivity of equality*) for each term a, b, each predicate $P \in \Omega_n$, each n-tuple of terms $\vec{s}$ and each $p \in \mathcal{P}os(P\vec{s}) \setminus \{\Lambda\}$
  if $(a \doteq b), P\vec{s}[a]_p, P\vec{s}[b]_p \in W_\alpha$ and $\Im(a \doteq b) = 1$,
  then $\Im(P\vec{s}[a]_p) = \Im(P\vec{s}[b]_p)$.

$\Im$ *is a partial* $\approx$*-interpretation on* W_α iff

- (*reflexivity of similarity*) for each term a, if $(a \approx a) \in W_\alpha$, $\Im(a \approx a) = 1$
- (*similar terms are almost indiscernible*) for each term a, b, each predicate $P \in \Omega_n$, each n-tuple of terms $\vec{s}$ and each $p \in \mathcal{P}os(P\vec{s}) \setminus \{\Lambda\}$
  if $(a \approx b), P\vec{s}[a]_p, P\vec{s}[b]_p \in W_\alpha$, then $1 - \Im(a \approx b) \geq |\Im(P\vec{s}[a]_p) - \Im(P\vec{s}[b]_p)|$.

An *interpretation* is a partial interpretation defined on W_λ. Every E-interpretation is actually an $\approx$-interpretation too.

Definition 3. Let $w[s]_p$ and $w[t]_p$ be two atoms, and $<$ be a *CSO* on the Herbrand base.

Let $\Im$ a partial E-interpretation on W_α $(\alpha < \lambda)$. $w[s]_p$ *is* $\Im$*-reduced into* $w[t]_p$, if $s \doteq t \in W_\alpha$, $s > t$, $\Im(s \doteq t) = 1$. $\Im$ forces the truth value of $w[s]_p$ to be that of $w[t]_p$.

Let $\Im$ be now a partial $\approx$-interpretation W_α, $w[s]_p$ *is* $\Im$*-reduced into* $w[t]_p$, if $s \approx t \in W_\alpha$, $s > t$, $\Im(s \approx t) > \Im(w[t]_p)$ or $\Im(w[t]_p) + \Im(s \approx t) > 1$.[2] Or stated otherwise $\Im$ bounds the truth values assignable to $w[s]_p$ in every extension of $\Im$.

[2] the two last conditions use $1 - \Im(s \approx t) \geq |\Im(w[s]_p) - \Im(w[t]_p)|$

2.2 $\approx$–interpretations and Metrics on the Individuals

Definition 4. Let $\Im$ be an $\approx$–interpretation with domain $\mathcal{I}$. $\Im$ is *a standard $\approx$–interpretation* iff for all $a, b \in \mathcal{I}$, $\Im(a \approx b) = 1 \Leftrightarrow \Im(a) = \Im(b)$.

It is easy to check that for each $\approx$–interpretation $\Im$ there exists a standard $\approx$–interpretation $\Im'$ such that for all WFFs F $\Im(F) = \Im'(F)$. (As for standard E-interpretations of the classical logic, the domain of $\Im'$ can be taken as being $\mathcal{I}/\equiv$ where $a \equiv b$ iff $\Im(a \approx b) = 1$, there remains to check the well-definiteness of $\Im(P)/\equiv$ and $\Im(f)/\equiv$ for $f \in \Sigma$ and $P \in \Omega$.)

Definition 5. Let $\Im$ be a standard $\approx$–interpretation with domain $\mathcal{I}$. We define $d_\Im$, the distance on individuals induced by $\Im$ as:

$$\begin{aligned} d_\Im : \mathcal{I} \times \mathcal{I} &\longrightarrow \mathcal{V}al \\ \langle a, b \rangle &\longmapsto 1 - \Im(a \approx b) \end{aligned}$$

Theorem 6. *Let $\Im$ be a standard $\approx$–interpretation with domain $\mathcal{I}$. $d_\Im$ is a distance on $\mathcal{I}$.*

Proof. Let a, b, c be individuals of $\mathcal{I}$. As usual, we define the sum of two truth values a and b,$a+b$ as the truth values denoted by the rational number $\min(1, a+b)$.

$d_\Im(a, a) = 0$: obvious since $\Im(a \approx a) = 1$ because of the reflexivity of $\approx$. If $d_\Im(a, b) = 0$, then $a = b$ because $\Im$ is a standard $\approx$–interpretation.

$d_\Im(a, b) = d_\Im(b, a)$: follows from the symmetry of $\approx$.

$d_\Im(a, c) \leq d_\Im(a, b) + d_\Im(b, c)$: follows from $1 - \Im(a \approx b) \geq |\Im(a \approx c) - \Im(b \approx c)|$ ("similar terms are almost indiscernible") i.e. $d_\Im(a, b) \geq |d_\Im(a, c) - d_\Im(b, c)|$.

For $M = 3$, $\approx$ has an appealing semantics, when $\frac{1}{2}$ is understood as *unknown*. Let $\Im$ be a standard $\approx$–interpretation with domain $\mathcal{I}$ that contains a and b.

$\Im(a \approx b) = 1$ iff $a = b$

Let a' and b' be any terms such that $\Im(a') = a$ and $\Im(b') = b$. If a and b are known to be discernible in $\Im$, i. e. there is an atom $P\vec{s}$ such that $\Im(P\vec{s}[a']_p) = 1$ and $\Im(P\vec{s}[b']_p) = 0$ where $p \in \mathcal{P}os(P\vec{s}) \setminus \{\Lambda\}$, then $1 - \Im(a \approx b) \geq 1$, i. e. $\Im(a \approx b) = 0$. That is to say whenever a and b are *not* known to be discernible, $\Im(a \approx b) = \frac{1}{2}$.

Definition 7. Let $\langle E, d \rangle$ and $\langle F, d' \rangle$ be metric spaces and $f : E^n \longrightarrow F$ be a mapping. We shall say that f is *non divergent on its ith argument* iff for each $\langle a_1, \ldots, a_n, b \rangle \in E^{n+1}$ $d'(f(a_1 ... a_i ... a_n), f(a_1 ... b ... a_n)) \leq d(a_i, b)$. f is *non divergent* iff f is non divergent on every argument.

If we take as distance between the truth values v and w $|v - w|$, the condition "similar terms are almost indiscernible" says that a standard $\approx$–interpretation maps predicates to non divergent mappings.

Actually this condition constrains also the functions to be interpreted by non divergent mappings as easily seen when applying this condition on $f \in \Sigma_n$ and $a_1, \ldots, a_n, a, b \in \mathcal{G}(\Sigma)$:

$$1 - \Im(a \approx b) \geq |\Im(f(...a...) \approx f(...a...)) - \Im(f(...a...) \approx f(...b...))| \quad (1)$$
$$\geq |1 - \Im(f(...a...) \approx f(...b...))| \quad (2)$$
$$d_\Im(a,b) \geq d_\Im(f(...a...), f(...b...)) \quad (3)$$

The converse holds too:

Theorem 8. *Let d be a $\mathcal{V}al$-valued distance on a set of individuals $\mathcal{I}$, If $\Im$ is an interpretation with domain $\mathcal{I}$, such that:*

- *$\Im(a \approx b) = 1 - d(a,b)$ and*
- *every function and predicate symbol is interpreted as a non divergent mapping by $\Im$,*

then $\Im$ is a standard $\approx$-interpretation.

Proof. The reflexivity of the similarity is verified in $\Im$. The condition "similar terms are almost indiscernible" is proved for the case in which P is $\approx$, and then for the general case. Both are proofs by induction on the length of the position p. They use the inequalities (1–3) given above.

This brief mathematical analysis of the $\approx$–interpretations made clear what the $\approx$–interpretations really are, and bears witness to the plausibility of the way we included function symbols in many-valued logics with a similarity relation (cf. [6]).

2.3 E–semantic and $\approx$–semantic Trees

Definition 9. A *transfinite E-semantic tree ($\approx$-semantic)* is the set of all partial E–interpretations (respectively $\approx$–interpretations) ordered by the embedding relation of the mappings noted $\lhd$.

In [15, page 42], the following properties are stated:

- $\lhd$ is a well-founded ordering.
- If $\Im$ is a E–interpretation on $W_{\alpha+1}$, then $\Im|_{W_\alpha}$ (i.e. the restriction of $\Im$ to the literals in W_α) is the predecessor of $\Im$ for the $\lhd$ ordering.

This is also true for $\approx$–interpretations.

Definition 10. Let S be a set of clauses. *The maximally consistent tree* of S, noted $MCT(S)$, is the maximal subtree of a transfinite E–semantic ($\approx$–semantic) tree of S such that:

> if for all c of S and all nodes $\Im$ of $MCT(S)$ and all ground substitutions σ such that $c\sigma$ contains only ground terms, then $\Im(c\sigma) = 1$, when $\Im(c\sigma)$ is defined.

The following terminology is used:

- if a node $\Im$ of the (E– or $\approx$–)semantic tree (for a given ordering) falsifies a clause $c \in S$, but none of the $\lhd$–predecessor of $\Im$ does falsify any clause of S, $\Im$ is called a *failure node*.
- a node of $MCT(S)$ is a *deduction node* if all of its $\lhd$–successors are called failure nodes.
- a *increasing sequence of nodes* of an semantic tree is a totally $\lhd$–ordered set $\{\Im_i\}_{i<\alpha}$ $(\alpha < \lambda)$ of partial interpretations.

Semantic trees are pictured as ordered trees, their edges are labeled with the truth values in an decreasing order, the rightmost being labeled with 0 (from the "more true" to the "more false").

The following lemma (cf. [15, page 44]) is still valid for the many-valued maximally consistent trees:

Lemma 11. *(closure lemma) The limit of an increasing sequence of elements of $MCT(S)$ belongs to $MCT(S)$.*

Proof. Let us assume that there is a ground instance C of clause of S that falsifies the limit. Since C is finite and $<$ is total on ground terms, there is a greatest literal in C. Let A_α be this greatest literal in C. α is a successor ordinal. Therefore there is an element of the sequence considered which also falsifies C, what contradicts the fact that this element belongs to $MCT(S)$.

3 A Resolution and Paramodulation Calculus

For proving that F is a theorem, i.e. F takes only designated truth values, one proves that $\bigvee_{k=S+1}^{M} J_k(F)$ cannot be satisfied. It is equivalent to prove that none of $J_k(F)$ (for $S < k \leq M$) can be satisfied, since $\vee$ is a standard disjunction; the problem is naturally split into smaller ones, and hence decreases the size of the search space. The clauses obtained from $J_k(F)$ can only take the truth values 0 or 1, for every literal has the form $J_k(\ell)$. Indeed, only the restrictions of $\forall$, $\wedge$ and $\vee$ to $\{0, 1\}$ are relevant and moreover this restriction is just the two–valued logic the many-valued aspect of the logic is only observed at atomic level. The soundness of the inference rules of the proposed calculus uses this essential fact. In this sense, the Resolution and Paramodulation calculus presented in the sequel is very close to the classical case. As for the classical logic, Ordered Resolution and Paramodulation [17, 15] shrinks efficiently the size of the search space for a refutation. The soundness and completeness arguments for unrestricted Resolution and Paramodulation are easily deduced from the ordered case. Hence we shall present here only the latter.

We shall handle only the similarity of Morgan [6], which is more general than the bivalent equality. [6] introduces a Hilbert style calculus for many valued logics without function symbols, but with a similarity relation.

For the sake of completeness, let us say that the bivalent equality can be handled by adding the bivalence axiom to the set of clauses at hand: $\forall x J_1(x \approx x) \vee J_M(x \approx x)$ (see also [6]).

We shall assume that $\approx$ and $\doteq$ are symmetric predicates, i.e. we shall not distinguish $a \approx b$ from $b \approx a$, etc.

Definition 12. Our calculus $\mathcal{O}RP_M$ applies on clauses; a clause is a set of literals of the form $J_k(\ell)$ with ℓ being an atom, in fact a clause is the universal closure of the (finite) disjunction of the elements of this set. A clause is *ground* iff there is no variables in any literal of the clause.

In the sequel, $\mathcal{V}ar(C)$ denotes the set of all the variables occurring in C and $<$ is a CSO on the atoms of $\mathcal{A}(\Omega, \Sigma, \mathcal{V})$ and the terms of $\mathcal{T}(\Sigma, \mathcal{V})$.

- the renaming.

$$\frac{C}{C\sigma}$$ with σ is such that $\mathcal{V}ar(C) = \mathcal{D}om(\sigma)$ and $\mathcal{V}Ran(\sigma)$ contains only "fresh" variables

  The conclusion of this inference is noted by $\mathrm{REN}(C, \sigma)$.
  Tacitly before applying the other rules, a renaming step is applied to all the premises in such a way that the premises of the inference rules do not share any variable.
- the binary resolution.

$$\frac{J_i(\ell\vec{s}) \vee C \quad J_j(\ell\vec{t}) \vee C'}{C\sigma \vee C'\sigma}$$ with
  - $i \neq j$
  - $\sigma = \vec{s} \sqcap \vec{t}$ is defined
  - for each $A \in C\sigma \cup C'\sigma$, $A \not\leq \ell\vec{s}\sigma$
  - for each $i' \leq i$, $J_{i'}(\ell\vec{s}\sigma) \notin C\sigma$
  - for each $j' \leq j$, $J_{j'}(\ell\vec{t}\sigma) \notin C'\sigma$

  O-RES($J_i(\ell\vec{s}) \vee C$, $J_i(\ell\vec{s})$, $J_j(\ell\vec{t}) \vee C'$, $J_j(\ell\vec{t})$, σ) denotes its conclusion.
- the binary factorisation.

$$\frac{J_i(\ell\vec{s}) \vee J_i(\ell\vec{t}) \vee C}{J_i(\ell\vec{s}\sigma) \vee C\sigma}$$ with
  - $\sigma = \vec{s} \sqcap \vec{t}$ is defined
  - for each $A \in C\sigma \cup C'\sigma$ $A \not\leq \ell\vec{s}\sigma$

  O-FACT($J_i(\ell\vec{s}) \vee J_i(\ell\vec{t}) \vee C$, $J_i(\ell\vec{s})$, $J_i(\ell\vec{t})$, σ) denotes its conclusion.

Fig. 2. Inference Rules of $\mathcal{O}RP_M$(Part 1)

Definition 13. The inference rules of the calculus are those listed in Figures 2 and 3.

– the paramodulation.

$$\frac{J_i(s \approx s') \vee C \qquad J_j(\ell\vec{t}) \vee C'}{\bigvee_{|k-j|<i} J_k(\ell\vec{t}[s']_p)\sigma \vee C\sigma \vee C'\sigma}$$

with
- $\sigma = s \sqcap \ell\vec{t}|_p$ is defined
- $s\sigma \not\leq s'\sigma$
- for each $A \in C\sigma$ $A \not\leq (s\sigma \approx s'\sigma)$
- for each $i' \leq i$, $J_{i'}(s\sigma \approx s'\sigma) \notin C\sigma$
- for each $A \in C'\sigma$, $A \not\leq \ell\vec{t}\sigma$
- for each $j' \leq j$, $J_{j'}(\ell\vec{t}\sigma) \notin C'\sigma$

O-PAR($J_j(\ell\vec{t}) \vee C'$, $J_j(\ell\vec{t}), p$, $J_i(s \approx s') \vee C$, $J_i(s \approx s'), \sigma$) denotes its conclusion.

– the negated reflexivity rule

$$\frac{J_i(s \approx t) \vee C}{C\sigma}$$

with
- $\sigma = s \sqcap t$ is defined
- $i \neq 1$
- for each $A \in C\sigma$, $A \not\leq (s\sigma \approx t\sigma)$
- for each $i' \leq i$, $J_{i'}(s\sigma \approx t\sigma) \notin C\sigma$

O-REFL($J_i(s \approx t) \vee C$, $J_i(s \approx t), \sigma$) denotes its conclusion.

Fig. 3. Inference Rules of ORP_M(Part 2)

We call $ORP^1_M(S)$ the set of clauses obtained by adding to S the clauses deduced by applying the inference rules of ORP_M to clauses in S. We define $ORP^0_M(S) =_{df} S$. For each $n > 1$, let $ORP^{n+1}_M(S) =_{df} ORP^1_M(ORP^n_M(S))$ and $ORP_M(S) =_{df} \bigcup_{n \geq 0} ORP^n_M(S)$

The paramodulation rule is a bit surprising: the inference blurs the knowledge of the truth value of the literal obtained by the paramodulation step: we obtain the disjunction $\bigvee_{|k-j|<i} J_k(\ell\vec{t}[s']_p)\sigma$ instead of $J_j(\ell\vec{t}[s']_p)\sigma$ we would have obtained in presence of bivalent equality.

To see what happens, let us look at the following simple inference:

$$\frac{J_i(a \approx b) \; J_j(b \approx c)}{\bigvee_{|k-j|<i} J_k(a \approx c)}$$

Which simply expresses the triangle inequality $d_{\mathfrak{I}}(a,c) \leq d_{\mathfrak{I}}(a,b) + d_{\mathfrak{I}}(b,c)$.

Paramodulating into an argument of an $\approx$-atom is a reasoning based on the triangle inequality. Paramodulating into a deeper position of an $\approx$-atom or into an atom with a predicate symbol different from $\approx$ is a reasoning based on the fact that the functions and the predicates are non divergent.

Lemma 14. *REC, O-FAC, O-RES, O-PAR, O-REFL are sound rules of* ORP_M.

Proof. The proof goes exactly as for the classical case. Let $\Im$ be an arbitrary $\approx$–interpretation, whose domain is $\mathcal{I}$. If $\Im$ does not satisfy all the premises of a given inference step, this step is trivially sound. In the sequel we consider only the case in which all the premises are satisfied by $\Im$. Let $\vec{x}$ be the n variables of $\mathcal{V}Ran(\sigma) \cup \mathcal{V}ar(C\sigma) \cup \mathcal{V}ar(C'\sigma)$ and $\vec{c}$ be an n–tuple of elements of $\mathcal{I}$. Let us write $\Im'$ instead of $\Im^{\vec{x}}_{\vec{c}}$.

- the renaming. Its soundness is obvious.
- the resolution. The three cases we have to consider are:
  - $\Im'(J_i(\ell\vec{s}\sigma)) = 1$. Hence $\Im'(J_j(\ell\vec{t}\sigma)) = 0$, whence $\Im'(C'\sigma) = 1$
  - $\Im'(J_j(\ell\vec{t}\sigma)) = 1$. Hence $\Im'(J_i(\ell\vec{s}\sigma)) = 0$, whence $\Im'(C\sigma) = 1$
  - $\Im'(J_i(\ell\vec{s}\sigma)) \neq 1$ and $\Im'(J_j(\ell\vec{t}\sigma)) \neq 1$. Hence $\Im'(J_i(\ell\vec{s}\sigma)) = \Im'(J_j(\ell\vec{t}\sigma)) = 0$, whence $\Im'(C'\sigma) = \Im'(C\sigma) = 1$.

  A fortiori $\Im'(C\sigma \vee C'\sigma) = 1$. This reasoning holds for each $\vec{c}$. Hence $\Im \models C\sigma \vee C'\sigma$
- the factorization. $\Im'(J_i(\ell\vec{s}\sigma))$ can take only the values 0 or 1.
  - If $\Im'(J_i(\ell\vec{s}\sigma)) = 0$, then $\Im'(C\sigma) = 1$, because the premise has been assumed to be satisfied by $\Im$ and is composed only by literals like $J_k(\ell)$.
  - If $\Im'(J_i(\ell\vec{s}\sigma)) = 1$, then the conclusion is satisfied by $\Im$.

  Hence $\Im'(J_i(\ell\vec{s}\sigma) \vee C\sigma) = 1$. This reasoning holds for each $\vec{c}$, hence $\Im \models J_i(\ell\vec{s}\sigma) \vee C\sigma$.
- the paramodulation. The case $\Im'(C\sigma \vee C'\sigma) = 1$ is immediate; there remains only to check the case $\Im'(C\sigma) = \Im'(C'\sigma) = 0$. In this case $\Im'(J_i(s\sigma \approx s'\sigma)) = \Im'(J_j(\ell\vec{t}\sigma)) = 1$. As $\Im'$ is a $\approx$–interpretation, $1 - \Im'(s\sigma \approx s'\sigma) \geq |\Im'(\ell\vec{t}\sigma) - \Im'(\ell\vec{t}[s']_p\sigma)|$, hence $\Im'(\bigvee_{|k-j|<i} J_k(\ell\vec{t}[s']_p\sigma)) = 1$, a fortiori $\Im'(\bigvee_{|k-j|<i} J_k(\ell\vec{t}[s']_p\sigma) \vee C\sigma \vee C'\sigma) = 1$. This is true for each choice of $\vec{c}$, therefore $\Im \models \bigvee_{|k-j|<i} J_k(\ell\vec{t}[s']_p\sigma) \vee C\sigma \vee C'\sigma$.
- the negated reflexivity rule. Let us assume that $\Im'(C\sigma) = 0$ hence $\Im'(J_i(s\sigma \approx t\sigma)) = 1$, but as $s\sigma = t\sigma$, $\Im'(s\sigma \approx t\sigma) = 1$, whence a contradiction. Therefore $\Im'(C\sigma) = 1$, and as previously, $\Im \models C\sigma$.

***Theorem 15** (soundness). If a set of clauses S is satisfiable in an $\approx$–interpretation, then the empty clause (written $\Box$) cannot be derived using ORP_M.*

Proof. We prove the converse. Let $\Im$ be an arbitrary $\approx$–interpretation of S. From the soundness of the inference rules, we deduce that if a derived clause is not satisfied by the interpretation $\Im$, then at least one of the premises is not satisfied by $\Im$.

Let us assume that there exists a derivation of $c \in ORP^{n+1}_M(S)$, such that $\Im \not\models c$. As all the inference rules are sound, at least one of the premises is falsified by $\Im$ hence there exists $c' \in ORP^n_M(S)$, such that $\Im \not\models c'$. Hence there exists $c_0 \in ORP^0_M(S) = S$ such that $\Im \not\models c_0$ i.e. $\Im \not\models S$. Hence S is $\approx$–unsatisfiable.

***Lemma 16.** Let S be a set of ground clauses and $\Im$ a node of $MCT(ORP_M(S))$ with k successors, then there exists an i such that the successors are labeled with $J_{i+1}(\ell), \ldots, J_{i+k}(\ell)$.*

Proof. Let A be the successor of the literal labeling the deduction node considered. If $s > t, s \approx t, B < A$ are such that $A|_p = s$ and $B|_p = t$ for some position $p \in \mathcal{P}os(A) \cap \mathcal{P}os(B)$, and $A[t]_p = B$ and B $\Im$-reduces A using $s \approx t$, then in any $\approx$-interpretation $\Im'$ extending $\Im$:

$$\Im(B) - sm \leq \Im'(A) \leq sm + \Im(B)$$

where $sm = 1 - \Im(s \approx t)$, and either $0 < \Im(B) - sm$ or $sm + \Im(B) < 1$. Considering the set $\mathcal{R}$ of all the pairs of atoms $s \approx t, B < A$ suited for $\Im$-reducing A

$$\max_{\langle s\approx t, B\rangle \in \mathcal{R}} (\Im(B) - \Im(s \approx t)) + 1 \leq \Im'(A) \leq \min_{\langle s\approx t, B\rangle \in \mathcal{R}} (\Im(s \approx t) + \Im(B)) - 1$$

Note that since there are only a finite number of truth values, a finite subset of $\mathcal{R}$ (with at most $M - k$ elements) is enough to get the boundaries above. These inequalities are just another way of stating the announced result.

Lemma 17. *Let S be a set of ground clauses and $\Im$ a deduction node of $MCT(ORP_M(S))$, whose k successors are labeled with $J_{i+1}(\ell), \ldots, J_{i+k}(\ell)$. We note $\Im_1, \ldots, \Im_k$ the corresponding extensions of $\Im$. There exist k clauses $C_1, \ldots, C_k$ in $ORP_M(S)$ falsified respectively by $\Im_1, \ldots, \Im_k$, such that none of the $J_{i+1}(\ell), \ldots, J_{i+k}(\ell)$ occurs in none of them, i.e. every clause of $C_1, \ldots, C_k$ is falsified by every $\Im_1, \ldots, \Im_k$.*

Proof. As $\Im_{i+1}, \ldots, \Im_{i+k}$ are failure nodes, there exist $C'_{i+1}, \ldots, C'_{i+k}$ in $ORP_M(S)$, such that for $1 \leq j \leq k$ $\Im_j \not\models C'_j$ We eliminate the $J_{i+1}(\ell), \ldots, J_{i+k}(\ell)$ by a method similar to the Gaussian elimination for solving linear systems.

Let j be the smallest index in $\{i+1, \ldots, i+k\}$ such that $J_j(\ell)$ occurs in at least one of $C'_{i+1}, \ldots, C'_{i+k}$, and let j' be the smallest index of the clause that contains $J_j(\ell)$. Furthermore, because $\Im_j$ is a failure node, C'_j does not contain the literal $J_j(\ell)$. Let n be the least integer such that $J_n(\ell)$ is in C'_j. By construction, none of $C'_i, \ldots, C'_{i+k}$ contain a literal like $J_p(\ell)$ whose index p is less than j.

We replace the clause C'_m by a clause C''_m that also proves that $\Im_m$ is a failure node, but does not contain $J_j(\ell)$.

- if $n > i + k$, we eliminate the $J_j(\ell)$ by Ordered Resolution. For $m \in \{i + 1, \ldots, i+k\}$, $C''_m = \text{O-RES}(C'_m, J_j(\ell), C'_j, J_n(\ell), \iota)$ if $J_j(\ell) \in C'_m$, else $C''_m = C'_m$.
- else if $J_j(\ell)$ belongs to C'_n, We replace C'_n and C'_j by their ordered resolvent which contains neither $J_j(\ell)$ nor $J_n(\ell)$. $C''_n = C''_j = \text{O-RES}(C'_n, J_j(\ell), C'_j, J_n(\ell), \iota)$, and for $m \in \{i+1, \ldots, i+k\} \setminus \{n, j\}, C''_m = C'_m$.
- else we replace C'_j by C'_n which contains neither $J_j(\ell)$ nor $J_n(\ell)$. $C''_j = C'_n$ and for $m \in \{i+1, \ldots, i+k\} \setminus \{j\}, C''_m = C'_m$.

It is easy to check that C''_m is well defined and falsified by $\Im_m$ (for each $i+1 \leq m \leq i+k$).

Furthermore the number of occurrences of ℓ has decreased strictly. We rerun the procedure until the elimination of $J_{i+k}(\ell)$. The termination of the iteration can be proved trivially.

Theorem 18. *If S a set of ground clauses unsatisfiable, then $MCT(ORP_M(S)) = \emptyset$.*

Proof. The proof schema is the same as in [15]; some of the arguments are specific to the similarity relation $\approx$ and to the calculus ORP_M. Here $\{A_i\}_{i<\alpha}$ stands again for the Herbrand base ordered by a CSO. We build a sequence of nodes $\{\Im_\alpha : \alpha \leq \lambda\}$ in $MCT(ORP_M(S))$, and we show that this sequence is empty. Let us assume $MCT(ORP_M(S)) \neq \emptyset$ and consider the rightmost branch in $MCT(ORP_M(S))$, more precisely, this branch is built as follows:

If $\alpha = 0$, $\Im_\alpha = \emptyset$.

If α is a successor ordinal, with predecessor κ: If $\Im_\kappa$ has no successor in $MCT(ORP_M(S))$, then the sequence is complete, else $\Im_\alpha$ is the rightmost $\approx$-interpretation successor of $\Im_\kappa$, which is not a failure node (we keep the literal A_κ "as false as possible").

If α is a limit-ordinal, $\Im_\alpha(A_i) = \Im_{i+1}(A_i)$ for $i < \alpha$. $\Im_\alpha$ is the limit of an increasing sequence of $\approx$-interpretations, therefore $\Im_\alpha$ is also an $\approx$-interpretation. By the closure lemma (lemma 11), $\Im_\alpha$ is in $MCT(ORP_M(S))$.

This sequence is not empty, for $MCT(ORP_M(S))$ is neither empty (by hypothesis). Its ordinality $\alpha < \lambda$ because S is unsatisfiable. Let us note $\Im_\alpha$, the last element of this sequence, simply $\Im$.

We shall proceed by a reductio ad absurdum and shall show that this node is actually a failure node; we shall distinguish three cases.

- $A_\alpha = (a \approx a)$. As $\Im(a \approx a) = 1$ and as $\Im$ is a deduction node, there exists in $ORP_M(S)$ a clause of the following form: $J_i(a \approx a) \vee C$ with $i > 1$ being the least index of a literal in this clause whose atom is A_α, but then O-REFL($J_i(a \approx a) \vee C, J_i(a \approx a), \iota$) is also in $ORP_M(S)$. If O-REFL($J_i(a \approx a) \vee C, J_i(a \approx a), \iota$) contains other occurrences of $a \approx a$, we reapply this inference rule as long as the conclusion contains no longer $a \approx a$. This happens after a finite number of steps. Hence $\Im$ is a failure node, but this contradicts the fact that $\Im$ belongs to $MCT(ORP_M(S))$.
- $\Im$ has exactly M $\lhd$-successors in $MCT(ORP_M(S))$. By lemma 17, there exists in $ORP_M(S)$ a clause all the literals of which are strictly less than A_α, and that is falsified by $\Im$. As in the previous case this leads to a contradiction.
- $\Im$ has strictly less than M $\lhd$-successors in $MCT(ORP_M(S))$. This means that A_α is $\Im$-reducible. Let the successors be labeled with $J_{i+1}(A_\alpha), \ldots, J_{i+k}(A_\alpha)$. By lemma 17, there exists in $ORP_M(S)$ a clause C which does not contain any of $J_{i+1}(A_\alpha), \ldots, J_{i+k}(A_\alpha)$, and that is falsified by all the successors of $\Im$.

If C would neither contain any literal of $J_1(A_\alpha),\ldots,J_i(A_\alpha)$, $J_{i+k+1}(A_\alpha),\ldots,J_M(A_\alpha)$, then C were also falsified by $\Im$, which were hence a failure node and could therefore not be an element of $MCT(\mathcal{O}RP_M(S))$. Let $J_j(A_\alpha)$ be the literal of A_α with the smallest index in C. Since A_α cannot take the truth value $\frac{M-j}{M-1}$ in any extension of $\Im$, there exist $s \approx t < A_\alpha$, and $B < A_\alpha$ as described in lemma 16 (the notations used there keep their meaning here), such that furthermore $j < \Im(B) - sm$ or $j > sm + \Im(B)$. Obviously $sm < 0$ (otherwise B would not $\Im$-reduce A_α using $s \approx t$). Let γ be such that $A_\gamma = sm$ ($\gamma < \alpha$) and consider for a while the successors of $\Im|_{W_\gamma}$. Since the branch built was the rightmost all the successors $\Im'$ of $\Im|_{W_\gamma}$ with $\Im'(s \approx t) < sm$ are failure nodes.
By lemma 17, there exists in $\mathcal{O}RP_M(S)$ a clause D falsifying all these nodes. D contains at least one literal $J_m(s \approx t)$ $\frac{M-m}{M-1} \leq sm$ (otherwise $\Im$ would not belong to $MCT(\mathcal{O}RP_M(S))$. Let n be the smallest such index in D. As $C, D \in \mathcal{O}RP_M(S)$, $C' = \text{O-PAR}(C, J_j(A_\alpha), p, D, J_m(s \approx t), \iota) \in \mathcal{O}RP_M(S)$. If C' contains some occurrences of A_α, we reapply the same argument on C' instead of C. Since at each iteration the number of A_α decreases strictly, at the end we have got a clause C'' falsified by $\Im$. We conclude as in the previous cases.

Therefore we must conclude that $MCT(\mathcal{O}RP_M(S))$ is empty.

This proof is interesting since we can derive easily strong strategies.

Theorem 19. *If S is a set of ground clauses unsatisfiable,* $\square \in \mathcal{O}RP_M(S)$.

Proof. $MCT(\mathcal{O}RP_M(S))$ is empty hence $\square \in \mathcal{O}RP_M(S)$.

The lifting lemmas are exactly as in the classical case (see [15]).

Theorem 20. *If S is a set of clauses unsatisfiable,* $\square \in \mathcal{O}RP_M(S)$.

4 Conclusion and Future Work

The $\approx$-interpretations have been characterized and the paper has presented a new and efficient proof procedure based on Ordered Resolution and Paramodulation for first-order many-valued logics. Its completeness has been proved by using the maximally consistent tree technique due to Rusinowitch. This work can be seen as a first step on considering clausal calculi for a wide class of many-valued logics.

The next step consists in extending it to some (semi-decidable) infinite valued logics like Orlowska's resolution calculus for ω^+-valued Post logic [8] or Dummett's LC [1], since there are very natural ω^+-valued distances on terms (see for ex. [3, chap. 6]).

Acknowledgement

The author whishes to thank Ricardo Caferra and Thierry Boy de la Tour for stimulating discussions on these topics and for reading and commenting earlier drafts of this paper.

References

1. Michael Dummett. A propositional calculus with denumerable matrix. *Journal of Symbolic Logic*, 24(2):97–106, June 1959.
2. Jean H. Gallier. *Logic for Computer Science*. Harper & Row, 1986.
3. J. W. Lloyd. *Foundations of Logic Programming*. Springer Verlag, second, extended edition, 1987.
4. Donald W. Loveland. *Automatic Theorem Proving: a logical basis*, volume 6 of *Fundamental Studies in Computer Science*. North Holland, 1978.
5. Charles Morgan. A theory of equality for a class of many valued predicate calculi. *Zeitschrift für mathematische Logik und Grundlagen der Mathematik*, 20:427–432, 1974.
6. Charles Morgan. Similarity as a theory of graded equality for a class of many-valued predicate calculi. In *Proceedings of the IEEE Symposium of many-valued logics*, pages 436–449, Bloomington, Long Beach, 1975.
7. Charles Morgan. A resolution principle for a class of many valued logics. *Logique et Analyse*, 74–76:311–339, 1976.
8. Ewa Orlowska. The resolution principle for ω^{+}-valued logic. *Fundamenta Informaticae*, II(1):1–15, 1978.
9. Ewa Orlowska. The resolution systems and their applications I. *Fundamenta Informaticae*, III:235–268, 1979.
10. Ewa Orlowska. The resolution systems and their applications II. *Fundamenta Informaticae*, III(3):333–362, 1980.
11. John Pais and Gerald E. Peterson. Using forcing to prove completeness of resolution and paramodalution. *Journal of Symbolic Computation*, 11(1 & 2):3–19, 1991.
12. G. A. Robinson and L. Wos. Paramodulation and theorem proving in first order theories with equality. In B. Meltzer and D. Michie, editors, *Machine Intelligence*, volume 4, pages 135–150. Edinburgh University Press, 1968.
13. James A. Robinson. A machine oriented logic based resolution principle. *Journal of the Association for Computing Machinery*, 12:23–41, 1965.
14. J. Barkley Rosser and Atwell R. Turquette. *Many-valued Logics*. North Holland, 1952.
15. Michaël Rusinowitch. *Démonstration automatique par des techniques de réécriture*. PhD thesis, Nancy-1 France, November 1987. Thèse d' Etat — also available as textbook (Inter Editions, Paris 1989).
16. Dana S. Scott. Background to formalization. In Hugues Leblanc, editor, *Truth, syntax and modality — proceedings of the Temple University conference on alternative semantics*, pages 244–273, Philadelphia, Pa, 29–30.12.70, 1973. North Holland.
17. James R. Slagle. Automatic theorem proving with renamable and semantic resolution. *Journal of the Association for Computing Machinery*, 14(4):687–697, October 1967.

18. Zbigniew Stachniak and Peter O'Hearn. Resolution in the domain of strongly finite logics. *Fundamenta Informaticae*, XIII:333–351, 1990.

This article was processed using the LaTeX macro package with LLNCS style

An Efficient Constraint Language for Polymorphic Order-sorted Resolution

Christian Prehofer*

Technische Universität München**

Abstract. In recent years various sorted logics have been developed, mostly to facilitate knowledge representation and to speed up automated deduction. We present a polymorphic order-sorted logic that can be implemented efficiently. Because the polymorphism is almost unrestricted, it is possible for two terms to have an exponential number of maximally general unifiers. To guarantee a single most general unifier, we embed the sorted logic into a more general constraint logic and create a distinct constraint satisfaction search space. This separates the total search space into two orthogonal ones and facilitates many optimizations. The main complexity gains are that the unnecessary generation of unifiers can be avoided and that the primary resolution search space remains constant if the complexity of the unification grows.

Keywords: Constraints, Order-sorted Logic, Polymorphism, Resolution, Unification.

1 Introduction

In recent years various sorted logics have been developed, mostly to facilitate knowledge representation and to speed up automated deduction. In sorted logics, a sort represents a subset of the domain and variables range only over a specific sort. If sorts can be ordered by set inclusion, the logic is called *order-sorted*. It is also desirable to specify the sortal behavior of functions. A sorted logic is *polymorphic* if a function can have different result sorts, depending on the argument sorts.

One motivation for sorts is, roughly speaking, that proofs can be kept at a more general level compared to unsorted logic, thus reducing the search space [Cohn, 1985, Frisch, 1985, Walther, 1985]. It has also been argued that deduction with sorts can be implemented more efficiently by specialized procedures [Cohn, 1989]. For example, for order-sorted logics, there are efficient methods to find the greatest lower bound (glb) of two sorts in a sort hierarchy [Aït-Kaci *et al.*, 1989, Huber and Varsek, 1987].

Our contribution is an efficient unification method for a very general class of polymorphism. Sorted logics only constrain variables to be of a certain sort which means that a most general unifier often cannot be expressed by the logic. Instead, we attach constraints on terms in order to represent multiple unifiers. For instance, a term $t = sum(X, Y)$ is of sort EVN if X and Y are both of sort EVN or both of

* Research carried out at the University of Illinois at Urbana-Champaign, USA.

** Full Address: Institut für Informatik, Technische Universität München, Postfach 20 24 20, W-8000 München 2, Germany. E-mail: prehofer@informatik.tu-muenchen.de

sort ODD. Then t and a variable Z:EVN have the two maximally general instances $sum(X{:}\text{EVN}, Y{:}\text{EVN})$ and $sum(X{:}\text{ODD}, Y{:}\text{ODD})$. In general, in this order-sorted logic, the number of maximally general instances grows exponentially with the term size and to decide if two terms unify is NP-complete [Schmidt-Schauß, 1989].

We express the above two terms as one term by attaching a constraint C on t to ensure that t is of sort EVN. This constrained term, t/C, represents all corresponding sorted common instances. If unification fails for t/C, an up to exponential number of maximally general sorted instances would fail also. Instead of an exponential number of unifiers, we now have a NP-complete constraint satisfaction problem, for which we create a distinct search space. This means that the primary resolution search space remains constant if the complexity of the unification grows. Furthermore, in our approach an exponential number of maximal sorted unifiers does not automatically require an exponential effort; we show that it suffices to find a single representative unifier (in the constraint satisfaction search space) and to delay further search.

The following section presents the above order-sorted logic formally and shows the problem of multiple unifiers. Then Section 3 introduces a more general logic where sort constraints are attached to terms. This is used in Section 4 to embed the order-sorted logic into the constraint logic CP, where a most general unifier always exists for unifiable terms. In Section 5, we discuss the advantages of CP-logic in the light of two examples. The first example shows the effect of CP-logic on the search space graphically, the second shows details of a derivation with CP-terms.

2 Order-sorted Logic

The syntax of our sorted logic differs from standard first-order logic only in variables that are quantified over sorts, written as

$$\forall X{:}s\ \phi,$$

where X is a variable and s is a sort. We specify sorts and polymorphic functions by a sort theory Σ in unsorted first-order logic. Then we define the meaning of a sorted sentence by an unsorted sentence.

Σ contains only unary predicate symbols from a distinct set of sort predicate symbols $\mathcal{S}$; $s \in \mathcal{S}$ is called a *sort*. An atomic formula with a sort predicate is called *characteristic literal*. These literals define the sorts. A term denotes an element of a sort s if the interpretation of $s(t)$ is true. These characteristic literals appear in the sort theory Σ only. In examples, we write sorts typographically as small capitals, but not characteristic literals, for example X:ODD and ODD(X). Σ contains only sentences of the following two forms:

- Subsort declarations: $\forall X\ s_1(X) \rightarrow s_2(X)$
- Function declarations: $\forall X_1, \ldots, X_n\ s_1(X_1) \wedge \ldots \wedge s_n(X_n) \rightarrow s_{n+1}(f(X_1, \ldots, X_n))$

A sort s_1 is subsort of s_2, written as $s_1 \preceq s_2$, iff $\Sigma \models \forall X\ s_1(X) \rightarrow s_2(X)$. We require that $\prec$ is a partial order on the sorts, i.e. for two sorts s_1 and s_2, we disallow that Σ entails both $s_1 \prec s_2$ and $s_2 \prec s_1$. Then we can represent the sorts in a diagram (see Fig. 1), where $s \prec s'$ if there is a downwards path from s' to s.

We abbreviate a function declaration (or declaration for short) $d = \forall X_1, \ldots, X_n$ $s_1(X_1) \wedge \ldots \wedge s_n(X_n) \rightarrow s_{n+1}(f(X_1, \ldots, X_n))$ by $f(s_1, \ldots, s_n) : s_{n+1}$. The sort s_{n+1}

Fig. 1. Sort Hierarchy Σ_1

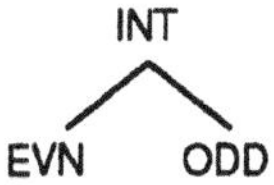

Fig. 2. Sort Hierarchy Σ_2

is called the *result sort*, denoted as $\mathcal{R}(d)$. The set of function declarations in Σ is required to be monotone, i.e. if Σ contains two declarations of the form

$$f(s_1, \ldots, s_n) : s_{n+1} \text{ and } f(s'_1, \ldots, s'_n) : s'_{n+1}$$

and $s_i \preceq s'_i$ for all $i \in \{1 \ldots n\}$, then $s_{n+1} \preceq s'_{n+1}$.

We require that the sorts are non-empty, i.e. for all $s \in \mathcal{S}$ there is a ground term t such that $\Sigma \models s(t)$. We also assume that there is at least one declaration for each function symbol.

Now we can define the meaning of a sentence: $\forall X{:}s\ \phi = \forall X\ (s(X) \rightarrow \phi)$. To ensure that substitutions comply with the restriction of a variable we define well-sorted substitutions with respect to Σ. A substitution θ is *well-sorted* iff $\Sigma \models \overline{\forall} s(X\theta)$ for all variables $X{:}s$.[3] A term t is a *Σ-instance* of t' iff $t = t'\theta$ for a well-sorted substitution θ. A set of terms I is a *complete set of incomparable common Σ-instances* (or CII_Σ) of a set of terms E, iff:

- Each element of I is a Σ-instance of all elements of E.
- Every term which is a Σ-instance of all elements of E is a Σ-instance of at least one element of I.
- No element of I is a Σ-instance of another element of I.

It can be shown that a CII_Σ is uniquely defined except for variants of its elements [Frisch, 1991b]. A well-sorted substitution θ is a *Σ-unifier* of a set of terms E iff $e\theta = e'\theta$ for all $e, e' \in E$ and $e\theta$ is a Σ-instance of all elements of E. We say a term t is of sort s if $\Sigma \models \overline{\forall} s(t)$. A term is *well-sorted* if it is of some sort. This corresponds to syntax-based notions of well-sortedness, e.g. [Schmidt-Schauß, 1989].

Example 1. Assuming Σ_1 (Fig. 1), a CII_Σ of the two variables $X{:}\text{A}$ and $Y{:}\text{B}$ consists of $X{:}\text{C}$ and $X{:}\text{D}$.

Example 2. Assuming Σ_2 (Fig. 2) with the function declarations

- $d_1 = sum(\text{INT}, \text{INT}) : \text{INT}$
- $d_2 = sum(\text{ODD}, \text{EVN}) : \text{ODD}$
- $d_3 = sum(\text{EVN}, \text{ODD}) : \text{ODD}$
- $d_4 = sum(\text{ODD}, \text{ODD}) : \text{EVN}$
- $d_5 = sum(\text{EVN}, \text{EVN}) : \text{EVN}$,

a CII_Σ of $sum(X{:}\text{INT}, Y{:}\text{INT})$ and $Z{:}\text{EVN}$ is

$$\{sum(X{:}\text{ODD}, Y{:}\text{ODD}), sum(X{:}\text{EVN}, Y{:}\text{EVN})\}.$$

[3] Here, and in general, an expression of the form $\overline{\forall}\phi$ denotes the universal closure of ϕ—that is, the formula $\forall x_1 \cdots x_n\ \phi$ where $x_1, \ldots, x_n$ are the freely-occurring variables of ϕ. Similarly, $\overline{\exists}\phi$ denotes the existential closure of ϕ.

A sort theory Σ is *regular* if every well-sorted term has one least sort. A sort s is the least sort of a term t if $\Sigma \models \overline{\forall} s(t)$ and for all $s' \in S$ $\Sigma \models \overline{\forall} s'(t)$ only if $s \preceq s'$. If Σ is regular, Schmidt-Schauß shows that in a sort theory as defined[4] above, a CII_Σ is at most exponential in the size of the unified terms. If Σ is not regular, CII_Σ's can be infinite. In the following, we assume that the sort theories are regular. We show in [Prehofer, 1992] that every monotone sort theory can be transformed into a regular one.

We confine ourselves to sentences in Skolem Normal Form (SNF). For translating sentences into SNF see [Frisch and Scherl, 1991, Frisch, 1991b]. We often take the view that a sort is attached to a variable instead of to the quantifier. In particular, we often omit the universal quantifiers for sentences in SNF and simply attach the sort to a variable, for instance $sum(X{:}\text{ODD}, X{:}\text{ODD})$.

3 Sort-Constraint Logic

We introduce a constraint logic where sort-constraints can be attached to every (sub)term. It is an instance of the frameworks in [Bürkert, 1991, Frisch, 1991a] and will be used as the formal basis for CP-logic in the next section. The only novel feature, attaching constraints to subterms, will be important for implementing polymorphic order-sorted logic by CP-logic.

Definition 1 Sort-Constrained Terms. Sort-constraints are quantifier-free formulas with the connectives $\wedge$ and $\vee$ and only contain positive characteristic literals. Every unconstrained term is a sort-constrained term with the implicit constraint TRUE. Assume a term t is a variable or $t = f(t_1, \ldots, t_n)$, where $t_1, \ldots, t_n$ are sort-constrained terms. Then t/C is a sort-constrained term, if the sort-constraint C only contains literals of the form $s(t')$, where t' is a subterm of t.[5]

For example, $t_0 = f(X/\text{INT}(X), g(Y/\text{EVN}(Y)))/(\text{ODD}(X) \wedge \text{ODD}(g(Y)))$ is a sort-constrained term. Let t^C denote the conjunction of all constraints that occur in a sort-constrained formula t and let t^N denote t without all constraints. Then we define the meaning of a constrained formula t as $t \equiv \overline{\forall}(t^C \rightarrow t^N)$. For example, the atom $p(t_0)$ is logically equivalent to $\overline{\forall}(\text{INT}(X) \wedge \text{EVN}(Y) \wedge (\text{ODD}(X) \wedge \text{ODD}(g(Y)))) \rightarrow p(f(X, g(Y)))$.

To ease our notation, we often write $p(t)/C$ instead of $p(t/C)$. Notice that C only contains characteristic literals and does not contain the predicate symbol p. An expression t is in *normal form* if $t = t^N/t^C$. An expression t is *Σ-satisfiable* if $\Sigma \models \overline{\exists} t^C$.

Definition 2 Σ-Instance. Let Σ be a sort theory. e_1 is a Σ-instance of e (often written $e_1 \leq_\Sigma e$) if, and only if, for some substitution θ, $e^N\theta = e_1^N$ and $\Sigma \models \overline{\forall}\,(e_1^C \rightarrow (e^C\theta))$.[6]

Theorem 3. *Let e_1/C_1 and e_2/C_2 be variable-disjoint expressions in normal form such that θ is a most general unifier of e_1 and e_2. Then $(e_1/C_1 \wedge C_2)\theta$ is a most general common Σ-instance (or MGCI_Σ) of e_1 and e_2, if $(C_1 \wedge C_2)\theta$ is Σ-satisfiable.*

---

[4] Finite and elementary in Schmidt-Schauß's terminology.

[5] Notice that t is a subterm of itself.

[6] This definition and the following theorem are adapted from [Frisch, 1991a]

> *Input:* $(\emptyset, \{t = t'\})$
> *Output:* $(\theta, \emptyset)$, such that $t\theta = t'\theta = \text{MGCI}_\Sigma(t, t')$ or fail
> Apply the following rules in any order:
>
> 1. $(\vartheta, U \cup \{X/C = Y/C'\}) \Longrightarrow (\vartheta\theta, U\theta)$, where
> $\theta = \{ X/C \mapsto X/(C \wedge C''),$
> $Y/C' \mapsto X/(C \wedge C'')\}$ and $C'' = C'\{Y \mapsto X\}$
> IF $(t\vartheta\theta)^C$ is not Σ-satisfiable THEN fail
> 2. $(\vartheta, U \cup \{f(t_1, \ldots, t_n)/C' = X/C\}) \Longrightarrow (\vartheta, U \cup \{X/C = f(t_1, \ldots, t_n)/C'\})$
> 3. $(\vartheta, U \cup \{X/C = f(t_1, \ldots, t_n)/C'\}) \Longrightarrow (\vartheta\theta, U\theta)$, where
> $\theta = \{ X/C \mapsto f(t_1, \ldots, t_n)/(C' \wedge C''),$
> $f(t_1, \ldots, t_n)/C' \mapsto f(t_1, \ldots, t_n)/(C' \wedge C'')\}$ and
> $C'' = C\{X \mapsto f(t_1, \ldots, t_n)\}$
> IF $(t\vartheta\theta)^C$ is not Σ-satisfiable OR X occurs in $f(\ldots)$ THEN fail
> 4. $(\vartheta, U \cup \{f(t_1, \ldots, t_n)/C = f'(t'_1, \ldots, t'_n)/C'\}) \Longrightarrow$
> $(\vartheta\theta, U\theta \cup \{t_1 = t'_1, \ldots, t_n = t'_n\})$, where
> $\theta = \{ f(t_1, \ldots, t_n)/C \mapsto f(t_1, \ldots, t_n)/(C \wedge C'),$
> $f'(t'_1, \ldots, t'_n)/C' \mapsto f(t_1, \ldots, t_n)/(C \wedge C')\}$
> IF $(t\vartheta\theta)^C$ is not Σ-satisfiable OR $f \neq f'$ THEN fail

Fig. 3. Constraint Unification Algorithm

To unify two sort-constrained terms, a generalized notion of substitution is required. Usually, a substitution is a mapping from variables to terms. We present T-substitutions as mappings from terms to terms.

Definition 4 T-substitution. A T-substitution is a function from sort-constrained terms to sort-constrained terms, usually written as a set of mappings $t \mapsto t'$, where t and t' are sort-constrained terms. The composition of two T-substitutions is defined as $\theta\vartheta(t) = \vartheta(\theta(t))$ for all t.

A T-substitution θ is *well-sorted*, if $t\theta \leq_\Sigma t$ for every sort-constrained term t. For two well-sorted T-substitutions θ and ϑ, θ is more general than ϑ, if there is a well-sorted T-substitution θ' such that $\theta\theta' = \vartheta$. A well-sorted T-substitution θ is a *Σ-T-unifier* of two terms e and e', if $e\theta = e'\theta$, $e\theta \leq_\Sigma e$ and $e\theta \leq_\Sigma e'$.

In the style of Martelli and Montanari [Martelli and Montanari, 1982], we describe unification in the form of equations. The unification rules in Figure 3 compute a MGCI_Σ of two terms, not necessarily in normal form. The rules work on a tuple (θ, U), where θ is a T-substitution and U is a set of equations. To obtain a MGCI_Σ of two variable-disjoint terms t and t', we start with $(\emptyset, \{t = t'\})$. Then we apply the conditional rules in any order until U is empty or until one rule fails. If we succeed and U is empty, then θ is a most general Σ-T-unifier of t and t', and $t\theta = \text{MGCI}_\Sigma(t, t')$ is Σ-satisfiable.

If we compute a MGCI_Σ of two terms in normal form with Theorem 3, all constraints are attached to the outermost term. The rules in Figure 3 conjoin the constraints of the subterms locally instead of conjoining all constraints at once. By locally, we mean that the constraints are closer to the terms they constrain to a certain sort. As we require the Σ-satisfiability of $(t\vartheta\theta)^C$, where $t\vartheta\theta$ is equal to $\mathrm{MGCI}_\Sigma(t,t')$ if the rules succeed, the correctness of the rules follows easily from Theorem 3.

Let L, L' be positive literals and let M_i, N_i be literals. Then the (binary) resolution inference rule for sort-constrained terms is defined as:

$$\frac{\begin{array}{c} L \vee M_1 \vee \ldots \vee M_m \\ \neg L' \vee N_1 \vee \ldots \vee N_n \end{array}}{(M_1 \vee \ldots \vee M_m \vee N_1 \vee \ldots \vee N_n)\theta}$$

if θ is a most general Σ-T-unifier of L and L', and $((L \vee M_1 \vee \ldots \vee M_m \vee N_1 \vee \ldots \vee N_n)\theta)^C$ is Σ-satisfiable

The satisfiability test in this rule assures that the derived clause represents some ground instance. The constraints in M_i and N_i are not necessarily Σ-satisfiable if $\mathrm{MGCI}_\Sigma(L,L')$ is Σ-satisfiable. For instance, if we resolve $p(X/\mathrm{INT}(X)) \vee q(f(X)/\mathrm{EVN}(X))$ with $\neg p(X/\mathrm{ODD}(X))$, then $q(f(X/\mathrm{ODD}(X) \wedge \mathrm{INT}(X))/\mathrm{EVN}(X))$ is not Σ-satisfiable.

One problem of computing MGCI_Σ's is to simplify the conjunctions of constraints $(C \wedge C')$, otherwise the terms grow larger and larger without limit. In the above rules, terms do not have to be in normal form and these conjunctions can be simplified locally, because constraints only contain subterms of the terms to which they are attached.

4 CP-Logic

CP-logic is a sort-constraint logic with a specific class of constraints and will be used to implement the sorted logic of Section 2. In Section 4.1, we translate well-sorted terms of the sorted logic into CP-terms. Then Section 4.2 shows that constraint simplification in CP can be performed by set operations. To solve the problem of Σ-satisfiability of CP-constraints, we create a distinct search space, as shown in Section 4.3

4.1 Embedding the Sorted Logic in the Constraint Logic CP

We first present examples of the class of constraints that is used to translate the sorted logic into CP-logic and give examples for unification of CP-terms.

The CP-constraint of a sorted variable $X{:}s$ is the disjunction of all subsorts of s, including s.[7] For instance, in Example 1, the variables $X{:}\mathrm{A}$ and $Y{:}\mathrm{B}$ translate into $X/(A(X) \vee C(X) \vee D(X))$ and $Y/(B(Y) \vee C(Y) \vee D(Y))$, respectively. Then these two variables have the MGCI_Σ $Z/(C(Z) \vee D(Z))$, where the sorts in the constraint $C(Z) \vee D(Z)$ are obtained by intersecting the sorts in the constraints of X and Y. If possible, we will use the notation for sorted variables in examples for brevity, i.e. $X{:}\mathrm{A}$ instead of $X/(A(X) \vee C(X) \vee D(X))$.

Similarly, we attach a disjunction to every non-variable term, where each disjunct stands for one declaration. For a declaration $d = f(s_1, \ldots, s_n) : s_{n+1}$ and a term $t = f(t_1, \ldots, t_n)$, we define $d(t)$ as $s_1(t_1) \wedge \ldots \wedge s_n(t_n)$. Let $\mathcal{U}(t)$ be the set of all declarations d such that $d \in \Sigma$ and $\Sigma \models \overline{\forall} d(t')$ for some $t' \leq_\Sigma t$.

Then a CP-constraint of a term $t = f(\ldots)$ is the disjunction of all $d_i(t)$'s with $d_i \in \mathcal{U}(t)$. For instance, in Example 2, translating

$$t_0 = sum(X{:}\mathrm{INT}, Y{:}\mathrm{INT})$$

[7] We choose this representation to motivate constraints on non-variable terms. More efficient versions be found in [Aït-Kaci *et al.*, 1989].

into CP-logic gives

$$sum(X{:}\mathrm{INT}, Y{:}\mathrm{INT})/(d_1(t_0) \vee d_2(t_0) \vee d_3(t_0) \vee d_4(t_0) \vee d_5(t_0)).$$

In examples, we often omit the argument t in $d(t)$ and, for example, write the above term as $t_0/d_1 \vee \ldots \vee d_5$. Then the logical meaning of the constraints is given implicitly by their position in the term.

A MGCI$_\Sigma$ of $Z/\mathrm{EVN}(Z)$ and $t_0/d_1 \vee \ldots \vee d_5$, using Theorem 3, is $t_0/(d_1(t_0) \vee \ldots \vee d_5(t_0)) \wedge \mathrm{EVN}(t_0)$, and can be simplified to

$$sum(X{:}\mathrm{INT}, Y{:}\mathrm{INT})/d_4 \vee d_5.$$

Only the two declarations with result sort EVN remained. We eliminated d_2 because $d_2(t_0) \wedge \mathrm{EVN}(t_0)$ is not Σ-satisfiable and d_3 for the same reason. Furthermore, d_1 is redundant, because whenever Σ entails $d_1(t_0) \wedge \mathrm{EVN}(t_0)$, Σ must entail $d_4(t_0) \vee d_5(t_0)$ also. In general, to obtain the constraint of the MGCI$_\Sigma$ of a variable X/C_X and a term t/C_t, we simply intersect the sorts in C_X with the result sorts of the declarations in C_t.

Instead of a CII$_\Sigma$ of size two, we now have only one CP-term. In the general case, a single CP-MGCI$_\Sigma$ represents a possibly exponential large CII$_\Sigma$ in the sorted logic. A MGCI$_\Sigma$ of $sum(X{:}\mathrm{INT}, Y{:}\mathrm{INT})/d_4 \vee d_5$ and $sum(X'{:}\mathrm{INT}, Y'{:}\mathrm{EVN})/d_1 \vee d_2 \vee d_5$ is

$$sum(X'{:}\mathrm{INT}, Y'{:}\mathrm{EVN})/d_5.$$

Again, we only have to intersect the declarations of both constraints.

The next step is to make the constraints minimal. For example, $t_1 = sum(X{:}\mathrm{EVN}, Y{:}\mathrm{ODD})$ has the constraint $d_1 \vee d_3$. For all Σ-instances t_1' of t_1, whenever Σ entails $d_1(t_1')$, Σ must entail $d_3(t_1')$ also and, therefore, we can omit $d_1(t_1)$. Then we can often detect that two terms have no MGCI$_\Sigma$ because they have no common declarations. For example, $sum(X'{:}\mathrm{INT}, Y'{:}\mathrm{EVN})/(d_1 \vee d_2 \vee d_5)$ and t_1/d_3 have no MGCI$_\Sigma$, and we can see this by examining the two constraints only.

In the remainder of this section, we show that these observations hold if Σ is regular. Because we reduce the operations on constraints to set operations, we treat constraints as sets, where each element stands for one disjunct, e.g. $t_1/\{d_1, d_3\}$ instead of $t_1/d_1 \vee d_3$. A set C is *downwards (upwards) closed* w.r.t. an ordering $<$ and a set D, if $d \in C$ entails $d' \in C$ for all $d' \in D$ with $d' < d$ ($d' > d$).

Definition 5 CP-Variable. Assume $C = \{s_1, \ldots, s_n\} \subseteq \mathcal{S}$ is downwards closed w.r.t. $\prec$ and $\mathcal{S}$. Then the variable $X/(s_1(X) \vee \ldots \vee s_n(X))$ is a CP-term. We let $\widehat{C}$ denote the translation of C into $(s_1(X) \vee \ldots \vee s_n(X))$ and write $X/\widehat{C}$.

Translating a sorted variable $X{:}s$ into an equivalent CP-variable gives $X/\widehat{C}$, with $C = \{s' \mid s' \preceq s\}$. Two constrained variables $X/\widehat{C}$ and $Y/\widehat{C}'$ have a MGCI$_\Sigma$[8]

[8] Notice that this MGCI$_\Sigma$ and the following ones obtained by set operations are no MGCI$_\Sigma$'s in the strict logical sense, because they have slightly stronger constraints. As they preserve the set of ground instances, this can be justified by Herbrand's Theorem since Σ has a minimal model. Versions of Herbrand's Theorem for constraint logics can be found in in [Bürkert, 1991, Frisch, 1991a]. In the following we do not differentiate between true MGCI$_\Sigma$'s and terms with these stronger constraints.

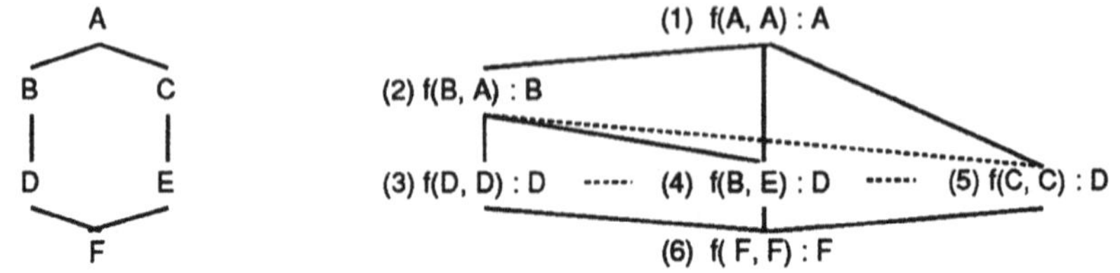

Fig. 4. Sort Theory Σ_3

$Z/(\widehat{C \cap C'})$, if $C \cap C' \neq \emptyset$. This follows from Theorem 3 because C and C' are downwards closed and the sorts are non-empty.

For each function symbol f, let $\mathcal{D}(f)$ denote the set of function declarations for f in Σ. The partial order of the sorts can be used to order the function declarations. Two declarations are ordered, denoted by

$$f(s_1, \ldots, s_n) : s_{n+1} \preceq f(s'_1, \ldots, s'_n) : s'_{n+1},$$

if $s_i \preceq s'_i$, $\forall i \in \{1, \ldots, n\}$. Observe that $\mathcal{U}(t)$ is upwards closed w.r.t. $\prec$ and $\mathcal{D}(f)$, but not downwards. For a declaration $d = f(s_1, \ldots, s_n) : s_{n+1}$, the *corresponding term* is defined as $f(X_1{:}s_1, \ldots, X_n{:}s_n)$. We often write d instead of d's corresponding term, for instance we write "instance of d" or "$\text{MGCI}_\Sigma(d, t)$".

Definition 6 CP-Term. Assume $t_1, \ldots .t_n$ are CP-terms and $t = f(t_1, \ldots, t_n)$. Assume further that $C = \{d_1, \ldots, d_n\} \subseteq \mathcal{D}(f)$ is non-empty and downwards closed w.r.t. $\prec$ and $\mathcal{U}(t)$. Then $t' = f(t_1, \ldots, t_n)/(d_1(t) \vee \ldots \vee d_n(t))$ is a CP-term. We let $\widehat{C}$ denote the translation of C into $(d_1(t) \vee \ldots \vee d_n(t))$ and write t' as $t/\widehat{C}$.

In examples, we often write C instead of $\widehat{C}$. To translate a sorted term t into CP, it would be possible to to use $\mathcal{U}(t)$ as the constraint of t. For further optimization, we show that the smaller set $\mathcal{I}(t)$, defined below, can be used instead. Recall that using smaller constraints facilitates the optimization that two non-variable terms do not unify if their constraints have no common declarations. The idea is to describe a term more accurately with the declarations in the constraint. We write $t \approx t'$, if t and t' are variants.

Definition 7 Initial Set. The initial set $\mathcal{I}(t)$ of a sort-constrained term $t = f(\ldots)$ is defined as $\{d \in \mathcal{U}(t) \mid$ there exists no $d' \in \mathcal{U}(t)$ with $d' \prec d$ and $\text{MGCI}_\Sigma(t, d) \approx \text{MGCI}_\Sigma(t, d')\}$.

Example 3. The following figure shows the initial sets of the terms in the left column w.r.t. Σ_3 (Fig. 4) and two variations of Σ_3. The numbers in the figure refer to the declarations in Figure 4, where the solid lines represent the relation $\prec$.

	Σ_3	$\Sigma_4 = \Sigma_3 - F$ and (6)	$\Sigma_5 = \Sigma_3 -$ (3) and (6)
$f(X{:}\text{B}, Y{:}\text{A})$	$\{2,3,4,6\}$	$\{2,3,4\}$	$\{2,4,5\}$
$f(X{:}\text{A}, X{:}\text{A})$	$\{1-6\}$	$\{1,2,3,5\}$	$\{1,2,4,5\}$
$f(X{:}\text{B}, Y{:}\text{B})$	$\{2,3,6\}$	$\{2,3\}$	$\{2,4,5\}$

The following theorem shows why $\mathcal{I}(t)$ can be used instead of $\mathcal{U}(t)$.

Theorem 8. *Let $t' \leq_\Sigma t$ be two non-variable* CP*-terms and let $d \in \mathcal{U}(t)$ be a declaration such that $\Sigma \models \overline{\forall} d(t')$. Then there exists $d' \in \mathcal{I}(t)$ with $d' \preceq d$ and $\Sigma \models \overline{\forall} d'(t')$.*[9]

Definition 9 Σ-valid Terms. All CP-variables are Σ-valid. A CP-term $f(\ldots)/\widehat{C}$ is Σ-valid if $\Sigma \models \overline{\forall}\widehat{C}$ and all subterms of $f(\ldots)$ are Σ-valid.

To translate a well-sorted term t of the sorted logic into a CP-term, we first translate all variables in t into CP-variables, as shown in the previous section, and obtain a sort-constrained term t'. Then we attach the constraint $\widehat{\mathcal{I}(t'')}$ on each non-variable subterm t'' of t', including t'. For instance, in Example 3 with Σ_3, $f(X{:}\text{B}, Y{:}\text{A})$ is translated into $f(X/B(X), Y/A(Y))/\widehat{\{2,3,4,6\}}$.

Since t is a well-sorted non-variable term, $\Sigma \models s(t)$ for some sort s. Hence, there must be a declaration $d \in \Sigma$, such that Σ entails $d(t)$. Then $d \in \mathcal{U}(t')$ and $\Sigma \models \overline{\forall}\widehat{\mathcal{U}(t')}$, and from Theorem 8 follows $\Sigma \models \overline{\forall}\widehat{\mathcal{I}(t')}$. Hence t and $t'/\widehat{\mathcal{I}(t')}$ have the same Σ-instances and the CP-term can be used for the sorted term and $t'/\widehat{\mathcal{I}(t')}$ is Σ-valid. Observe that a Σ-valid, non-variable CP-term t/C ensures the satisfiability of $C\theta$ for a well-sorted T-substitutions θ if $t\theta^C$ is Σ-satisfiable.

4.2 Reducing Constraint Simplification in CP to Set Operations

We first show by examples under what conditions the logical conjunction of two CP-constraints can be simplified by set operations. If t/C' is a MGCI$_\Sigma$ of a term $t/\widehat{C}$ and a variable, it is possible that $C' \subset \mathcal{I}(t)$, as $sum(X{:}\text{INT}, Y{:}\text{INT})/(d_4 \vee d_5)$ above. Can we still get a MGCI$_\Sigma$ by set intersection? For instance, assuming Σ_5 of Example 3, the two CP-terms $t_2 = f(X{:}\text{B}, Y{:}\text{A})/\{2,4\}$ and $t_3 = f(X{:}\text{C}, Y{:}\text{C})/\{5\}$ have the MGCI$_\Sigma$ $f(X{:}\text{F}, Y{:}\text{F})/\{4,5\}$, but no common declarations.

We need another invariant which will be preserved by our unification rules. Two declarations can be ordered by their result sorts, written as $d \preceq_\mathcal{R} d'$, if $\mathcal{R}(d) \preceq \mathcal{R}(d')$. Because Σ is monotone, $d \preceq d'$ entails $d \preceq_\mathcal{R} d'$. The dotted lines in Figure 4 indicate where $d \preceq_\mathcal{R} d'$, but not $d \preceq d'$. Observe that $\preceq_\mathcal{R}$ is not anti-symmetric, for example (3) $\preceq$ (4) and (4) $\preceq$ (3) for the declarations in Figure 4.

Definition 10 Result-Complete. A non-variable CP-term $t/\widehat{C}$ is result-complete if C is downwards closed w.r.t. $\preceq_\mathcal{R}$ and $\mathcal{I}(t)$.

Can a similar case happen if two terms are result-complete? For example, if we change $\mathcal{R}(5)$ to C in Σ_5, t_2 and t_3 become result-complete, but Σ_5 is no longer regular, because $f(X{:}\text{F}, Y{:}\text{F})$ no longer has a least sort. We show that this observation holds for constraint simplification in the general case:

Lemma 11. *Let $t' \leq_\Sigma t$ be two non-variable* CP*-terms. Then $\mathcal{I}(t') \subseteq \mathcal{I}(t)$.*

Theorem 12. *$t/\widehat{C}\theta$ is a result-complete* MGCI$_\Sigma$ *of the two result-complete* CP*-terms $t' = f(t_1,\ldots,t_n)/\widehat{C'}$ and $t'' = f(t'_1,\ldots,t'_n)/\widehat{C''}$, iff θ is a most general Σ-T-unifier of $f(t_1,\ldots,t_n)$ and $f(t'_1,\ldots,t'_n)$, $t = f(t_1,\ldots,t_n)\theta$, and $C = C' \cap C'' \neq \emptyset$.*

[9] All missing proofs can be found in [Prehofer, 1992].

Proof. We show that $t/\widehat{C}\theta$ and $t/(\widehat{C}' \wedge \widehat{C})\theta$ have the same gound Σ-instances. Assume $C = \{d_1, \ldots, d_n\}, C' = \{d'_1, \ldots, d'_m\}, C'' = \{d''_1, \ldots, d''_p\}$. Using distributivity, $(\widehat{C}' \wedge \widehat{C})$ gives $\vee_{i=1,\ldots,m;j=1,\ldots,p}(d'_i(t) \wedge d''_j(t))$. All disjuncts with $d'_i = d''_j$ are in $\widehat{C}$. Hence $\widehat{C} \rightarrow (\widehat{C}' \wedge \widehat{C}'')$ and we only have to show that all disjuncts with $d'_i \neq d''_j$ are redundant:

- If $d'_i \prec d''_j$ ($d''_i \prec d'_j$ is symmetric) then $(d'_i(t) \wedge d''_j(t)) \leftrightarrow d'_i(t)$. Now $d'_i \in \mathcal{U}(t)$ implies $d'_i \in \mathcal{U}(t')$, and the same for t''. Hence, if $d'_i \in \mathcal{U}(t)$ then $d'_i \in \mathcal{U}(t'')$ because C'' is downwards closed. Then $d'_i \in C''$ follows.
- If $d'_i \not\prec d''_j$ and $d''_j \not\prec d'_i$, then, if d'_i, d''_j, t have no common Σ-instance, $(d'_i(t) \wedge d''_j(t))$ is not Σ-satisfiable and is therefore redundant.
  Otherwise, let s be the unique least sort of $\text{MGCI}_\Sigma(d_i, d'_j)$. Then there has to be at least one declaration $d_0 \in \Sigma$ with $\mathcal{R}(d_0) = s$ and $\text{MGCI}_\Sigma(d_i, d'_j) \leq_\Sigma d_0$. Because s is the least sort, $s \preceq \mathcal{R}(d'_i)$ and $s \preceq \mathcal{R}(d''_j)$.
  Let $I_0 = \mathcal{I}(t)$ and $I_1 = \{d \in I_0 \mid d \preceq d_0\}$. We show first that $I_1 \subseteq C$. Because t is a Σ-instance of t' and of t'', Lemma 11 entails that $I_0 \subseteq \mathcal{I}(t')$ and $I_0 \subseteq \mathcal{I}(t'')$. Then the result-completeness of C' and C'' entails $I_1 \subseteq C$, because $d \preceq d_0 \preceq_\mathcal{R} d'_i$ and $d \preceq d_0 \preceq_\mathcal{R} d''_j$ for all $d \in I_1$.
  It remains to show that $\Sigma \models \overline{\forall}(d'_i(t) \wedge d''_j(t) \rightarrow \widehat{I}_1)$. Assume $t_0 \leq_\Sigma t$ with $\Sigma \models \overline{\forall}(d'_i(t_0) \wedge d''_j(t_0))$. Then $t_0 \leq_\Sigma \text{MGCI}_\Sigma(d'_i, d''_j)$ and hence $t_0 \leq_\Sigma d_0$ and Σ entails $\overline{\forall} d_0(t_0)$. From Theorem 8 follows that there is a $d'_0 \in I_0$ with $d'_0 \prec d_0$ (i.e. $d'_0 \in I_1$) such that Σ entails $\overline{\forall} d'_0(t_0)$. □

For $S \subseteq \mathcal{S}, D \subseteq \mathcal{D}$, we define $S \sqcap D = \{d \in D \mid \mathcal{R}(d) \in S\}$.[10]

Theorem 13. *$f(t_1, \ldots, t_n)/\widehat{C}''$ is a* MGCI_Σ *of the* CP*-term $t = f(t_1, \ldots, t_n)/\widehat{C}$ and a* CP*-variable $X/\widehat{C}'$ iff $C'' = C' \sqcap C'' \neq \emptyset$, X does not occur in t and $\widehat{C}''$ is Σ-satisfiable. Furthermore, $f(t_1, \ldots, t_n)/\widehat{C}''$ is result-complete if $f(t_1, \ldots, t_n)/\widehat{C}$ is result-complete.*

It is easy to see that Theorem 12 and Theorem 13 preserve result-completeness. Using Theorem 8, it can be shown that applying Theorem 12 preserves Σ-validity for terms $t/\widehat{C}$ with $C \supseteq \mathcal{I}(t)$.

4.3 Constraint Satisfaction in CP

In the previous section we simplified two CP-constraints locally. Now we show how to ensure global Σ-satisfiability for the rules in Figure 3 with CP-constraints. In the original sorted logic we had the problem of multiple unifiers. Each one of these corresponds to exactly one solution of the corresponding constraint satisfaction problem in CP-logic. Compared to sorted logic, we can now separate the problem of multiple unifiers (i.e. the constraint satisfaction problem in CP) out into a distinct search space.

After translating the sorted logic into CP-logic, all terms are Σ-valid. The problem is that computing MGCI_Σ's with Theorem 13 does not preserve Σ-validity. For

---

[10] Efficient implementations of this set operation are discussed in [Prehofer, 1992].

instance, in Example 2, Theorem 13 gives the MGCI$_\Sigma$

$$t_0 = sum(X/\{\text{INT}, \text{EVN}, \text{ODD}\}, Y/\{\text{INT}, \text{EVN}, \text{ODD}\})/\{d_4, d_5\}.$$

Assume that X and Y have additional occurrences. Now if X is unified with a variable $Z/\{\text{EVN}\}$ and Y with $Z'/\{\text{ODD}\}$, the constraint $\{d_4, d_5\}$ is not Σ-satisfiable. One way to solve this problem is to check all constraints on terms whenever a term is substituted for another term, but these constraint checks are costly and often redundant.

Instead, to guarantee Σ-satisfiability, we maintain a Σ-valid Σ-instance for all CP-terms. As we see later on, this Σ-valid Σ-instance can be seen as a representative unifier from the corresponding CII$_\Sigma$ in the sorted logic. If we, in the unification rules, solve an equation $s/C_s = t/C_t$, we also solve an equation with the two corresponding Σ-valid Σ-instances, $s'/C'_s = t'/C'_t$, where $s'/C'_s \leq_\Sigma s/C_s$ and $t'/C'_t \leq_\Sigma t/C_t$. Since there can be many Σ-valid terms s'/C'_s with $s'/C'_s \leq_\Sigma s/C_s$, we call s'/C'_s the *choice term* of s/C_s. If s/C_s is Σ-valid, it is sufficient to use s/C_s as the choice term of s/C_s.

Consider solving an equation $X/C = t/C_t$ with Theorem 13, where X is a variable and t is a non-variable term. Let t'/C'_t, with $t' = f(t_1, \ldots, t_n)$, be the MGCI$_\Sigma$ of X/C and t/C_t and let $t'' = f(t'_1, \ldots, t'_n)/C''_t$ be the MGCI$_\Sigma$ of the choice terms of X and t. Then t'' is a choice term of t'/C'_t.

Since Theorem 13 does not preserve Σ-validity, we have to assure that the choice term t'' is Σ-valid. For that purpose, the unification rules select a declaration $d = f(s_1, \ldots, s_n) : s_0$ from C''_t and add the equations $t'_1 = X_1/s_1(X_1), \ldots, t'_n = X_n/s_n(X_n)$, where all X_i are new variables, to the set of equations. This assures that $\Sigma \models d(t'')$ and t'' is Σ-valid, or the added equations are not solvable.

Since we have a choice for selecting d, we call t'' a *choice point*. To see how choice terms relate to choice points, recall that a term t is Σ-valid after t has been translated from the sorted logic into CP-logic. Therefore, t itself is initially used as the choice term of t. If the unification rules, as in the case above, add equations that only affect the choice terms, the choice terms may become true Σ-instances of their corresponding CP-terms. Since the added equations depend on the declarations chosen at some choice terms, these choice terms are called choice points.

For instance, solving the equation $X'{:}\text{EVN} = t_1$, where

$$t_1 = sum(X{:}\text{INT}, sum(Y{:}\text{INT}, Z{:}\text{INT})/\{d_1, d_2, d_3, d_4, d_5\})/\{d_1, d_2, d_3, d_4, d_5\},$$

with Theorem 13 gives

$$t'_1 = \text{MGCI}_\Sigma(X', t_1) = sum(X{:}\text{INT}, sum(Y{:}\text{INT}, Z{:}\text{INT})/\{d_1, d_2, d_3, d_4, d_5\})/\{d_4, d_5\}.$$

Since X' and t_1 are Σ-valid, their choice terms are identical to X' and t_1, respectively. To obtain a choice term t''_1 for t'_1, we arbitrarily select d_5 from $\{d_4, d_5\}$. This adds two new equations, $X_1{:}\text{EVN} = X{:}\text{INT}$ and $X_2{:}\text{EVN} = sum(Y{:}\text{INT}, Z{:}\text{INT})/\{d_1 - d_5\}$. To solve these, we have to make $sum(Y, Z)$ a choice point. We select d_4 at $sum(Y, Z)$ and add equations similar to above. After solving all these equations, we obtain

$$t''_1 = sum(X{:}\text{EVN}, sum(Y{:}\text{ODD}, Z{:}\text{ODD})/\{\overline{d_4}, d_5\})/\{d_4, \overline{d_5}\},$$

where the selected declarations are indicated by overlining.

Then all the subterms of the choice point t'', including t'', are Σ-valid. Notice that t'' without constraints on terms, $sum(X{:}\textsc{evn}, sum(Y{:}\textsc{odd}, Z{:}\textsc{odd}))$, is a common instance of X' and t in the original sorted language.

Obviously, a term differs from its choice term only in the constraints. Therefore, to represent the CP-term and its choice term, it is sufficient to use two separate constraints on one term. The first one, called *cp*, is the constraint of the CP-$\textsc{mgci}_\Sigma$. The second, called *ch*, represents a choice term. We write *ch*-constraints as subscripts. For instance, we write t_1' and t_1'' as one term:

$$sum(X{:}\text{INT}_{\text{EVN}}, sum(Y{:}\text{INT}_{\text{ODD}}, Z{:}\text{INT}_{\text{ODD}})/\{d_1 - d_5\}_{\{\overline{d_4}, d_5\}})/\{d_4, d_5\}_{\{d_4, \overline{d_5}\}}$$

Observe that the first choice point, t_1'', limits the choice for the second, $sum(Y, Z)$, to $\{d_4, d_5\}$. Selecting d_4 at t_1'', for example, restricts the choices at $sum(Y, Z)$ to $\{d_2, d_3\}$.

We now have a constraint satisfaction search space where every declaration of every choice point is a possible choice.[11] Therefore, it is desirable to have small constraint sets, as pursued in Section 4.1.

If unification fails for the current choice term, we have to search for another choice term; we will implement this search with backtracking. If there is no other choice term, unification fails for the CP-term. To improve backtracking, we build up a dependency structure between choice points and their immediate subterms, for instance the subterms X and $sum(Y, Z)$ will keep a reference to the choice point t_1'. One term can be subject to several choices, so a set of references is maintained in the CP-unification rules in the next section. We define $\mathcal{P}$ to be a data structure that contains the set of all choice points that have been created so far. Summarizing the above observations, we obtain the following dual terms, each representing two CP-terms:

- Dual Variables: $X_{cp,ch,bt}$, where $cp, ch \subseteq \mathcal{S}, bt \subseteq \mathcal{P}$
- Dual Terms: $f_{cp,ch,cd,bt}(t_1, \ldots, t_n)$, where $cp, ch \subseteq \mathcal{D}(f), cd \in ch, bt \subseteq \mathcal{P}$

The parameter *bt* of a term t will always be the set of all choice points which are a direct superterm of some occurrence of t. The parameter *cd* indicates the chosen declaration if t is a choice point, otherwise, if $cd = nil$, t is not a choice point. Initially, after translating sorted terms into CP-terms, *ch* is set to *cp*, $cd = nil$ and $bt = \emptyset$. The parameters *cd* and *bt* are used to organize backtracking and have no (logical) meaning. They are attached to terms (in substitutions) for notational convenience.

As long as a term t/cp and all superterms of t/cp are Σ-valid, t's *bt* set is empty and $cd = nil$. Only if $bt \neq \emptyset$ and $ch \subseteq cp$, t is affected by a choice and we say that t has a choice term, as described above. A T-substitution with dual terms, such as $\{t_{cp,ch,cd,bt} \mapsto t'_{cp',ch',cd',bt'}\}$, stands for two T-substitutions $\{t/cp \mapsto t'/cp'\}$ and $\{t/ch \mapsto t'/ch'\}$, if we represent t and the choice term of t separately. Furthermore, we must assure for the above T-substitution on dual terms that $cd' \in ch'$ and that bt' is the set of all choice points which are direct superterms of t'.

---

[11] It is sufficient to try only maximal declarations from t's constraint C [Prehofer, 1992].

In examples, as above, we attach only the *cp* and *ch* constraints to the terms and show the chosen declarations and the dependence structure, represented by the *bt*'s, separately.

The Unification Rules

The rules in Figure 5 are a refinement of the rules in Figure 3 and apply the theorems of Section 4.2 to dual terms. The rules compute the CP-MGCI$_\Sigma$ and its choice term at the same time.

We showed that we can represent CP-constraints as sets and implement the logical conjunctions as set operations. Therefore, in the rules in Figure 5, we treat constraints as sets.

All four rules unify two terms t, t' and both are throughout replaced by their MGCI$_\Sigma$, hence the *bt* set of their MGCI$_\Sigma$ is obtained by the union of the *bt* sets of t and of t'. To assure Σ-satisfiability, the rules have to check that no constraints are empty. The requirement $cd \in ch$ assures that the choice term is a Σ-instance of the CP-MGCI$_\Sigma$.

The rules operate on arbitrary equations between dual terms which possibly represent an exponential number of sorted unifiers. In practice most terms do not represent multiple unifiers. Then the parameters *ch*, *bt* and *cd* are not needed and simpler terms and simpler rules can be used [Prehofer, 1992].

Rule 3 creates choice points and adds new equations as shown above. The new variables in the added equations have $\mathcal{S}$ as *cp*-constraint to assure that the *cp*-constraints of the subterms of the choice points are not affected. These variables also have a *bt* set with a reference to the newly created choice point. This reference is propagated by the equations to all direct subterms of the new choice term.

Rule 3 makes a choice by arbitrarily selecting a maximal declaration. Whenever a failure occurs because of the choices at the choice points in $\mathcal{P}$ (i.e. $cd \notin ch$ or $ch = \emptyset$), the procedure *backtracking*, explained in the next section, is invoked.

Rule 4 is complicated because both terms are possible choice points. If neither of them is a choice point, the rule is simple. If one of them is, we have to make sure that its choice term is also a choice term of the resulting MGCI$_\Sigma$. If both are choice terms, their chosen declarations, *cd* and *cd'*, must be identical, if not, the previously made choices fail and the procedure *backtracking* is invoked. Furthermore, if both are choice points, Rule 4 collapses them into one choice point.

Unification Backtracking

To get a new Σ-valid choice term, we have to search through the choice points and find new consistent choices. We perform this search in the constraint satisfaction search space by specialized chronological backtracking. If we apply the rule of resolution, as presented in Section 3,

$$\frac{\begin{array}{c} L \vee M_1 \vee \ldots \vee M_m \\ \neg L' \vee N_1 \vee \ldots \vee N_n \end{array}}{(M_1 \vee \ldots \vee M_m \vee N_1 \vee \ldots \vee N_n)\theta,}$$

we first compute a most general Σ-T-unifier θ for L and L', and apply it to the remaining literals, M_i and N_i, in the resolvent. Assume M_i and N_i contain only Σ-valid terms. Then applying θ preserves Σ-satisfiability. If M_i and N_i have choice

1. $(\vartheta, U \cup \{X_{cp,ch,bt} = Y_{cp',ch',bt'}\}) \Longrightarrow (\vartheta\theta, U\theta)$, where
$cp'' = cp \cap cp'$
$ch'' = ch \cap ch'$
$bt'' = bt \cup bt'$
$\theta = \{X_{cp,ch,bt} \mapsto X_{cp'',ch'',bt''}, Y_{cp',ch',bt'} \mapsto X_{cp'',ch'',bt''}\}$
IF $cp'' = \emptyset$ OR ($ch'' = \emptyset$ AND not $backtrack(max(\mathcal{P}), max(bt''))$) THEN fail
2. $(\vartheta, U \cup \{f_{cp',ch',cd',bt'}(t_1, \ldots, t_n) = X_{cp,ch,bt}\}) \Longrightarrow$
$(\vartheta, U \cup \{X_{cp,ch,bt} = f_{cp',ch',cd',bt'}(t_1, \ldots, t_n)\})$
3. $(\vartheta, U \cup \{X_{cp,ch,bt} = f_{cp',ch',cd',bt'}(t_1, \ldots, t_n)\}) \Longrightarrow (\vartheta\theta, U\theta \cup W)$, where
$cp'' = cp \sqcap cp'$
$ch'' = ch \sqcap ch'$
$bt'' = bt \cup bt'$
$\theta = \{X_{cp,ch,bt} \mapsto f_{cp'',ch'',cd',bt''}(t_1, \ldots, t_n),$
$f_{cp',ch',cd',bt'}(t_1, \ldots, t_n) \mapsto f_{cp'',ch'',cd',bt''}(t_1, \ldots, t_n)\}$
$W = \emptyset$
IF $cp'' = \emptyset$ OR X occurs in $f_{cp',ch',cd',bt'}(t_1, \ldots, t_n)$ THEN fail
IF $ch'' = \emptyset$ AND not $backtrack(max(\mathcal{P}), max(bt''))$ THEN fail
IF $cd' \neq nil$ AND $cd' \notin ch''$ AND not $backtrack(max(\mathcal{P}), f_{cp'',ch'',cd',bt''}(\ldots))$ THEN fail
IF $cd' = nil$ AND $cp'' \neq cp'$ THEN make $f_{cp'',ch'',cd',bt''}(\ldots)$ a choice point:
$\mathcal{P} = \mathcal{P} \cup f_{cp'',ch'',cd',bt''}(t_1, \ldots, t_n)$
Select a maximal declaration $cd' = f(s_1, \ldots, s_n) : s_{n+1}$ from ch'' and $W = \{X^i_{S,s_i,\{\uparrow f\}} = t_i \mid i \in \{1 \ldots n\}\}$, where $\{\uparrow f\}$ is a one element set containing a reference to $f_{cp'',ch'',cd',bt''}(t_1, \ldots, t_n)$ and X^i are new variables.
4. $(\vartheta, U \cup \{f_{cp,ch,cd,bt}(t_1, \ldots, t_n) = f'_{cp',ch',cd',bt'}(t'_1, \ldots, t'_n)\}) \Longrightarrow$
$(\vartheta\theta, U\theta \cup \{t'_1 = t_1, \ldots, t'_n = t_n\})$, where
$cp'' = cp \cap cp'$
$ch'' = ch \cap ch'$
$bt'' = bt \cup bt'$
$$cd'' = \begin{cases} cd \text{ IF } cd' = nil \\ cd' \text{ ELSE} \end{cases}$$
$\theta = \{f_{cp,ch,cd,bt}(t_1, \ldots, t_n) \mapsto f_{cp'',ch'',cd',bt''}(t_1, \ldots, t_n),$
$f'_{cp',ch',cd',bt'}(t'_1, \ldots, t'_n) \mapsto f_{cp'',ch'',cd',bt''}(t_1, \ldots, t_n)\}$
IF $f \neq f'$ OR $cp'' = \emptyset$ THEN fail
IF $ch'' = \emptyset$ AND not $backtrack(max(\mathcal{P}), max(bt''))$ THEN fail
IF $cd \neq nil$ AND $cd' \neq nil$ AND $cd \neq cd'$
AND not $backtrack(max(\mathcal{P}), f_{cp'',ch'',cd',bt''}(\ldots))$ THEN fail
IF $cd'' \neq nil$ AND $cd'' \notin ch''$ AND not $backtrack(max(\mathcal{P}), f_{cp'',ch'',cd',bt''}(\ldots))$ THEN fail

Fig. 5. CP-Unification Rules

points, then these choice points must have been created at previous resolution steps. Therefore, we assume that $\mathcal{P}$ includes all choice points of all previous unifications in the current derivation. These choice points can affect the current choices even if they do not appear in the currently resolved clauses. Hence, backtracking covers all choice points that have been created in the whole derivation of the two resolved clauses.

```
FUNCTION backtrack(Pterm, min_backtrack)
IF min_backtrack < Pterm THEN skip = TRUE
ELSE skip = FALSE
found = FALSE
Assume Pterm = f_{cp,ch,cd,bt}(t_1, ..., t_n)
REPEAT
    WHILE (not found AND not skip AND has_more_choices(ch, cd))
        cd = next_choice(ch, cd)
        found = try_declaration(Pterm, cd)
    ENDWHILE
    skip = FALSE
    IF (not found) THEN
        IF (backtrack(previous_choice_point(Pterm), min_backtrack)) THEN cd = nil
        ELSE RETURN FALSE
UNTIL(found)
RETURN TRUE
```

```
FUNCTION try_declaration(Pterm, d)
Undo all choices at choice points t with Pterm ≤ t
Assume d = f(s_1, ..., s_n) : s_{n+1} and Pterm = f_{cp,ch,cd,bt}(t_1, ..., t_n)
FOR EACH t_i DO
    IF t_i = X_{cp',ch',bt'} THEN
        ch' = ch' ⊓ s_i
        IF ch' = ∅ THEN RETURN FALSE
    ELSE IF t_i = f_{cp',ch',cd',bt'}(t'_1, ..., t'_n) THEN
        ch' = ch' ⊓ s_i
        IF ch' = ∅ THEN RETURN FALSE
        IF cd' ≠ NIL AND t_i < Pterm AND cd' ∉ ch' THEN RETURN FALSE
RETURN TRUE
```

Fig. 6. Backtracking Algorithm

In the following, we restrict ourselves to SLD-resolution such as in Prolog. We show in [Prehofer, 1992] how to extend unification backtracking to general resolution. This unification backtracking is independent of the ordinary Prolog backtracking. Unification backtracking could be performed like ordinary chronological backtracking. After a failure, the last choice point is selected and all substitutions since this point are undone.

In this specialized backtracking, we will find the first choice point that can affect the failure, using the dependency structure. Also, we will not undo substitutions and *cp*-constraints. Because choice terms differ only in the *ch*-constraints, only these have to be undone. Thus we solely work in the constraint satisfaction search space. Furthermore, the dependency structure is constant during a search in this search space, because it depends only on the primary resolution search space.

We use the time of creation to order the choice points $\mathcal{P}$. This ordering, denoted by $<$, does not necessarily reflect the term structure, that is for two choice points, $t < t'$, that t' can be a subterm of t or vice versa.

Then the search space can be described by a list of choice points, $c_1, c_2, c_3, \ldots, c_n$, where the c_i's appear in the order they have been created in the application of the above rules. At each c_i a choice is made that is — if possible — consistent with the subterms of c_i. If a variable X/C is a subterm of several c_i's, the constraint of its choice term must be the intersection of the sorts that have been selected for X at each occurrence of X. Also, choice points can be subterms of each other and limit choices of other choice terms. Therefore, if we make a choice at c_i, this choice may invalidate the current choice at another choice point c', if $cd \notin ch$ for c'.

The algorithm in Figure 6 modifies *ch* constraints and the chosen declarations (*cd*) to obtain a Σ-satisfiable choice term.[12] The function ***backtrack*** handles choice points recursively in descending order with respect to $<$ and is called with the maximal element of $\mathcal{P}$. The second argument of the function backtrack, *min_backtrack*, determines the minimal backtracking depth. It indicates the maximal choice point that can change the choice for a term t with $ch = \emptyset$ or $cd \notin ch$. If $ch = \emptyset$, the maximal element in the *bt* set of t is the first choice point that affects *ch*. In the second case, $cd \notin ch$, we have to backtrack at least to t and make a new valid choice.

The following functions are used to iterate over the maximal elements of *cd*: First, in Rule 3, a maximal declaration is selected. If backtracking is invoked, the function *next_choice*(*ch*, *cd*) iterates over the remaining maximal elements of *cd* until it reaches the last element. Only then the function *has_more_choices*(*ch*, *cd*) returns false. In that case, *cd* can be set to *nil* and the next invocation of the function *next_choice*(*ch*, *cd*) starts over again with the first maximal declaration.

The function *try_declaration*(t, d) takes a choice point t and a declaration d as its arguments. It intersects the *ch*-constraints of the subterms of t with the sorts of d and returns true if the choice is consistent with the choices at all choice points t' with $t' < t$. If this succeeds at all choice points we obtain a new choice term.

5 Examples and Comparisons to Sorted Logic

The first example is intended to show the restructuring of the search space by using CP-logic. We assume Σ_2 as above and the two program clauses:

1. $p(X{:}\text{EVN})$
2. $p(X{:}\text{ODD})$

When solving the goal

$$\neg p(sum(X_1{:}\text{INT}, X_2{:}\text{INT})) \vee \neg p(sum(X_3{:}\text{INT}, X_4{:}\text{INT}))$$

with SLD-resolution (Prolog), there are two choices at each resolution step and the at each of these there are, in sorted logic, two unifiers. This creates the search space in Figure 7. The choices caused by multiple unifiers are circled.

The search space for CP-logic (see Fig. 8) shows the constraint satisfaction search space in the form of the curved lines. Each curved line corresponds to one unifier in the sorted logic. At every resolution step, we only have to select one of these unifiers for a deduction. To see the advantages from this 2-dimensional search space, assume the number of multiple unifiers doubles at one unification. Then in the order-sorted case, the whole subtree doubles; in CP-logic only the constraint search space doubles.

[12] We assume that if it modifies one occurrence of a term t, it also modifies all occurrences of t.

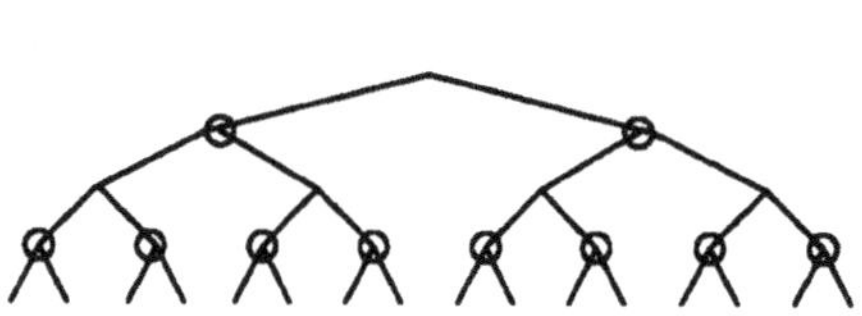

Fig. 7. Search Space with Sorted Logic

Fig. 8. Search Space with CP-Logic

Furthermore, a CP-term can be seen as an upper bound (w.r.t $\leq_\Sigma$) of all its choice terms, which correspond to common instances in sorted logic. If this single CP-term fails, all sorted common instances would fail also. That means, in the sorted case, we would have to generate each one of them and each single one would fail. Also, search in the constraint satisfaction search space can be highly optimized, as shown in the next example.

The second example shows the details of a derivation with CP-logic, in particular backtracking. We again use SLD-resolution and assume Σ_6 (Fig. 9) and the two program clauses:

1. $p(X{:}\text{EVN}_{\text{EVN}})$
2. $q(Y{:}\text{POS}_{\text{POS}})$

Recall that the sorts and declarations in subscripts indicate the choice terms, which correspond to sorted unifiers. The top section of Figure 10 shows the initial goal list. The lower five sections show the goals after each of the five resolution steps with the the five goals. Since unification backtracking includes the already resolved goals, only these are shown.

The second section of Figure 10 shows the first goal after unification with Fact 1. Both *sum*(...) terms have become choice points. At the outer *sum*(...) term, to which we refer as choice point (a), declaration (4) has been selected. Therefore, the second choice point, $sum(Y, Z)$, to which we refer as (b), is restricted to the declarations $\{2, 3\}$, because declaration (4) at (a) restricts (b) to sort ODD. At choice point (b), declaration (3) has been selected.

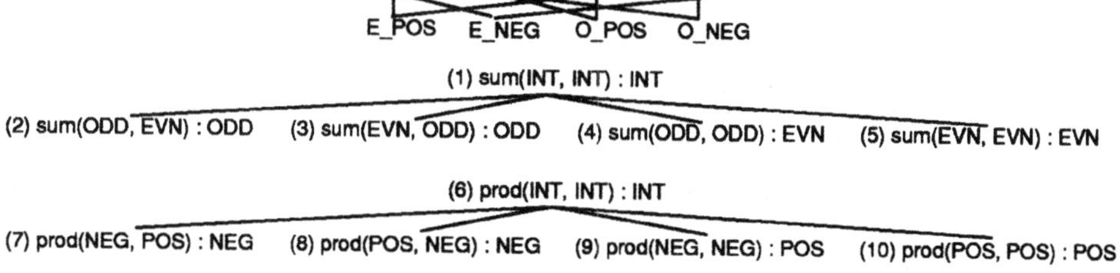

Fig. 9. Sort Theory Σ_6

	Initial Goals
1	$p(sum(X{:}\mathrm{INT}_{\mathrm{INT}}, sum(Y{:}\mathrm{INT}_{\mathrm{INT}}, Z{:}\mathrm{INT}_{\mathrm{INT}})\{1-5\}_{\{1-5\}})/\{1-5\}_{\{1-5\}}))$
2	$q(prod(Y{:}\mathrm{INT}_{\mathrm{INT}}, Z{:}\mathrm{INT}_{\mathrm{INT}})\{6-10\}_{\{6-10\}})$
3	$q(Z{:}\mathrm{INT}_{\mathrm{INT}})$
4	$p(Y{:}\mathrm{INT}_{\mathrm{INT}})$
5	$p(X{:}\mathrm{INT}_{\mathrm{INT}})$
1	$p(sum(X{:}\mathrm{INT}_{\mathrm{ODD}}, sum(Y{:}\mathrm{INT}_{\mathrm{EVN}}, Z{:}\mathrm{INT}_{\mathrm{ODD}})\{1-5\}_{\{2,\bar{3}\}})/\{4,5\}_{\{\bar{4},5\}}))$
1	$p(sum(X{:}\mathrm{INT}_{\mathrm{ODD}}, sum(Y{:}\mathrm{INT}_{\mathrm{E\_POS}}, Z{:}\mathrm{INT}_{\mathrm{O\_POS}})\{1-5\}_{\{2,\bar{3}\}})/\{4,5\}_{\{\bar{4},5\}}))$
2	$q(prod(Y{:}\mathrm{INT}_{\mathrm{E\_POS}}, Z{:}\mathrm{INT}_{\mathrm{O\_POS}})\{9,10\}_{\{9,\overline{10}\}})$
1	$p(sum(X{:}\mathrm{INT}_{\mathrm{ODD}}, sum(Y{:}\mathrm{INT}_{\mathrm{E\_POS}}, Z{:}\mathrm{INT}_{\mathrm{O\_POS}})\{1-5\}_{\{2,\bar{3}\}})/\{4,5\}_{\{\bar{4},5\}}))$
2	$q(prod(Y{:}\mathrm{INT}_{\mathrm{E\_POS}}, Z{:}\mathrm{POS}_{\mathrm{O\_POS}})\{9,10\}_{\{9,\overline{10}\}})$
3	$q(Z{:}\mathrm{POS}_{\mathrm{O\_POS}})$
1	$p(sum(X{:}\mathrm{INT}_{\mathrm{ODD}}, sum(Y{:}\mathrm{EVN}_{\mathrm{E\_POS}}, Z{:}\mathrm{INT}_{\mathrm{O\_POS}})\{1-5\}_{\{2,\bar{3}\}})/\{4,5\}_{\{\bar{4},5\}}))$
2	$q(prod(Y{:}\mathrm{EVN}_{\mathrm{E\_POS}}, Z{:}\mathrm{POS}_{\mathrm{O\_POS}})\{9,10\}_{\{9,\overline{10}\}})$
3	$q(Z{:}\mathrm{POS}_{\mathrm{O\_POS}})$
4	$p(Y{:}\mathrm{EVN}_{\mathrm{E\_POS}})$
1	$p(sum(X{:}\mathrm{EVN}_{\mathrm{EVN}}, sum(Y{:}\mathrm{EVN}_{\mathrm{E\_POS}}, Z{:}\mathrm{INT}_{\mathrm{E\_POS}})\{1-5\}_{\{4,\bar{5}\}})/\{4,5\}_{\{4,\bar{5}\}}))$
2	$q(prod(Y{:}\mathrm{EVN}_{\mathrm{E\_POS}}, Z{:}\mathrm{POS}_{\mathrm{E\_POS}})\{9,10\}_{\{9,\overline{10}\}})$
3	$q(Z{:}\mathrm{POS}_{\mathrm{E\_POS}})$
4	$p(Y{:}\mathrm{EVN}_{\mathrm{E\_POS}})$
5	$p(X{:}\mathrm{EVN}_{\mathrm{EVN}})$

Fig. 10. Goal-Reduction for the Prolog Example

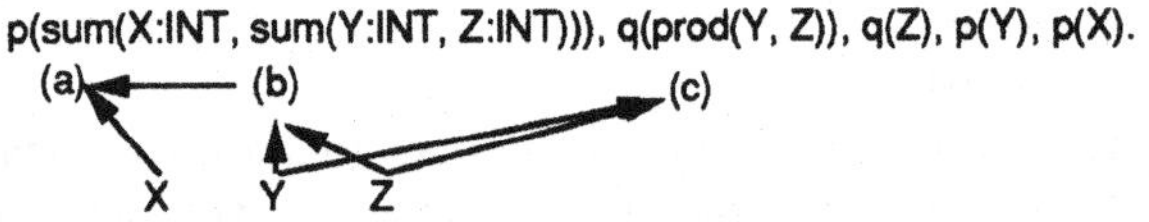

Fig. 11. Query with a Possible Dependency Structure

Then, in the third section, the second literal in the goal has been unified with the second program clause, making *prod*(Y, Z) the third choice point, called (c). At (c) declaration (10) has been selected. Figure 11 shows the references from the subterms to the three choice points (a)-(c). The *bt* set of a term in Figure 11 contains all the terms to which it has an outgoing arrow. For instance, the *bt* set of Z is $\{b, c\}$.

Now there are two choices at each point and, because there are no conflicts between any choices, this corresponds to a CII_{Σ} of size eight in the sorted language.

Unifying the third and fourth goals with the above program clauses changes the *cp*-constraints of Y and Z. To resolve the last goal, unification with the first program clause fails for the choice term of X, which is of sort ODD, but not for the *cp*-constraint.

Then the function *backtracking* is invoked with the parameters (c) and (a), hence skips the choice points (c) and (b). At (a), backtracking changes *cd* to declaration (5), which restricts the *ch*-constraint of (b) to $\{4, 5\}$. Then at (b) and (c), because the *cp*-constraints of Y and Z have have been reduced, (5) and (10), respectively, are the only possible choices, which give the final instances of the five goals.

Observe that Y and Z have become more restricted in the primary resolution search space and that these additional *cp*-constraints limit our choices at (b) and (c), thus pruning the constraint search space. This can be seen as constraint propagation from the resolution search space to the constraint search space.

The claim is that if there are many sorted unifiers, it is not necessary to generate all of them. With sorted logic, an exponential number of maximal unifiers directly causes an exponential blowup of the resolution search space. Instead, we simply select a representative unifier in the form of a choice term to ensure Σ-satisfiability and delay further search. Then, if further constraints make the search space smaller, many sorted unifiers would fail. When the representative fails, we use intelligent backtracking and these additional constraints to effectively reduce the search space.

6 Conclusions

Most other implementation-oriented approaches to order-sorted resolution often avoid polymorphism or impose strong restrictions on it, e.g. [Beierle *et al.*, 1991, Huber and Varsek, 1987, Smolka, 1989]. The approaches by Beierle and Smolka are based on parametric polymorphism (see also [Hanus, 1989, Mycroft and O'Keefe, 1984]) and allow for sort constructors, which are not supported here. Apart from this, our approach is much more general, since the sortal behavior of functions can be specified with minimal restrictions. It our view, the existence of multiple unifiers is not an obstacle, but an opportunity for efficient implementations.

To handle an exponential number of maximally general unifiers in a polymorphic sort theory, we propose constraining non-variable terms by disjunctions of declarations. For unification, the logical conjunction of these constraints can be reduced to set operations. The embedding in CP-logic assures a single most general unifier and creates a separate constraint satisfaction search space. The complexity of the constraint search space does not affect the complexity of the primary resolution search space. Informally speaking, with sorted logic, the total search space doubles if the number of maximal unifiers doubles; with CP-logic, only the constraint search space doubles.

Also, we commit ourselves to only one solution of the constraint satisfaction problem, which can be seen as selecting a representative unifier instead of generating all maximal unifiers a priori. Furthermore, a CP-term can be seen as an upper bound of all corresponding unifiers in the sorted case. This means that if unification fails for the CP-term, an exponential number of sorted unifiers would fail also.

The restructuring of the search space allows for additional optimizations. For instance, constraints can be propagated from the resolution search space to the constraint satisfaction search space. We can build up a dependency structure, which only depends on the primary resolution search space, to perform unification backtracking more intelligently and to propagate constraints within the constraint satisfaction search space.

These results motivate shifting search from the main reasoner to the constraint search space by expressing knowledge with sorts and sortal behavior of functions. It is an open question if this method of restructuring the search space can be applied to other logics in a similar way.

Acknowledgement

The author is indebted to Alan Frisch for numerous valuable discussions. Also Rich Scherl, Andreas Werner, and Christoph Brzoska contributed to this work.

References

[Aït-Kaci *et al.*, 1989] H. Aït-Kaci, R. Boyer, P. Lincoln, and R. Nasr. Efficient lattice operations. *ACM Transactions on Programming Languages and Systems*, 11(1):115–146, January 1989.

[Beierle *et al.*, 1991] C. Beierle, G. Meyer, and H. Semle. Extending the Warren abstract machine to polymorphic order-sorted resolution. In *Logic Programming: Proceedings of the 1991 International Symposium*, pages 272–286, October 1991.

[Bürkert, 1991] Hans-Jürgen Bürkert. *A Resolution Principle for a Logic with Restricted Quantifiers*. Springer-LNAI 568, 1991.

[Cohn, 1985] Anthony G. Cohn. On the solution of Schubert's Steamroller in many sorted logic. In *Proceedings of the Ninth International Joint Conference on Artificial Intelligence*, pages 1169–1174, August 1985.

[Cohn, 1989] Anthony G. Cohn. Taxonomic reasoning with many-sorted logics. *Artificial Intelligence Review*, 3:89–128, 1989.

[Frisch and Scherl, 1991] Alan M. Frisch and Richard B. Scherl. A general framework for modal deduction. In *Principles of Knowledge Representation and Reasoning: Proceedings of the Second International Conference*, pages 196–207. Morgan Kaufman, San Mateo, CA, 1991.

[Frisch, 1985] Alan M. Frisch. An investigation into inference with restricted quantification and a taxonomic representation. *SIGART Newsletter*, (91):28–31, 1985.

[Frisch, 1991a] Alan M. Frisch. The substitutional framework for hybrid reasoning. Working Notes of the 1991 Fall Symposium on Principles of Hybrid Reasoning. Asilomar, CA., November 1991.

[Frisch, 1991b] Alan M. Frisch. The substitutional framework for sorted deduction: Fundamental results on hybrid reasoning. *Artificial Intelligence*, 49:161–198, 1991.

[Hanus, 1989] Michael Hanus. Horn clause programs with polymorphic types: semantics and resolution. pages 225–240. TAPSOFT'89, Springer Verlag, 1989.

[Huber and Varsek, 1987] H. Huber and L. Varsek. Extended Prolog for order-sorted resolution. In *Proceedings of the 4th IEEE Symposium on Logic Programming*, pages 34–45, 1987.

[Martelli and Montanari, 1982] Alberto Martelli and Ugo Montanari. An efficient unification algorithm. *ACM Transactions on Programming Languages and Systems*, 4(2):258–282, April 1982.

[Mycroft and O'Keefe, 1984] A. Mycroft and U. O'Keefe. A polymorphic type system for Prolog. *Artificial Intelligence*, (23):295–307, 1984.

[Prehofer, 1992] Christian F. Prehofer. A constraint language for order-sorted polymorphic resolution. Master's thesis, Univ. of Illinois at Urbana-Champaign, January 1992.

[Schmidt-Schauß, 1989] Manfred Schmidt-Schauß. *Computational Aspects of an Order-Sorted Logic with Term Declarations*. Springer-LNAI 395, 1989.

[Smolka, 1989] Gert Smolka. *Logic programming over polymorphically order-sorted types*. PhD thesis, Universität Kaiserslautern, May 1989.

[Walther, 1985] Christoph Walther. A mechanical solution of Schubert's Steamroller by many-sorted resolution. *Artificial Intelligence*, 26(2):217–224, 1985.

This article was processed using the LaTeX macro package with LLNCS style

Default theory for Well Founded Semantics with Explicit Negation

Luís Moniz Pereira   José Júlio Alferes   Joaquim Nunes Aparício
CRIA, Uninova and DCS, U. Nova de Lisboa, 2825 Monte da Caparica, Portugal
{lmp|jja|jna}@fct.unl.pt

Abstract

One aim of this paper is to define a default theory for Well Founded Semantics of logic programs which have been extended with explicit negation, such that the models of a program correspond exactly to the extensions of the default theory corresponding to the program.

To do so we must introduce a new default theory semantics that satisfies principles of modularity, enforcedness, and uniqueness of minimal extension (if it has an extension), which have a natural counterpart in the program semantics. Other default theories, with which we compare our own, do not satisfy all these principles, namely Reiter's default theory, Baral and Subrahmanian's well founded extensions, and Przymusinski's stationary extensions.

The relationship between logic programs and defaults theories opens the way for a mutual fertilization, which we have enhanced:

On the one hand, we widen the class of programs which can be given a semantics by a suitable default theory, show that explicit negation can be translated into classical negation in such a default theory, and that it clarifies the use of rules in extended logic programs.

On the other hand, since our logic program semantics is definable by a monotonic fixpoint operator, it has desirable computational properties, including top–down and bottom–up procedures. As our semantics is sound with respect to Reiter's default semantics, whenever an extension exists, we thus provide sound methods for computing the intersection of all extensions for an important subset of Reiter's default theories.

1 Introduction

A relationship between logic programs and default theories was first proposed in [3] and [4]. The idea is to translate every program rule into a default one and then compare extensions of the default theory with the semantics of the corresponding program.

In [4], stable model semantics [5] was shown equivalent to a special case of default theories in the sense of Reiter [15]. This result was generalized in [6] to programs with explicit negation and answer-set semantics, where they claim that explicit negation corresponds in fact to classical negation used in default theories.

Well Founded Semantics for Default Theories [1] extends Reiter's semantics of default theories, resolving some issues of the latter, namely that some theories have no extension and that some theories have no least extension. Based on the way such issues were resolved in [2], the well founded semantics for programs without explicit negation was shown equivalent to a special case of the extension classes of default theories in the sense of [1]. It turns out that in attempting to extend this result to extended logic programs with explicit negation we get some unintuitive results and no semantics of such logic programs relates to known default theories.

To overcome that, here we identify principles a default theory semantics should enjoy, and introduce a default theory semantics that extends that of [2] to the larger class of logic programs, and complies with those principles.

Such relationship to a larger program class improves the cross-fertilization between logic programs and default theories, since we generalize previous results concerning their relationship [1, 2, 3, 4, 6, 13], and also because there is an increasing use of logic programming with explicit negation for nonmonotonic reasoning [6, 9, 10, 11, 17]. It also clarifies the meaning of logic programs combining both explicit negation and negation by default. In particular, it shows in what way explicit negation corresponds to classical negation in our default theory, and elucidates the use of rules in extended logic programs. Like defaults rules are unidirectional, so their contrapositives are not implicit; the rule connective $\leftarrow$ is not material implication but has rather the flavour of an inference rule, as in defaults.

On the other hand, since our logic program semantics is definable by a monotonic fixpoint operator, it has desirable computational properties, including top–down and bottom–up procedures. As our semantics is sound with respect to Reiter's default semantics, whenever an extension exists, we thus provide sound methods for computing the intersection of all extensions for an important subset of Reiter's default theories .

2 Language Used

Given a first order language *Lang* [12], an extended logic program is a set of rules of the form $H \leftarrow B_1, \ldots, B_n, \sim C_1, \ldots, \sim C_m \quad m \geq 0, n \geq 0$, where $H, B_1, \ldots, B_n, C_1, \ldots, C_m$ are classical literals. A (syntactically) classical literal (or explicit literal) is either an atom A or its explicit negation $\neg A$. We also use the symbol $\neg$ to denote complementary literals in the sense of explicit negation. Thus $\neg\neg A = A$. The symbol $\sim$ stands for negation by default[1]. $\sim L$ is called a default literal. Literals are either classical or default literals. A set of rules stands for all its ground instances wrt *Lang*.

Concerning default theories, *Lang*(AT) is the propositional language generated by considering the ground atoms of *Lang* to be propositional symbols, where *Lits* is the set of its ground literals.

Definition 2.1 (Default rule) *A propositional default is a triple $d = (p(d), j(d), c(d))$ where $p(d)$ and $c(d)$ are formulas of Lang(AT) and $j(d)$ is a finite subset of*

[1]This designation has been used in the literature instead of the more operational *"negation as failure (to prove)"*. Another appropriate designation is *"implicit negation"*, in contradistinction to explicit negation.

Lang(AT). p(d) (resp. j(d), resp. c(d)) is called the prerequisite *(resp.* justification, *resp.* consequence*) of default d. The default d is also denoted* $\frac{p(d):j(d)}{c(d)}$.

Definition 2.2 (Default theory) *A default theory Δ is a pair (D, W) where $W \subseteq$ Lang and D is a set of default rules.*

3 *WFSX* overview

In this section we briefly review $WFSX$ semantics for logic programs extended with explicit negation. For full details the reader is referred to [8].

$WFSX$ follows from one basic "coherence" requirement: $\neg L$ implies $\sim L$ (if L is explicitly false, L must be false) for any explicit literal L.

Example 1 *[8]* Consider program $P = \{a \leftarrow \sim b,\ \ b \leftarrow \sim a,\ \ \neg a \leftarrow\}$.

If $\neg a$ were to be simply considered as a new atom symbol, say a', and WFS used to define the semantics of P (as suggested in [14]), the result would be $\{\neg a, \sim\neg b\}$, so that $\neg a$ is true and a is undefined. We insist that $\sim a$ should hold, and a not, because $\neg a$ does. Accordingly, the WFSX of P is $\{\neg a, b, \sim a, \sim\neg b\}$, since b follows from $\sim a$.

We begin by providing a definition of interpretation for programs with explicit negation which incorporates coherence from the start.

Definition 3.1 (Interpretation) *[8] By an interpretation I of a language Lang we mean any set $T \cup \sim F$[2], where T and F are disjoint subsets of classical literals over the Herbrand base, and if $\neg L \in T$ then $L \in F$ (coherence)*[3]*. The set T contains all ground classical literals* true *in I, the set F contains all ground classical literals* false *in I. The truth value of the remaining classical literals is* undefined *(The truth value of a default literal $\sim L$ is the 3-valued complement of L.)*

We next extend with an additional rule the P modulo I transformation of [12], itself an extension of the Gelfond-Lifschitz modulo transformation, to account for coherence.

Definition 3.2 ($\frac{P}{I}$ transformation) *[8] Let P be an extended logic program and let I be an interpretation. By $\frac{P}{I}$ we mean a program obtained from P by performing the following three operations for every atom A :*

- *Remove from P all rules containing a negative premise $L = \sim A$ such that $A \in I$.*
- *Remove from P all rules containing a premise L (resp. $\neg L$) such that $\neg L \in I$ (resp. $L \in I$).*
- *Remove from all remaining rules of P their negative premises $L = \sim A$ such that $\sim A \in I$.*
- *Replace all the remaining negative premises by proposition* **u**[4].

[2] By $\sim\{a_1, \ldots, a_n\}$ we mean $\{\sim a_1, \ldots, \sim a_n\}$.

[3] For any literal L, if L is explicitly false L must be false. Note that the complementary condition "if $L \in T$ then $\neg L \in F$" is implicit.

[4] The special proposition **u** is *undefined* in all interpretations.

The resulting program $\frac{P}{I}$ is by definition non-negative, and it always has a unique $least(\frac{P}{I})$, where $least(\frac{P}{I})$ is:

Definition 3.3 (Least-operator) *We define $least(P)$, where P is a non-negative program, as the set of literals $T \cup \sim F$ obtained as follows:*

- *Let P' be the non-negative program obtained by replacing in P every negative classical literal $\neg L$ by a new atomic symbol, say $'\neg\_L'$.*
- *Let $T' \cup \sim F'$ be the least 3-valued model of P'.*
- *$T \cup \sim F$ is obtained from $T' \cup \sim F'$ by reversing the replacements above.*

The least 3-valued model of a non-negative program can be defined as the least fixpoint of the following generalization of the van Emden–Kowalski least model operator Ψ for definite logic programs:

Definition 3.4 Ψ^* *operator*
Suppose that P is a non-negative program, I is an interpretation of P and A is a ground atom. Then $\Psi^*(I)$ is an interpretation defined as follows:

- *$\Psi^*(I)(A) = 1$ iff there is a rule $A \leftarrow A_1, \ldots, A_n$ in P such that $I(A_i) = 1$ for all $i \leq n$.*
- *$\Psi^*(I)(A) = 0$ iff for every rule $A \leftarrow A_1, \ldots, A_n$ there is an $i \leq n$ such that $I(A_i) = 0$.*
- *$\Psi^*(I)(A) = 1/2$, otherwise.*

To avoid incoherence, a partial operator is defined that transforms any non-contradictory set of literals into an interpretation, whenever contradiction[5] is not present.

Definition 3.5 (The *Coh* operator) *[8] Let $I = T \cup \sim F$ be a set of literals such that T is not contradictory. We define $Coh(I) = I \cup \sim\{\neg L \mid L \in T\}$.*

Definition 3.6 (The Φ operator) *[8] Let P be a logic program and I an interpretation, and let $J = least(\frac{P}{I})$. If $Coh(J)$ exists we define $\Phi_P(I) = Coh(J)$. Otherwise $\Phi_P(I)$ is not defined.*

Definition 3.7 (WFS with explicit negation) *[8] An interpretation I of an extended logic program P is called an Extended Stable Model (XSM) of P iff $\Phi_P(I) = I$. The F-least Extended Stable Model is called the Well Founded Model. The semantics of P is determined by the set of all $XSMs$ of P.*

4 A Default Theory for Extended Logic Programs

The relationship between the semantics of the default theories of [2] and the semantics of the class of logic programs extended with explicit negation has not been defined. We will now introduce a default theory corresponding to this broader class of programs (whose $WFSX$ semantics was reviewed above).

[5] We say a set of literals S is contradictory iff for some literal L, $L \in S$ and $\neg L \in S$.

4.1 Principles Required of Default Theories

Next we argue about principles a default theory semantics should enjoy, and relate it to logic programs extended with explicit negation, where the said principles can also be considered desireable.

Definition 4.1 (Uniqueness of minimal extension) *Whenever a default theory has an extension there is a minimal one.*

It is well known that Reiter's default theories do not comply with this principle.

Definition 4.2 (Union of theories) *By the union of two default theories $\Delta_1 = (D_1, W_1)$ and $\Delta_2 = (D_2, W_2)$ with languages $L(\Delta_1)$ and $L(\Delta_2)$ we mean the theory $\Delta = \Delta_1 \cup \Delta_2 = (D_1 \cup D_2, W_1 \cup W_2)$ with language $L(\Delta) = L(\Delta_1) \cup L(\Delta_2)$, whenever $W_1 \cup W_2$ is consistent.*

Example 2 Consider the two default theories:

$$\Delta_1 = (\{\tfrac{:\neg a}{\neg a}, \tfrac{:a}{a}\}, \{\}) \quad \Delta_2 = (\{\tfrac{:b}{b}\}, \{\})$$

Classical default theory, well–founded semantics, and stationary semantics all identify $\{b\}$ as the single extension of Δ_2.

Since the languages of the two theories are disjoint, one would expect their union to include b in all its extensions. However, both the well founded semantics as well as the least stationary semantics give the value unknown to b in the union theory, and therefore are not modular (cumulative). There is an objectionable interaction among the default rules of both theories when put together. Classical default theory is modular but has two extensions: $\{\neg a, b\}$ and $\{a, b\}$, failing to give a unique minimal extension to the union.

Definition 4.3 (Modularity) *Let Δ_1, Δ_2 be two default theories such that $L(\Delta_1) \cap L(\Delta_2) = \{\}$ and let $\Delta = \Delta_1 \cup \Delta_2$, with extensions $E^i_{\Delta_1}$ $E^j_{\Delta_2}$ and E^k_{Δ}. A semantics for default theories is* modular *iff:*

$$\forall_A (\forall_i A \in E^i_{\Delta_1} \Rightarrow \forall_k A \in E^k_{\Delta})$$
$$\forall_A (\forall_j A \in E^j_{\Delta_2} \Rightarrow \forall_k A \in E^k_{\Delta})$$

Consider now the following examples:

Example 3 Let $(D_1, W_1) = (\{d_1 = \frac{:\neg b}{a}, d_2 = \frac{:\neg a}{b}\}, \{\})$. The two classical extensions are $\{a\}$ and $\{b\}$. Stationary default semantics has one more extension, namely $\{\}$.

Example 4 Let $(D_2, W_2) = (\{d_1 = \frac{:\neg b}{a}, d_2 = \frac{:\neg a}{b}\}, \{\neg a\})$. The only classical extension is $\{\neg a, b\}$. In the least stationary extension, $E = \Gamma^2_{\Delta}(E) = \{\neg a\}$, $j(d_2) \in E$ but $c(d_2) \notin E$.

Definition 4.4 *Given an extension E:*

- *a default d is* applicable *in E iff $p(d) \subseteq E$ and $\neg j(d) \cap E = \{\}$*

- *an applicable default d is* applied in E *iff* $c(d) \in E$

In classical default semantics every applicable default is applied. This prevents the uniqueness of a minimal extension. In example 3, because one default is always applied, one can never have a single minimal extension. In [2, 13], in order to guarantee a unique minimal extension, it becomes possible to apply or not an applicable default. However, this abandons the notion of maximality of application of defaults of classical default theory. But, in example 4, we argue that at least rule d_2 should be applied.

We want to retain the principle of uniqueness of minimal extension coupled with a notion of maximality of application of defaults which we call enforcedness.

Definition 4.5 (Enforcedness) *Given a theory Δ with extension E, a default d is* enforceable in E *iff $p(d) \subseteq E$ and $j(d) \subseteq E$. An extension* is enforced *if all enforceable defaults in D are applied.*

We argue that, whenever E is an extension, if a default is enforceable then it must be applied. Note that an enforceable default is always applicable.

Another way of viewing enforcedness is that if d is an enforceable default, and E is an extension, then the default rule d must be understood as an inference rule $p(d), j(d) \rightarrow c(d)$ and so $c(d) \in E$ must hold.

The well founded semantics and stationary semantics both sanction minimal extensions where enforceable defaults are not applied, viz. example 4. However, in this example they still allow an enforced extension $\{b, \neg a\}$. This is not the case in general:

Example 5 Let $(D, W) = (\{\frac{:\neg b}{c}, \frac{:\neg a}{b}, \frac{:\neg a}{a}\}, \{\neg b\})$. The only stationary extension is $\{\neg b\}$, which is not enforced.

4.2 Ω–Default Theory

In this section we introduce a default theory semantics which is modular and enforced. Moreover, when it is defined it has a unique minimal extension.

When relating non-disjunctive logic programs to defaults it is customary to restrict prerequisites, justifications and conclusions to ground literals. Also, each program rule corresponds to a such default, and the theory W is empty [1, 2, 3, 4, 6, 13]. We shall do the same, though in our case, we also allow W to contain ground literals.

If this may seem an excessive restriction, we prove below that the extensions of such default theories correspond to models of normal logic programs extended with explicit negation, and so have comparable expressibility.

In order to relate default theories to extended logic programs, we must provide a modular semantics for default theories, except if they are contradictory as in the example below:

Example 6 In the default theory $(\{\frac{:}{\neg a}, \frac{:}{a}\}, \{\})$ its two defaults with empty prerequesites and justifications should always be applied, which clearly enforces a contradiction. Note that this would also be the case if the default theory were to be written as $(\{\}, \{a, \neg a\})$.

Consider now example 2, which alerted us about nonmodularity in stationary or well founded default semantics, where $D = \{\frac{:\neg a}{\neg a}, \frac{:a}{a}, \frac{:b}{b}\}$ and $\{\}$ is the least extension. This result obtains because $\Gamma_\Delta(\{\})$, by having a and $\neg a$ forces, via the deductive closure operator Cn, $\neg b$ (and all the other literals) to belong to it. This implies the nonapplicability of the third default in the second iteration. For that not to happen one should inhibit $\neg b$ from belonging to $\Gamma_\Delta(\{\})$, which can be done by preventing the application of Cn. In the language to which we restrict ourselves this is not problematic because, as formulae are literals, Cn does not introduce anything except in the contradictory case. Next we define $\Gamma'_\Delta(E)$, similar to $\Gamma_\Delta(E)$ but where the Cn operator is absent.

Definition 4.6 ($\Gamma'_\Delta(E)$) *Suppose (D, W) is a default theory and E is a* context. *The operator $r^{E,D}(S)$, mapping sets of literals S into sets of literals, is defined as follows:*

$$r^{E,D}(S) = S \cup \{c(d) | d \in D, p(d) \subseteq S, \neg j(d) \cap E = \{\}\} \tag{1}$$

We define the sequence:

$$\begin{aligned} r_0^{E,D}(W) &= W \\ r_{n+1}^{E,D}(W) &= r^{E,D}(r_n^{E,D}(W)) \\ r_\infty^{E,D}(W) &= \bigcup_{n=0}^{\infty} r_n^{E,D}(W) \end{aligned}$$

and $\Gamma'_\Delta(E) = r_\infty^{E,D}(W)$.

Reconsider now example 4, that showed that stationary default extensions are not always enforced. The nonenforced extension is (the least extension) $E = \Gamma^2(E) = \{\neg a\}$, where $\Gamma(E) = \{\neg a, a, b\}$. The semantics obtained is that $\neg a$ is true and a is undefined.

To avoid this counterintuitive result we want to ensure that, for an extension $E : \forall d \in D \;\; \neg c(d) \in E \Rightarrow c(d) \notin \Gamma(E)$, i.e. if $\neg c(d)$ is true then $c(d)$ is false[6].

It is easily recognized that this condition is satisfied by seminormal default theories: if $\neg c(d)$ belongs to an extension then any seminormal rule with conclusion $c(d)$ cannot be applied. This principle is exploited in our semantics.

Definition 4.7 *Given a default theory Δ, we dub Δ^s its seminormal version*[7] *obtained as follows: There is a rule $d^s = \frac{p(d):j(d),c(d)}{c(d)}$ in Δ^s for each default rule $d = \frac{p(d):j(d)}{c(d)}$ in Δ.*

Definition 4.8 (Ω–extension) *For a theory Δ we define $\Omega_\Delta(E) = \Gamma'_\Delta(\Gamma'_{\Delta^s}(E))$. E is an Ω-extension of Δ iff:*

- *E contains no pair of complementary literals A and $\neg A$*
- $E = \Omega_\Delta(E)$

[6] cf. definition the semantics of [2].

[7] In Reiter's formalization a default is seminormal if it is of the form $\frac{p(d):j(d)\wedge c(d)}{c(d)}$. Since we are only considering ground versions of the defaults the definitions are equivalent.

- $E \subseteq \Gamma'_{\Delta^\bullet}(E)$

The need for this last condition is related to the fact that the operator Ω is the composition of two anti-monotonic operators, that in some cases are equivalent[8]. Thus, in those cases, for every fixpoint E of Ω_Δ there is one other fixpoint $E' = \Gamma'_{\Delta^\bullet}(E)$. The last condition chooses the least of E and E'.

Example 7 For the default theory $\Delta = (\{\frac{:\neg a}{a}\}, \{\})$ there are two fixpoints of Ω_Δ : $E = \{\}$ and $E' = \Gamma'_{\Delta^\bullet}(E) = \{a\}$. Only E obeys the last condition.

Definition 4.9 (Ω–Default semantics) *Let Δ be a default theory, $E = \Omega_\Delta(E)$ an extension, and L a literal.*

- *L is* true *w.r.t. extension E iff $L \in \Omega_\Delta(E)$*
- *L is* false *w.r.t. extension E iff $L \notin \Gamma'_{\Delta^\bullet}(E)$*
- *Otherwise L is* unknown *(or* undefined*)*

Example 8 Consider the default theory $\Delta = (\{\frac{:\neg c}{c}, \frac{:\neg b}{a}, \frac{:\neg a}{b}, \frac{:}{\neg a}\}, \{\})$. Its only extension is $\{\neg a, b\}$. In fact $\Gamma'_{\Delta^\bullet}(\{\neg a, b\}) = \{c, b, \neg a\}$ and $\Gamma_\Delta(E)(\{c, b, \neg a\}) = \{\neg a, b\}$. Thus: $\neg a$ and b are true, c is undefined and a and $\neg b$ are false.

It is easy to see that some theories may have no Ω–extension.

Definition 4.10 (Contradictory theory) *A default theory Δ is contradictory iff it has no Ω–extension.*

Example 9 The theory $\Delta = (\{\frac{:}{a}, \frac{:}{\neg a}\}, \{\})$ has no Ω–extension.

Theorem 4.1 (Ω is monotonic) *If Δ is a noncontradictory theory then Ω_Δ is monotonic.*

Proof: We begin by stating a lemma:

Lemma 4.2 *Let $\Delta = (D, W)$ be a noncontradictory default theory, and $\Delta' = (D \cup \{\frac{:}{L} \mid L \in W\}, \{\})$. E is an Ω–extension of Δ iff is an Ω–extension of Δ'.*

Proof: It is easy to see that every Ω–extension of Δ and of Δ' contains W. Thus for each Ω–extension of one of the theories the set of rules in D applied is the same in the other theory. $\diamond$

Now we prove that if Δ is a noncontradictory default theory then Γ'_Δ is anti-monotonic.

Without loss of generality (cf. lemma 4.2 above) we consider $\Delta = (D, \{\})$. Let A and B be sets of literals such that $A \subseteq B$.

We want to prove $A \subseteq B \Rightarrow \Gamma'_\Delta(B) \subseteq \Gamma'_\Delta(A)$, i.e.:

$$A \subseteq B \Rightarrow \forall d(c(d) \in \Gamma'_\Delta(B) \Rightarrow c(d) \in \Gamma'_\Delta(A)).$$

[8]Example of such cases are theories that are already seminormal, theories without negative conclusion, etc.

Let $c(d)$ be a conclusion of a default rule, such that $c(d) \in \Gamma'_\Delta(B)$. We prove that $c(d) \in \Gamma'_\Delta(A)$ given that $A \subseteq B$.

$$\begin{aligned} c(d) \in \Gamma'_\Delta(B) &\implies \exists_\lambda c(d) \in r^{B,D}_{\lambda+1}(\{\}) \\ &\implies \exists_{\lambda_b \leq \lambda} p(d) \in r^{B,D}_{\lambda_b}(\{\}) \wedge \neg j(d) \cap B = \{\} \end{aligned}$$

We must prove now:

$$\exists_{\lambda_b} p(d) \in r^{B,D}_{\lambda_b}(\{\}) \wedge \neg j(d) \cap B \Rightarrow \exists_{\lambda_a} p(d) \in r^{A,D}_{\lambda_a}(\{\}) \wedge \neg j(d) \cap A$$

Since $\neg j(d) \cap B = \{\} \Rightarrow \neg j(d) \cap A = \{\}$, it remains to prove that:

$$\exists_{\lambda_b} p(d) \in r^{B,D}_{\lambda_b}(\{\}) \Rightarrow \exists_{\lambda_a} p(d) \in r^{A,D}_{\lambda_a}(\{\})$$

$$\begin{array}{lll} \text{if } p(d) = \{\} & \text{then} & \lambda_a = \lambda_b = 0 \text{ and } c(d) \in r^{B,D}_1(\{\}) \subseteq \Gamma'_\Delta(A) \\ \text{if } p(d) \neq \{\} & \text{then} & \exists_{\lambda' < \lambda_b} p(d) \in r^{B,D}_{\lambda'+1}(\{\}) \end{array}$$

For the last case, and given that λ strictly decreases, the same proof applies recursively[9].

Since Ω_Δ is the composition of two antimonotonic operators, it is monotonic. ◇

Definition 4.11 (Iterative contruction of the least Ω–extension) *In order to obtain a constructive definition for the least (in the set inclusion order sense) Ω–extension of a theory we define the following transfinite sequence $\{E_\alpha\}$:*

$$\begin{aligned} E_0 &= \{\} \\ E_{\alpha+1} &= \Omega(E_\alpha) \\ E_\delta &= \bigcup \{E_\alpha \mid \alpha < \delta\} \quad \textit{for a limit ordinal } \delta \end{aligned}$$

By theorem 4.1 and according to the properties of monotonic operators, there must exist a smallest λ for the sequence above, such that E_λ is the smallest fixpoint of Ω. If E_λ is a Ω–extension then it is the smallest one. Otherwise, by the proposition below, there are no Ω–extensions for the theory.

Proposition 4.1 *If the least fixpoint E of Ω_Δ is not a Ω–extension of Δ then Δ has no Ω–extensions.*

Proof: Two cases are conceivable for E not to be a Ω–extension: either E has a pair of literals A and $\neg A$, or $\Gamma'_{\Delta^\bullet}(E) \subset E$.

In the first case every fixpoint of Ω_Δ also has that same pair of literals, thus no extension exists for Δ. The second case cannot occur because $\Gamma'_{\Delta^\bullet}$ is antimonotonic. ◇

[9]Note that the expression is similar to the initial one, but with $\lambda' < \lambda_b \leq \lambda$.

Example 10 Consider the default theory Δ of example 8. In order to obtain the least (and only) extension of Δ we build the sequence:

$$\begin{array}{lllllll} E_0 & = & \{\} \\ E_1 & = & \Gamma'_\Delta(\Gamma'_{\Delta^\bullet}(\{\})) & = & \Gamma'_\Delta(\{c, a, b, \neg a\}) & = & \{\neg a\} \\ E_2 & = & \Gamma'_\Delta(\Gamma'_{\Delta^\bullet}(\{\neg a\})) & = & \Gamma'_\Delta(\{c, b, \neg a\}) & = & \{\neg a, b\} \\ E_3 & = & \Gamma'_\Delta(\Gamma'_{\Delta^\bullet}(\{\neg a, b\})) & = & \Gamma'_\Delta(\{c, b, \neg a\}) & = & \{\neg a, b\} \;=\; E_2 \end{array}$$

As E_2 does not contain any pair of complementary literals, and $E_2 \subseteq \Gamma'_{\Delta^\bullet}(E_2)$, it is the least Ω–extension of Δ.

Example 11 Let $\Delta = (\{\frac{:}{a}, \frac{:}{\neg a}\}, \{\})$. Let us build the sequence:

$$\begin{array}{lllllll} E_0 & = & \{\} \\ E_1 & = & \Gamma'_\Delta(\Gamma'_{\Delta^\bullet}(\{\})) & = & \Gamma'_\Delta(\{a, \neg a\}) & = & \{a, \neg a\} \\ E_2 & = & \Gamma'_\Delta(\Gamma'_{\Delta^\bullet}(\{a, \neg a\})) & = & \Gamma'_\Delta(\{\}) & = & \{a, \neg a\} \;=\; E_1 \end{array}$$

As E_1 has a pair of complementary literals Δ has no Ω–extensions.

This new default semantics satisfies all the principles required above (section 4.1).

Theorem 4.3 (Uniqueness of minimal extension) *If Δ has an extension then there is one least extension E_m.*

Proof: The set 2^{AT} is a complete lattice under set inclusion. Since $\Omega_\Delta(L)$ is defined from $2^{AT} \rightarrow 2^{AT}$ and is monotonic then $lfp(\Omega_\Delta)$ always exists. $\diamond$

Theorem 4.4 (Enforcedness) *If E is a Ω–extension then E is enforced.*

Proof: We want to prove that for any default rule $\{p(d), j(d)\} \subseteq E \Rightarrow c(d) \in E$.

Given that E is a Ω–extension, by definition $\Omega(E) \subseteq \Gamma'_{\Delta^\bullet}(E)$ holds. Thus for any default rule:

$$\{p(d), j(d)\} \subseteq E \Rightarrow \{p(d), j(d)\} \subseteq \Gamma'_{\Delta^\bullet}(E), \text{ and } \neg j(d) \notin E.$$

By definition $p(d) \in \Gamma'_{\Delta^\bullet}(E) \Leftrightarrow \exists_\lambda p(d) \in r_\lambda^{\Gamma'_{\Delta^\bullet}(E),D}(\{\})$.

Since $\neg j(d) \notin E$, it follows that: $c(d) \in r_{\lambda+1}^{\Gamma'_{\Delta^\bullet}(E),D}(\{\})$. Thus:

$$c(d) \in r_\infty^{\Gamma'_{\Delta^\bullet}(E),D}(\{\}) = \Gamma'_\Delta(\Gamma'_{\Delta^\bullet}(E)) = E$$

$\diamond$

Corollary 1 *If E is an Ω–extension of Δ then for any $d = \frac{:}{c(d)} \in \Delta$, $c(d) \in E$.*

Proof: Follows directly from enforcedness for true prerequisites and justifications. $\diamond$

Theorem 4.5 (Modularity) *Let $L(\Delta_1)$ and $L(\Delta_2)$ be the languages of two default theories. If $L(\Delta_1) \cap L(\Delta_2) = \{\}$ then, for any corresponding extensions E_1 and E_2, there always exists an extension E of $\Delta = \Delta_1 \cup \Delta_2$ such that $E = E_1 \cup E_2$.*

Proof: Since the languages are disjoint, the rules of Δ_1 and Δ_2 do not interact on that count. Additionally, since the Cn operator is not applied, we never obtain the whole set of literals as a result of $\Gamma'_{\Delta^\bullet}$, and hence they do not interact on that count either. ◇

4.3 Comparison with Reiter's semantics

Comparing ous with Reiter's semantics for default theories, we prove that for theories whose language is such that our semantics can be applied the former is a generalization of the latter, in the sense that whenever Reiter's semantics (Γ–extension) gives a meaning to a theory (i.e. the theory has at least one Γ–extension), our semantics also provides one.

Moreover, when both semantics give meaning to a theory our semantics is sound w.r.t. the intersection of all Γ–extensions. Thus we provide a monotonic fixpoint operator for computing a subset of the intersection of all Γ–extensions. With that purpose we begin by stating a theorem:

Theorem 4.6 *Consider a theory Δ such that our semantics is defined. Then every Γ–extension is a Ω–extension.*

Proof: We begin by stating two lemmas

Lemma 4.7 $E = \Gamma_\Delta(E) \Rightarrow E = \Gamma'_{\Delta^\bullet}(E)$.

Proof: By definition of Γ_Δ, $E = \Gamma_\Delta(E) \Leftrightarrow (\forall_{d \in D}\, p(d) \in E \wedge \neg j(d) \cap E = 0$.
But:

$$\begin{array}{lll} \text{Since} & (\forall_{d \in D}\, p(d) \in E \wedge \neg j(d) \cap E = 0 & \Rightarrow c(d) \in E) \\ \text{then} & (\forall_{d \in D}\, p(d) \in E \wedge \neg j(d) \cap E = 0 \wedge \neg c(d) \cap E = 0 & \Rightarrow c(d) \in E) \end{array}$$

i.e., by definition, $E = \Gamma'_{\Delta^\bullet}(E)$. ◇

Lemma 4.8 $E = \Gamma_\Delta(E) \Rightarrow E = \Gamma'_\Delta(E)$.

Proof: Similar to the proof of lemma 4.7. ◇

Now we prove that for an E such that $E = \Gamma_\Delta(E)$, $E = \Omega_\Delta(E)$ holds.

By definition, $\Omega_\Delta(E) = \Gamma'_\Delta(\Gamma'_{\Delta^\bullet}(E))$. By lemma 4.7, $\Omega_\Delta(E) = \Gamma'_\Delta(E)$, and by lemma 4.8, $\Gamma'_\Delta(E) = E$.

For E to be a Ω–extension two more conditions must hold: E cannot have a pair of complementary literals, which holds because it is a Γ–extension; and $E \subseteq \Gamma'_{\Delta^\bullet}(E)$. It is easy to see that the latter condition holds given that $E = \Gamma_\Delta(E)$. ◇

Theorem 4.9 (Generalization of Reiter's semantics) *If a theory Δ has at least one Γ–extension, it has at least one Ω–extension.*

Theorem 4.10 (Soundness wrt to Reiter's semantics) *If a theory Δ has a Γ-extension, whenever L belongs to the least Ω-extension it also belongs to the intersection of all Γ-extensions.*

Proof: Follows directly from the theorem and the property that the least Ω–extension is the intersection of all Ω–extensions. $\diamond$

4.4 Comparison with Przymusinski's semantics

We now draw some comparisons with stationary extensions [13]. It is not the case that every stationary extension is a Ω–extension since, as noted above, nonmodular or nonenforced stationary extensions are not Ω–extensions. As shown in the example below, it is also not the case that every Ω–extension is a stationary extension.

Example 12 Let $\Delta = (\{\frac{:\neg b}{c}, \frac{:\neg a}{b}, \frac{:\neg a}{a}, \frac{:}{\neg b}\}, \{\})$. The only Ω–extension of Δ is $\{c, \neg b\}$. Note that this is not a stationary extension.

However, in a large class of cases these semantics coincide. In particular:

Proposition 4.2 *If for every default $d = \frac{p(d):j(d)}{c(d)}$ $c(d)$ is a positive literal then Ω coincides with Γ^2_Δ.*

5 Relation between the Semantics of Default Theories and Logic Programs with Explicit Negation

In this section we prove the exact correspondence among the Ω–extensions and the XSMs of extended logic programs.

Definition 5.1 (Program corresponding to a default theory) *Let $\Delta = (D, \{\})$ be a default theory. We say that an extended logic program P corresponds to Δ iff:*

- *For every default of the form $\frac{\{a_1,\ldots,a_n\} \,:\, \{b_1,\ldots,b_m\}}{C} \in \Delta$ there exists a rule $c \leftarrow a_1, \ldots, a_n, \sim\neg b_1, \ldots, \sim\neg b_m \in P$, where b_j denotes the complement of $\neg b_j$.*
- *No rules other than these belong to P.*

Definition 5.2 (Interpretation corresponding to a default context) *An interpretation I of a program P corresponds to a context E of the corresponding default theory T iff for every classical literal L of P (and literal L of T):*

- *$I(L) = 1$ iff $L \in E$ and $L \in \Gamma'_{\Delta^*}(E)$.*
- *$I(L) = \frac{1}{2}$ iff $L \notin E$ and $L \in \Gamma'_{\Delta^*}(E)$.*
- *$I(L) = 0$ iff $L \notin E$ and $L \notin \Gamma'_{\Delta^*}(E)$.*

Theorem 5.1 (Correspondence) *Let $\Delta = (D, \{\})$ be a default theory corresponding to the program P. E is a Ω-extension of Δ iff the interpretation I corresponding to E is a XSM of P. (Proof ahead)*

According to this theorem we can say that explicit negation is nothing but classical negation in default theories. As Ω default semantics is a generalization of Γ default semantics (in the sense of theorems 4.9 and 4.10), and since answer-sets semantics correspond to Γ default semantics [6], it turns out that answer-sets semantics (and hence the semantics defined in [17]) are special cases of $WFSX$ semantics in the same sense.

Example 13 Consider program $P = \{c \leftarrow \sim c, \quad a \leftarrow \sim b, \quad b \leftarrow \sim a, \quad \neg a \leftarrow\}$. The corresponding default theory is $\Delta = (\{\frac{:\neg c}{c}, \frac{:\neg b}{a}, \frac{:\neg a}{b}, \frac{:}{\neg a}\}, \{\})$.

As calculated in example 8, the only Ω–extension of Δ is $E = \{\neg a, b\}$ and $\Gamma'_{\Delta^s}(E) = \{\neg a, b, c\}$. The XSM corresponding to this extension is:

$$M = \{\neg a, \sim a, b, \sim\neg b, \sim\neg c\}^{10}.$$

It is easy to verify that M is the only XSM of P.

Proof of Theorem 5.1: We begin by stating some propositions useful in the sequel.

Proposition 5.1 *Let $\Delta = (D, \{\})$ be a default theory and E a context such that $\Gamma'_\Delta(E)$ is noncontradictory. Then:*

$$L \in \Gamma'_\Delta(E) \Leftrightarrow \exists \frac{\{b_1, \ldots, b_n\} \; : \; \{c_1, \ldots, c_m\}}{L} \in D, \; \forall i, j \; b_i \in \Gamma'_\Delta(E) \wedge \neg c_j \notin E$$

Proof: It is easy to see that under these conditions $\Gamma'_\Delta(E) = \Gamma_\Delta(E)$. Thus the proof follows from properties of the Γ_Δ operator. ◇

Proposition 5.2 *Let E be an extension of a default theory $\Delta = (D, \{\})$. Then:*

$$L \in \Omega(E) \Leftrightarrow \exists \frac{\{b_1, \ldots, b_n\} \; : \; \{\neg c_1, \ldots, \neg c_m\}}{L} \in D \textit{ such that}$$

$$\forall i, j, \; b_i \in E \wedge b_i \in \Gamma'_{\Delta^s}(E) \wedge c_j \notin \Gamma'_{\Delta^s}(E).$$

Proof: By definition of Γ'_Δ, and given that $W = \{\}$, it follows from proposition 5.1 that for $L \in \Omega(E)$ there must exist at least one default in D applied in the second step, i.e. with all prerequesites in $\Omega(E)$ and all negations of justifications not in $\Gamma'_{\Delta^s}(E)$. By hypothesis E is an extension; thus $E = \Omega(E)$ and $E \subseteq \Gamma'_{\Delta^s}(E)$; so for such a rule all prerequesites are in E and in $\Gamma'_{\Delta^s}(E)$, and all negations of justifications are not in $\Gamma'_{\Delta^s}(E)$. ◇

Proposition 5.3 *Let E be an extension of a default theory $\Delta = (D, \{\})$. Then:*

$$L \notin E \Rightarrow \forall \frac{\{b_1, \ldots, b_n\} \; : \; \{\neg c_1, \ldots, \neg c_m\}}{L} \in D, \; \exists i, j, \; b_i \notin E \vee c_j \in \Gamma'_{\Delta^s}(E)$$

---

[10] Note that c is undefined in M.

Proof: If $L \notin E$ then, given that E is an extension, $L \notin \Omega_\Delta(E)$. Thus no default rule for L is applicable in the second step, i.e. given that $W = \{\}$, and by proposition 5.1, no rule with conclusion L is such that all its prerequisites are in $\Omega_\Delta(E)$ and no negation of a justification is in $\Gamma'_{\Delta^s}(E)$. ◇

Proposition 5.4 *Let E be an extension of a default theory $\Delta = (D, \{\})$. Then:*

$$L \notin \Gamma'_{\Delta^s}(E) \Leftrightarrow \forall \frac{\{b_1,\ldots,b_n\} : \{\neg c_1,\ldots,\neg c_m\}}{L} \in D,$$

$$\exists i,j,\ b_i \notin \Gamma'_{\Delta^s}(E) \vee c_j \in E \vee \neg L \in E$$

Proof: Similar to the proof of 5.3 but now applied to the first step, which imposes the use of seminormal defaults. Thus the need for $\neg L \in E$. ◇

We now prove the main theorem:

($\rightarrow$) E is a Ω-extension of $\Delta \Rightarrow I$ is a XSM of P.

Here we must prove that for any (classical and default) literal F, $F \in I \Leftrightarrow F \in \Phi(I)$. We do this in three parts: for any classical literal L, $L \in I \Rightarrow L \in \Phi(I)$; $L \notin I \Rightarrow L \notin \Phi(I)$; $\sim L \in I \Leftrightarrow \sim L \in \Phi(I)$.

Each of these proofs proceeds by: translating conditions in I into conditions in E via correspondence; finding conditions in Δ given the conditions in E, and the fact that E is an extension; translating conditions in Δ into conditions in P via correspondence; using those conditions in P to determine the result of operator Φ.

$L \in I \Leftrightarrow I(L) = 1 \Rightarrow L \in E \Leftrightarrow L \in \Omega_\Delta(E)$, because I corresponds to E and E is a Ω-extension.

By proposition 5.2 $L \in \Omega_\Delta(E) \Leftrightarrow \exists \frac{\{b_1,\ldots,b_n\} : \{\neg c_1,\ldots,\neg c_m\}}{L} \in D$ such that $\forall i, b_i \in E \wedge b_i \in \Gamma'_{\Delta^s}(E)$ and $\forall j, c_j \notin \Gamma'_{\Delta^s}(E)$.

By translating, via the correspondence definitions, the default and the conditions on E into a rule and conditions on I:

$L \in E \Rightarrow \exists L \leftarrow b_1,\ldots,b_n,\sim c_1,\ldots,c_m \in P$ such that $\forall i, I(b_i) = 1$ and $\forall j, I(c_j) = 0 \Rightarrow L \in least(\frac{P}{I})$, by properties of $least(\frac{P}{I})$.

Given that the operator Coh does not delete literals from I, $L \in I \Rightarrow L \in \Phi(I)$.

$L \notin I \Leftrightarrow L \notin E$, because I corresponds to E.

$L \notin E \Rightarrow \forall \frac{\{b_1,\ldots,b_n\} : \{\neg c_1,\ldots,\neg c_m\}}{L} \in D$ then either a $b_i \notin E$ or a $c_j \in \Gamma'_{\Delta^s}(E)$, by proposition 5.3.

Translating, via the correspondence definitions, the default and the conditions on E into a rule and conditions on I:

$L \notin E \Rightarrow \forall L \leftarrow b_1,\ldots,b_n,\sim c_1,\ldots,c_m \in P$ then $\exists i,j\ I(b_i) \neq 1 \vee I(c_j) \neq 0 \Rightarrow L \notin least(\frac{P}{I})$, by properties of $least(\frac{P}{I})$.

Given that the operator Coh does not add classical literals to I, $L \notin I \Rightarrow L \notin \Phi(I)$.

$\sim L \in I \Leftrightarrow L \notin \Gamma'_{\Delta^s}(E)$, since E corresponds to I.

$L \notin \Gamma'_{\Delta^s}(E) \Leftrightarrow \forall \frac{\{b_1,\ldots,b_n\} : \{\neg c_1,\ldots,\neg c_m\}}{L} \in D$ then $\exists i,j\ b_i \notin \Gamma'_{\Delta^s}(E) \vee c_j \in E \vee \neg L \in E$, by proposition 5.4.

$L \notin \Gamma'_{\Delta}\bullet(E) \Leftrightarrow (\forall L \leftarrow b_1, \ldots, b_n, \sim c_1, \ldots, c_m \in P$ then $\exists i, j\ I(b_i) = 0 \vee I(c_j) = 1) \vee \neg L \in E$.

$\sim L \in I \Leftrightarrow \sim L \in least(\frac{P}{I}) \vee \neg L \in E$ (*), by properties of the least operator.

It was proven before that $\neg L \in E \Leftrightarrow \exists \neg L \leftarrow b_1, \ldots, b_n, \sim c_1, \ldots, \sim c_m \in P$ then $\exists i, j\ I(b_i) = 1 \vee I(c_j) = 0)$.

By properties of $least(\frac{P}{I})$, $\neg L \in E \Leftrightarrow \neg L \in least(\frac{P}{I})$.

Using correspondence, we can simplify the equivalence (*) to: $\sim L \in I \Leftrightarrow \sim L \in least(\frac{P}{I}) \vee \neg L \in least(\frac{P}{I}) \Leftrightarrow \sim L \in \Phi(I)$, this last equivalence being due to the definitions of operators Coh and Φ.

($\leftarrow$) I is a XSM of $P \Rightarrow E$ is a Ω-extension of T.

By definition of correspondence between interpretations and contexts, it is easy to see that E is consistent and $E \subseteq \Gamma'_{\Delta}\bullet(E)$. So we only have to prove that $E = \Omega_{\Delta}(E)$. We do this by proving that $\forall L\ L \in E \Leftrightarrow L \in \Omega_{\Delta}(E)$.

$L \in E \Leftrightarrow I(L) = 1$ by definition of corresponding context.

$I(L) = 1 \Leftrightarrow \exists L \leftarrow b_1, \ldots, b_n, \sim c_1, \ldots, \sim c_m \in P$, where $n, m \geq 0$ such that $\forall i\ I(b_i) = 1$ and $\forall j\ I(c_j) = 0$, because I is a XSM of P.

By translating, via the correspondence definitions, the rule and the conditions on I into a default and conditions on E:

$I(L) = 1 \Leftrightarrow \exists \frac{\{b_1, \ldots, b_n\} : \{\neg c_1, \ldots, \neg c_m\}}{L} \in D$ such that $\forall i\ b_i \in E \wedge b_i \in \Gamma'_{\Delta}\bullet(E)$ and $\forall j\ c_j \notin E \wedge c_j \notin \Gamma'_{\Delta}\bullet(E)$

Given that such a rule exists under such conditions, it follows easily from theorem 5.1 that $L \in E \Leftrightarrow L \in \Omega_{\Delta}(E)$. $\diamond$

Acknowledgements

We thank ESPRIT BRA COMPULOG (no. 3012), Instituto Nacional de Investigação Científica, Junta Nacional de Investigação Científica e Tecnológica, and Gabinete de Filosofia do Conhecimento for their support.

References

[1] C. Baral and V. S. Subrahmanian. Stable and extension class theory for logic programs and default logics. In *International Workshop on Nonmonotonic Reasoning*, 1990.

[2] C. Baral and V. S. Subrahmanian. Dualities between alternative semantics for logic programming and nonmonotonic reasoning. In A. Nerode, W. Marek, and V. S. Subrahmanian, editors, *Logic Programming and NonMonotonic Reasoning'91*. MIT Press, 1991.

[3] N. Bidoit and C. Froidevaux. Minimalism subsumes default logic and circumscription in stratified logic programming. In *Symposium on Principles of Database Systems*. ACM SIGACT-SIGMOD, 1987.

[4] N. Bidoit and C. Froidevaux. General logic databases and programs: default logic semantics and stratification. *Journal of Information and Computation*, 1988.

[5] M. Gelfond and V. Lifschitz. The stable model semantics for logic programming. In R. A. Kowalski and K. A. Bowen, editors, *5th International Conference on Logic Programming*, pages 1070–1080. MIT Press, 1988.

[6] M. Gelfond and V. Lifschitz. Logic programs with classical negation. In Warren and Szeredi, editors, *7th International Conference on Logic Programming*, pages 579–597. MIT Press, 1990.

[7] A. Marek and M. Truszczynski. Stable semantics for logic programs and default theories. In *North American Conference on Logic Programming'89*. MIT Press, 1989.

[8] L. M. Pereira and J. J. Alferes. Well founded semantics for logic programs with explicit negation. In *European Conference on Artificial Inteligence'92*. John Wiley & Sons, Ltd, 1992. *To appear.*

[9] L. M. Pereira, J. N. Aparício, and J. J. Alferes. Counterfactual reasoning based on revising assumptions. In Ueda and Saraswat, editors, *International Logic Programming Symposium'91*. MIT Press, 1991.

[10] L. M. Pereira, J. N. Aparício, and J. J. Alferes. Hypothetical reasoning with well founded semantics. In B. Mayoh, editor, *Scandinavian Conference on AI'91*. IOS Press, 1991.

[11] L. M. Pereira, J. N. Aparício, and J. J. Alferes. Nonmonotonic reasoning with well founded semantics. In Koichi Furukawa, editor, *8th International Conference on Logic Programming'91*, pages 475–489. MIT Press, 1991.

[12] H. Przymusinska and T. Przymusinski. *Semantic Issues in Deductive Databases and Logic Programs*. Formal Techniques in Artificial Intelligence. North Holland, 1990.

[13] H. Przymusinska and T. Przymusinski. Stationary default extensions. Technical report, California Polytechnic at Pomona and University of California at Riverside, 1991.

[14] T. Przymusinski. Extended stable semantics for normal and disjunctive programs. In Waren and Szeredi, editors, *7th International Conference on Logic Programming*, pages 459–477. MIT Press, 1990.

[15] R. Reiter. A logic for default reasoning. *Artificial Intelligence*, 13:68–93, 1980.

[16] A. Van Gelder, K. A. Ross, and J. S. Schlipf. The well-founded semantics for general logic programs. *Journal of ACM*, pages 221–230, 1990.

[17] G. Wagner. A database needs two kinds of negation. In B. Thalheim, J. Demetrovics, and H-D. Gerhardt, editors, *MFDBS'91*, pages 357–371. Springer-Verlag, 1991.

A Default Theory Review

Here we review some known default theory semantics. We begin by reviewing Reiter's classical default logic [15]. We then review (partly following [2]) the well-founded [2] and stationary [13] default logics, which correspond respectively to the well founded and stationary semantics of (nonextended) logic programs.

A.1 Classical Default Theories

Definition A.1 ($R^{E,D}(S)$) *Suppose* (D, W) *is a default theory,* S *is a set of formulas, and* E *is a set of formulas called the* context. *The operator* $R^{E,D}(S)$ *mapping formulas into formulas is defined as:*

$$R^{E,D}(S) = Cn(S \cup c(d) | d \in D, p(d) \in S, \neg j(d) \cap E = \{\}) \qquad (2)$$

where Cn *denotes the Tarskian consequence operator, and* $\neg j(d)$ *denotes the set* $\{\neg\beta | \beta \in j(d)\}$.

Note that the operator above enables to add the consequences of those defaults whose justifications are consistent with context E.

Definition A.2 ($\Gamma_\Delta(E)$) *Suppose* (D, W) *is a default theory and* E *is a context. Then*

$$\begin{aligned} R_0^{E,D}(W) &= Cn(W) \\ R_{n+1}^{E,D}(W) &= R^{E,D}(R_n^{E,D}(W)) \\ R_\infty^{E,D}(W) &= \bigcup_{n=0}^{\infty} R_n^{E,D}(W) \end{aligned}$$

As noted in [7], if (D, W) is a default theory $R_\infty^{E,D}(W)$ is identical to Reiter's operator $\Gamma_\Delta(E)$ [15].

Definition A.3 (Default extension) E *is a classical extension of a default theory* (D, W) *iff* $E = R_\infty^{E,D}(W)$ *or, equivalently, iff* $E = \Gamma_\Delta(E)$.

One problem of classical default theory is that it may have multiple extensions and then no single minimal extension gives meaning to the theory, or no extensions at all (and no meaning is given), in cases where a definite meaning is expected.

Example 14 The default theory $W = \{p\}, D = \{\frac{:\neg q}{q}\}$ has no extensions. However p is a fact, and we would expect it to be true.

A.2 Well Founded and Stationary Default Semantics of General Logic Programs

We review here two approaches which relate general logic programs with default theories.

Baral et. al. [2] introduce the well founded semantics for default theories giving a meaning to default theories with multiple extensions. Furthermore, the semantics is defined for all theories, identifying a single extension for each theory.

Let $\Delta = (D, W)$ be a default theory, and let $\Gamma_\Delta(E)$ be as above. Since $\Gamma_\Delta(E)$ is antimonotonic $\Gamma_\Delta^2(E)$ is monotonic [2].

Definition A.4 (Well founded semantics) *[2]*[11]

- *A formula F is* true *in a default theory* Δ *with respect to the well-founded semantics iff* $F \in lfp(\Gamma^2)$.
- *F is* false *in* Δ *w.r.t. the well founded semantics iff* $F \notin gfp(\Gamma^2)$.
- *Otherwise F is said to be* unknown *(or* undefined*).*

This semantics is defined for all theories and is equivalent to the Well Founded Model semantics [16] of general logic programs.

Definition A.5 (Stationary extension) *[13]*
Given a default theory Δ*, E is a stationary default extension iff:*

- $E = \Gamma^2_\Delta(E)$
- $E \subseteq \Gamma_\Delta(E)$

Definition A.6 (Stationary default semantics) *[13]*

- *A formula L is true in E iff* $L \in \Gamma^2_\Delta(E) = E$.
- *A formula L is false in E iff* $L \notin \Gamma_\Delta(E)$.
- *Otherwise L is said to be* unknown *(or* undefined*).*

Remark A.1 *Note that every default theory has at least one stationary default extension. The least stationary default extension always exists, and corresponds to the well founded semantics for default theories above. Indeed,* $\Gamma^2_\Delta = lfp(\Gamma^2_\Delta)$*, and* $\Gamma_\Delta = gfp(\Gamma^2_\Delta)$*, whenever* $E \subseteq \Gamma_\Delta(E)$*(i.e.* $lfp(\Gamma^2_\Delta) \subseteq gfp(\Gamma^2_\Delta)$*). Moreover, the least stationary default extension can be computed by iterating the monotonic operator* Γ^2_Δ.

Example 15 Consider the default theory of example 14. We have $\Gamma_\Delta(\{p\}) = \{p, q\}$ and $\Gamma^2_\Delta(\{p\}) = \{p\}$. p is true in the theory Δ.

[11] In [2] the notation used is slightly different but the definitions are equivalent.

Computing answers for disjunctive logic programs

Ulrich Furbach

University of Koblenz, Rheinau 3 – 4,
D 5400 Koblenz, Germany
uli@uni-koblenz.de

Abstract. Various approaches to computing answers for disjunctive logic programs are presented. Emphasis is put on splitting based approaches; a naive application of the splitting rule is taken as a starting point for both, the development of a fixpoint semantics and the SPLIT-SLD-calculus. This calculus is proven correct and complete and it is compared with model elimination and InH-Prolog.

Logic programming by means of non-Horn clauses is a field which is investigated in database research, theorem proving and logic programming. There is work aiming at a semantic characterization of those "disjunctive logic programs", e.g. Minker and Rajasekar ([MR90, MR88]) investigated fixpoint semantics and proof-procedures, the latter was done also by others like Herre ([Her88]), Decker ([Dec91]), Casanova et.al. ([CGS89]) and the author ([Fur91a]). As another somewhat different approach Loveland came with the idea of near-Horn programs and of using the splitting rule as a base for efficient implementations ([Lov78, Lov87]). A very similar idea was presented in [CEFB84], where the matrix reduction method from Prawitz was used to construct Horn programs out of disjunctive logic programs.

In this paper we want to discuss three kinds of operational semantics for disjunctive logic programs, namely *V-resolution*, as it is introduced already in [CL73], *SETHEO*, a model elimination prover and *splitting based evaluation models* together with a *fixpoint semantics*. This fixpoint semantics is derived very naturally from the splitting rule and does not reflect any details from an interpreter, as it is the case in [RLS91].

This is organized as follows: In the first section we informally present the calculus of V-resolution just to demonstrate the query answering procedure which is introduced elsewhere ([Fur91a]). In section 2 it is demonstrated, that this concept can be implemented very easily by using the theorem prover SETHEO.
Section 3 contains the main part of this paper. First a basic model for the use of the splitting rule as a basis for evaluation is introduced and motivated by this, we then define a fixpoint semantics and investigate its relation to the basic model and to a minimal model semantics. Finally the SPLIT-SLD-proof procedure is derived from these ideas, it is proven correct and complete and it is compared with model elimination and InH-Prolog.

Before we get into these items let us fix what we are talking about.

Disjunctive logic programs A *disjunctive logic program* is a consistent set of *program clauses* derived from a first order predicate formula which is transformed in conjunctive normal form. Program clauses are written as $A_1, \ldots, A_n \leftarrow B_1, \ldots, B_m$, where the A_i's and B_i's are atoms.

A *query* is a clause of the form $\leftarrow B_1, \ldots, B_m$. The set of variables, occurring in a query $\leftarrow Q$ are called the *query variables.* The query variables are assumed to be existentially quantified. As usually an *answer substitution* for a query $\leftarrow Q$ is a substitution for the query variables of $\leftarrow Q$. An *answer* for a query $\leftarrow Q$ is a formula of the form $Q\sigma_1 \vee \cdots \vee Q\sigma_n$ where each σ_i is an answer substitution. Hence, answers are disjunctions of the query-clause Q, which is a conjunction. And finally we have a *correct answer* C for a program P and a query Q, if C is an answer for $\leftarrow Q$ and C is a logical consequence of P.

1 V-resolution

V-resolution is a kind of Herbrand proof procedure, where the variables occurring in a clause are *not* treated to be universally quantified. Instead one has to work with a sufficient number of copies for each clause. The interesting point is, that V-resolution memorizes the substitutions performed during a deduction; this is very much like the concept of computed answer substitutions as defined in SLD-resolution. In the first order case we have the additional complication that during "ancestor resolution steps" one has to check whether the substitutions computed until then are consistent with the mgu of the actual step. In SLD-resolution such a case can never occur, since one always uses a new variant of an input clause. V-resolution only introduces new variants when it is necessary, which allows us to use this concept to derive indefinite answers.

Instead of formally defining the calculus, we sketch it by using an example taken from [WP88]:

	DP_0		
$symptom(S)$	$\leftarrow$		(1)
$cause(C_1), cause(C_2)$	$\leftarrow$	$symptom(S)$	(2)
$treatment(T_0)$	$\leftarrow$	$cause(C_1)$	(3)
$treatment(T_1)$	$\leftarrow$	$cause(C_1)$	(4)
$treatment(T_0)$	$\leftarrow$	$cause(C_2)$	(5)
$treatment(T_2)$	$\leftarrow$	$cause(C_2)$	(6)

Together with the goal $\leftarrow t(x)$[1], we get the following refutation, where the substitutions corresponding to a clause are written besides each clause:

$$\leftarrow t(x') \mid \emptyset$$

$$\downarrow \quad (4)$$

$$\leftarrow c(C_1) \mid \{x'/T_1\}$$

$$\downarrow \quad (2)$$

$$c(C_2) \leftarrow s(S) \mid \{x'/T_1\}$$

[1] Predicate symbols are abbreviated by taking just the first letter

$$\downarrow \quad (1)$$

$$c(C_2) \leftarrow\mid \{x'/T_1\}$$

$$\downarrow \quad (6)$$

$$t(T_2) \leftarrow\mid \{x'/T_1\}$$

$$\downarrow \quad \text{goal with new query variable } x''$$

$$\Box \mid \{x'/T_1, x''/T_2\}$$

In the last step we see, that a "naive" ancestor resolution step with the goal $\leftarrow t(x')$ would employ the mgu $\{x'/T_2\}$ which is inconsistent with the substitution $\{x'/T_1\}$. Therefore a new variant of the original goal clause is used. It is straightforward how to construct the *computed (indefinite) answer* $t(T_1), t(T_2) \leftarrow$ out of tho substitutions attached to the empty clause.

The reader verifies easily, that alternative choices of input clauses during the deduction would result in the refutation: $\leftarrow t(x') \mid \emptyset \xrightarrow{*} \Box \mid \{x'/T_0\}$ and, hence, the *computed (definite) answer* $t(T_0) \leftarrow$.

We want to point out that V-resolution is not intended as a practical procedure for computing answers of disjunctive logic programs, moreover it is a means for a formal investigation and indeed its correctness and completeness with respect to the introduced concept of answers is proven in [Fur91a].

2 SETHEO

The above idea of storing substitutions can be implemented very easily in SETHEO, a first order predicate logic prover, based on the model elimination calculus ([LSBB89]), which brings us to a practical alternative for computing answers.

Besides the fact that SETHEO is efficient due to its Prolog-technology design, it is of particular interest, that it provides a number of data-structures, such as strings, lists and sets. The input language of SETHEO can be understood as a first order logic programming language. Another construct, which turns out to be useful for our purpose are global variables with destructive assignment. These non-logical variables can be used very much as the variable concept in procedural programming languages, however, with the "logical property" that they are updated through backtracking and that a precise semantics can be defined through transformation.

Applying the fact, that indefinite answers can occur only if the query of the program is used more than once during a refutation (cf. [WP88]), we transform a given query into a new one, which stores substitutions for query variables in global variables. This transformation is the following:

Given a disjunctive logic program DP and a query $\leftarrow Q$ with variables $x_1, \ldots, x_n$. Then

- transform the query into

```
goal  <-  Q, $X1:=[X1|$X1],...,$Xn:=[Xn|$Xn].
```

thus getting a new program clause and
– generate the new query

```
<-  $X1:=[],...,$Xn:=[], goal, write($X1), ..., write($Xn).
```

where global variables are marked with prefix **$**.

Whenever the modified goal, i.e. the new program clause, is used during a model elimination proof of SETHEO, the instantiations of the query variables $x_1, \ldots, x_n$ are stored in the corresponding global variables. The new goal performs the initialisations of the global variables and after calling the query, outputs the lists of substitutions. [2]

In the above example we get

```
goal <-  t(X), $X := [X|$X].
     <-  $X :=[], goal, write($X).
```

where the value of **$X** is a list, which is appended to the substitution for **X**, each time the query is used during a refutation. Note that assignments to global variables are backtracked, i.e. the above list contains only those substitutions which are really used during a refutation, namely $[T_1 T_2]$ or $[T_0]$. Of course there is a close relationship to the substitution-storing component in V-resolution, and hence we construct in the same manner answers out of these substitutions.

3 Splitting-based evaluation

The splitting rule as a means to derive Horn clauses from disjunctive logic programs is introduced and discussed by Loveland and his co-workers [Lov78, RLS91]. In the context of logic programming the rule reads as follows:

> **Splitting:** $DP \cup \{A_1, \ldots, A_n \leftarrow B\}$ is an unsatisfiable set of ground clauses **iff**
> $DP \cup \{A_i \leftarrow B\}$ and each set from $\{DP \cup \{A_j \leftarrow\} \mid 1 \leq j \leq n \wedge j \neq i\}$ are unsatisfiable sets of ground clauses **iff**
> $DP \cup \{A_i \leftarrow B\}$ and each set from $\{DP \cup \{A_j \leftarrow B\} \mid 1 \leq j \leq n \wedge j \neq i\}$ are unsatisfiable sets of ground clauses.

These two variants of the splitting rule, although being logically equivalent, show significant differences when used within the context of programming language semantics. We will use in the following only the second form of splitting, i.e. each of the clauses which are the result of an application of the splitting rule, contain the body of the original disjunctive clause. This will be discussed later on.

We propose to use the splitting rule as a basis for the parallel evaluation of disjunctive logic programs. We believe that this approach is interesting for several reasons:

---

[2] The substitutions should be processed further to get an answer instead of simply writing them.

- It can be seen as a very moderate, i.e. coarse grain form of and-parallelism. Thus, problems concerning the choice of subgoals to be processed in parallel are avoided; here it is a priori clear, where to split the program clauses.
- Besides the possibility of parallel evaluation this approach offers the advantage that SLD-based interpreters can be used to process disjunctive (and hence first-order) logic programs.
- NearHorn-PROLOG proof-procedures can be understood as a sequential implementation of our model.

3.1 A basic model

The following basic model is intended as a starting point and motivation for subsequent discussions of fixpoint and dynamic splitting semantics.

The splitting rule as formulated above is valid only for the ground case, therefore we have to additionally take care for variables. Instead of formulating a rule for the first order level, we give a completeness proof of our basic model in theorem 4 later in this section.

Given a disjunctive logic program DP and a query Q, we use the splitting rule to derive a set of Horn clause programs; for this we apply the above rule corresponding to non-ground clauses. The variables of such a split clause are called splitted variables. Each of the Horn clause programs, together with the query, can be understood as a process:

The Basic Model

- Use the splitting rule to derive Horn clause programs $HP_1, \ldots, HP_n$ from DP, together with a set of variables V which are split variables.
- Find all solutions S_i for each program HP_i and the query, where S_i contains the answer substitutions and the substitutions for variables from V.
- Delete all solutions, where the substitutions for variables from V are not compatible.

It is important to note that variants of clauses which contain splitted variables inherit this property for corresponding clauses.

Take as an example the disjunctive logic program DP_1

DP_1		
$P(x), Q(y,x)$	$\leftarrow$	$R(y,z)$
$Q(x,x)$	$\leftarrow$	$P(x)$
$R(a,b)$	$\leftarrow$	
$R(b,b)$	$\leftarrow$	

We derive two Horn clause programs

HP_1	HP_2
$P(x) \leftarrow R(y,z)$	$Q(y,x) \leftarrow R(y,z)$
$Q(x,x) \leftarrow P(x)$	$Q(x,x) \leftarrow P(x)$
$R(a,b) \leftarrow$	$R(a,b) \leftarrow$
$R(b,b) \leftarrow$	$R(b,b) \leftarrow$

When these two Horn sets are evaluated independently with the query $\leftarrow Q(a, x')$ we can get two proofs with substitutions $\{x'/a, y/b, z/b\}$ and $\{y/a, x'/x, z/b\}$. y and z are both split variables and therefore must be replaced consistently by substitutions. Since the substitutions for y are not compatible, one has to look for an alternative proof, yielding compatible answer substitutions. Another proof is possible only for the left-hand program with substitutions $\{x'/a, y/a, z/b\}$, which corresponds to a synchronous substitution for the *splitted variable* y. The basic model with this example can be depicted as follows.

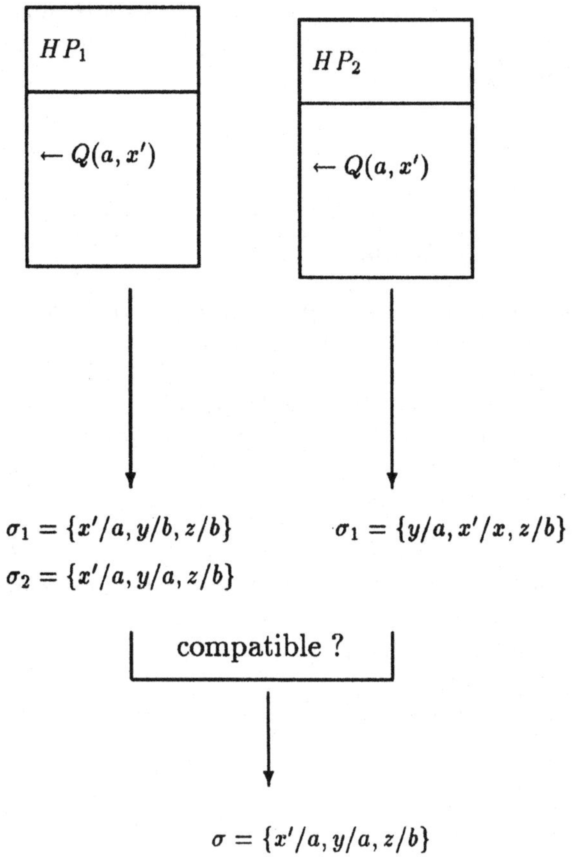

We are currently experimenting with a Unix-simulation using tasks, consisting of one SETHEO to process one Horn clause program; another SETHEO-task is used as well for checking the consistency of answer substitutions. One improvement of this basic model, which uses a shared memory for the splitted variables is in work and is discussed in more detail in [Fur91b].

3.2 A fixpoint characterization

Let us now give a fixpoint characterization which is motivated by the above basic model. This is in contrast to the semantics in [RLS91], which, too, is motivated by the splitting rule, but is intended to mirror the InH-Prolog proof-procedure.

Our operator is aiming at simply combining the unique least models of the Horn clause programs obtained through splitting. However, there are two problems: One is the aforementioned consistency of split variables, which would get lost, by computing the least models separately. The second problem stems from the fact that in general one has to provide a sufficient number of copies of a non-Horn clause *before* splitting.

The first problem could be avoided by splitting only ground instances, but of course we would still have the problem, that without a goal it is impossible to know how much copies are needed. We therefore propose an operator for the original disjunctive logic program, which generates models for the splitted programs:

The operator $\mathcal{T}_{DP}$ maps sets of interpretations $\mathcal{I}$ into sets of interpretations:

A fixpoint operator

$$\mathcal{T}_{DP}(\mathcal{I}) = \{I \in split(\{\{A\} \mid A \in I'\} \cup T_{DP}(I')) \mid I' \in \mathcal{I}\}$$

where

$T_{DP}(I)$ is an operator which maps interpretations, i.e. sets of atoms, into sets of sets of atoms:

$$T_{DP}(I) = \{\{A_1, \ldots, A_n\} \mid A_1, \ldots, A_n \leftarrow B_1, \ldots, B_m \text{ is a ground instance of a clause in } P \text{ and } B_1, \ldots, B_m \in I \}$$

and

$$split(S) = \{A \mid \{A\} \in S\} \text{ iff } S \text{ contains only singletons}$$

$$split(\{\{A_1, \ldots, A_n\}\} \cup S) = split(\{\{A_1\}\} \cup S) \cup \ldots \cup split(\{\{A_n\}\} \cup S)$$

Let us discuss this construction with a very small example:

DP_2	
a, b	$\leftarrow$
c	$\leftarrow a$
c	$\leftarrow b$

In a first step the operator T_{DP} when applied to the empty set gives $\{\{a, b\}\}$, which can be interpreted as "There are two alternatives for building an interpretation, namely a or b". This is exactly, what the operator *split* does, giving us the two possible interpretations $\{a\}$ and $\{b\}$. These interpretations are enlarged and new ones are generated when T_{DP} is iterated:

$$T^1_{DP} = \{\{a\}, \{b\}\}$$
$$T^2_{DP} = \{\{a, c\}, \{b, c\}, \{a, b, c\}\}$$
$$T^3_{DP} = T^2_{DP}$$

This example is handled in [RLS91] as well. There the authors define two operators, which have to be ω-iterated: The first iteration of the *case operator* T^c_P gives

$$T^c_P\uparrow\omega = \{(a, \{b\}, \emptyset), (b, \{a\}, \emptyset), (a, \emptyset, \{a\}), (b, \emptyset, \{b\}), (c, \emptyset, \{c\})$$
$$(c, \{b\}, \emptyset), (c, \{a\}, \emptyset), (c, \emptyset, \{a\}), (c, \emptyset, \{b\}), \ldots\}$$

The second iteration with the *join operator* T^j_P uses the result of the case operator and gives:

$$T^j_P(T^j_P(T^j_P(\emptyset))) = \{(a, \{a\}), (b, \{b\}), (c, \{c\}), (c, \{a\}), (c, \{b\}), \ldots\} \cup$$
$$\{(a \vee b, \emptyset), (a \vee c, \emptyset), (b \vee c, \emptyset), (c, \emptyset), \ldots\} \cup$$
$$\{(a \vee b \vee c, \emptyset), \ldots\}$$

The close relation of this semantics to the InH-Prolog interpreter is obvious with the result, which states that a tuple (L, A) is in $T^j_P \uparrow \omega$ iff there exists an InH-Prolog derivation from $P \cup A$. Thus the block-structure of the InH-Prolog procedure is modeled directly in this semantics. In contrast, our semantics is concerned with generating models.

To demonstrate this with a more complicated example take

DP_3		
wobbly wheel	$\leftarrow$	
flat tyre , broken spokes	$\leftarrow$	wobbly wheel
punctured tube , leaky valve	$\leftarrow$	flat tyre

Iteration of T_{DP} yields:

$$T^1_{DP} = \{\{ww\}\}$$
$$T^2_{DP} = \{\{ww, ft\}, \{ww, bs\}\}$$
$$T^3_{DP} = \{\{ft, ww, pt\}, \{ft, ww, lv\},$$
$$\{bs, ww, pt, ft\}, \{bs, ww, lv, ft\},$$
$$\{ww, ft, bs\}, \{ww, bs\}\}$$

We have to point out, that the above construction is very similar to the one given in [Dec91]. We reinvented this transformation from a very different starting point: As discussed previously we wanted to model the use of the splitting rule within a fixpoint semantics, while Deckers motivation was to get a close relationship to the usual closure operator for Horn clauses.

Decker uses in his approach conjunctions of ground atoms to denote Herbrand interpretations and disjunctions of conjuncts to denote alternative interpretations resp. models. Transforming our set structures into this clause-oriented approach allows to use Deckers results to characterize the resulting interpretations from $\mathcal{T}_{DP}\uparrow\omega(\{\emptyset\})$.

For this the following two definitions are in order: A model M of DP is *supported* if there is a mapping *value*, assigning each atom of the Herbrand base a non-negative integer, such that, for each atom in M there exists a ground instance $A_1,\ldots,A_n \leftarrow B_1,\ldots,B_m$ of a clause in DP such that $\exists i : A = A_i$ and $B_1,\ldots,B_m \in M$ and $\forall 1 \leq i \leq m : value(B_i) < value(A)$.
The operator *min* is used to filter out all interpretations I from a set of interpretations $\mathcal{I}$ for which there exists $I' \in \mathcal{I}$ such $I' \subset I$.

Theorem 1 (Decker). *Let MM_{DP} be the set of minimal models of DP, then the following holds:*

- $M \in \mathcal{T}_{DP}\uparrow\omega(\{\emptyset\})$ *iff* M *is supported*
- $M \in min(\mathcal{T}_{DP}\uparrow\omega(\{\emptyset\}))$ *iff* $M \in MM_{DP}$

Let us now discuss the relationship of of $\mathcal{T}_{DP}\uparrow\omega(\{\emptyset\})$ and our basic splitting rule approach.

3.3 Fixpoint semantics and the basic model

To get rid, at least on a theoretical level, of the "splitted variable problem", we apply our basic model to the set of all ground instances of clauses from the disjunctive logic program DP: This gives us first a set of ground Horn clause programs $\mathcal{HP}$. Then, to each of the programs $HP \in \mathcal{HP}$ we apply the usual consequence operator T_{HP}, which gives us a set $\{T_{HP}\uparrow\omega(\emptyset) \mid HP \in \mathcal{HP}\}$ of minimal models for the Horn clause programs from $\mathcal{HP}$.

Lemma 2. *Each model from $\{T_{HP}\uparrow\omega(\emptyset) \mid HP \in \mathcal{HP}\}$ is a supported model of DP.*

We do not get all supported models of DP: take as an example the program $DP = \{a, b \leftarrow\}$. Splitting gives us two Horn clause programs with $\{a\}$ and $\{b\}$ as minimal models; clearly both are supported models of DP, but $\{a, b\}$ is a supported model of DP as well. At least we get the minimal models:

Lemma 3. *Each model from $MM(DP)$, i.e. each minimal model of DP, is in $\{T_{HP}\uparrow\omega(\emptyset) \mid HP \in \mathcal{HP}\}$.*

Proof: Let M be an arbitrary but fixed model from $MM(DP)$. Hence, for each ground clause $A_1,\ldots,A_n \leftarrow B_1,\ldots,B_m$ if $B_1,\ldots,B_m \in M$ then $\exists i : A_i \in M$.

Let A be such an atom; from the construction of the Horn programs in $\mathcal{HP}$ we conclude, that there must exist a Horn clause set HP which contains the clause $A \leftarrow B_1, \ldots, B_m$ and therefor M is a model of HP as well.
Since minimal models of DP are supported models of DP we can further proof by induction on $value(A)$, that there must be an $HP \in \mathcal{HP}$ which contains A in its minimal model, i.e. in $T_{HP}\uparrow\omega(\emptyset)$.

qed

Note that the proof of this lemma is based on the fact that we used the version of the splitting rule, which duplicates the body of a split clause.

Together with the operator *min* we get the

Theorem 4. $MM(DP) = min\{T_{HP}\uparrow\omega(\emptyset) \mid HP \in \mathcal{HP}\}$

Let us finally note that in [FM91] procedures for constructing minimal models are discussed, which are based on the same idea as our consequence operator. Since in that paper Fernandez and Minker concentrate on hierarchical disjunctive deductive databases only, their model construction can be done systematically by handling one disjunctive clauses (rsp. all of its ground clauses) after another and they don't have to iterate their model construction operator.

3.4 A model with dynamic splitting

In this section we give an evaluation model which generates the Horn clause programs dynamically, i.e. every time a disjunctive clause should be used by a resolution procedure, the splitting with dynamic process-creation occurs.

Informally, this is defined by using SLD-resolution and whenever a variant of a disjunctive clause should be used, the SPLIT-SLD-procedure ignores all but one positive clauses. As a side-effect it generates a new proof task, which contains a modified program; this can be interpreted as a "call-by-need splitting"

A *proof task* is a tuple (G, DP), where G is a goal and DP is a disjunctive logic program, possibly containing protected clauses.

Dynamic splitting – SPLIT-SLD

Let DP be a disjunctive logic program, G_0 and $G = \leftarrow D_1, \ldots, D_i, \ldots, D_l$ be goals and $C = A_1, \ldots, A_n \leftarrow B_1, \ldots, B_m$ a variant of a clause from DP. Then $\langle G', (G_0, DP \cup \{C'\})\rangle$ is derived from G and C using mgu θ if the following condition holds:

- D_i is an atom, called the selected atom in G
- There is an j, such that θ is a mgu of A_j and D_i.
- G' is the goal $\leftarrow (D_1, \ldots, D_{i-1}, B_1, \ldots, B_m, D_{i+1}, \ldots, D_l)\theta$.
- $(G_0, DP \cup \{C'\})$ is a proof task where $C' = (A_1, \ldots, A_{j-1}, A_{j+1}, \ldots, A_n \leftarrow B_1, \ldots, B_m)\theta$.
  In case of $C = A \leftarrow B_1, \ldots, B_m$, there is only G' and no proof task derived; in this case we also say there is an empty proof task ϵ.

Let (G, DP) be a proof task. A sequence of pairs $\langle G, \epsilon\rangle = \langle G_0, T_0\rangle, \ldots, \langle G_n, T_n\rangle = \langle \Box, \epsilon\rangle$ is a SPLIT-SLD-refutation for (G, DP), iff

> $\forall 0 \leq i \leq n-1 : (G_{i+1}, T_{i+1})$ is derived from G_i and an appropriate C from DP and if $T_{i+1} = (G, DP') \neq \epsilon$, there is a SPLIT-SLD-refutation for (G, DP')

Figure 1 depicts a SPLIT-SLD-refutation with the program DP_2.

Finally we introduce the concept of a set of *case-refutations* associated to a SPLIT-SLD-refutation $\langle G, \epsilon\rangle = \langle G_0, T_0\rangle, \ldots, \langle G_n, T_n\rangle = \langle \Box, \epsilon\rangle$ for (G, DP):

> If $\forall 1 \leq i \leq n : T_i = \epsilon$,
> then $G_0, \ldots, G_n$ is the associated case-refutation with respect to DP
> else $G_0, \ldots, G_n$ with respect to DP and the case-refutations of SPLIT-SLD-refutations for all $T_i \neq \epsilon$ are the associated case refutations of T_i.

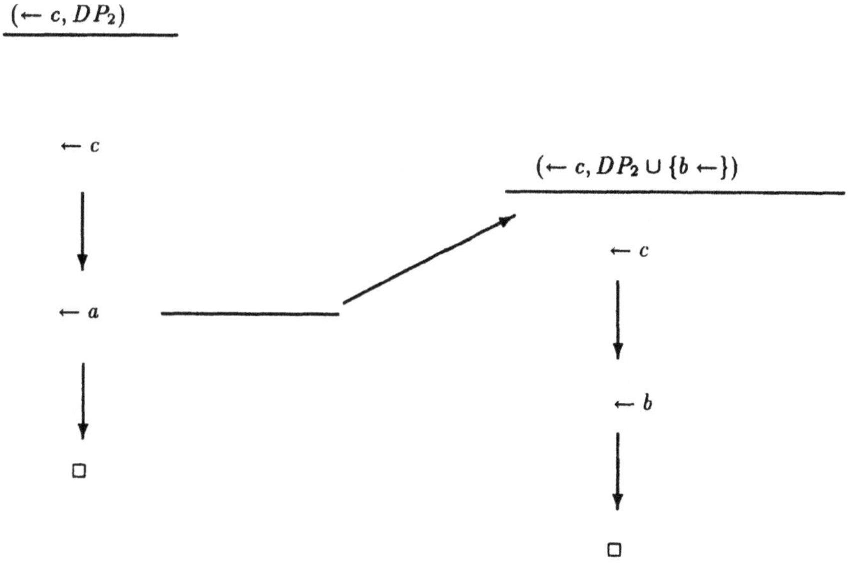

Fig. 1. An SPLIT-SLD-refutation

The SPLIT-SLD-refutation in figure 1 has two case refutations, namely $\leftarrow c, \leftarrow a, \Box$ with respect to DP_2 and $\leftarrow c, \leftarrow b, \Box$ with respect to $DP_2 \cup \{b \leftarrow\}$.

At this point it is obvious that we came up with a proof procedure which is nearly a "modular" version of InH-Prolog ([LR90]). Instead of having restart blocks we have the notion of proof tasks and instead of the cancellation rule we perform input resolution steps with split clauses. The semantics, however, seems to be closer to the Horn clause concepts in our case and admits easier correctness and completeness proofs.

In InH-Prolog the unifiers of each resolution step has to be applied to the deferred heads as well. In SPLIT-SLD-resolution the unifiers within a case-refutation are without effect in other case-refutations and thus we have to model this in addition.

Obviously, each case-refutation is a SLD-refutation. As in the basic model we want to perform these SLD-refutations in parallel. To this end we have to define the necessary communication between these tasks, i.e. the splitted variables.

Let $\langle G', (G_0, DP \cup \{C'\})\rangle$ be derived from G and C by SPLIT-SLD-resolution, then $Var(C) \cap Var(C')$ are splitted variables. We assume, that if a splitted variable x is renamed within a case-refutation, it still can be associated with its variant; we further assume that this relation is transitive and that variables are unique throughout all case-refutations. Altogether we use a function *svar* which gives for a splitted variable x

$$svar(x) = \{x' \mid x' \text{ is a variant of the splitted variable } x\}$$

Answers can be extracted from those SPLIT-SLD-refutations, which make consistent substitutions for splitted variables:

Let (G, DP) be a SPLIT-SLD-refutation with m associated case refutations $\{G_0^i, \ldots, G_{n_i}^i$ with respect to $DP_i \mid 1 \leq i \leq m\}$ and $\theta^i = \theta_1^i \cdots \theta_{n_i}^i$ the compositions of the mgu's used in the i-the case-refutation. Then

$$\begin{aligned}
&\{\theta^i \mid 1 \leq i \leq m\} \text{ is consistent if} \\
&\forall x(x \in Dom(\theta^1) \cup \cdots \cup Dom(\theta^m) \wedge x \text{ is split} \\
&\qquad \rightarrow \{x'\theta^i \mid x' \in (svar(x) \cup \{x\}) \wedge 1 \leq i \leq m\} \text{ is unifiable })
\end{aligned}$$

A SPLIT-SLD-*computed answer* is $G\theta^1 \vee \cdots \vee G\theta^m$.

Theorem 5 (Correctness). *Let* (G, DP) *be a proof task and* $G\theta^1 \vee \cdots \vee G\theta^m$ *a* SPLIT-SLD-*computed answer. Then*

$$DP \models G\theta^1 \vee \cdots \vee G\theta^m$$

Proof: By transformation into the basic model. Induction on the number of case-refutations yields, that for every case-refutation i and every ground instance $G\theta^i\sigma$ of $G\theta^i$, there exists a Horn clause program $HP \in \mathcal{HP}$, such that $G\theta^i\sigma$ is in $T_{HP}\uparrow\omega(\emptyset)$. Together with theorem 4 we get the correctness.

qed

Theorem 6 (Completeness). *Let* (G, DP) *be a proof task and* $G\theta^1 \vee \cdots \vee G\theta^m$ *a correct answer. Then there exists a* SPLIT-SLD-*computed answer* $G\mu^1 \vee \cdots \vee G\mu^r$ *which implies* $G\theta^1 \vee \cdots \vee G\theta^m$.

Proof: From $DP \models G\theta^1 \vee \cdots \vee G\theta^m$ we conclude, that for every ground substitution σ and for every minimal model M of DP the disjunction $(G\theta^1 \vee \cdots \vee G\theta^m)\sigma$ is true in M. From theorem 4 we learn that it is true in every set from $\{T_{HP}\uparrow\omega(\emptyset) \mid HP \in \mathcal{HP}\}$ as well, i.e. in every success-set of the Horn clause programs from $\mathcal{HP}$. Hence, we know that

$$\forall M \in \{T_{HP}\uparrow\omega(\emptyset) \mid HP \in \mathcal{HP}\} : \exists i : G\theta^i\sigma \in M$$

Since $(G\theta^1 \vee \cdots \vee G\theta^m)\sigma$ is ground

$$\exists n \forall M \in \{T_{HP}\uparrow n(\emptyset) \mid HP \in \mathcal{HP}\} : \exists i : G\theta^i\sigma \in M$$

That is, for every Horn clause program $HP \in \mathcal{HP}$, there is a disjunct $G\theta^i\sigma$ with $G\theta^i\sigma \xrightarrow{*} \Box$. From the lifting lemma for SLD-resolution we conclude, that there exists a SLD-refutation $G \xrightarrow{*} \Box$ of the same length with computed answer substitution μ, which is more general than $\theta^i\sigma$.

Whenever a clause $A \leftarrow B_1, \ldots, B_l$ is used during that refutation, which is derived from a disjunctive clause $A_1, A_2 \leftarrow B_1, \ldots, B_l$ through applying a ground substitution and subsequent splitting [3] we generate a proof task $(G, DP \cup \{A_2 \leftarrow B_1, \ldots, B_l\})$. By construction of $\mathcal{HP}$, there exists a $HP \in \mathcal{HP}$, with $A_2 \leftarrow B_1, \ldots, B_l$ as a program clause and a j such that $G\theta^j\sigma \in T_{HP}\uparrow n(\emptyset)$. We now argue in the same manner, noting, that only those answers of the SLD-refutations which are consistent with respect to the splitted variables, contribute to the computed answer of the SPLIT-SLD-refutation .

Since for any $G\theta^i\sigma$ there is only a finite number of Horn clause programs $HP \in \mathcal{HP}$ with different proofs, we conclude the construction of an SPLIT-SLD-refutation by induction on that number. Since during that construction not all instantiations of the goal are necessarily used, we can conclude only, that there exists a computed answer $G\mu^1 \vee \cdots \vee G\mu^r$ which implies the given correct answer.

qed

Discussion

We have discussed various possibilities to compute answers to disjunctive logic programs. In particular we gave a fixpoint semantics which is based on a splitting rule approach and which allows to derive a proof procedure as its backward-chaining counterpart very naturally. Numerous connections to other work have been discussed throughout the paper.

Let us give a final example, the interpretation of which is then discussed using three calculi, namely SPLIT-SLD-resolution, InH-Prolog and model elimination.

The example is

DP_3	
$a(x), b(x) \leftarrow$	
$c(1)$	$\leftarrow a(1)$
$c(1)$	$\leftarrow b(1)$
$c(2)$	$\leftarrow b(x)$

In figure 2 there is a SPLIT-SLD-refutation of $\leftarrow c(x)$.

Note that in this case the substitution for the splitted variable x is consistent, namely 1 in both case-refutations. The computed answer is $c(1)$; however, it would be possible to give a case-refutation for $(\leftarrow c(x), DP_3 \cup \{\leftarrow b(1)\})$ where the splitted

[3] Without loss of generality we assume that there are at most two positive literals in each clause.

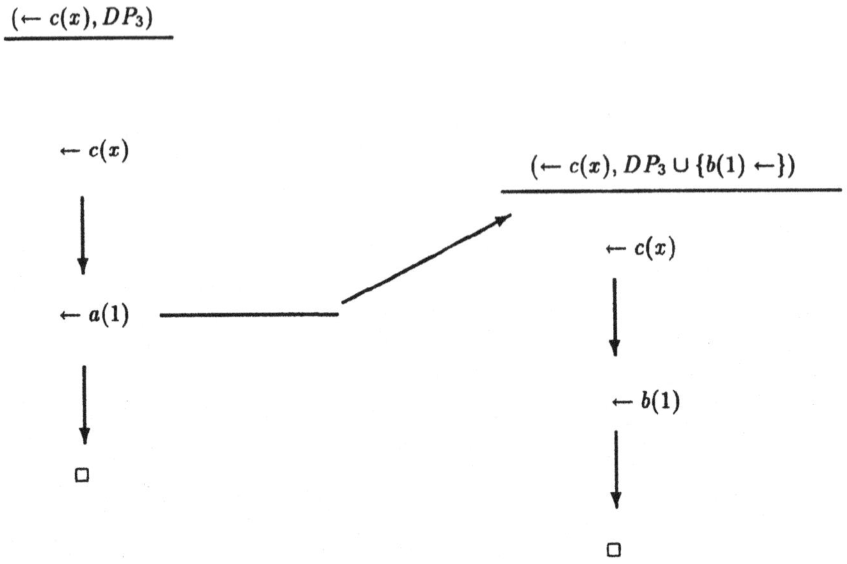

Fig. 2. SPLIT-SLD-refutation of DP_3

variable, resp. a variant of it, is substituted by 2, which does not lead to a computed answer.

A refutation in the model elimination calculus a la SETHEO is given in figure 3. The tree is constructed by extension steps, whereas the nodes marked by asterix are closed by reduction steps. Note that every unification during one of these steps has to be applied to the entire tree. As a consequence variables which are splitted ones in SPLIT-SLD-refutation are always substituted in a consistent way. The disadvantage of this calculus is, that one always has to take care for reduction steps with nodes, which are in the path up to the root of the tree. In the example the path marked with ** is closed by reduction with the root. This reduction step with an ancestor literal corresponds to the resolution step with $\leftarrow b(1)$ in the second case-refutation of the above SPLIT-SLD-refutation.

A derivation using InH-Prolog according to [RLS91] is given finally:

```
?- c(x)
:- a(1)
:-[] # {b(1)}     deferring a disjunctive head

?- c(x) # b(1)    restart
:- b(1) # b(1)
:- []   # b(1)    cancellation
```

In this calculus only positive heads of a non-Horn clause are deferred, while in SPLIT-SLD-resolution the body of such a clause is duplicated as well. We have discussed this before; aiming at an efficient proof procedure this could be done with SPLIT-SLD-resolution in the same manner,

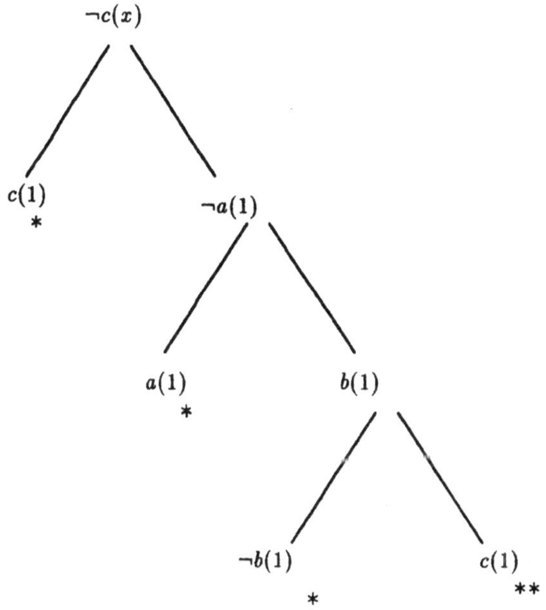

Fig. 3. A model elimination refutation of DP_3

Again, like with model elimination, the unifications have to be applied to the deferred heads (right of the #-wall), as well. Thus the case of inconsistent substitutions for splitted variables cannot occur. On the other hand our calculus seems to be better suited for implementation on a distributed system.

This is clearly a topic for further work, together with the following to points: On the base of our splitting based model it should be possible to bring in non-monotonic aspects very easily, by using negation by finite failure within the derived Horn clause programs. Secondly, the concept of answers certainly has to be generalized, to allow more intensional answers.

References

[CEFB84] R. Caferra, E. Eder, B. Fronhöfer, and W. Bibel. Extension of prolog through matrix reduction. In *Proc. ECAI-84*, pages 101–104. North-Holland, 1984.

[CGS89] M.A. Casanova, R. Guerreiro, and A. Silva. Logic programming with general clauses and defaults based on model elimination. In *Proc. of the 11th IJCAI*, pages 395–400. Morgan Kaufmann, 1989.

[CL73] C. L. Chang and R. C. T. Lee. *Symbolic Logic and Mechanical Theorem Proving.* Academic Press, 1973.

[Dec91] H. Decker. On the declarative, operational and procedural semantics of disjunctive computational theories. In *Proc. Workshop on the Deductive Approach to Information Systems and Databases, Spain*, 1991.

[FM91] J.A. Fernandez and J. Minker. Bottom-up evaluation of hierarchical disjunctive deductive databases. In Koichi Furukawa, editor, *Proc. 8th International Conference on Logic Programming*, pages 660–675, 91.

[Fur91a] U. Furbach. Answers for disjunctive logic programs. In T. Christaller, editor, *Proc. of the GWAI'91*, pages 23 - 32. Springer, Informatik Fachberichte 285, 1991.

[Fur91b] U. Furbach. Splitting as a source of parallelism in disjunctive logic programs. Technical Report 8/91, Universität Koblenz, 1991. Also in Proc. of the IJCAI-Workshop PPAI'91.

[Her88] H. Herre. Negation and constructivity in logic programming. *J. New Gener. Comput. Syst.*, 1(4):295–305, 1988.

[Lov78] D.W. Loveland. *Automated Theorem Proving: A logical basis*. North-Holland, 1978.

[Lov87] D.W. Loveland. Near-Horn Prolog. In J.-L. Lassez, editor, *Proc. of the 4th Int. Conf. on Logic Programming*, pages 456–469. The MIT Press, 1987.

[LR90] D.W. Loveland and D.W. Reed. A near-horn prolog for compilation. Technical report cs-1989-14, Duke University, 1990.

[LSBB89] R. Letz, J. Schumann, S. Bayerl, and W. Bibel. SETHEO: A High-Performance Theorem Prover. Technical report, ATP–Report, Technische Universität München, 1989. To appear in Journal of Automated Reasoning.

[MR88] J. Minker and A. Rajasekar. Procedural interpretation of non-horn logic programs. In Lusk and Overbeek, editors, *Proc. of the 9th CADE*, pages 279–293. Springer LNCS 310, 1988.

[MR90] J. Minker and A. Rajasekar. A fixpoint semantics for disjunctive logic programs. *J. Logic Programming*, 9:45–74, 1990.

[RLS91] D.W. Reed, D.W. Loveland, and B.T. Smith. An alternative characterization of disjunctive logic programs. Technical Report CS-1991-11, Duke Univerity, 1991.

[WP88] T. Wakayama and T.H. Payne. Case inference in resolution-based languages. In Lusk and Overbeek, editors, *Proc. of the 9th CADE*, pages 313–322. Springer LNCS 310, 1988.

Expanding Logic Programs

Cees Witteveen

Delft University of Technology,
Dept. of Mathematics and Computer Science,
Julianalaan 132, 2628 BL Delft, The Netherlands.
witt@dutiab.tudelft.nl

Abstract. We discuss the problem of finding acceptable models for (propositional) logic programs with negation and constraints.
It is well-known that although the well-founded semantics is defined for every general program, not every program with constraints has a well-founded model. We will argue that, in case the program is consistent but has no well-founded model, we should look for an *expansion* of the current program having a well-founded model.
We will discuss some properties these expansions and the methods generating them should have. In particular we show that there are tractable and complete expansion methods, i.e., efficient methods returning an expansion having a well-founded model whenever the original program is consistent.
Furthermore, we will investigate the complexity of expansion minimization problems, showing that in general they are NP-hard. If, however, we restrict these problems to local minimization problems, they can be solved efficiently.
This work can be viewed as a logical reconstruction of ideas presented (procedurally) in truth-maintenance (e.g. dependency-directed backtracking), auto-epistemic logic and abductive reasoning.

1 Introduction

1.1 Background

For general logic programs, the *well-founded* (wf) model semantics ([13, 14, 15]) has been proposed as a universally defined 3-valued stable model semantics.
If, however, *(integrity) constraints* are added to a program, the wf-semantics looses its universality: not for every program P there exists a wf-model satisfying all the constraints. Sometimes this defect is caused by the fact that P has no (3-valued) model at all. In other cases, however, there exist 3-valued models of P, but none of them is stable.

In the former case, it seems reasonable to *retract* information contained in the program in order to restore consistency. In the latter case, it seems more appropriate to follow an alternative approach, which will be called the *expansion* approach. Loosely speaking, this approach comes down to expanding the current program P to a larger program P', whenever P does not have a wf-model satisfying all the constraints. Such a P' will be called an *expansion* of P.

In principle, this approach is justified by the way the negation operator $\sim$ is interpreted in the stable model semantics. For, in this semantics, negation is interpreted as *negation by default*, i.e., a negated literal b occurring in the body of some rule is assumed to be false, unless we have a (well-founded) reason to believe b. Having no reason for b implies that the negation of b will be assumed to be true and therefore it might occur that the head of a rule containing such a negated literal will be evaluated as true. In the presence of constraints, however, the adoption of such a principle can have serious repercussions, if considering b to be false would have the effect that one or more constraints are violated. In that case it is natural to reconsider the current interpretation and to reverse the argument: *the occurrence of such a violation is seen as a convincing reason for b to be true.*
Such a reason for b can be provided by *adding a rule* for b to P and computing the wf-model of the expansion of P, hoping that this model indeed will satisfy the constraints.

1.2 Goal of the paper

In this paper, we will explore the use and limits of program expansion for general programs with constraints. We will only consider finite propositional logic programs. Most of the results, however, can be easily generalized to the general case.
In particular, we will develop a general theory of program expansion, analyzing such properties as *completeness* and *minimality* of expansion methods and their complexity profiles. Our aim is to give a characterization of feasible expansion methods.

1.3 Related research

In truth maintenance -a closely related formalism- program expansion has been applied in the context of *dependency-directed backtracking methods* ([2, 12, 16]). As these methods mainly have been stated informally and in an procedural way, there are little or no formal results.
In Auto-Epistemic Logic (AEL), Morris ([10]) has suggested something like program expansion for auto-epistemic theories that have no AE-extension. The simple idea is: if there is no AE-extension for a set of premises S, then a set-inclusion minimal set of ordinary (i.e. modal-operator-free) premises is added to S such that an AE-extension exists.
In logic programming, the work of Pereira et al. ([11]) on Contradiction Removal Semantics can be seen as a special expansion method, allowing for revision of assumptions.
Also the *abduction problem* in logic programming is closely related to the expansion problem: an abduction problem can be reformulated as the search for a suitable expansion of a program if some constraints are added restricting the admissible truth-value of some atoms.
In a subsequent paper we will deal with such a reformulation of abduction as

program expansion.

In none of the references cited above, however, a general characterization of program expansion has been given and also the aspects of completeness and tractability of expansion methods have not been discussed as we will do in this paper.

2 Preliminaries

2.1 Programs and rules

By a finite propositional program P we mean a finite set of rules of the form

$$c \leftarrow a_1 \wedge a_2 \wedge \ldots \wedge a_m \wedge \sim a_{m+1} \wedge \sim a_{m+2} \wedge \ldots \wedge \sim a_n$$

where $c, a_1, a_2, \ldots, a_m, a_{m+1}, a_{m+2} \ldots, a_n$ are positive literals and $\sim$ stands for negation by default. We will often abbreviate such a rule by $c \leftarrow \alpha$ and we will use $\alpha^+ = a_1 \wedge \ldots \wedge a_m$ to stand for the positive literals in the body α and $\alpha^- = \sim a_{m+1} \wedge \ldots \wedge \sim a_n$ to stand for the conjuction of the negative literals in α.
$\top$ will denote the empty conjunction. Both α^+ and α^- may be empty; if α^- is empty (denoted by $\alpha^- = \top$), we say that the rule is *positive*.
The *body* of a rule r is denoted by $\alpha(r)$. We will use $\alpha^+(r)$ and $\alpha^-(r)$ to denote, respectively, the α^+-part and the α^--part of $\alpha(r)$. Likewise, $hd(r)$ will denote the head of r and $hds(P) = \{hd(r) \mid r \in P\}$. For a given wff ϕ, At(ϕ) will denote the set of atoms occurring in ϕ.
The *Herbrand Base* B_P of a program P is the set of all ground atoms belonging to P. In the propositional case, B_P simply is the set of all atoms $At(P)$ mentioned in P.

The *positive sub-program* P^+ of P is the sub-program consisting of the set of the positive rules occurring in P, i.e.,

$$P^+ = \{c \leftarrow \alpha^+ \mid c \leftarrow \alpha \in P \wedge \alpha^- = \top\}$$

The *positive program derived from* P is the program P^{++} obtained from P by neglecting the negative part α^- in the body of the rules of P, i.e.,

$$P^{++} = \{c \leftarrow \alpha^+ \mid c \leftarrow \alpha \in P\}$$

Note that for every P, $P^+ \subseteq P^{++}$.
The Herbrand Base belonging to both P^+ and $P^+ +$ is equal to B_P.

A program with *constraints* is a program P, containing a special subset $\Delta(P)$ of rules, called *constraints*. Constraints are used to declare that a given conjunction ϕ of literals forms a *nogood*, i.e., that it is inconsistent to consider all the literals of ϕ to be true simultaneously.
Constraints are represented as rules of the form

$$\bot \leftarrow \alpha^+ \wedge \alpha^-$$

where, like in ordinary rules, α^+ is a conjunction of atoms and α^- a conjunction of negated atoms. The special atom $\bot$ does not occur as antecedent of any rule in P.

2.2 Interpretations

Let S be a set of literals. Then $\sim S = \{\sim a \mid a \in S\}$ is also a set of literals, where $\sim\sim x = x$. Let $Lit(P) = At(P) \cup \sim At(P)$.
An *interpretation* I *of* P is a consistent subset of $Lit(P)$, i.e., it can be represented by the union $I_t \cup \sim I_f$ of two disjunctive sets $I_t, I_f \subseteq B_P$, where I_t is the *true-set* and I_f the *false-set* of I. If $I_t \cup I_f = B_P$, I is a 2-valued interpretation, else I is a *3-valued* interpretation, where atoms occurring in $I_u = B_P - (I_t \cup I_f)$ are evaluated as *unknown*.
We will also use I as a *truth-assignment* $B_P \rightarrow \{\mathbf{f}, \mathbf{u}, \mathbf{t}\}$, defining for $a \in B_P$, $I(a) = x$ iff $a \in I_x$, $x \in \{f, u, t\}$.

Usually, two partial orderings for the set $\{\mathbf{f}, \mathbf{u}, \mathbf{t}\}$ of truth-values are distinguished. The *truth ordering* $<_t$ defined as $\mathbf{f} <_t \mathbf{u} <_t \mathbf{t}$, is used to evaluate the truth-value of rules. The *knowledge ordering* $<_k$ defined as $\mathbf{u} <_k \mathbf{f}$, $\mathbf{u} <_k \mathbf{t}$ is used to compare interpretations and models.

Using the truth-ordering. Conjunction is interpreted as the finite meet under the $<_t$-ordering and $\sim$ is defined by $\sim \mathbf{t} = \mathbf{f}$, $\sim \mathbf{f} = \mathbf{t}$ and $\sim \mathbf{u} = \mathbf{u}$, i.e. $\wedge$ and $\sim$ are defined using the *strong Kleene truth tables* ([8, 17]). For the operator $\leftarrow$ the so-called *weak implication table* is used: $I(c \leftarrow \beta) = \mathbf{t}$ iff $I(\beta) \leq_t I(c)$ and $\mathbf{f}$ otherwise. We will sometimes expand our propositional language to include the special atoms $\mathbf{u}, \mathbf{f}$ and $\mathbf{t}$. We will assume that every interpretation I of a program containing these symbols satisfies $I(\mathbf{u}) = \mathbf{u}$, $I(\mathbf{t}) = \mathbf{t}$ and $I(\mathbf{f}) = \mathbf{f}$.

Using the knowledge-ordering. The partial *knowledge* ordering $<_k$ is defined by $\mathbf{u} <_k \mathbf{f}$, $\mathbf{u} <_k \mathbf{t}$. This ordering can be extended in a natural way to 3-valued interpretations by defining $I \leq_k I'$ iff $I(a) \leq_k I'(a)$ for every $a \in B_P$. Of course then, $\leq_k$ reduces to the set-inclusion relation for interpretations, since $I \leq_k I'$ iff $I \subseteq I'$. This allows us to define k(nowledge)-minimal interpretations as set-inclusion minimal interpretations: Given a set $\mathcal{I}$ of partial interpretations, a *k-minimal* interpretation is an inclusion minimal interpretation $I \in \mathcal{I}$.

We will sometimes expand our propositional language to include the special atoms $\mathbf{u}, \mathbf{f}$ and $\mathbf{t}$. We will assume that every interpretation I of a program containing these symbols satisfies $I(\mathbf{u}) = \mathbf{u}$, $I(\mathbf{t}) = \mathbf{t}$ and $I(\mathbf{f}) = \mathbf{f}$.

Given an interpretation I the *partial evaluation* E_I *associated with* I is defined as

$$E_I(a) = \begin{cases} \mathbf{t} \text{ if } a \in I \\ \mathbf{f} \text{ if } \sim a \in I \\ a \text{ else} \end{cases} \qquad E_I(\sim a) = \begin{cases} \mathbf{t} \quad \text{if} \sim a \in I \\ \mathbf{f} \quad \text{if } a \in I \\ \sim a \text{ else} \end{cases}$$

for $a \in B_P$ and

$$E_I(\alpha \wedge \beta) = E_I(\alpha) \wedge E_I(\beta)$$

for conjunctions of literals.

Connected with interpretations and partial evaluations we distinguish two *reductions* of programs:

1. *the reduction of P w.r.t. I* defined as

   $$P_I = \{c \leftarrow \alpha^+ \wedge I(\alpha^-) \mid c \leftarrow \alpha \in P\}$$

   Note that P_I is always a *positive* program. The Herbrand Base of P_I is identical to B_P.
2. *the reduction of P w.r.t.* E_I defined as

   $$P_{E_I} = \{c \leftarrow E_I(\alpha) \mid c \leftarrow \alpha \in P\}$$

The Herbrand Base of both P_I and P_{E_I} is identical to B_P. We will need both reductions in describing 3-valued stable models and the fixed point characterization of the well-founded model.

2.3 Models and stable models

I is called a *model* of P iff $I(r) = \mathbf{t}$ for every $r \in P$[1]. For programs P with a set of constraints, we define I to be a *Δ-model* of P iff I is a model of P and $I(\bot) <_t \mathbf{t}$. Note that for every Δ-model M of P, $M(\alpha) <_t \mathbf{t}$ for every constraint $\bot \leftarrow \alpha \in \Delta(P)$.
P is called *Δ-inconsistent* if, for every model I of P, $I(\bot) = \mathbf{t}$.
Distinguishing ordinary models from Δ-models satisfying the constraints allows us to use standard semantics for logic programs with constraints, remembering that Δ-models have to satisfy an additional criterion.

A model M is a $<_t$-minimal model of P iff for every model M', $M'(a) \leq_t M(a)$ implies that $M(a) = M'(a)$. We will use $MIN_t(P)$ to denote a $<_t$-minimal model of P.
If P is a positive program, $MIN_t(P)$ is uniquely defined (cf. [9]). Using $<_t$-minimal models, two-valued *stable* models can be defined as follows (cf. [6]):

Definition 2.1 *M is a 2-valued* **stable** *model of P if $M = MIN_t(P_M)$ and $At(M) = B_P$.*

[1] We require that, for every interpretation I, $I(\top) = \mathbf{t}$. So $I(a \leftarrow \top) = \mathbf{t}$ iff I is a model of a program containing such a rule.

Note that $P_M = \{c \leftarrow \alpha^+ \wedge M(\alpha^-) \mid c \leftarrow \alpha \in P\}$ will contain the fixed propositions $\mathbf{t}$ and $\mathbf{f}$[2].
Three-valued stable models can be easily defined using a generalization of the least model semantics to the 3-valued case ([14]):

Definition 2.2 *Then M is a (3-valued) stable model of P, denoted as $M \in STAB3(P)$, iff $M = MIN_t(P_M)$.*

Note that, in general, for 3-valued stable models P_M will contain the fixed atoms $\mathbf{f}, \mathbf{u}, \mathbf{t}$.

Naturally, for programs P with constraints, a *Δ-stable model M* of P is a stable model of P satisfying $M(\perp) <_t \mathbf{t}$.

2.4 Well-founded models

The intersection $I \cap I'$ of 3-valued interpretations I and I', represented as sets, again is a 3-valued interpretation.
It is well-known that every program has a unique k-minimal 3-valued model. This model will be denoted by $MIN_k(P)$.
It is also well-known that intersections also preserve (stable) model properties: if M and M' be 3-valued (stable) models of P then $M \cap M'$ is also a 3-valued (stable) model of P.
Since every P has at least one 3-valued (stable) model, this implies that for every program P there exists a unique k-minimal *stable* model of P. This model is called the *well-founded* (wf-) model $WF(P)$ of P.

There is also a constructive fix-point characterization of the well-founded model of propositional programs (see [18]), which, using our terminology, can be stated very succinctly:
Let $T_P : 2^{Lit(P)} \rightarrow 2^{Lit(P)}$ be defined as:

$$T_P(I) = (\ MIN_t((P_{E_I})^+) \cap At(P)\ ) \cup (\ MIN_t((P_{E_I})^{++}) \cap \sim At(P)\ )$$

Then $lfp(T_P) = WF(P)$.

To characterize the set of atoms evaluated as true in the wf-model, we introduce the notion of a *well-founded proof*. Intuitively, such a proof constitutes a non-circular argument for establishing an atom as true in the wf-model.
Let P be a program and M be the wf-model of P. We will say that $c \in B_P$ has a *well-founded proof*, abbreviated *wf-proof*, w.r.t. P iff there exists a sequence $\sigma_P(c) = (r_1, r_2, \ldots, r_m)$ of rules in P such that

1. $hd(r_m) = c$

[2] It is immediate that P_M is equivalent to the program $\{c \leftarrow \alpha^+ \mid c \leftarrow \alpha \in P\ ,\ M(\alpha^-) = \mathbf{t}\}$ since rules $c \leftarrow \alpha^+ \wedge \mathbf{f}$ can be deleted from the program, without changing the $<_t$-minimal model of the program. The latter form is nothing more than a reformulation of the definition given in [6].

2. for every r_i, r_j in $\sigma_P(c)$,
   (a) $i \neq j$ implies $hd(r_i) \neq hd(r_j)$
   (b) $M(\alpha(r_i)) = t$ and
   (c) $At(\alpha^+(r_i)) \subseteq hds(\{r_1, \ldots, r_{i-1}\}$
3. no subsequence of $\sigma_P(c)$ satisfies 1 and 2.

Lemma 2.3 *Let M be the wf-model of a program P. Then every $c \in M_t$ has a wf-proof w.r.t. P.*

Finally, given a program P, will call $b \in B_P$ an *assumption* if b occurs in the α^--part of some rule and $WF(P)(b) = \mathbf{f}$.

2.5 Problems with the wf-semantics

Generalizing the well-founded semantics to programs with constraints, we define M to be the *wf_Δ model of P* if M is the wf-model of P and $M(\bot) <_t \mathbf{t}$. Although the wf-model is uniquely defined for every general program P, not every program P with constraints does have a wf_Δ model.
For example, the program

$$P_1 = \{a \leftarrow \top \ , \ c \leftarrow a \wedge \sim b \ , \ \bot \leftarrow c\}$$

does not have a wf_Δ-model, and the same is true for the the program

$$P_2 = \{c \leftarrow \top \ , \ \bot \leftarrow c\}$$

We argue that in the first case there is a very natural solution to the problem: consider the wf-model $M = \{a, c, \bot, \sim b\}$ for P_1. Here, the stable semantics makes the assumption b false, since there seems to be no grounded reason to make b true. This leads to the undesirable consequence that the constraint $\bot \leftarrow c$ is violated.
We can repair such a defect of the wf-semantics by recognizing these *repercussions* of the stability requirements in the presence of constraints by (i) adding some reason (clause) for one or more such assumptions b and (ii) determining the wf-model of the obtained *expansion* of the original program. We will call such an approach a program expansion approach.
On the other hand, in case of P_2, the program itself is Δ-inconsistent and the stable model semantics cannot be blamed for a defect. Here, in fact, every expansion of P_2 is Δ-inconsistent, so there exists no expansion having a wf_Δ-model. In this latter case, general theory revision methods could be applied, including the application of *contraction* operations (cf. e.g. [4]). However, as mentioned in the introduction, we will restrict our attention to expansion methods.

In the next section, we will discuss a framework for program expansion, discussing some abstract properties of methods generating expansions and analyzing some complexity issues.

3 Program expansion: general characteristics and results

An **expansion** of a program P is a program P' such that $P \subseteq P'$ and $\Delta(P) = \Delta(P')$.
A **wf-expansion** of a program P is an expansion P' of P such that P' has a wf_Δ-model.
We will study expansions of programs by studying properties of expansion *methods* that are used to generate them.

Definition 3.1 *Given a class of programs $\mathcal{P}$, a* **wf-expansion method** *is a (partial) computable mapping E, whose domain is a class of programs, assigning to every $P \in dom(E)$ a wf-expansion P' of P.*
If $P \in dom(E)$, $M_E(P)$ denotes the **wf-expansion model** *$WF(E(P))$ returned by E.*

Example 3.2 Consider the program

$$P = \{a \leftarrow \top\ ,\ c \leftarrow a \wedge \sim b\ ,\ \bot \leftarrow c\}$$

Suppose we have an expansion method E, which applied to a program P, expands P by adding a rule $b \leftarrow \top$ for every assumption b occurring in some wf-proof $\sigma(\bot)$ in P.
Then E will return the expansion $P' = P \cup \{b \leftarrow \top\}$ and finds the wf_Δ model $M' = M_E(P) = \{a, b, \sim c, \sim \bot\}$ as the wf-expansion model for P.
If, however, the rule $c \leftarrow b$ is added to P, E is not defined for $P \cup \{c \leftarrow b\}$: adding $b \leftarrow$ to the program now results in a program P' having no wf_Δ-model, since the wf-model of P' now contains $\bot$.
If we take another method E' that always adds a rule $b \leftarrow \sim b$ for every assumption b occurring in some wf-proof of $\bot$, E' is also defined for this latter program.
■

As the example given above shows, some expansion methods may succeed for some programs but not for others. Given some program P, however, we would like to know whether there exists *some* method for successful expansion of *this* P. This motivates the following definition:

Definition 3.3 *A program P is called* **revisable** *iff there exists a wf-expansion method E, such that $P \in dom(E)$.*

Intuitively, we would expect that program expansion should be possible whenever P is Δ-consistent. The following result shows that the formalization of program-expansion by wf-expansion methods corresponds to our intuitions:

Theorem 3.4 *A program P is revisable iff P has a Δ-model.*

Proof. The only-if part is trivial. To prove the if-part, note that for every P, the inclusion minimal model $MIN_k(P)$ of P is uniquely defined. Of course, we have $MIN_k(P)(\bot) = u$ iff P has a 3-valued Δ-model.
Now it is easy to show that $MIN_k(P)$ equals the wf_Δ-model of the expansion $P' = P \cup \{a \leftarrow \sim a \mid MIN_k(P)(a) = \mathbf{u},\ a \neq \bot\}$ of P and the proposition follows.
□

Now we can decide exactly *when* program expansion is applicable, it's time to ask *how efficiently* such a decision can be made.

Theorem 3.5 *Let P be a program. Then there exists a linear algorithm to decide whether or not P is revisable.*

The proof is based on the fact that the inclusion-minimal model $MIN_k(P)$ of P is a subset of $At(P)$ and equals the positive part $MIN_t(P^+) \cap At(P)$, of the unique $<_t$-minimal model $MIN_t(P^+)$, where P^+ is the positive subprogram derived from P.
Since P^+ is positive, by [1], P^+ can be computed in linear time. Then, again in linear time, we can check whether or not $\perp \in MIN_k(P)$. If so, clearly, P is not revisable, else P is revisable.

3.1 Complete expansion methods.

We have seen that it is easy to decide whether or not a program can be expanded in order to find a suitable model. So, in principle, it would make sense to ask for methods that given a program P would return an expansion for P, if possible.

Among such expansion methods that may be defined for a class of programs, we are interested in most powerful expansion methods, i.e. expansion methods that have the property that they can be applied successfully on every revisable program. Such expansion methods we will call *complete*.
More exactly, let $\mathcal{P}$ be a class of programs and let us define

$$CONS(\mathcal{P}) = \{P \in \mathcal{P} \mid \text{P is a } \Delta\text{-consistent program } \}$$

Then we can define completeness w.r.t. $\mathcal{P}$ as

Definition 3.6 *A program expansion method E is said to be called* **complete** *w.r.t. a class $\mathcal{P}$ iff $dom(E) = CONS(\mathcal{P})$.*

The question is, do there exist complete expansion methods for the class of programs with constraints and if so, are they tractable. As an easy consequence of the theorems 3.4 and 3.5, we can easily prove that tractable and complete expansion methods do exist:

Proposition 3.7 *There exists a tractable complete expansion method for general programs with constraints.*

Proof. Let $WF(P) = M$ and for any P, let

$$P' = P \cup \{a \leftarrow \sim a \mid MIN_k(P)(a) = \mathbf{u}\}$$

We define an expansion method E_{k-min} as follows:
for any P,

$$E_{k-min}(P) = \begin{cases} (P, M) & \text{if } M \text{ is the } wf_\Delta\text{-model of } P \\ (P', MIN_k(P)) & \text{if } P \text{ has no } wf_\Delta \text{ model, but is revisable} \\ \text{undefined} & \text{else} \end{cases}$$

Since the wf-model of P can be computed in quadratic time and the expansion of P in linear time, $E_{k-min}(P)$ is a tractable expansion method. Since $E_{k-min}(P)$ is defined whenever P is revisable, it follows that E_{k-min} is complete. □

Remark. There is no result analogous to Proposition 3.7 for expansion methods in a classical 2-valued setting.
This may come as a surprise since, for programs without constraints, expansions of a program having a 2-valued stable model can be easily found: let P be an arbitrary program. Then the expansion

$$P' = P \cup \{ b \leftarrow \top \mid \exists r \in P \,.\, \sim b \in \alpha^-(r) \}$$

has a unique stable model M' which equals the $<_t$-minimal model $MIN_t(P^+)$ of the positive subtheory P^+ of P. Clearly, P^+ can be constructed in linear time and by the result of [1], $M' = MIN_t(P^+)$ can be determined in linear time as well.

For programs with constraints, however, even finding an ***arbitrary*** expansion having a 2-valued stable model is NP-hard, as can be shown by a polynomial turing-reduction of the NP-complete 3-SAT problem to this problem ([19]).

3.2 Simple Expansion Methods

We want to study properties of expansions in a simple setting. Therefore, we would like to restrict the form of rules that can be added in expansions of programs, but without loosing generality, of course.

Let us call an expansion of P *simple* if it does not introduce atoms outside $At(P)$ and every rule added has at most one antecedent. Simple expansions have the advantage that they do not enlarge the Herbrand universe of the program by adding arbitrary elements. The question, however, is, don't we loose expansion power in restricting expansions to simple expansions?
The following proposition shows that the effect of any expansion method can be simulated by a simple expansion method:

Proposition 3.8 *Let P' be an expansion of P having a wf-model M' Then there exists a simple expansion P'' of P having a wf-model $M'' = M' \cap Lit(P)$.*

Proof. (Sketch.) Let $A = hds(P' - P) \cap At(P)$ be the set of heads of rules added to P which do also occur in $At(P)$.
Construct P'' as follows:

$$P'' = P \cup \{ a \leftarrow \top \mid a \in A \wedge M'(a) = \mathbf{t} \} \cup \{ a \leftarrow \sim a \mid a \in A \wedge M'(a) = \mathbf{u} \}$$

Then P'' is a simple expansion of P. Let $M'' = M' \cap Lit(P)$ and let N be the wf-model of P''.
By construction of P'' it follows that $M''_t \subseteq N_t$.
On the other hand, if $a \in N_t$ then there exists a wf-proof $\sigma(a)$ in P'' and it is easy to see that this proof can be expanded to a wf-proof of a in P'; so $a \in M''_t$.
Therefore $M''_t = N_t$.
Analogously, we can show that $M''_u = N_u$. Thus, $N = M''$ is the wf-model of P''. □

This result implies that without loss of generality we can restrict our attention to simple expansions.

4 Minimality of Expansion Methods

Obviously, we are not interested in expansion methods returning an arbitrary expansion, even if these methods are complete and tractable. Besides completeness, we should investigate whether an expansion method is *minimal* in some sense. The following minimization criteria will be distinguished:

1. *minimality of knowledge*
   E is a minimal-knowledge method if for every $P \in dom(E)$ such that P does not have a wf_Δ model, E returns an inclusion or knowledge-minimal wf-model, i.e., for every E' and $P \in dom(E) \cap dom(E')$, $WF_E(P) \cap Lit(P) \subseteq WF_{E'}(P) \cap Lit(P)$.
2. *minimality of change*
   E is a minimal-change method if it minimizes the difference between the (inconsistent) model of P and the expansion model returned.
3. *minimality of expansion*
   E is a minimal-expansion method if it minimizes the amount of information added to P in order to find an expansion model.

It is not difficult to find examples of theories in which minimal knowledge, minimal change and minimal expansion methods differ:

Example 4.1 Let P be a program containing the rules:

$$\{b \leftarrow a_1 \ , \ b \leftarrow a_2 \ , \ c_1 \leftarrow b \ , \ c_2 \leftarrow b \ , \ \bot \leftarrow\sim c_1 \ , \ \bot \leftarrow\sim c_2\}$$

Then $M = WF(P) = (\emptyset, \{a_1, a_2, b, c_1, c_2\})$ and M is not a wf_Δ model.
The following minimal expansion methods and their outcomes can be distinguished:

1. A knowledge-minimal expansion method will return an expansion $P' = P \cup \{b \leftarrow \top\}$ and a expansion model $M' = \{b, c_1, c_2, \sim a_1, \sim a_2\}$.
2. A minimal change method will result in $P' = P \cup \{c_1 \leftarrow \top \ , \ c_2 \leftarrow \top\}$ and $M' = \{c_1, c_2, \sim a_1, \sim a_2, \sim b\}$.
3. A minimal knowledge method will produce $P' = P \cup \{a_1 \leftarrow\sim a_1 \ , \ a_2 \leftarrow\sim a_2\}$ and $M' = \emptyset$.

■

The following proposition shows that E_{k-min} in fact is a minimal knowledge method.

Proposition 4.2 *E_{k-min} is a minimal knowledge expansion method.*

Proof. Let P be a program having no wf_Δ model, $P \in dom(E_{k-min})$ and suppose E' is an arbitrary expansion method defined for P. Let $M' = M_{E'}(P)$. Since M' is a model of P', $M' \cap Lit(P)$ is a propositional model of $P \subset P'$.
Now the proposition follows, since $M' \cap Lit(P) \supseteq MIN_k(P)$, and $MIN_k(P)$ is the wf-expansion model returned by E_{k-min}. □

As an immediate consequence, we have the result that knowledge minimal methods can be tractable and complete:

Theorem 4.3 *E_{k-min} is a tractable and complete minimal knowledge expansion method.*

There is little hope, however, that complete and efficient minimal change or minimal expansion methods will be found:
Let MIN-EXPANSION and MIN-CHANGE denote, respectively, the problem to find a minimal expansion model and the problem to find a minimal change model.

Proposition 4.4 *MIN-CHANGE and MIN-EXPANSION are NP-Hard search problems.*

Proof. With a polynomial turing-reduction from the NP-complete Hitting Set problem (cf. [5]). Let $(S, \mathcal{C}, K)$ be an instance of the Hitting Set problem, where S is a finite set, $\mathcal{C} \subseteq 2^S$ is a set of nonempty subsets of S and K is a positive integer. The instance $(S, \mathcal{C}, K)$ has a solution if there is a subset $S' \subset S$, $|S'| \leq K$, such that for every $C \in \mathcal{C}$, $C \cap S' \neq \emptyset$.
Such an instance can be solved by solving the MIN-EXPANSION problem for the program

$$P_{\mathcal{C}} = \{\bot \leftarrow \sim s_{i_1} \wedge \ldots \wedge \sim s_{i_k} \mid \{s_{i_1}, \ldots, s_{i_k}\} \in \mathcal{C}\}$$

Note that every such a program $P_{\mathcal{C}}$ has at least one expansion $P'_{\mathcal{C}}$ having a wf_Δ-model: just select from each $C_j \in \mathcal{C}$ an element s_j and add the rules $s_j \leftarrow$ to $P_{\mathcal{C}}$. Hence, for every $P_{\mathcal{C}}$ there exists a minimal expansion $P'_{\mathcal{C}}$ having a wf_Δ-model. In general, every expansion $P'_{\mathcal{C}}$ having a wf_Δ-model must contain rules of the form $s \leftarrow \beta$ with $s \in S$ such that for every $\bot \leftarrow \gamma \in P_{\mathcal{C}}$, $At(\gamma) \cap hds(P') \neq \emptyset$. Note that if $P'_{\mathcal{C}}$ is such a minimal expansion, the program $P''_{\mathcal{C}} = P_{\mathcal{C}} \cup \{s \leftarrow \top \mid s \in hds(P'_{\mathcal{C}}) \cap S\}$ is also a suitable and minimal expansion. So without loss of generality, we can assume that every minimal expansion consists of this kind of rules.

Claim: $P'_{\mathcal{C}}$ is a minimal expansion revision of $P_{\mathcal{C}}$ such that $|P'_{\mathcal{C}} - \Delta| \leq K$ iff $hds(P'_{\mathcal{C}} - \Delta)$ is a solution of the Hitting Set instance $(S, \mathcal{C}, K)$.

Proof of the claim. Without loss of generality, we may assume that $P'_{\mathcal{C}}$ contains only normal rules of the form $s \leftarrow \top$ where $s \in S$. Since $P'_{\mathcal{C}}$ has a wf_Δ-model, for every $\bot \leftarrow \beta \in \Delta \subset P'_{\mathcal{C}}$, $hds(P'_{\mathcal{C}} - \Delta) \cap At(\beta) \neq \emptyset$. But note that $\{At(\beta) \mid \bot \leftarrow \beta \in \Delta\} = \mathcal{C}$. Hence, if $|hds(P'_{\mathcal{C}} - \Delta)| = |P'_{\mathcal{C}} - \Delta| \leq K$, then

$hds(P'_C - \Delta) \subseteq S$ is a solution of Hitting Set.
Conversely, If S', $|S'| \leq K$ is a solution of the Hitting Set instance (S, C, K), $P'_C = \{s \leftarrow \top \mid s \in S'\} \cup P_C$ is an expansion having a wf_Δ-model. Clearly then, a minimal expansion P''_C should satisfy $|P''_C - \Delta| \leq |S'| \leq K$.

This clearly shows that Hitting Set is polynomial turing reducible to the minimal expansion problem.

Using this reduction, note that the minimal expansion solution equals the minimal change solution for the transformed Hitting Set instances. Hence, the minimal change problem must be NP-Hard, too. □

These results are rather disappointing. On the one hand, although knowledge minimal methods may be complete and tractable, one can argue that such methods are *too skeptical*, in the sense that they also expand parts of the program that are not involved in the violation of the constraints.

Example 4.5 Consider the program

$$P = \{b \leftarrow a \ , \ \bot \leftarrow\sim c\}$$

P does not have a wf_Δ-model but is revisable since $MIN_k(P) = \emptyset$. So

$$E_{k-min}(P) = (\emptyset \ , \ P \cup \{a \leftarrow\sim a \ , \ b \leftarrow\sim b \ , \ c \leftarrow\sim c\})$$

Note, however, that the addition of rules for a and b is not necessary at all, since they have nothing to do with the evaluation of $\bot$ as true. Clearly, then, only the addition of $c \leftarrow\sim c$ seems to be *relevant*. ■

On the other hand, minimal change and minimal expansion methods do not suffer from irrelevant expansions, but they are hard to compute.
We can solve these problems -at least partially - by, firstly, restricting expansion methods to *relevant* expansions and, secondly, by introducing the notion of a *locally minimal expansion*. In particular, we will show that locally minimal expansions can be computed in polynomial time.

4.1 Relevant expansion methods

Note that $\bot$ is established as true in the wf-model of P iff there exists a wf-proof of $\bot$ in P. Such a program P will be revisable if every wf-proof $\sigma(\bot)$ in P contains at least one rule with one or more negative antecedents (assumptions). Now clearly, all a *relevant* expansion method has to do is to *prevent* wf-proofs of $\bot$ to occur. So, all a relevant expansion method has to add are rules for atoms occurring in such wf-proofs of $\bot$.
Since adding a rule (in an expansion) can only make a difference for atoms evaluated **f**, it seems reasonable to concentrate on assumptions b by adding rules $b \leftarrow \gamma$ for them.

Definition 4.6 *Let M be the wf-model of P, $c \in B_P$ and $M(c) = \mathbf{t}$. Then b occurs in the* **foundations** *of c, denoted as $b \in F_P(c)$, if there is a rule $c \leftarrow \alpha \in P$, $M(\alpha) = \mathbf{t}$ and*

1. *$\sim b$ occurs in α or*
2. *$\sim b$ occurs in $F_P(a)$ for some $a \in At(\alpha^+)$.*

Definition 4.7 *A expansion method E is* **relevant** *iff, for every P, $E(P) = (M', P')$ implies that for every rule $r' \in P' - P$, $hd(r')$ occurs in the foundations $F_P(\perp)$ of $\perp$ in P.*

Let A be an arbitrary subset of $F_P(\perp)$. Such a set will be called a *set of foundations of P*.
A is called **saturated** iff every wf-proof σ of $\perp$ in P contains an assumption b occurring in A.
Given a saturated set of foundations A of P, let $P(A) = \{b \leftarrow\sim b \mid b \in A\}$ and let $P'(A)$ denote the expansion $P'(A) = P \cup P(A)$.
Let us call a relevant expansion method E to be saturated if, given a program P having no wf_Δ model, E returns an expansion $P'(A)$ for some saturated set of foundations A.

Saturated relevant expansion methods are successful:

Lemma 4.8 *Let $A \neq \emptyset$ be a saturated set of foundations for P. Then $P'(A) = P \cup P(A)$ has a wf_Δ model.*

Proof. Let M be the wf-model of P and let N be the wf-model of the expansion $P' = P'(A)$ of P. Remember that the well-founded model can be defined as the least fixpoint of the operator T_P:

$$T_P(I) = (\ MIN_t((P_{E_I})^+) \cap At(P)\ ) \cup (\ MIN_t((P_{E_I})^{++}) \cap \sim At(P)\ ) \qquad (1)$$

Let us define $M^0, N^0 = \emptyset$ and for $k > 0$, $M^k = T_P(M^{k-1})$ and $N^k = T_{P'}(N^{k-1})$.

With induction over k, it is proved that for $k \geq 0$, $N^k \subseteq M^k$. Hence, $N = lfp(T_{P'}) \subseteq lfp(T_P) = M$.

Now assume $N(\perp) = \mathbf{t}$. Then there exists a wf-proof $\sigma'(\perp)$ of $\perp$ in P' and, since $N \subseteq M$, $\sigma'(\perp)$ must also be a wf-proof of $\perp$ in P. Therefore, A contains some assumption b occurring in $\sigma'(\perp)$ and $N(a) = \mathbf{f}$. But then N cannot be a model of P', since $a \leftarrow\sim a \in P'$ implies that $N(a) \geq_t \mathbf{u}$. Since $\sigma'(\perp)$ was chosen arbitrarily, a contradiction has been derived. Hence, $N(\perp)$ cannot be equal to $\mathbf{t}$. Therefore, N is the wf_Δ model of P'. □

It is not difficult to see that $MIN_k(P)(\perp) = \mathbf{u}$ iff $F_P(\perp)$ is saturated. Thus, we have:

Proposition 4.9 *P is revisable iff $F_P(\perp)$ is saturated.*

It is immediate from the preceding proposition that saturated relevant expansion methods are complete. Moreover, as an almost trivial consequence we have:

Proposition 4.10 *Let E_{rel} be the relevant expansion method which, for every revisable P that has no wf-model, returns the expansion $P'(F_P(\perp))$. Then E_{rel} is a knowledge minimal relevant expansion method which is tractable and complete.*

Proof. By Lemma 4.8 and Proposition 4.9, E_{rel} is complete. Clearly, $F_P(\perp)$ is efficiently computable. □

One of the objections (irrelevancy) against E_{k-min} has been met by introducing E_{rel}. But still E_{rel} does have a disadvantage: it may add rules that are not really necessary to form a wf-expansion. Clearly, such a problem would disappear if we could solve the minimal expansion problem for relevant expansion methods. But, using the same reduction as we have shown for the general minimal expansion problem, it can be seen that the relevant minimal expansion problem remains NP-Hard.
There is, however, a possibility to derive tractable and complete relevant expansion methods that are *locally* expansion minimal.

Remark. Instead of the set of foundations we can easily define a subset called the *maximal foundations* of $\perp$ by slightly altering the definition of a set of foundations of an atom (cf. [7]):
b occurs in the *maximal foundations* of c, denoted as $b \in MaxF_P(c)$, if there is a rule $c \leftarrow \alpha \in P$, $M(\alpha) = \mathbf{t}$ and

1. if $\alpha^- \neq \top$ then $\sim b$ occurs in α^-
2. if $\alpha^- = \top$ then $\sim b$ occurs in $F_P(a)$ for some $a \in At(\alpha^+)$.

It is not difficult to see that $MaxF_P(\perp)$ is saturated iff $F_P(\perp)$ is saturated.

Remark. Instead of choosing the foundations, the reader might wonder why we didn't restrict expansions to assumptions that occur in wf-proofs of $\perp$. Since all a relevant expansion method has to do is to prevent such wf-proofs to occur, this would seem a better motivated choice.
There is a simple reason for *not* choosing such a set of assumptions as the basis for an expansion method: determining such a set of assumptions turns out to be NP-Hard (cf. [19]).

4.2 Locally minimal relevant expansions

In general, a wf-expansion P' of P is *locally minimal* if there exists no subset P'' of P' such that P'' is also a wf-expansion of P.

Applied to complete and relevant expansion methods the idea is to make use of the saturated set of foundations A the method generates for each program P. This set is used to search for a minimal subset A' of A such that

$P'(A') = P \cup P(A')$ has a wf-expansion.

To develop such a minimization method, we can use an anti-monotonic property of the WF operator:

Observation 4.11 *Let P be a program and A, B be subsets of $F_P(\bot)$. Then $A \subseteq B$ implies $WF(P'(B)) \subseteq WF(P'(A))$.*

Now suppose we have a saturated set of foundations A and let $a \in A$.
If $P'_{A-\{a\}}$ has no wf_Δ-model, according to the proposition above, for no subset A' of $A - \{a\}$, $P_{A'}$ will have a wf_Δ-model. Hence a has to be included in every subset A' of A having a wf-expansion $P'_{A'}$. On the other hand, if $P'_{A-\{a\}}$ has a wf_Δ-model, $\{a\}$ can be removed from A and we continue with $A = A - \{a\}$ until every element of A has been checked. The result will be a locally minimal set of foundations to be used in a locally minimal expansion method.

The following algorithm LOCMIN computes such a locally minimal set of foundations, given some arbitrary saturated set of foundations A:

LOCMIN(P, A):
begin
    $A' := A$;
    $i := 1$;
    while $i \leq |A|$
        if $P'_{A'-\{a_i\}}$ is a wf-expansion of P **then** $A' := A' - \{a_i\}$ **fi**
        $i := i + 1$
    wend
    LOCMIN$(P, A) := A'$
end

If P is revisable, $F_P(\bot)$ is saturated. Hence, LOCMIN$(P, F_P(\bot))$ will return a locally minimal set of foundations. Since $F_P(\bot)$ can be computed in $O(||P||)$ time and the wf-model of a propositional theory can be computed in roughly $O(N \times ||P||)$ time, where $N = |At(P)|$, it follows that the total time to compute a locally minimal expansion is

$$O(N(N \times (||P|| + N)))) = O(N^2 \times ||P||)$$

which is bounded above by $O(||P||^3)$.

Example 4.12 *Consider the following program:*

$$P = \{e \leftarrow \sim a \ , \ e \leftarrow \sim a \wedge \sim b \ , \ d \leftarrow e \ , \ e \leftarrow d \wedge \sim c \ , \ \bot \leftarrow e\}$$

The wf-model of P is

$$M = \{\sim a, \sim b, \sim c, e, d, \bot\}$$

According the definition of $F_P(\bot)$ we have:

$$F_P(\bot) = \{a, b, c\}$$

Since P is revisable, $F_P(\bot)$ is saturated. Computing $LOCMIN(P, F_P(\bot))$ we find $LOCMIN(P, F_P(\bot)) = \{a\}$. Hence, $P \cup \{a \leftarrow\!\sim a\}$ is a successful locally minimal expansion of P.

5 Conclusion

We have investigated the use of expansions in finding acceptable models of logic programs with constraints. Although there exists a straight-forward extension of the wf-model in the form of the 3-valued propositionally least model of the program which can be computed easily, we feel that this model contains too less information.
Relevant expansions could remedy this defect, but we also remarked that they may contain superfluous rules. Local minimization of relevant expansions seem to be the best we can hope for, since the minimal change and minimal expansion problems we have distinguished are NP-Hard.

Acknowledgements. I would like to thank the anonymous referees for their suggestions and insightful remarks on a preliminary version of this work.

References

1. W. Dowling, J. Gallier, Linear-Time Algorithms for Testing the Satisfiability of Propositional Horn Formulae, *Journal of Logic Programming*,**3**, (1984), 267–284.
2. J. Doyle, A Truth Maintenance System, Artificial Intelligence 12, 1979
3. Ch. Elkan, Logical Characterizations of Nonmonotonic TMSs, in: A. Kreczmar and G. Mirkowska (eds), *Mathematical Foundations of Computer Science 1989*, Springer-Verlag, Heidelberg, 1990, pp. 218–224.
4. P. Gärdenfors, *Knowledge in Flux*, MIT Press, Cambridge, MA, 1988.
5. M. R. Garey, D. S. Johnson, *Computers and Intractability*, Freeman, New York, 1979.
6. M. Gelfond and V. Lifschitz, The Stable Model Semantics for Logic Programming, in: *Fifth International Conference Symposium on Logic Programming*, pp. 1070-1080, MIT Press 1988.
7. C. M. Jonker, Cautious Backtracking and Well-Founded Semantics in Truth Maintenance Systems. Technical report RUU-CS-91-26, Department of Computer Science, Utrecht University, 1991.
8. S. Kleene, Introduction of metamathematics, Van Nostrand, 1952.
9. J. W. Lloyd, *Foundations of Logic Programming*, Springer Verlag, Heidelberg, 1987
10. P. H. Morris, Autoepistemic Stable Closures and Contradiction Resolution, in: M. Reinfrank et al. (eds), *Non-Monotonic Reasoning*, LNAI 346, Springer-Verlag, 1988, pp. 60–73.

11. L. M. Pereira, J. J. Alferes and J. N. Aparicio. Contradiction Removal within well-founded semantics. In: A. Nerode, W. Marek and V. S. Subrahmanian, (eds.), *First International Workshop on Logic Programming and Non-monotonic Reasoning*, MIT Press, 1991
12. C. J. Petrie, Revised Dependency-Directed Backtracking for Default Reasoning, Proc. AAAI, 1987.
13. H. Przymusinska and T. Przymusinski, Semantic Issues in Deductive Databases and Logic Programs, in: R.B. Banerji (ed), *Formal Techniques in Artificial Intelligence, A Sourcebook*, Elsevier, Amsterdam, 1990, pp. 321-367.
14. T. Przymusinski, Well-founded semantics coincides with three-valued stable semantics, *Fundamenta Informaticae*, XIII:445–463, 1990
15. T. Przymusinski, Three-valued nonmonotonic formalisms and semantics of logic programs, *Artificial Intelligence*, 49, (1991), 309–343.
16. M. Reinfrank, Fundamentals and Logical Foundations of Truth Maintenance, Linköping Studies in Science and Technology. Dissertations no. 221, Linköping University, 1989.
17. R. Turner, *Logics for Artificial Intelligence*, Ellis Horwood, Chichester, 1987.
18. C. Witteveen, Skeptical Reason Maintenance is Tractable, in: J. Allen and E. Sandewall, Proceedings of the Second International Congress of Knowledge Representation, Morgan Kaufmann, Los Altos, CA, 1991.
19. C. Witteveen, Theory Revision and Expansions, to appear.

This article was processed using the LaTeX macro package with LLNCS style

Disjunctive Logic Programming, Constructivity and Strong Negation

Heinrich Herre*and David Pearce†

Abstract

Logic programming research has been based largely on Horn-clause logic, because definite programs can be interpreted efficiently using SLD-resolution. However, there are several reasons to extend the ideas and concepts of Horn-clause logic programming to more general formulas. The paper offers a framework for discussing questions of constructivity and completeness that arise in the field of clause logic programming. Constructive properties of different calculi are investigated and their relation to a certain family of constructive logics with strong negation is established.

1 Introduction

Constructivity and completeness are important properties of SLD-resolution. Combined with algorithmic completeness, they make the domain of definite theories a natural choice for a programming language. Logic programming research has been based chiefly on Horn-clause logic, because definite programs can be interpreted efficiently. However, there are several reasons to extend the ideas and concepts of Horn-clause logic programming to more general sets of formulas.

In the areas of deductive databases and knowledge representation it is important to handle negative information. Explicit negative information plays an important role in natural discourse, and hence in the representation of knowledge. A logical system containing explicit negation and the usual propositional connectives can be considered as a kind of machine language for knowledge representation. In many cases knowledge representation languages can be translated into full predicate logic and predicate logic into open logic using Skolem functions.

If one admits negation in formulas without any restriction on the underlying deductive system then important properties of Horn-clause logic are forfeited. At the heart of Horn-clause logic, representing the basis of Prolog, are the following two properties of SLD-resolution (where $\models$ is the classical consequence relation):

(1) If $S \models \exists x F(x)$, where S is a set of Horn-clauses, and $F(x)$ is a conjunction of atomic formulas, then there is a substitution σ such that $S \models F\sigma$; moreover, σ can be generated by an SLD-refutation of $S \cup \{\neg\exists x F(x)\}$. This property is sometimes

*FB Mathematik/Informatik, Universität Leipzig, Augustusplatz 10-11, D-7010 Leipzig

†Gruppe Logik, Wissenstheorie & Information, Institut für Philosophie, Freie Universität Berlin, Habelschwerdter Allee 30, D-1000 Berlin 33

called the *existential property*. In [He 86], [Go 87] deductive systems $(L, \vdash)$ satisfying this property are called *constructively correct* to indicate that the calculus in question is sound with respect to the idea of constructive existence as it arises in constructive mathematics and logic.

(2) If $S \models F\sigma$, where S a set of Horn-clauses, then $S \vdash \exists x F(x)$ and there is an SLD-refutation of $S \cup \{\neg\exists x F(x)\}$ generating (essentially) the substitution σ. This property is called completeness. But one has to distinguish two kinds of completeness: classical completeness and *constructive* completeness. If S is a set of Horn-clauses then the two notions coincide. A substitution σ is said to be a classical *witness* for F in S, if $S \models F\sigma$. A deductive system $\vdash_C$ is called constructively complete over classical logic CL (or simply constructively complete) if for every open formula F and every classical witness σ for F in S, σ can be generated by a suitable proof diagram which derives $(\exists)F$ from S, ie. $S \vdash_C F\sigma$.

If non-Horn clauses are admitted in a program S, then completeness and constructive completeness no longer coincide and either or both of the properties (1) and (2) may be lost. It is an important task, therefore, to develop proof systems which are applicable to arbitrary (or at least non-Horn) sets of open formulas and which satisfy properties (1) and (2). The problem is of considerable practical significance since, where practical applications are concerned, most problem-solving tasks have a constructive character. Examples abound in areas like deductive question-answering, automated programming, automated design, automated planning and intelligent robots, cf. [Gr 69], [Lu 71]. Surprisingly, however, the automated theorem-proving community has so far paid little attention to the concept of constructivity and its relevance for applications.

The aim of the present paper is to outline a rather broad framework for dealing with issues of constructivity arising in generalised (ie. non-Horn clause) logic programming. Within this framework we go on to discuss various constructive properties of different deductive systems or *calculi*. The calculi in question fall roughly into two groups: those which apply to general (disjunctive) clauses, and those whose 'domains' are restricted to formulas of a special class (called here simply *program formulas*); in both cases negation may be present. Consequently, the former category subsumes what are normally called disjunctive logic programs (or databases), and the latter includes what one might term *general* logic programs, ie. where negated atoms can appear in both the heads and bodies of program clauses. We do not deal here with nonmonotonic calculi eg. of the kind which arise through the presence of negation-as-failure.

We shall not discuss philosophical questions of constructivism in any detail. However, it should be mentioned at the outset that our construal of "constructivity" is broader than that associated with intuitionism. Even for a rather restricted class of queries the intuitionistic derivability relation, '$\vdash_I$', does not deliver up all classically correct answers. Thus, for instance, it is easy to find a literal $L(x)$, a set of formulas S and a substitution σ such that σ is a classical but not an intuitionistic witness for L in S, ie. $S \models L\sigma$ but $S \not\vdash_I L\sigma$. It is arguable, however, that the notion of constructive method extends that of being an intuitionistic method. In particular, therefore, we want to allow for the possibility that, if IW is the set of intuitionistic witnesses, CW the set of constructive witnesses, and CLW the set of classical witnesses, then $IW \subset CW \subseteq CLW$. Whether one should go still further and assume that $CW = CLW$ seems to depend a good deal on the context involved, ie. on the syntax of formulas in S, on the area of application and the types of queries to be answered.

1.1 Plan of the Paper

In section 2 the preliminaries are summarised. Section 3 offers a framework for constructive calculi; we define the notions of constructively correct and constructively complete systems. In section 4 several correctness and completeness results for definite and indefinite answers are surveyed and classified. The main approaches are due to Casanova [Ca 89], Furbach [Fu 91], Hofbauer [Ho 86], Herre [He 86, 88] and Wakayama [WP 88]; being based chiefly on Herbrand's theorem. In section 5 we turn to calculus restrictions that are constructively correct. In particular, we discuss special input resolution and its relation to a suitable calculus of forward chaining. Finally, in section 6, the connections with constructive logics with strong negation are studied.

2 Preliminaries

We assume familarity with the first-order predicate calculus and the standard notions of set theory.

2.1 Syntax

A signature, denoted by Σ, is a set of relational, functional and constant symbols. $\mathcal{L}(\Sigma)$ is the first-order language based on Σ. If Σ is not specified we write simply $\mathcal{L}$. Atoms, literals and ground literals are defined in the usual way. The set OF of open formulas is the smallest set containing the atomic formulas and being closed with respect to the propositional connectives $\neg, \wedge, \vee$, and $\rightarrow$. Formulas of the form

$$L_1 \wedge \ldots \wedge L_m \rightarrow L$$

where $L, L_1, \ldots, L_m$ are literals, are called *program formulas.*[1] A program formula is said to be Horn if all its literals are atoms. A *clause* C is a formula of the form

$$C := A_1 \vee \ldots \vee A_m \vee \neg B_1 \vee \ldots \vee \neg B_n,$$

where A_i, B_j are atomic formulas.[2] C is said to be Horn if $m = 1$. The existential closure of an open formula F is denoted by $\exists(F)$ and is called an *existential sentence.*

[1] To simplify matters we shall assume whenever appropriate that *program clauses* of the form $L \leftarrow L_1, \ldots, L_n$ are replaced by their corresponding program formulas in the obvious way. This assumption is harmless since we are not imposing any fixed logic on the propositional connectives, eg. in general we do not suppose that $(A \rightarrow B)$ is the same as $(\neg A \vee B)$. Accordingly, we shall treat those deduction methods that are defined for sets of program clauses as applying *mutatis mutandis* to the appropriate program formulas. This applies in particular to the generalised form of SLD-resolution discussed in §4 as well as to the approaches of [Wag 91] and [GL 90, 91] discussed in §6 below.

[2] If the underlying logic is assumed to be classical, then disjunctive logic programs, whose formulas have the shape $L_1 \wedge \ldots \wedge L_m \rightarrow M_1 \vee \ldots \vee M_n$ (where L_i, M_j are literals) can be regarded as sets of clauses.

A *universal theory* is a set of open formulas. For a set of clauses S let $cl(S)$ be the set of program formulas generated from S as follows. Let $C := L_1 \vee \ldots \vee L_n$ be a clause; then we stipulate

$$cl_i(C) = \neg L_1, \wedge \ldots \wedge \neg L_{i-1} \wedge \neg L_{i+1} \wedge \ldots \neg \wedge L_n \rightarrow L_i$$

$$cl(C) : \{cl_i(C) : i \in \{1, \ldots, n\}\}$$

$$cl(S) = \bigcup_{C \in S} cl(C).$$

For later use we adopt the following abbreviations (where S is a set of clauses or program formulas): $GL(S)$, denotes the set of all ground literals in the language of S, $KGL(S)$ denotes the set of all conjunctions of elements of $GL(S)$, and $KL(S)$ the set of all conjunctions of literals of S. Further, we write $Ex(S)$ for the set of all prenex existential formulas (including the set $KL(S)$) whose matrix belongs to $KL(S)$, and $Ex^*(S)$ for the set of all formulas from $Ex(S)$ without free variables. Finally, we set $E(S) := KGL(S) \cup Ex^*(S)$, and write $PF(S)$ for the set of all program formulas of the language $\mathcal{L}(S)$.

2.2 Notions from model theory

Let P be a universal theory. $B(P)$ denotes the *Herbrand base* of P, ie. the set of all ground atoms of the signature of P. $U(P)$ is the set of all variable-free terms of $\Sigma(P)$. A substitution is an operation which replaces occurrences of variables by a term throughout an expression E. The result is denoted by $E\lambda$ and is called an *instance* of E. If the substitution λ is injective and the values of λ are variables then $E\lambda$ is said to be a *variant* of E. An interpretation for P is, as usual, a relational structure $\mathcal{A}$, which associates a declarative meaning to the symbols of $\Sigma(P)$. A structure $\mathcal{A}$ is a *model* for P if every formula A in P is true in A, denoted by $\mathcal{A} \models A$. Let $Mod(P)$ be the class of all models of P. An *Herbrand model* for P is one in which the universe equals $U(P)$. Herbrand models can be represented by subsets $I \subseteq B(P)$. For a class K of structures let $Th(K) = \{F : F \in L(K) \;\&\; K \models F\}$. The classical consequence relation is denoted by $\models$. For universal theories S we write $S \models_H F$ if every Herbrand model of S is a model of F. An Herbrand model I is said to be *minimal* for S if I is a model of S and no proper subset of I represents a model of S. Analogously, we define the notion of a *maximal* model. $Min(S)$ (resp. $Max(S)$) is the set of all minimal (resp. maximal) Herbrand models of S. A universal theory is said to be consistent if it has a model; then it has always an Herbrand model. Partial Herbrand structures and models are defined in §5 below.

2.3 Notions from proof theory

We assume that a calculus $\mathcal{C}$ is equipped with an appropriate concept of proof. If X is a set of formulas, let $\mathcal{D}(\mathcal{C}, X)$ be the set of all proof diagrams over X. A formula F is derivable from the set X in $\mathcal{C}$, denoted by $X \vdash_{\mathcal{C}} F$, if there is a derivation $D \in \mathcal{D}(\mathcal{C}, X)$ with root F and initial formulas from X. $\mathcal{D}(\mathcal{C}, X, F)$ is the set of all proof diagrams over X proving F. A restriction of $\mathcal{C}$ is defined by a function Ω

satisfying the condition $\Omega(X) \subseteq \mathcal{D}(\mathcal{C}, X)$ for every set X of formulas. The restricted calculus is denoted by $\mathcal{C}(\Omega)$. A calculus $\mathcal{C}$ can be applied to sets of formulas, thereby defining a closure operator $Cl_{\mathcal{C}}$, where $Cl_{\mathcal{C}}(X) = \{F : X \vdash_{\mathcal{C}} F\}$. We also write $Cl_{\mathcal{C}}(X) = X^{\vdash_{\mathcal{C}}}$. Since a calculus $\mathcal{C}$ need not always be defined for arbitrary sets of formulas and syntactic restrictions may also be placed on the set of formulas to be derived, certain refinements need to be considered. Accordingly, a calculus $\mathcal{C}$ will be be defined here as a triple

$$\mathcal{C} = (\mathbf{S}, \mathbf{Q}, \vdash_{\mathcal{C}}),$$

where:

1. $\mathbf{S}$ is a collection of sets of formulas, ie. $\mathbf{S} \subseteq Pow(L(\Sigma))$, called the *domain* of $\mathcal{C}$;

2. $\mathbf{Q}$, the *range* of $\mathcal{C}$, is a set of formulas representing goals, problems or queries to be proved (or refuted); we suppose further that $\mathbf{Q} = \mathbf{Q_0} \cup (\exists)\mathbf{Q_0}$, where $\mathbf{Q_0}$ is a set of open formulas, and we refer to elements of $\mathbf{Q}$ as *query formulas*;

3. lastly, $\vdash_{\mathcal{C}}$ is the derivability relation.

An example of a calculus of this kind is the resolution system which is defined only for sets of clauses and whose provable sentences can be understood as existential sentences. We shall also consider more restricted calculi, eg. whose domains comprise sets of program formulas and whose query formulas are literals or conjunctions of literals.

We recall some features of the resolution calculus of [Ro 65]. Let C and D be clauses, σ and τ renaming substitutions such that $var(C\sigma) \cap var(D\tau) = \emptyset$. Let $X \subseteq C\sigma, Y \subseteq D\tau$ be subsets of literals and assume that X and Y are complementary. If λ is a unifier for $at(X) \cup at(Y)$ then the clause $E = ((C\sigma \backslash X\sigma) \vee (D\tau \backslash Y\tau))\lambda$ is said to be a *resolvent* of C and D, in symbols $Res(C, D, E)$. The relation $Res(C, D, E)$ will also be denoted by $C \xrightarrow{D,\lambda} E$.

The notion of a proof diagram or a resolution tree over a set S of clauses is defined as usual. A resolution tree $\mathcal{T}$ over S is said to be a refutation of S if the values of the initial nodes of $\mathcal{T}$ are formulas from S and the value of the root of $\mathcal{T}$ is the empty clause, denoted by $\Box$. Let C be a clause and S a set of clauses. A refutation $\mathcal{T}$ of the pair (S, C) is called an *n-refutation* if there are exactly n initial nodes in $\mathcal{T}$ whose values are variants of the clause C. For an existential sentence $G := \exists x F(x)$, $F(x)$ a conjunction of literals, we write $S \vdash_R G$ if there is a refutation over $S \cup \{\neg G\}$ and $S \vdash_R^n G$ iff there is an n-refutation for $(S, \neg G)$.

3 Constructive Calculi

From the above a calculus is understood to be a system $\mathcal{C} = (\mathbf{S}, \mathbf{Q}, \vdash_{\mathcal{C}})$, where $\mathbf{S}$ is a collection of sets of formulas, $\mathbf{Q}$ is a set of query formulas and $\vdash_{\mathcal{C}}$ a derivability relation. One may also regard a calculus as a subsystem of a logic $L = (Fm(L), \vdash_L, \models_L)$ given by the set $Fm(L)$ of formulas, by a derivability relation $\vdash_L$ and a truth relation $\models_L$. In the following all the calculi we shall consider can be regarded as subsystems of classical logic, CL; but in many cases they can also be represented as subsystems of a suitable nonclassical logic. We now introduce the main concepts of *c-correctness* and *c-completeness* for a calculus, where 'c' stands for 'constructive'.[3] Let $\mathcal{C} = (\mathbf{S}, \mathbf{Q}, \vdash_{\mathcal{C}})$, be a calculus.

Definition 1 *$\mathcal{C}$ is said to be* c-correct *if for every theory $S \in \mathbf{S}$ and formula $F \in (\exists)\mathbf{Q}_0$, where $F := \exists x G(x)$, the following holds: if $S \vdash_{\mathcal{C}} F$ then there is a substitution σ such that*

$$S \vdash_{\mathcal{C}} G\sigma \ \ \& \ \ S \models_{\mathcal{C}} G\sigma.$$

We assume in this case that the substitution σ can be generated from a suitable proof diagram.

Definition 2 *$\mathcal{C}$ is said to be* c-complete over *a logic L if for every theory $S \in \mathbf{S}$, open query formula $Q \in \mathbf{Q}_0$ and substitution σ satisfying $S \models_L Q\sigma$, there is a proof of $\exists(Q)$ from S generating a substitution λ at least as general as σ such that*

$$(1) \quad S \vdash_{\mathcal{C}} Q\lambda.$$

$\mathcal{C}$ is called simply *c-complete* if it is c-complete over classical logic. Moreover, we say that $\mathcal{C}$ is *c-complete for S (over L)* if (1) is assumed for S only. Evidently, if $\mathcal{C} = L$ then c-completeness (over L) amounts to completeness in the customary sense. Weaker forms of c-correctness and c-completeness can be defined to take account of indefinite solutions; see Proposition 1 below.

No calculus $\mathcal{C}$ which is complete with respect to classical first-order logic and whose domain $\mathbf{S}_{\mathcal{C}}$ contains all open theories is constructively correct. The main problem which concerns us, therefore, is to classify and investigate calculi $\mathcal{C}$ which are subsystems of a logic L, and which satisfy appropriate constructivity conditions. This general approach was formulated in [Go 87], cf. also [Mi 91].

We conclude this section with some familiar examples.

Prolog Here $\mathbf{S}$ comprises all sets of Horn formulas; $\mathbf{Q}$ is the set of existentially closed formulas whose matrix is a conjunction of atomic formulas; $\vdash$ is SLD-resolution.

Intuitionistic calculus $(\{\emptyset\}, \mathbf{Q}, \vdash_I)$, where $\mathbf{Q}$ is the set of existential sentences.

Harrop-calculus $(Pow(H), \mathbf{Q}, \vdash_I)$; where H is the set of Harrop formulas and $\mathbf{Q}$ is as before.

[3] The terms constructive correctness and completeness were introduced by first-named author in [He 88]. Constructive correctness is a variant of the well-known Existential Property whose presence in a logical system is usually taken to indicate that the system is constructive. In [He 88] constructive completeness was defined only wrt classical logic. However, as work on this paper progressed it became clear that many calculi that are not c-complete over CL are indeed c-complete over an appropriate constructive logic. This prompts the more general definition (Def 2) given here. The definitions remain in a certain sense restricted, however, since in this paper we are dealing primarily with existential queries. They should be suitably extended if one wishes to consider other important classes of queries, such as $\forall\exists$-formulas.

Complete theories $(\{S\}, \mathbf{Q}, \vdash)$; where S is a (classically) complete theory, $\mathbf{Q}$ is the set of all existential formulas in the language of S, and $\vdash$ is a complete classical derivability relation. Furthermore, it assumed that S has an Herbrand model.

Clearly, the first of these is c-complete and correct, the second and third are c-correct, and the fourth is c-complete for S.

4 Constructivity and Completeness in Clause Logic

SLD-resolution is defined for definite programs and can be used to compute answer substitutions for queries. There is a natural generalisation of this concept to clause logic. Let Q be an open query formula and S a set of clauses. A *definite solution for Q in S* is a substitution σ such that $S \models Q\sigma$; an *indefinite solution for Q in S* is a finite number of substitutions $\sigma_1, \ldots, \sigma_n$ satisfying $S \models \bigvee_{i \leq n} Q\sigma_i$. A finite set M of substitutions is an n-indefinite solution if it is an indefinite solution and $card(M) = n$. No calculus which is complete for CL is correct for definite solutions.

Proposition 1 (Correctness and completeness for indefinite solutions)
Let S be a set of clauses and let $Q \in KL(S)$ be a a query formula. (1) If T is a refutation of $S \cup \{\neg Q\}$ then there are substitutions $\tau_1, \ldots, \tau_n$ derived from T such that $S \models \bigvee_{i<n} Q\tau_1$.
(2) If $S \models \bigvee_{i \leq n} Q\tau_i$ then there is refutation T of $S \cup \{\neg Q\}$ generating substitutions $\tau_1, \ldots, \tau_m, m \leq n$, such that every τ_i is more general than a $\sigma_j, j \leq n$, and $S \models \bigvee_{i \leq m} Q\tau_i$.

The substitutions $\tau_1, \ldots, \tau_m$ can be extracted from T in a natural way. Take all leaf nodes whose values are a variant of $\neg Q$, compose the substitutions along the path from the leaf to the root, then restrict to $var(Q)$. The substitutions $\tau_1, \ldots, \tau_m$ are called *answer substitutions*. Proposition 1 is basically Herbrand's theorem. It appears in the literature in several variations, cf. [Si 91], [Fu 91], [Ca 89], [Ho 86], [Gr 69], [Lu 71].

Proposition 2 (c-correctness for 1-refutations)
Let S be a consistent set of clauses and $Q \in KL(S)$ be a query formula. Let T be a refutation of $S \cup \{\neg Q\}$. Assume that in T there is exactly one leaf node whose value is a variant of $\neg Q$. Then there is a substitution σ derived from T such that $S \models Q\sigma$.

Proof. Immediate from Proposition 1.

The converse of Proposition 2 can be proved by way of some lemmata. We introduce the following notation: if L is a literal, then $(L)^1 = L$, $(L)^{n+1} = (L)^n \vee L$.

Lemma 1 *Let S be a consistent set of clauses and L a ground literal. If $S \models L$ then there is a resolution tree R over S whose root is $(L)^n$ for a certain $n < \omega$.*

Proof. Let $S \models L$, then $GI(S) \models L$. By the compactness theorem there is a finite subset $X \subseteq GI(S)$ such that $X \models L$. We show that there is a resolution tree R over

X having the root $(L)^n$ for a certain $n < \omega$. By the transformation lemma [Ro 65] this implies the existence of a resolution tree over S having the root $(L)^n$.

Obviously, it is sufficient to prove the following: Let X be a non-empty consistent set of ground clauses and L be a ground literal, $\neg L \notin S$. Let R be a resolution tree over $S \cup \{\neg L\}$ of height > 0 having C as root and $\neg L$ as an initial node. Then there is a resolution tree R^* over S and a subclause $C_1 \subseteq C$ such that $C_1 \vee (L)^S$ is the root of R^* for a certain $s < \omega$. We prove this inductively on the height of R, denoted by $h(R)$.

(1) The case $h(R) = 1$ is trivial.

(2) Assume $h(R) = n > 1$. Then C has two immediate predecessors D_1, D_2 which are the roots of resolution trees R_1, R_2. Let A be the resolvent atom used in the step $Res(D_1, D_2, C)$. R_1, R_2 are resolution trees over $S \cup \{\neg L\}$. By assumption $\neg L$ is an initial node of R. From this follows that $\neg L$ is an initial node of R_1 or R_2. Without loss of generality let us assume that $\neg L$ is an initial node of R. We distinguish two cases (a) and (b).

(a) $h(R_1) = 0$, then $D_1 = \neg L$ and $at(L) = A$.

(a1) $\neg L$ is not an initial node of R. Then there is a subclause $D_2' \subseteq C$ and $s < \omega$ such that $D_2' \vee (L)^S = D_2$ is the root of R. D_2' and R_2 satisfy the claimed conditions.

(a2) $\neg L$ is an initial node of R. We may assume that $h(R_2) > 0$ (otherwise $h(R) = 1$). Since $h(R) < n$ we can use the induction hypothesis. We obtain a resolution tree R_2^* over S and $D_2' \subseteq D_2$ with root $D_2' \vee (L)^t$ for a certain $t < \omega$. Assume that $L \notin D_2'$, then $D_2' \subseteq C$ and we are done. Assume $L \in D_2'$, then $D_2' \backslash \{L\} \subseteq C$. Let $D_2^* = D_2' \backslash \{L\}$ and $D_2' \vee (L)^t = D_2^* \vee (L)^s \vee (L)^t$. Then R_2^* and D_2^* satisfy the desired properties.

(b) $0 < h(R_1) < n$.

Since $h(R) < n$, by the induction hypothesis there is a resolution tree R_1^* over S and a subclause $D_1' \subseteq D_1$ such that $D_1' \vee (L)^s$ is the root of R_1^* for a certain $s < \omega$. If $D_1^* = D_1' \backslash \{L\} \subseteq C$ then R_1^* and D_1^* satisfy the desired conditions. Assume $D_1' \backslash \{L\} \not\subseteq C$. Then $A \in at(D_1' \backslash \{L\})$ (in general: if $D_1' \subseteq D_1$ and $A \notin at(D_1')$ then $D_1' \subseteq C$ because A is the resolvent atom). Without loss of generality let $A \in D_1'$. We distinguish two subcases.

(b1) $\neg L$ is not an initial node of R_2. By assumption $\neg A \in D_2$. Let $D_1^* = D_1' \backslash \{A\}$ (all occurrences of A in D_1' are cancelled) and $D_2' = D_2 \backslash \{\neg A\}$ (all ocurrences of $\neg A$ are cancelled in D_2), and $E = D_1^* \vee (L)^s v D_2'$. Then $Res(D_1' \vee (L)^s, D_2, E)$. E is the root of a resolution tree R^* over S constructed from R_1^* and R_2. Then $E = D_1^* \vee D_2' \vee (L)^s$ and $D_1^* \vee D_2' \subseteq C$, ie. the desired properties are satisfied.

(b2) Let $\neg L$ be an initial node of R_2. By the induction hypothesis there is a resolution tree R_2^* over S and a subclause $D_2' \subseteq D_2$ such that $D_2' \vee (L)^s$ is the root of R_2^* for a certain $s < \omega$. We distinguish two cases.

(b2.1) $A \notin at(D_2')$. Then $D_2' \subseteq C$ and R_2^*, $D_2' \vee (L)^s$ satisfy the desired conditions.

(b2.2) $A \in at(D_2')$. Then $\neg A \in D_2'$ (A is the resolvent atom). We take the following resolvent of D_1' and $D_2' \vee (L)^s$, denoted by E, $E = (D_1' \backslash A) \vee (D_2' \backslash \{\neg A\}) \vee (L)^s$. It is clear that $D^* = (D_1' \backslash \{A\}) \vee (D_2' \backslash \{\neg A\}) \subseteq C$, hence we may construct (using R_1^* and R_2^*) a resolution tree R^* over S with root $D^* \vee (L)^s$. □

Lemma 2 *Let S be a consistent set of clauses, $C = L_1 \vee \ldots \vee L_k$ and $S \models \neg L_1 \wedge \ldots \wedge \neg L_k$. Then there is a 1-refutation for (S, C) generating an answer substitution* *which is identical on var(C).*

Proof. Assume $S \models \bigwedge_{i \leq k} \neg L_i$ and $var(C) = \{x_1, \ldots, x_m\}$. $\{c_1, \ldots, c_m\}$ are pairwise distinct constants not appearing in $S \cup \{C\}$ and let $\theta = \{x_1/c_1, \ldots, x_m/c_m\}$. Then $S \models \neg L_i\theta$, $i = 1, 2, \ldots, k$. By lemma 1 there is a resolution tree R_i over S whose root is $(\neg L_i\theta)^{s_i}$, $i = 1, 2, \ldots, k$. We may assume that none of the variables $x_1, \ldots, x_m$ appears in R_i. Using $R_i, i \leq k$, it is easy to construct a 1-refutation tree R for $(S, C\theta)$. In the tree R we substitute back the constants c_i for x_i at any occurrence of $c_i, i \leq m$. In this way we get a 1-refutation tree for (S, C) whose answer substitution associated to C is identical on $var(C)$. □

Proposition 3 (c-completeness of 1-refutations)
Let S be a consistent set of clauses, $C = L_1 \vee \ldots \vee L_m$, and assume $S \models (\neg L_1 \wedge \ldots \wedge \neg L_m)\theta$. Then there is a 1-refutation tree R for (S, C) having an answer substitution τ such that $\theta = \tau$ on $var(C)$.

Proof. By Lemma 2 there is a 1-refutation tree R_1 for $(S, C\theta)$ with answer substitutions $\mu_1, \ldots, \mu_n$ which are identical on $var(C\theta)$. We may assume that $dom(\mu_i) \cap (dom(\theta) \cup var(C\theta) \cup var(C)) = \emptyset$, $i \leq n$. We consider the following subtree R_2 of R_1 determined by the $C\theta$-branch, $C_0 = C\theta$

$$C_0 \xrightarrow{D_1,\mu_1} C_1 \xrightarrow{D_2,\mu_2} \ldots \xrightarrow{D_n,\mu_n} \square = C_n.$$

$D_1, D_2, \cdots, D_n$ are the associated neighbour-clauses. Let $\sigma = \theta \cup \mu$. Then

$$C \xrightarrow{D_1,\sigma} C_1.$$

We may construct a 1-refutation tree R for (S, C) such that the C-branch of R is defined by:

$$C \xrightarrow{D_1,\sigma} C_1 \xrightarrow{D_2,\mu_2} \ldots \xrightarrow{D_n,\mu_n} \square = C_n.$$

Then $\tau = \sigma \circ \mu_2 \circ \ldots \circ \mu_n$ is the answer substitution of R associated with C. Let $x \in var(C)$. Then $\sigma(x) = \theta(x)$. Since μ_i is identical on $var(\theta(x))$ it follows that $\tau(x) = \theta(x)$. □

Proposition 3 expresses the property that 1-refutations characterise c-completeness over classical logic. Variations of Propositions 2 and 3 can be found in [He 86], [Ca 89], [Fu 91] and [WP 88]. In [Ca 89] an extended version of weak model-elimination [Lo 78], denoted by $\mathcal{C}(WME)$, is introduced. The restriction for $\mathcal{C}(WME)$ uses the notion of an initial support from the query formula Q. The results in [WP 88] are based on the calculus of LO-resolution, [CL 73]. Then the definite solutions of a query formula E in S are characterised by the single-entry LO-proofs of E over S. Other forms of calculus restrictions are defined in [Ke 88] and [Gi 87]. It is intuitively clear that in most situations the full power of the classical predicate calculus is not used. For this reason it is important to study c-correct restrictions $\mathcal{C}(\Omega)$ of a complete calculus $\mathcal{C}$ that are not necessarily c-complete (over classical logic) but are sufficient for the solution of a given class of problems. It is to this issue that we now turn.

5 Input Resolution and Forward Chaining

1-refutations are too complex to be implemented very efficiently. For this reason it is important to consider stronger restrictions that are c-correct. In [He 86] the notion of Special Input or SI-resolution was introduced. Let S be a set of clauses and C a clause. A 1-refutation $\mathcal{T}$ of C over S is called an *SI-refutation* of (S,C) if $\mathcal{T}$ is a binary input resolution tree without factoring. In [WP 90] SI-resolution is called SEI-resolution (SEI standing for Single Entry Input). SI-resolution is not constructively complete as the following example shows.

Example 1 Let $S = \{\neg P(a) \vee \neg Q(a), P(a) \vee \neg Q(a), P(a) \vee Q(a)\}$. Then $S \models P(a)$ but there is no SI-refutation for $(S, \neg P(a))$.

There is a close connection between SI-resolution and SLD-resolution: SI-resolution for $(S, L_1 \vee \ldots \vee L_n)$ is the same as SLD-resolution for $(cl(S), \leftarrow \neg L_1, \ldots, \neg L_n)$. We introduce the following deduction relations: if S is a set of clauses and $F := \exists x(L_1 \wedge \ldots L_n)$, we write $S \vdash_{SI} F$ iff there is an SI-refutation of $(S, \neg F)$; if S is a set of program formulas and $Q := L_1 \wedge \ldots \wedge L_n$, we write $S \vdash_{SLD} F$ iff there is an SLD-refutation of $(S, \neg Q)$. The associated calculi are denoted by $\mathcal{C}(SI)$ and $\mathcal{C}(SLD)$ respectively.

Proposition 4 *(1)* $S \vdash_{SI} F$ iff $cl(S) \vdash_{SLD} F$.
(2) $\mathcal{C}(SI)$ *and* $\mathcal{C}(SLD)$ *are c-correct.*

(1) is immediately clear and (2) follows from Proposition 2.

We now introduce a calculus $\mathcal{C}(FC)$ that describes forward chaining trees. The concept of forward chaining plays a fundamental role in the field of expert systems. Our treatment extends and generalises some of the ideas presented in [MD 91]. Forward chaining is defined for sets of program formulas, ie. formulas having the form

$$L_1 \wedge \ldots \wedge L_n \rightarrow L,$$

where the L_i and L are literals. The calculus of forward chaining, $\mathcal{C}(FC)$, is based on the following rules:

$$\text{(substitution)} \quad \frac{D}{D\sigma}, \quad D \in KL(S) \cup PF(S), \sigma \text{ a substitution;}$$

$$\text{(conjunction)} \quad \frac{E \quad D}{E \wedge D}, \quad \text{if } E, D \in KL(S);$$

$$\text{(modus ponens)} \quad \frac{K \quad K \rightarrow L}{L}, \quad K \in KL(S), K \rightarrow L \in PF(S);$$

$$\text{(}\exists\text{-rule)} \quad \frac{F(x/t)}{\exists x F(x)}, \quad F \in Ex(S), t \text{ a term.}$$

Furthermore, contraction and commuting of literals in formulas from $KL(S)$ are admitted. An FC-proof d over S, S being a set of program formulas, is defined as usual; the initial formulas have to be taken from S. Let $\mathcal{D}(FC, S)$ denote the set of all FC-proofs over S. Then for a formula F we write $S \vdash_{FC} F$ iff there is a proof $d \in \mathcal{D}(FC, S)$ whose root has the value F. The following proposition is straightforward.

Proposition 5 (constructive correctness of $\mathcal{C}(FC)$)
Let S be a set of program formulas and $F \in Ex^*(S)$, $F := \exists\vec{x}G(\vec{x})$. If $S \vdash_{FC} F$ then there is a substitution σ such that $S \vdash_{FC} G\sigma$ and $S \models G\sigma$. σ can be extracted from a proof d of F.

The relation $\vdash_{FC}$ can be semantically characterised. A 3-valued interpretation is defined by a partial Herbrand structure, ie. a set J of ground literals. J is said to be consistent (or to be a *proper* partial Herbrand structure) if there is no ground atom A such that $\{A, \neg A\} \subseteq J$. $val_J(F) \in \{0,1,u\}$ is defined for arbitrary sentences F. It is assumed that $0 < u < 1$ and $\neg(0) = 1, \neg(1) = 0, \neg(u) = u$. $val_J(F)$ is defined inductively on the complexity of F. If F is a ground literal then $val_J(F) = 1$ iff $F \in J$. In addition,

$$val_J(\neg F) = \neg val_J(F)$$

$$val_J(F \vee G) = max\{val_J(F), val_J(G)\}$$

$$val_J(F \wedge G) = min\{val_J(F), val_J(G)\}$$

$$val_J(F \rightarrow G) = \begin{cases} 1 & \text{if } val_J(G) \geq val_J(F) \text{ or } val_J(F) = u \\ 0 & \text{if } val_J(F) = 1 \text{ and } valJ(G) \neq 1 \end{cases}$$

$$val_J(\exists x F(x)) = max\{val_J(F(x/t) : t \in U(S)\}$$

$$val_J(\forall x F(x)) = min\{val_J(F(x/t) : t \in U(S)\}.$$

Define $J \models_3 F$ iff $val_J(F) = 1$; then if J is consistent val_J is called a 3-valued (or partial Herbrand) model of F. $Mod_3(S) = \{J : J \models_3 F\}$ is the set of all 3-valued models of S; and finally $S \models_3 F$ iff $Mod_3(S) \subseteq Mod_3(\{F\})$.

Proposition 6 *Let S be a consistent set of program formulas. Then S has a least 3-valued model.*

Proof. As in [He 88]. Let U be a set of 3-valued models of S. We show that $\bigcap U = J_0$ is a 3-valued model of S. Let $C \in S$, $C := L_1 \wedge \ldots \wedge L_n \rightarrow L$, $K := L_1 \wedge \ldots \wedge L_n$ and for all $J \in U : J \models_3 C$, ie. $val_J(\forall x C(x)) = 1$. It follows that for every variable-free substitution σ whose domain contains the variables from K the condition $val_J(K\sigma) \leq val_J(L\sigma)$ or $val_J(K\sigma) = u$ is satisfied. Assume that $J_0 \not\models_3 C$, then there is a variable free substitution τ such that $val_{J_0}(C\tau) \in \{0, u\}$. By the truth definition for implication '$\rightarrow$' we have $val_{J_0}(C\tau) = 0$. Then $val_{J_0}(K\tau) = 1$ and $val_{J_0}(L\tau) \neq 1$. From $val_{J_0}(K\tau) = 1$ it follows that $val_{J_0}(L_1\tau) = \ldots = val_{J_0}(L_m\tau) = 1$ and this implies that $\{L_1\tau, \ldots, L_m\tau\} \subseteq J_0$. But then $\{L_1\tau, \ldots, L_m\tau\} \subseteq J$, ie. $val_J(K\tau) = 1$, for every $J \in U$. From this it follows that $val_J(L\tau) = 1$, hence $L\tau \in J$, for all $J \in U$. Therefore $L\tau \in \bigcap U = J_0$, contradicting that $val_{J_0}(L\tau) < 1$. □

Proposition 7 (Completeness)
Let S be a consistent set of program formulas and F an existential sentence. Then $S \vdash_{FC} F$ iff $S \models_3 F$.

Proof (sketch). The proof is based on the following remark. Let $GI(S)$ be the set of all ground instances of formulas from S. Let $T_S \uparrow \omega$ be the smallest set of ground literals containing the ground literals from $GI(S)$ and closed with respect to the rule of modus ponens applied to formulas of $GI(S)$. (In [He 88] $T_S \uparrow \omega$ is the least fixpoint of an operator T_S). It is easily verified that $T_S \uparrow \omega$ is the smallest 3-valued model of S, denoted by $J_0(S)$, and that for all ground literals L the following condition is satisfied: $S \vdash_{FC} L$ iff $L \in J_0(S)$. Assume $F := \exists x G(x)$, and $G(x) := L_1(x) \wedge \ldots \wedge L_m(x)$.
($\Rightarrow$) Let $S \vdash_{FC} F$, then by Proposition 5 there is variable free substitution σ such that $\{L_1\sigma, \ldots, L_m\sigma\} \subseteq GL(S)$ and $S \vdash_{FC} G\sigma$. From this it follows that $S \vdash_{FC} L_i\sigma$, $i \leq m$, and from the above remark $\{L_1\sigma, \ldots, L_m\sigma\} \subseteq J_0(S)$. Since $J_0(S)$ is the least model it follows that $S \models_3 G\sigma$, which implies that $S \models_3 F$.
($\Leftarrow$) Suppose that $S \models_3 F$, so that $J_0(S) \models_3 F$. Then there exists a variable-free substitution σ such that $J_0(S) \models_3 G\sigma$, whence $\{L_1\sigma, \ldots, L_m\sigma\} \subseteq J_0(S)$. From the above remark we obtain $S \vdash_{FC} L_1\sigma, \ldots, L_m\sigma$, and applying the conjunction rule yields $S \vdash_{FC} L_1\sigma \wedge \ldots \wedge L_m\sigma$. Finally, from the $\exists$-rule we obtain $S \vdash_{FC} F$. □

Proposition 8 *Let S be a consistent set of clauses and F a formula from $E(S)$. Then the following conditions (1),(2),(3) are equivalent.*

(1) $cl(S) \vdash_{SLD} F$;

(2) $S \vdash_{SI} F$;

(3) $cl(S) \vdash_{FC} F$.

Proof. In [He 91] it is shown that for ground literals L the following holds: $S \vdash_{SI} L$ iff $L \in T_{cl(S)} \uparrow \omega$. This can be used to show the equivalence of (2) and (3). □

The following proposition characterizes those theories S for which $C(SI)$ is constructively complete.

Proposition 9 *Suppose that S is a consistent set of clauses and*

$$\Delta(S) = \{L : L \text{ is a ground literal } \& \ S \models L\}.$$

Then $C(SI)$ is c-complete for S if and only if $\Delta(S)$ is the least 3-valued model of $cl(S)$.

(The proof uses results from [He 88] together with Propositions 7 and 8).

6 Constructive Calculi and Constructive Logics

Up to now we have discussed constructive properties of different calculi measured with respect to classical derivability, $\vdash$, or consequence, $\models$. The calculi considered are c-correct but not necessarily c-complete over classical logic. But one may also regard constructivity as a property or set of properties, present or absent, in a logic in general. In this sense one may regard classical logic as being non-constructive, intuitionistic logic as being (in a well-defined sense) constructive. An important family of constructive logics, investigated by Nelson [Ne 49] and others, can be obtained from intuitionistic logic by replacing (or augmenting) intuitionistic negation with a so-called *strong* negation. In these systems not only is truth constructively construed (in terms of provability) but so too is falsity (in terms of refutability). In this section we shall briefly discuss these logics and show how the calculi introduced earlier, in particular special input resolution and forward chaining, are closely related to them. In fact, we shall see that, whilst they are not c-complete over classical logic, these calculi are c-complete over suitable constructive logics. We begin, however, by comparing forward chaining with some recently proposed systems of logic programming.

6.1 Strong Negation and Answer Sets

Logic programming with strong negation was introduced by Pearce & Wagner [PW 90, PW 91] with the idea of using a suitable fragment of Nelson's constructive logic as an extended logic programming language. An extensive treatment of this system can be found in [Wag 91]. We can regard the system of [Wag 91] as determining a calculus, $\mathcal{C}(SN)$, of logic programming with Strong Negation. Though independently formulated, $\mathcal{C}(FC)$ and $\mathcal{C}(SN)$ are essentially the same, though the latter may be viewed as a slight generalisation of the former.[4] The main difference is that $\mathcal{C}(SN)$ admits an arbitrary formula F in $(\wedge, \vee, \neg)$ as a query and as the antecedent of a program formula. Quantifiers are not explicitly introduced; however in the usual manner the free variables of a (generalised) program formula

$$F \rightarrow L$$

can be thought to be universally quantified and those of a query F to be existentially quantified. Accordingly, the proof theory of $\mathcal{C}(SN)$ includes additional rules to handle complex formulas. However, it is easily seen that these rules conservatively extend those of forward chaining:

Proposition 10 *Let S be a set of program formulas and $F \in Ex^*(S)$. Then $S \vdash_{FC} F$ iff $S \vdash_{SN} F$.*

where $\vdash_{SN}$ is the derivability relation of $\mathcal{C}(SN)$.

A similar relationship holds between $\mathcal{C}(FC)$ and the system of logic programming developed by Gelfond & Lifschitz in [GL 90, GL 91]. The latter is defined for (extended) logic programs whose rules may contain an additional negation-as-failure operator, '*not*', (in their bodies). However, programs without '*not*' can be represented as finite sets of program formulas. Gelfond & Lifschitz do not present a full calculus in our sense, but their programs are interpreted by so-called *answer sets*. In the case of a noncontradictory program, its answer set is a consistent set of ground literals, in other words a proper partial Herbrand structure. The answer set of a contradictory program S is the (unique) general Herbrand structure comprising the set $GL(S)$ of all ground literals. The following is readily verified:

[4] $\mathcal{C}(FC)$ and its main properties expressed in Propositions 5-8 were independently investigated by the first-named author. The similarities with $\mathcal{C}(SN)$ became apparent during the preparation of the present paper. The analogs of Propositions 5-7 for $\mathcal{C}(SN)$ can be found in [Wag 91].

Proposition 11 *Let S be a consistent set of program formulas. The (unique) answer set of S defines the least 3-valued model of S.*

6.2 Constructive Logics

We consider three strong negation systems of constructive logic, which we call N, N^+ and NC, respectively. N^+ is the basic logic of Nelson [Ne 49], usually denoted by 'N'. However, we use 'N' for the weaker variant presented in [Lo 72] and [AN 84]. NC is the contrapositive system 'S' of [Ne 59].[5] As an axiomatic system, N is characterised by the usual axioms and rules for positive intuitionistic predicate logic together with additional axiom schemata governing strong negation:

$$\neg(A \wedge B) \leftrightarrow \neg A \vee \neg B$$

$$\neg(A \vee B) \leftrightarrow \neg A \wedge \neg B$$

$$\neg(A \rightarrow B) \leftrightarrow A \wedge \neg B$$

$$\neg\neg A \leftrightarrow A$$

$$\neg \forall x A(x) \leftrightarrow \exists x \neg A(x)$$

$$\neg \exists x A(x) \leftrightarrow \forall x \neg A(x)$$

where $A \leftrightarrow B$ abbreviates $(A \rightarrow B) \wedge (B \rightarrow A)$; the symbols '$\neg$' and '$\rightarrow$' now standing for strong negation and constructive implication, respectively. Nelson's original logic of [Ne 49], denoted here by N^+, is obtained from N by adding a further axiom

$$A \rightarrow (\neg A \rightarrow B).$$

We denote the derivability relations for N and N^+ by $\vdash_N$ and $\vdash_{N^+}$, respectively.

Following [Lo 72] we present a single-conclusion Gentzen-style sequent system N_s, equivalent to N. Multiple-conclusion systems are discussed in, eg., [Gu 77] [AN 84], [Ak 88]. Below, Γ and Δ range over finite (possibly empty) *sets* of formulas; thus a reduced stock of structural rules will suffice.

The System N_s

Axiom (1) $A \Rightarrow A$

Structural Rules

$$\text{(Weakening)}\ \frac{\Gamma \Rightarrow A}{\Gamma, B \Rightarrow A} \qquad \text{(Cut)}\ \frac{\Gamma \Rightarrow A \quad \Gamma, A \Rightarrow B}{\Gamma \Rightarrow B}$$

[5] To make the paper fairly self-contained, we reproduce here Hilbert and Gentzen systems for N and N^+. The remaining propositions and claims of this section are, however, given without proofs, which are usually straightforward.

Inference Rules

$$(\rightarrow\Rightarrow)\frac{\Gamma\Rightarrow A \quad \Delta, B\Rightarrow\alpha}{\Gamma,\Delta, A\rightarrow B\Rightarrow\alpha} \qquad (\Rightarrow\rightarrow)\frac{\Gamma, A\Rightarrow B}{\Gamma\Rightarrow A\rightarrow B}$$

$$(\wedge\Rightarrow)\frac{\Gamma, A, B\Rightarrow\alpha}{\Gamma, A\wedge B\Rightarrow\alpha} \qquad (\Rightarrow\wedge)\frac{\Gamma\Rightarrow A \; \Gamma\Rightarrow B}{\Gamma\Rightarrow A\wedge B}$$

$$(\vee\Rightarrow)\frac{\Gamma, A\Rightarrow\alpha \quad \Gamma, B\Rightarrow\alpha}{\Gamma, A\vee B\Rightarrow\alpha} \qquad (\Rightarrow\vee)\frac{\Gamma\Rightarrow A}{\Gamma\Rightarrow A\vee B} \quad \frac{\Gamma\Rightarrow B}{\Gamma\Rightarrow A\vee B}$$

$$(\forall\Rightarrow)\frac{\Gamma, A(t)\Rightarrow B}{\Gamma, \forall x A(x)\Rightarrow B} \qquad (\Rightarrow\forall)\frac{\Gamma\Rightarrow A(x)^{\star}}{\Gamma\Rightarrow\forall x A(x)}$$

$$(\exists\Rightarrow)\frac{\Gamma, A(x)^{\star}\Rightarrow B}{\Gamma, \exists A(x)\Rightarrow B} \qquad (\Rightarrow\exists)\frac{\Gamma\Rightarrow A(t)}{\Gamma\Rightarrow\exists x A(x)}$$

$$(\neg\rightarrow\Rightarrow)\frac{\Gamma, A, \neg B\Rightarrow\alpha}{\Gamma, \neg(A\rightarrow B)\Rightarrow\alpha} \qquad (\Rightarrow\neg\rightarrow)\frac{\Gamma\Rightarrow A \quad \Gamma\Rightarrow\neg B}{\Gamma\Rightarrow\neg(A\rightarrow B)}$$

$$(\sim\wedge\Rightarrow)\frac{\Gamma, \neg A\Rightarrow\alpha \quad \Gamma, \neg B\Rightarrow\alpha}{\Gamma, \neg(A\wedge B)\Rightarrow\alpha} \qquad (\Rightarrow\neg\wedge)\frac{\Gamma\Rightarrow\neg A}{\Gamma\Rightarrow\neg(A\wedge B)} \quad \frac{\Gamma\Rightarrow\neg B}{\Gamma\Rightarrow\neg(A\wedge B)}$$

$$(\neg\vee\Rightarrow)\frac{\Gamma, \neg A, \neg B\Rightarrow\alpha}{\Gamma, \neg(A\vee B)\Rightarrow\alpha} \qquad (\Rightarrow\neg\vee)\frac{\Gamma\Rightarrow\neg A \quad \Gamma\Rightarrow\neg B}{\Gamma\Rightarrow\neg(A\vee B)}$$

$$(\neg\neg\Rightarrow)\frac{\Gamma, A\Rightarrow B}{\Gamma, \neg\neg A\Rightarrow B} \qquad (\Rightarrow\neg\neg)\frac{\Gamma\Rightarrow A}{\Gamma\Rightarrow\neg\neg A}$$

$$(\sim\forall\Rightarrow)\frac{\Gamma, \neg A(x)^{\star}\Rightarrow B}{\Gamma, \neg\forall x A(x)\Rightarrow B} \qquad (\Rightarrow\neg\forall)\frac{\Gamma\Rightarrow\neg A(t)}{\Gamma\Rightarrow\neg\forall x A(x)}$$

$$(\sim\exists\Rightarrow)\frac{\Gamma, \neg A(t)\Rightarrow B}{\Gamma, \neg\exists x A(x)\Rightarrow B} \qquad (\Rightarrow\neg\exists)\frac{\Gamma\Rightarrow\neg A(x)^{\star}}{\Gamma\Rightarrow\neg\exists x A(x)}$$

In rules marked with a star, it is required that x does not occur free in Γ or B. The sequent system N_s^+, equivalent to N^+, is obtained by adding the further axiom

(2) $\quad A, \neg A\Rightarrow B.$

Lastly, for the contraposable strong negation system, denoted here by NC, we refer the reader to [Ne 59], where a sequent system (called 'S') is formulated. This system lacks axiom (2) above, has only a weaker variant of the weakening rule, but adds a contraposition axiom

$$(A\rightarrow B)\Leftrightarrow(\neg B\rightarrow\neg A).$$

In this logic, the principle $A\wedge\neg A\rightarrow B$ is no longer generally valid, though the weaker form $(A\wedge\neg A)\rightarrow(B\vee\neg B)$ holds. NC can also be characterised as an extension of N^+ containing an additional implication sign standing for strong, contraposable implication. Axioms governing this connective are given in [Ra 74]. Notice that in each of the sequent systems the rule *cut* is eliminable, and that some rules are interderivable: eg. the rules for $\vee$ and $\exists$ can be derived in virtue of the definability of $\vee$ in terms of $(\neg, \wedge)$ and of $\exists$ in terms of $(\neg, \forall)$.

Like intuitionistic logic, N and N^+ satisfy the principle of constructible truth, also known as the Disjunction Property (DP):

$$\vdash_L F \vee G \;\Rightarrow\; \vdash_L F \text{ or } \vdash_L G,$$

for $L = N, N^+$. Similarly the Existential Property (EP) holds for these systems, namely for any existential sentence $\exists x G(x)$,

$$\vdash_L \exists x G(x) \;\Rightarrow\; \vdash_L G(t)$$

for some closed term t, again where $L = N, N^+$. Unlike intuitonistic logic, in N and N^+ the corresponding principle of constructible falsity holds:

$$\vdash_L \neg(F \wedge G) \;\Rightarrow\; \vdash_L \neg F \text{ or } \vdash_L \neg G.$$

Likewise, in view of the interdefinability of $\forall$ and $\exists$, universal sentences are constructively refutable.

Unlike in classical and intuitionistic logic, N possesses EP even if one reasons from sets of program formulas. Thus, let $\mathcal{C}(N) = (\mathbf{S}, \mathbf{Q}, \vdash_N)$, where $\mathbf{S}$ is the collection of all sets of program formulas and let $\mathbf{Q}$ contain the formulas in $KL(S)$ for any $S \in \mathbf{S}$. We have:

Proposition 12 (c-correctness of $\mathcal{C}(N)$)
Let $S \in \mathbf{S}$ be a set of program formulas and let $F \in Ex^*(S)$ where $F := \exists x G(x)$. Then

$$S \vdash_N F \;\Rightarrow\; S \vdash_N G(t)$$

for some closed term t.

(Actually, we could choose a somewhat larger domain and range for $\mathcal{C}(N)$, including eg. at least those of $\mathcal{C}(SN)$.) Moreover, from from the completeness theorem for N it follows that

Proposition 13 *$\mathcal{C}(N)$ is c-complete over N.*

6.3 Strong Negation and C-Completeness

In [Pe 92] it is shown how N is related to the calculus of logic programming with strong negation, $\mathcal{C}(SN)$, and to the answer set semantics of [GL 90, 91], (see also [Wag 91]). For any set of formulas S in the domain of $\mathcal{C}(SN)$ and any query formula F of $\mathcal{C}(SN)$ in the language of S,

$$S \vdash_{SN} F \quad \text{iff} \quad S \vdash_N (\exists) F.$$

Similarly for the system of Gelfond & Lifschitz: if S is a consistent set of program formulas, then

$$\mathbf{A} \text{ is an answer set of } S \quad \text{iff} \quad \mathbf{A} = \{L \in GL(S) : S \vdash_N L\}.$$

For an arbitrary (possibly inconsistent) set of program formulas the same relation holds where N is replaced by N^+. From our earlier observations on $\mathcal{C}(FC)$ and $\mathcal{C}(SN)$ it follows that:

Proposition 14 *If S is a set of program formulas and $F \in Ex(S)$, then*

$$S \vdash_{FC} F \quad \text{iff} \quad S \vdash_N F.$$

Moreover, FC and N deliver the same answer substitutions. Consequently, we obtain

Proposition 15 *The calculi $\mathcal{C}(FC)$ and $\mathcal{C}(SN)$ are c-complete over N.*

Proposition 8 relates $\mathcal{C}(FC)$ (and hence also $\mathcal{C}(SN)$) to SLD- and SI-resolution. Consequently, the latter are also closely connected to constructive logics with strong negation. In the case of SLD-resolution, it is appropriate to take N^+. In fact, one can show that SLD-resolution is c-complete over N^+, therefore, on its domain, $\vdash_{SLD}$ coincides with $\vdash_{N^+}$ (for related claims concerning SLD-resolution on Horn clauses, see [Ak 89]). It is interesting to note, however, that N^+ is stronger than these restricted calculi when the domain consists of general clauses. Consider, eg. , the set S of clauses of Example 1 above. As remarked, $S \not\vdash_{SI} P(a)$, but it is easily verified that $S \vdash_{N^+} P(a)$. In fact one can show that for the domain of clauses and for query formulas comprising conjunctions of literals, N^+ is complete over classical logic. So, for clause logic programming 1-refutations correspond to derivability in N^+.[6]

In the case of special input resolution, there is a direct relationship to the contrapositive strong negation logic, NC. First, we have to convert any clause C to an appropriate program formula, by convention say $cl_1(C)$, which is classically (but not constructively) equivalent to C. If S is any set of clauses, let $S' = \{cl_1(C) : C \in S\}$ and let $F \in Ex(S)$.

Proposition 16 $\quad S \vdash_{SI} F \quad \text{iff} \quad S' \vdash_{NC} F.$

So, modulo the conversion to program formulas, the calculus $\mathcal{C}(SI)$ is c-complete over NC.

6.4 Summary of Calculi and Completeness Properties

calculus	description	domain	c-complete over
	1-refutations	clauses	CL
	(N^+) clause logic	clauses	CL, N^+
$\mathcal{C}(SLD)$	SLD-resolution	prog fmls	N^+
$\mathcal{C}(SI)$	Special Input Res	prog fmls	NC
$\mathcal{C}(FC)$	Forward Chaining	prog fmls	N
$\mathcal{C}(SN)$	Log Prog with Str Neg	gen prog fmls	N
$\mathcal{C}(N)$		prog fmls	N

[6] Notice, therefore, that on the domain of program formulas, N^+, like SLD-resolution, is c-correct, but not of course c-complete. On the domain of clauses, on the other hand, N^+ is c-complete (over CL) for literal query formulas, but c-correctness is lost. These differences are due to the constructive character of the implication connective in N^+, for which $A \rightarrow B$ is not the same as $\neg A \vee B$.

6.5 Remark on Intuitionistic Logic

We have seen how various c-correct calculi turn out to be c-complete over constructive logics in the strong negation family. Commonly, intuitionistic logic (IL) is viewed as the paradigm of constructive reasoning. In fact, intuitionistic negation '$-$' can be defined in N^+, eg. by $-A := A \rightarrow \neg A$, and under this definition N^+ conservatively extends IL. However, since the negation of N^+ is 'stronger' than intuitionistic negation, the correctness and completeness properties described above do not carry over to IL. Thus, for instance, though positive (intuitionistic) logic suffices for the derivability of atoms from purely positive programs (definite theories), there is a sense in which, even on Horn-clauses, SLD-resolution is not faithfully captured in IL (see [Ak 89] for a discussion). Moreover, on the domain of program formulas IL is not c-correct. In our view these (positive and negative) results suggest that strong negation logics have a greater potential than intuitionistic logic in the areas of both extended and clause-logic programming.

References

[Ak 88] Akama, S, Constructive logic with strong negation and model theory, *Notre Dame J Formal Logic* 29 (1988), 18-27

[Ak 89] Akama, S, Resolution in Constructivism, *Logique et Analyse* (1989), 385-399

[Ak 89a] Akama, S, *Constructive Falsity: Foundations and their Applications to Computer Science*, Dissertation, Keio University, Japan, 1989

[AN 84] Almukdad, A & Nelson, D, Constructible Falsity and Inexact Predicates, *J Symbolic Logic* 49 (1984), 231-233.

[Av 89] Avron, A, Simple Consequence relations, TR 145/89 (1989), The Eskanasy Institute of CS, Tel-Aviv University

[Be 88] Bellin, G, & J Ketonen, Notes on Direct Logic, Linear Logic and its Implementation, Preprint, 1988

[Ca 89] Casanova, M A, R Guereiro, & A Silva, Logic Programming with general clauses and defaults based on model elimination, in, *Proc of the 11th IJCAI*, Morgan Kaufmann, 1989

[CL 73] Chang, C L, & Lee, R C T, Symbolic Logic and Mechanical Theorem Proving, Academic Press, New York- San Francisco- London, 1973

[Fu 91] Furbach, U, Answers for disjunctive logic programs, in *GWAI'91: 15. Fachtagung für KI*, Informatik-Fachberichte 285, Springer-Verlag, 1991, 23-32

[Gi 87] Girard, J-Y, Linear Logic, *TCS* (1987), 1-101

[GL 90] Gelfond, M & Lifschitz, V, Logic Programs with Classical Negation, in D Warren & P Szeredi (eds.), *Proc. ICLP-90*, MIT Press, 1990.

[GL 91] Gelfond, M, & V Lifschitz, Classical Negation in Logic Programs and Disjunctive Databases, *New Gener Comp* (1991)

[Go 87] Goltz. H-J, & H Herre, *Mathematische Grundlagen der logischen Programmierung*, IIR Info-Reporte 5 (1987) 160pp.

[Gr 69] Green, C, Applications of theorem-proving to problem solving, *Proc IJCAI 69* (1969), 219- 238

[Gu 77] Gurevich, Y, Intuitionistic Logic with Strong Negation, *Studia Logica* 36 (1977), 49-59.

[He 86] Herre, H, Konstruktive Resolutionsmethoden, Preprint P-Math 21/86, Institut für Mathematik, AdW der DDR, Berlin 1986

[He 88] Herre, H, Negation and Constructivity in Logic Programming, *J New Gener Comput Syst* 1 (1988), 295 - 305

[He 91] Herre, H, Nonmonotonic Reasoning and Logic Programs; NIL (Workshop on Nonmonotonic and Inductive Logic), Karlsruhe 1990, LNCS vol. 543, Springer-Verlag, 38-58

[Ho 86] Hofbauer, D, R Kutsche & D Siefkes, On the Logic of Logic Programming, Preprint TU Berlin, 1986, 16pp.

[Lo 72] López-Escobar, E G K, Refutability and Elementary Number Theory, *Indag. Math.* 34 (1972), 362-374.

[Lo 78] Loveland, D, W, Automated Theorem Proving: A logical basis, North-Holland, 1978

[Ll 87] Lloyd,J.W. Foundations of Logic Programming, Springer 1987

[MD 91] Mathieu, P, & J-P Delahaye, The logical compilation of knowledge bases, LNCS, Springer-Verlag, 1991

[Mi 91] Miller, D, G Nadathur, F Pfenning, & A Scedŗov, Uniform proofs as a foundation for logic programming, *Ann Pure and Applied Logic* 51 (1991), 125-157

[Ne 49] Nelson, D, Constructible falsity, *JSL* 14 (1949), 16-26.

[Ne 59] Nelson, D, Negation and Separation of Concepts in Constructive Systems, in A Heyting (Ed), *Constructivity in Mathematics*, North-Holland, Amsterdam, 1959

[Pe 92] Pearce, D, Reasoning with Negative Information, II: Hard Negation, Strong Negation and Logic Programs, to appear in [PW 92].

[PW 90] Pearce, D & Wagner, G, Reasoning with Negative Information I: Strong Negation in Logic Programs, in L Haaparanta, M Kusch & I Niiniluoto (eds.), *Language, Knowledge, and Intentionality*, (*Acta Philosophica Fennica* 49), Helsinki, 1990.

[PW 91] Pearce, D & Wagner, G, Logic Programming with Strong Negation, in P Schroeder-Heister (ed.), *Extensions of Logic Programming*, LNAI, 475, Springer-Verlag, Berlin, 1991.

[PW 92] Pearce, D & Wansing, H, (eds.) *Nonclassical Logics and Information Processing*, LNAI 619, Springer-Verlag, Berlin, 1992.

[Ra 74] Rasiowa, H, *An Algebraic Approach to Non-classical Logics*, PWN, Warsaw, & North-Holland, Amsterdam, 1974.

[Ro 65] Robinson, J, A, A Machine-oriented Logic Based on the Resolution Principle, *J. ACM* 12 (1965), 23-41

[Si 91] Sieg, W, Herbrand Analyses, *Archive for ML* 30 (1991), 409-441

[St 84] Stickel, M, A Prolog technology theorem prover, *New Gener Computing* 2 (1984), 371 - 383

[Wag 91] Wagner, G, Logic Programming with Strong Negation and Inexact Predicates, *J. Logic and Computation* 1 (1991), 835-859

[WP 88] Wakayama, T, & T H Payne, Case inference in resolution-based laguages. In Lusk and Overbeek, Proc. of the 9th CADE, 313-322; Springer LNCS 310, 1988

[WP 90] Wakayama, T, & T H Payne, Case-Free Programs: An Abstraction of Definite Horn Programs, LNCS, 1990

Lecture Notes in Artificial Intelligence (LNAI)

Vol. 451: V. Marík, O. Stepánková, Z. Zdráhal (Eds.), Artificial Intelligence in Higher Education. Proceedings, 1989. IX, 247 pages. 1990.

Vol. 459: R. Studer (Ed.), Natural Language and Logic. Proceedings, 1989. VII, 252 pages. 1990.

Vol. 462: G. Gottlob, W. Nejdl (Eds.), Expert Systems in Engineering. Proceedings, 1990. IX, 260 pages. 1990.

Vol. 465: A. Fuhrmann, M. Morreau (Eds.), The Logic of Theory Change. Proceedings, 1989. X, 334 pages. 1991.

Vol. 475: P. Schroeder-Heister (Ed.), Extensions of Logic Programming. Proceedings, 1989. VIII, 364 pages. 1991.

Vol. 476: M. Filgueiras, L. Damas, N. Moreira, A.P. Tomás (Eds.), Natural Language Processing. Proceedings, 1990. VII, 253 pages. 1991.

Vol. 478: J. van Eijck (Ed.), Logics in AI. Proceedings. 1990. IX, 562 pages. 1991.

Vol. 481: E. Lang, K.-U. Carstensen, G. Simmons, Modelling Spatial Knowledge on a Linguistic Basis. IX, 138 pages. 1991.

Vol. 482: Y. Kodratoff (Ed.), Machine Learning – EWSL-91. Proceedings, 1991. XI, 537 pages. 1991.

Vol. 513: N. M. Mattos, An Approach to Knowledge Base Management. IX, 247 pages. 1991.

Vol. 515: J. P. Martins, M. Reinfrank (Eds.), Truth Maintenance Systems. Proceedings, 1990. VII, 177 pages. 1991.

Vol. 517: K. Nökel, Temporally Distributed Symptoms in Technical Diagnosis. IX, 164 pages. 1991.

Vol. 518: J. G. Williams, Instantiation Theory. VIII, 133 pages. 1991.

Vol. 522: J. Hertzberg (Ed.), European Workshop on Planning. Proceedings, 1991. VII, 121 pages. 1991.

Vol. 535: P. Jorrand, J. Kelemen (Eds.), Fundamentals of Artificial Intelligence Research. Proceedings, 1991. VIII, 255 pages. 1991.

Vol. 541: P. Barahona, L. Moniz Pereira, A. Porto (Eds.), EPIA '91. Proceedings, 1991. VIII, 292 pages. 1991.

Vol. 542: Z. W. Ras, M. Zemankova (Eds.), Methodologies for Intelligent Systems. Proceedings, 1991. X, 644 pages. 1991.

Vol. 543: J. Dix, K. P. Jantke, P. H. Schmitt (Eds.), Nonmonotonic and Inductive Logic. Proceedings, 1990. X, 243 pages. 1991.

Vol. 546: O. Herzog, C.-R. Rollinger (Eds.), Text Understanding in LILOG. XI, 738 pages. 1991.

Vol. 549: E. Ardizzone, S. Gaglio, F. Sorbello (Eds.), Trends in Artificial Intelligence. Proceedings, 1991. XIV, 479 pages. 1991.

Vol. 565: J. D. Becker, I. Eisele, F. W. Mündemann (Eds.), Parallelism, Learning, Evolution. Proceedings, 1989. VIII, 525 pages. 1991.

Vol. 567: H. Boley, M. M. Richter (Eds.), Processing Declarative Kowledge. Proceedings, 1991. XII, 427 pages. 1991.

Vol. 568: H.-J. Bürckert, A Resolution Principle for a Logic with Restricted Quantifiers. X, 116 pages. 1991.

Vol. 587: R. Dale, E. Hovy, D. Rösner, O. Stock (Eds.), Aspects of Automated Natural Language Generation. Proceedings, 1992. VIII, 311 pages. 1992.

Vol. 590: B. Fronhöfer, G. Wrightson (Eds.), Parallelization in Inference Systems. Proceedings, 1990. VIII, 372 pages. 1992.

Vol. 592: A. Voronkov (Ed.), Logic Programming. Proceedings, 1991. IX, 514 pages. 1992.

Vol. 596: L.-H. Eriksson, L. Hallnäs, P. Schroeder-Heister (Eds.), Extensions of Logic Programming. Proceedings, 1991. VII, 369 pages. 1992.

Vol. 597: H. W. Guesgen, J. Hertzberg, A Perspective of Constraint-Based Reasoning. VIII, 123 pages. 1992.

Vol. 599: Th. Wetter, K.-D. Althoff, J. Boose, B. R. Gaines, M. Linster, F. Schmalhofer (Eds.), Current Developments in Knowledge Acquisition - EKAW '92. Proceedings. XIII, 444 pages. 1992.

Vol. 604: F. Belli, F. J. Radermacher (Eds.), Industrial and Engineering Applications of Artificial Intelligence and Expert Systems. Proceedings, 1992. XV, 702 pages. 1992.

Vol. 607: D. Kapur (Ed.), Automated Deduction – CADE-11. Proceedings, 1992. XV, 793 pages. 1992.

Vol. 610: F. von Martial, Coordinating Plans of Autonomous Agents. XII, 246 pages. 1992.

Vol. 611: M. P. Papazoglou, J. Zeleznikow (Eds.), The Next Generation of Information Systems: From Data to Knowledge. VIII, 310 pages. 1992.

Vol. 617: V. Mařík, O. Štěpánková, R. Trappl (Eds.), Advanced Topics in Artificial Intelligence. Proceedings, 1992. IX, 484 pages. 1992.

Vol. 619: D. Pearce, H. Wansing (Eds.), Nonclassical Logics and Information Processing. Proceedings, 1990. VII, 171 pages. 1992.

Vol. 622: F. Schmalhofer, G. Strube, Th. Wetter (Eds.), Contemporary Knowledge Engineering and Cognition. Proceedings, 1991. XII, 258 pages. 1992.

Vol. 624: A. Voronkov (Ed.), Logic Programming and Automated Reasoning. Proceedings, 1992. XIV, 509 pages. 1992.

Vol. 627: J. Pustejovsky, S. Bergler (Eds.), Lexical Semantics and Knowledge Representation. Proceedings, 1991. XII, 381 pages. 1992.

Vol. 633: D. Pearce, G. Wagner (Eds.), Logics in AI. Proceedings. VIII, 410 pages. 1992.

Lecture Notes in Computer Science

Vol. 597: H. W. Guesgen, J. Hertzberg, A Perspective of Constraint-Based Reasoning. VIII, 123 pages. 1992. (Subseries LNAI).

Vol. 598: S. Brookes, M. Main, A. Melton, M. Mislove, D. Schmidt (Eds.), Mathematical Foundations of Programming Semantics. Proceedings, 1991. VIII, 506 pages. 1992.

Vol. 599: Th. Wetter, K.-D. Althoff, J. Boose, B. R. Gaines, M. Linster, F. Schmalhofer (Eds.), Current Developments in Knowledge Acquisition - EKAW '92. Proceedings. XIII, 444 pages. 1992. (Subseries LNAI).

Vol. 600: J. W. de Bakker, K. Huizing, W. P. de Roever, G. Rozenberg (Eds.), Real-Time: Theory in Practice. Proceedings, 1991. VIII, 723 pages. 1992.

Vol. 601: D. Dolev, Z. Galil, M. Rodeh (Eds.), Theory of Computing and Systems. Proceedings, 1992. VIII, 220 pages. 1992.

Vol. 602: I. Tomek (Ed.), Computer Assisted Learning. Proceedings, 1992. X, 615 pages. 1992.

Vol. 603: J. van Katwijk (Ed.), Ada: Moving Towards 2000. Proceedings, 1992. VIII, 324 pages. 1992.

Vol. 604: F. Belli, F. J. Radermacher (Eds.), Industrial and Engineering Applications of Artificial Intelligence and Expert Systems. Proceedings, 1992. XV, 702 pages. 1992. (Subseries LNAI).

Vol. 605: D. Etiemble, J.-C. Syre (Eds.), PARLE '92. Parallel Architectures and Languages Europe. Proceedings, 1992. XVII, 984 pages. 1992.

Vol. 606: D. E. Knuth, Axioms and Hulls. IX, 109 pages. 1992.

Vol. 607: D. Kapur (Ed.), Automated Deduction – CADE-11. Proceedings, 1992. XV, 793 pages. 1992. (Subseries LNAI).

Vol. 608: C. Frasson, G. Gauthier, G. I. McCalla (Eds.), Intelligent Tutoring Systems. Proceedings, 1992. XIV, 686 pages. 1992.

Vol. 609: G. Rozenberg (Ed.), Advances in Petri Nets 1992. VIII, 472 pages. 1992.

Vol. 610: F. von Martial, Coordinating Plans of Autonomous Agents. XII, 246 pages. 1992. (Subseries LNAI).

Vol. 611: M. P. Papazoglou, J. Zeleznikow (Eds.), The Next Generation of Information Systems: From Data to Knowledge. VIII, 310 pages. 1992. (Subseries LNAI).

Vol. 612: M. Tokoro, O. Nierstrasz, P. Wegner (Eds.), Object-Based Concurrent Computing. Proceedings, 1991. X, 265 pages. 1992.

Vol. 613: J. P. Myers, Jr., M. J. O'Donnell (Eds.), Constructivity in Computer Science. Proceedings, 1991. X, 247 pages. 1992.

Vol. 614: R. G. Herrtwich (Ed.), Network and Operating System Support for Digital Audio and Video. Proceedings, 1991. XII, 403 pages. 1992.

Vol. 615: O. Lehrmann Madsen (Ed.), ECOOP '92. European Conference on Object Oriented Programming. Proceedings. X, 426 pages. 1992.

Vol. 616: K. Jensen (Ed.), Application and Theory of Petri Nets 1992. Proceedings, 1992. VIII, 398 pages. 1992.

Vol. 617: V. Mařík, O. Štěpánková, R. Trappl (Eds.), Advanced Topics in Artificial Intelligence. Proceedings, 1992. IX, 484 pages. 1992. (Subseries LNAI).

Vol. 618: P. M. D. Gray, R. J. Lucas (Eds.), Advanced Database Systems. Proceedings, 1992. X, 260 pages. 1992.

Vol. 619: D. Pearce, H. Wansing (Eds.), Nonclassical Logics and Information Processing. Proceedings, 1990. VII, 171 pages. 1992. (Subseries LNAI).

Vol. 620: A. Nerode, M. Taitslin (Eds.), Logical Foundations of Computer Science – Tver '92. Proceedings. IX, 514 pages. 1992.

Vol. 621: O. Nurmi, E. Ukkonen (Eds.), Algorithm Theory – SWAT '92. Proceedings. VIII, 434 pages. 1992.

Vol. 622: F. Schmalhofer, G. Strube, Th. Wetter (Eds.), Contemporary Knowledge Engineering and Cognition. Proceedings, 1991. XII, 258 pages. 1992. (Subseries LNAI).

Vol. 623: W. Kuich (Ed.), Automata, Languages and Programming. Proceedings, 1992. XII, 721 pages. 1992.

Vol. 624: A. Voronkov (Ed.), Logic Programming and Automated Reasoning. Proceedings, 1992. XIV, 509 pages. 1992. (Subseries LNAI).

Vol. 625: W. Vogler, Modular Construction and Partial Order Semantics of Petri Nets. IX, 252 pages. 1992.

Vol. 626: E. Börger, G. Jäger, H. Kleine Büning, M. M. Richter (Eds.), Computer Science Logic. Proceedings, 1991. VIII, 428 pages. 1992.

Vol. 627: J. Pustejovsky, S. Bergler (Eds.), Lexical Semantics and Knowledge Representation. Proceedings, 1991. XII, 381 pages. 1992. (Subseries LNAI).

Vol. 628: G. Vosselman, Relational Matching. IX, 190 pages. 1992.

Vol. 629: I. M. Havel, V. Koubek (Eds.), Mathematical Foundations of Computer Science 1992. Proceedings. IX, 521 pages. 1992.

Vol. 630: W. R. Cleaveland (Ed.), CONCUR '92. Proceedings. X, 580 pages. 1992.

Vol. 631: M. Brynooghe, M. Wirsing (Eds.), Programming Language Implementation and Logic Programming. Proceedings, 1992. XI, 492 pages. 1992.

Vol. 632: H. Kirchner, G. Levi (Eds.), Algebraic and Logic Programming. Proceedings, 1992. IX, 457 pages. 1992.

Vol. 633: D. Pearce, G. Wagner (Eds.), Logics in AI. Proceedings. VIII, 410 pages. 1992. (Subseries LNAI).